Google Earth Engine Applications

Google Earth Engine Applications

Special Issue Editors

Lalit Kumar

Onisimo Mutanga

MDPI • Basel • Beijing • Wuhan • Barcelona • Belgrade

Special Issue Editors
Lalit Kumar
University of New England
Australia

Onisimo Mutanga
University of KwaZulu-Natal
South Africa

Editorial Office
MDPI
St. Alban-Anlage 66
4052 Basel, Switzerland

This is a reprint of articles from the Special Issue published online in the open access journal *Remote Sensing* (ISSN 2072-4292) from 2016 to 2019 (available at: https://www.mdpi.com/journal/remotesensing/special_issues/GEE)

For citation purposes, cite each article independently as indicated on the article page online and as indicated below:

LastName, A.A.; LastName, B.B.; LastName, C.C. Article Title. *Journal Name* **Year**, *Article Number*, Page Range.

ISBN 978-3-03897-884-8 (Pbk)
ISBN 978-3-03897-885-5 (PDF)

Cover image courtesy of Lalit Kumar.

Contents

About the Special Issue Editors ix

Onisimo Mutanga and Lalit Kumar
Google Earth Engine Applications
Reprinted from: *Remote Sens.* **2019**, *11*, 591, doi:10.3390/rs11050591 1

Lalit Kumar and Onisimo Mutanga
Google Earth Engine Applications Since Inception: Usage, Trends, and Potential
Reprinted from: *Remote Sens.* **2018**, *10*, 1509, doi:10.3390/rs10101509 5

Manuel Campos-Taberner,Álvaro Moreno-Martínez, Francisco Javier García-Haro, Gustau Camps-Valls, Nathaniel P. Robinson, Jens Kattge, and Steven W. Running
Global Estimation of Biophysical Variables from Google Earth Engine Platform
Reprinted from: *Remote Sens.* **2018**, *10*, 1167, doi:10.3390/rs10081167 20

Ate Poortinga, Nicholas Clinton, David Saah1, Peter Cutter, Farrukh Chishtie, Kel N.Markert, Eric R. Anderson, Austin Troy, Mark Fenn, Lan Huong Tran, Brian Bean,Quyen Nguyen, Biplov Bhandari, Gary Johnson and Peeranan Towashiraporn
An Operational Before-After-Control-Impact (BACI) Designed Platform for Vegetation Monitoring at Planetary Scale
Reprinted from: *Remote Sens.* **2018**, *10*, 760, doi:10.3390/rs10050760 37

Yu Hsin Tsai, Douglas Stow, Hsiang Ling Chen, Rebecca Lewison, Li An and Lei Shi
Mapping Vegetation and Land Use Types in Fanjingshan National Nature Reserve Using Google Earth Engine
Reprinted from: *Remote Sens.* **2018**, *10*, 927, doi:10.3390/rs10060927 50

Nathaniel P. Robinson, Brady W. Allred, Matthew O. Jones, Alvaro Moreno, John S. Kimball, David E. Naugle, Tyler A. Erickson and Andrew D. Richardson
A Dynamic Landsat Derived Normalized Difference Vegetation Index (NDVI) Product for the Conterminous United States
Reprinted from: *Remote Sens.* **2017**, *9*, 863, doi:10.3390/rs9080863 64

Ran Goldblatt, Alexis Rivera Ballesteros and Jennifer Burney
High Spatial Resolution Visual Band Imagery Outperforms Medium Resolution Spectral Imagery for Ecosystem Assessment in the Semi-Arid Brazilian Sertão
Reprinted from: *Remote Sens.* **2017**, *9*, 1336, doi:10.3390/rs9121336 78

Leandro Parente and Laerte Ferreira
Assessing the Spatial and Occupation Dynamics of the Brazilian Pasturelands Based on the Automated Classification of MODIS Images from 2000 to 2016
Reprinted from: *Remote Sens.* **2018**, *10*, 606, doi:10.3390/rs10040606 104

Dimosthenis Traganos, Bharat Aggarwal, Dimitris Poursanidis, Konstantinos Topouzelis, Nektarios Chrysoulakis and Peter Reinartz
Towards Global-Scale Seagrass Mapping and Monitoring Using Sentinel-2 on Google Earth Engine: The Case Study of the Aegean and Ionian Seas
Reprinted from: *Remote Sens.* **2018**, *10*, 1227, doi:10.3390/rs10081227 118

Jacky Lee, Jeffrey A. Cardille and Michael T. Coe
BULC-U: Sharpening Resolution and Improving Accuracy of Land-Use/Land-Cover Classifications in Google Earth Engine
Reprinted from: *Remote Sens.* **2018**, *10*, 1455, doi:10.3390/rs10091455 . . . . **132**

Roberta Ravanelli, Andrea Nascetti, Raffaella Valeria Cirigliano, Clarissa Di Rico, Giovanni Leuzzi, Paolo Monti and Mattia Crespi
Monitoring the Impact of Land Cover Change on Surface Urban Heat Island through Google Earth Engine: Proposal of a Global Methodology, First Applications and Problems
Reprinted from: *Remote Sens.* **2018**, *10*, 1488, doi:10.3390/rs10091488 . . . . **153**

Mingzhu He, John S. Kimball, Marco P. Maneta, Bruce D. Maxwell, Alvaro Moreno, Santiago Beguería and Xiaocui Wu
Regional Crop Gross Primary Productivity and Yield Estimation Using Fused Landsat-MODIS Data
Reprinted from: *Remote Sens.* **2018**, *10*, 372, doi:10.3390/rs10030372 . . . . **174**

Masoud Mahdianpari, Bahram Salehi, Fariba Mohammadimanesh, Saeid Homayouni and Eric Gill
The First Wetland Inventory Map of Newfoundland at a Spatial Resolution of 10 m Using Sentinel-1 and Sentinel-2 Data on the Google Earth Engine Cloud Computing Platform
Reprinted from: *Remote Sens.* **2019**, *11*, 43, doi:10.3390/rs11010043 . . . . **195**

Rosa Aguilar, Raul Zurita-Milla, Emma Izquierdo-Verdiguier and Rolf A. de By
A Cloud-Based Multi-Temporal Ensemble Classifier to Map Smallholder Farming Systems
Reprinted from: *Remote Sens.* **2018**, *10*, 729, doi:10.3390/rs10050729 . . . . **222**

Jun Xiong, Prasad S. Thenkabail, James C. Tilton, Murali K. Gumma, Pardhasaradhi Teluguntla, Adam Oliphant, Russell G. Congalton, Kamini Yadav and Noel Gorelick
Nominal 30-m Cropland Extent Map of Continental Africa by Integrating Pixel-Based and Object-Based Algorithms Using Sentinel-2 and Landsat-8 Data on Google Earth Engine
Reprinted from: *Remote Sens.* **2016**, *.9*, 1065, doi:10.3390/rs9101065 . . . . **240**

Eric A. Sproles, Ryan L. Crumley, Anne W. Nolin, Eugene Mar and Juan Ignacio Lopez Moreno
SnowCloudHydro—A New Framework for Forecasting Streamflow in Snowy, Data-Scarce Regions
Reprinted from: *Remote Sens.* **2018**, *10*, 1276, doi:10.3390/rs10081276 . . . . **267**

Cheng-Chien Liu, Ming-Chang Shieh, Ming-Syun Ke and Kung-Hwa Wang
Flood Prevention and Emergency Response System Powered by Google Earth Engine
Reprinted from: *Remote Sens.* **2018**, *10*, 1283, doi:10.3390/rs10081283 . . . . **282**

Nazmus Sazib, Iliana Mladenova and John Bolten
Leveraging the Google Earth Engine for Drought Assessment Using Global Soil Moisture Data
Reprinted from: *Remote Sens.* **2018**, *10*, 1265, doi:10.3390/rs10081265 . . . . **302**

Gonzalo Mateo-García, Luis Gómez-Chova, Julia Amorós-López, Jordi Muñoz-Marí, Gustau Camps-Valls
Multitemporal Cloud Masking in the Google Earth Engine
Reprinted from: *Remote Sens.* **2018**, *10*, 1079, doi:10.3390/rs10071079 . . . . **325**

Kel N. Markert, Calla M. Schmidt, Robert E. Griffin, Africa I. Flores, Ate Poortinga, David S. Saah, Rebekke E. Muench, Nicholas E. Clinton, Farrukh Chishtie, Kritsana Kityuttachai, Paradis Someth, Eric R. Anderson, Aekkapol Aekakkararungroj and David J. Ganz
Historical and Operational Monitoring of Surface Sediments in the Lower Mekong Basin Using Landsat and Google Earth Engine Cloud Computing
Reprinted from: *Remote Sens.* **2018**, *10*, 909, doi:10.3390/rs10060909 . . . . **343**

Felipe de Lucia Lobo, Pedro Walfir M. Souza-Filho, Evlyn Márcia Leão de Moraes Novo, Felipe Menino Carlos and Claudio Clemente Faria Barbosa
Mapping Mining Areas in the Brazilian Amazon Using MSI/Sentinel-2 Imagery (2017)
Reprinted from: *Remote Sens.* **2018**, *10*, 1178, doi:10.3390/rs10081178 . . . . **362**

Dimosthenis Traganos, Dimitris Poursanidis, Bharat Aggarwal, Nektarios Chrysoulakis and Peter Reinartz
Estimating Satellite-Derived Bathymetry (SDB) with the Google Earth Engine and Sentinel-2
Reprinted from: *Remote Sens.* **2018**, *10*, 859, doi:10.3390/rs10060859 . . . . **376**

Sean A. Parks, Lisa M. Holsinger, Morgan A. Voss, Rachel A. Loehman and Nathaniel P. Robinson
Mean Composite Fire Severity Metrics Computed with Google Earth Engine Offer Improved Accuracy and Expanded Mapping Potential
Reprinted from: *Remote Sens.* **2018**, *10*, 879, doi:10.3390/rs10060879 . . . . **394**

About the Special Issue Editors

Lalit Kumar a professor at the School of Environmental and Rural Science at The University of New England in Australia. His main research interests are environmental modelling, the utilization of high spatial and high spectral resolution data for vegetation mapping and understanding light interactions at the plant canopy level. Most of his recent work has investigated the impacts of climate change on coastal areas and the utilisation of time series remote sensing data and global climate models for mapping historical changes in coastal areas and projections of future changes.

Onisimo Mutanga is a Professor and SARChI Chair on land use planning and management. His expertise lies on vegetation (including agricultural crops) patterns and condition analysis in the face of global and land use change using remote sensing. He integrates ecology, biodiversity conservation and remote sensing to model the impacts of forest fragmentation, pests and diseases, and invasive species on agricultural and natural ecosystems.

Editorial

Google Earth Engine Applications

Onisimo Mutanga [1,*] and Lalit Kumar [2]

1 School of Agricultural, Earth and Environmental Sciences, University of KwaZulu Natal, P. Bag X01 Scottsville, Pietermaritzburg 3209, South Africa

2 Ecosystem Management, School of Environmental and Rural Science, University of New England, Armidale, NSW 2351, Australia; lkumar@une.edu.au

* Correspondence: Mutangao@ukzn.ac.za

Received: 8 March 2019; Accepted: 11 March 2019; Published: 12 March 2019

Keywords: cloud computing; big data analytics; long term monitoring; data archival; early warning systems

---

1. Introduction

The Google Earth Engine (GEE) is a cloud computing platform designed to store and process huge data sets (at petabyte-scale) for analysis and ultimate decision making [1]. Following the free availability of Landsat series in 2008, Google archived all the data sets and linked them to the cloud computing engine for open source use. The current archive of data includes those from other satellites, as well as Geographic Information Systems (GIS) based vector data sets, social, demographic, weather, digital elevation models, and climate data layers.

The easily accessible and user-friendly front-end provides a convenient environment for interactive data and algorithm development. Users are also able to add and curate their own data and collections, while using Google's cloud resources to undertake all the processing. The end result is that this now allows scientists, independent researchers, hobbyists, and nations to mine this massive warehouse of data for change detection, map trends, and quantify resources on the Earth's surface like never before. One does not need the large processing powers of the latest computers or the latest software, meaning that resource-poor researchers in the poorest nations of the world have the same ability to undertake analysis as those in the most advanced nations.

The purpose of this special issue was to solicit papers that take advantage of the Google Engine cloud computing geospatial tools to process large data sets for global applications. Special priority was given to papers from developing nations on how the availability of GEE data and processing has enabled new research that was difficult or impossible before. Key areas covered processing shortcomings, programming, and difficulties in handling data in the cloud atmosphere. We are pleased to report that a total of 22 papers were published in this special issue, covering areas around vegetation monitoring, cropland mapping, ecosystem assessment, and gross primary productivity, among others. A plethora of data sets used ranged from coarse spatial resolution data such as MODIS (Moderate Resolution Imaging Spectroradiometer) to very high-resolution data sets (Worldview -2) and the studies covered the whole globe at varying spatial and temporal scales.

Since its inception in 2010, Google Earth engine usage was investigated using articles drawn from a total of 158 journals. The study showed a skewed usage towards developed countries as compared to developing regions such as Africa [1], with Landsat being the most widely used data set.

This editorial categorized the papers into five main themes, whose contributions are summarized.

1.1. Vegetation Mapping and Monitoring

A number of articles examined the utility of GEE in vegetation mapping and monitoring. This includes the global estimation of key biodiversity variables such as Leaf Area Index (LAI), Fraction of Absorbed Photosynthetically Active Radiation (FAPAR), Fraction Vegetation Cover (FVC), and Canopy water content (CWC) using MODIS historical data [2]. At a planetary scale in Vietnam [3], mapped vegetation using MODIS derived EVI products and the GEE web based application. Using a user defined baseline period, they could monitor the degradation or improvement of vegetation and the impact of mitigation efforts by the Vietnam government. A related study in a Chinese nature reserve used multi seasonal Landsat TM composites to map vegetation and general landcover by minimizing cloud cover and terrain effects. A combination of spectral vegetation indices, terrain ancillary data, and simple illumination transformation could predict vegetation classification with accuracy above 70% [4]. A more refined and accurate 30 m NDVI composite, spanning the past 30 years, was developed for the United States of America using the Google Engine cloud-based planetary processing platform [5]. Missing data due to clouds was filled by using a climate driven modelling approach and data was produced at multiple scales [5]. Using Google Engine, an ecosystem assessment study in a Brazilian semi-arid landscape showed that high spatial resolution data (Worldview) could yield higher classification accuracy compared to medium resolution Landsat TM, with full spectral resolution information [6]. Trees, shrubs, and bare land were classified, with a clear distinction between trees and shrubs, a mammoth task using prior data sets.

Attention was also paid towards rangeland monitoring using long term satellite data in a cloud computing environment. Specifically, a total of 17 pastureland maps were produced in Brazil using MODIS data from 2000 to 2016 with an accuracy above 80% [7]. Results showed an increase in pasture area for most areas analyzed. In another study, [8] integrated GEE and Sentinel 2 data in a machine learning environment to map sea grasses in the Aegean and Iron seas. They could successfully map the seasonal and inter-annual variation of seagrasses up to a depth of 40 m.

1.2. Landcover Mapping

A number of studies assessed land cover dynamics at different spatial scales. Taking advantage of GEE, which provides data access and advanced analytical techniques on big data, [9] used the Bayesian Updating of Land Cover (BULC) algorithm to fuse Landsat data with GlobCover 2009, thereby improving the spatial resolution of the global output from 300 m to 30 m in Brazil. The approach is widely applicable, since it employs an unsupervised algorithm, which does not require intensive ground truthing data [9]. The potential of GEE was also demonstrated in handling huge long term data sets at a global scale to analyze the impact of land cover change on surface urban heat island, taking advantage of the already established climate tool Engine to extract huge land surface temperature data [10]. More than 6000 Landsat images from 2000 to 2011 were processed.

The problem of cloud cover is not new in remote sensing. The availability of time series data and GEE platform facilitated the development of algorithms that solve the cloud cover and terrain effects problems of land cover mapping in a Chinese protected area [4]. The cloud computing platform also facilitated computation of spectral vegetation indices [11] from multi-seasonal Landsat data as well as illumination normalization algorithms, yielding successful land cover classification results. Apart from Landsat data, another study employed high resolution Sentinel 1 and 2 satellite data to map wetland extent at a provincial scale in Newfoundland, Canada [12]. The study produced the first detailed regional wetland map, leveraging on high resolution Sentinel SAR and optical data, the GEE computational power, and advanced machine learning algorithms.

1.3. Agricultural Applications

Remotely sensed agricultural applications that include crop yield estimation, crop area mapping, pests and diseases vulnerability, and suitability assessments, among others, are critical for sustaining productivity and food security. This issue reports on a number of studies that used GEE cloud computing for agricultural applications across varying scales. The GEE platform provided an opportunity to fuse Terra MODIS data and Landsat to estimate Gross Primary Productivity of seven crops in Montana, USA from 2008–2015 at 30 m spatial resolution [11]. The estimated cropland productivity patterns and their seasonal variations compared favorably with the country level crop data. High spatial resolution Worldview 2 data was also used to map small holder heterogeneous cropland areas in the African environment of Mali, using ensemble rules [13]. The cloud platform, with high processing capabilities, allowed the computation of a number of ensemble rules to optimize classification accuracy. At a continental scale, cropland and non-cropland areas were mapped for the entire continent of Africa using a combination of 10-day Sentinel data and 16-day Landsat TM data. 30 m resolution composites were generated using the satellite data, together with elevation data yielding a 30-m slope layer derived from the Shuttle Radar Topographic Mission (SRTM) [14]. The data was then subjected to pixel-based (Random Forest) and object-based (Recursive Hierarchical Segmentation) classifications, yielding results comparable to FAO reports [14].

1.4. Disaster Management and Earth Sciences

Earth science related research, as well as studies that directly address disaster extent and response, were done across the globe and reported in this issue. A snowcloud hydro model, applicable to different environments in Chile, Spain, and the USA was developed using MOD10A1 [15]. The cloud-based model forecasts monthly stream flows in snow areas and mapped snow cover areas and is generally applicable. With respect to disasters, [16] developed a flood prevention and response system using the cloud based GEE platform. The system integrates a whole range of datasets from remote sensing and ancillary sources at each stage of flood events (before, during and after) including Formosat-2, Synthetic aperture radar, and GIS topographic data, and was successfully tested to manage the Typhoon Soudelor in August 2015. The GEE platform was also used to assess drought occurrence using soil moisture as an indicator at a global scale [17]. The cloud based engine facilitated an integration of soil moisture global data sets and web-based processing tools to forecast drought duration as well as intensity and the model was successfully tested in Ethiopia and South Africa [17].

Other applications in this issue include cloud masking using multi-temporal approaches [18], surface sediment monitoring [19], and mining area mapping using sentinel data [20].

In summary, this issue has demonstrated the power of GEE platform in handling huge data sets at various scales and building automated programs that can be used at an operational level. This is a great step in solving environmental problems affecting the earth and is critical in achieving the UN millennium development goals. The applications demonstrated are wide ranging, from mining, agriculture, ecosystem services, and drought monitoring, among others. Day to day, monthly, seasonal and long-term monitoring of phenomena at high spatial resolution and covering large extents is now possible with the availability of such platforms that can handle big data.

Acknowledgments: This work was supported by the DST/NRF Chair in Land use planning and management, Grant No. 84157.

Conflicts of Interest: The authors declare no conflict of interest.

References

1. Kumar, L.; Mutanga, O. Google Earth Engine Applications Since Inception: Usage, Trends, and Potential. *Remote Sens.* **2018**, *10*, 1509. [CrossRef]
2. Campos-Taberner, M.; Moreno-Martínez, Á.; García-Haro, F.J.; Camps-Valls, G.; Robinson, N.P.; Kattge, J.; Running, S.W. Global Estimation of Biophysical Variables from Google Earth Engine Platform. *Remote Sens.* **2018**, *10*, 1167. [CrossRef]
3. Poortinga, A.; Clinton, N.; Saah, D.; Cutter, P.; Chishtie, F.; Markert, K.N.; Anderson, E.R.; Troy, A.; Fenn, M.; Tran, L.H.; et al. An Operational Before-After-Control-Impact (BACI) Designed Platform for Vegetation Monitoring at Planetary Scale. *Remote Sens.* **2018**, *10*, 760. [CrossRef]
4. Tsai, Y.H.; Stow, D.; Chen, H.L.; Lewison, R.; An, L.; Shi, L. Mapping Vegetation and Land Use Types in Fanjingshan National Nature Reserve Using Google Earth Engine. *Remote Sens.* **2018**, *10*, 927. [CrossRef]

5. Robinson, N.P.; Allred, B.W.; Jones, M.O.; Moreno, A.; Kimball, J.S.; Naugle, D.E.; Erickson, T.A.; Richardson, A.D. A Dynamic Landsat Derived Normalized Difference Vegetation Index (NDVI) Product for the Conterminous United States. *Remote Sens.* **2017**, *9*, 863. [CrossRef]
6. Goldblatt, R.; Rivera Ballesteros, A.; Burney, J. High Spatial Resolution Visual Band Imagery Outperforms Medium Resolution Spectral Imagery for Ecosystem Assessment in the Semi-Arid Brazilian Sertão. *Remote Sens.* **2017**, *9*, 1336. [CrossRef]
7. Parente, L.; Ferreira, L. Assessing the Spatial and Occupation Dynamics of the Brazilian Pasturelands Based on the Automated Classification of MODIS Images from 2000 to 2016. *Remote Sens.* **2018**, *10*, 606. [CrossRef]
8. Traganos, D.; Aggarwal, B.; Poursanidis, D.; Topouzelis, K.; Chrysoulakis, N.; Reinartz, P. Towards Global-Scale Seagrass Mapping and Monitoring Using Sentinel-2 on Google Earth Engine: The Case Study of the Aegean and Ionian Seas. *Remote Sens.* **2018**, *10*, 1227. [CrossRef]
9. Lee, J.; Cardille, J.A.; Coe, M.T. BULC-U: Sharpening Resolution and Improving Accuracy of Land-Use/Land-Cover Classifications in Google Earth Engine. *Remote Sens.* **2018**, *10*, 1455. [CrossRef]
10. Ravanelli, R.; Nascetti, A.; Cirigliano, R.V.; Di Rico, C.; Leuzzi, G.; Monti, P.; Crespi, M. Monitoring the Impact of Land Cover Change on Surface Urban Heat Island through Google Earth Engine: Proposal of a Global Methodology, First Applications and Problems. *Remote Sens.* **2018**, *10*, 1488. [CrossRef]
11. He, M.; Kimball, J.S.; Maneta, M.P.; Maxwell, B.D.; Moreno, A.; Beguería, S.; Wu, X. Regional Crop Gross Primary Productivity and Yield Estimation Using Fused Landsat-MODIS Data. *Remote Sens.* **2018**, *10*, 372. [CrossRef]
12. Mahdianpari, M.; Salehi, B.; Mohammadimanesh, F.; Homayouni, S.; Gill, E. The First Wetland Inventory Map of Newfoundland at a Spatial Resolution of 10 m Using Sentinel-1 and Sentinel-2 Data on the Google Earth Engine Cloud Computing Platform. *Remote Sens.* **2018**, *11*, 43. [CrossRef]
13. Aguilar, R.; Zurita-Milla, R.; Izquierdo-Verdiguier, E.; de By, R.A. A Cloud-Based Multi-Temporal Ensemble Classifier to Map Smallholder Farming Systems. *Remote Sens.* **2018**, *10*, 729. [CrossRef]
14. Xiong, J.; Thenkabail, P.S.; Tilton, J.C.; Gumma, M.K.; Teluguntla, P.; Oliphant, A.; Congalton, R.G.; Yadav, K.; Gorelick, N. Nominal 30-m Cropland Extent Map of Continental Africa by Integrating Pixel-Based and Object-Based Algorithms Using Sentinel-2 and Landsat-8 Data on Google Earth Engine. *Remote Sens.* **2017**, *9*, 1065. [CrossRef]
15. Sproles, E.A.; Crumley, R.L.; Nolin, A.W.; Mar, E.; Lopez Moreno, J.I. SnowCloudHydro—A New Framework for Forecasting Streamflow in Snowy, Data-Scarce Regions. *Remote Sens.* **2018**, *10*, 1276. [CrossRef]
16. Liu, C.-C.; Shieh, M.-C.; Ke, M.-S.; Wang, K.-H. Flood Prevention and Emergency Response System Powered by Google Earth Engine. *Remote Sens.* **2018**, *10*, 1283. [CrossRef]
17. Sazib, N.; Mladenova, I.; Bolten, J. Leveraging the Google Earth Engine for Drought Assessment Using Global Soil Moisture Data. *Remote Sens.* **2018**, *10*, 1265. [CrossRef]
18. Mateo-García, G.; Gómez-Chova, L.; Amorós-López, J.; Muñoz-Marí, J.; Camps-Valls, G. Multitemporal Cloud Masking in the Google Earth Engine. *Remote Sens.* **2018**, *10*, 1079. [CrossRef]
19. Markert, K.N.; Schmidt, C.M.; Griffin, R.E.; Flores, A.I.; Poortinga, A.; Saah, D.S.; Muench, R.E.; Clinton, N.E.; Chishtie, F.; Kityuttachai, K.; et al. Historical and Operational Monitoring of Surface Sediments in the Lower Mekong Basin Using Landsat and Google Earth Engine Cloud Computing. *Remote Sens.* **2018**, *10*, 909. [CrossRef]
20. Lobo, F.D.L.; Souza-Filho, P.W.M.; Novo, E.M.L.d.M.; Carlos, F.M.; Barbosa, C.C.F. Mapping Mining Areas in the Brazilian Amazon Using MSI/Sentinel-2 Imagery (2017). *Remote Sens.* **2018**, *10*, 1178. [CrossRef]

Article

Google Earth Engine Applications Since Inception: Usage, Trends, and Potential

Lalit Kumar [1,*] and Onisimo Mutanga [2]

[1] School of Environmental and Rural Science, University of New England, Armidale, NSW 2351, Australia
[2] School of Agricultural, Earth and Environmental Sciences, University of KwaZulu- Natal, P. Bag X01 Scottsville, Pietermaritzburg 3209, South Africa; MutangaO@ukzn.ac.za
* Correspondence: lkumar@une.edu.au; Tel.: +61-2-6773-5239

Received: 30 July 2018; Accepted: 18 September 2018; Published: 20 September 2018

Abstract: The Google Earth Engine (GEE) portal provides enhanced opportunities for undertaking earth observation studies. Established towards the end of 2010, it provides access to satellite and other ancillary data, cloud computing, and algorithms for processing large amounts of data with relative ease. However, the uptake and usage of the opportunity remains varied and unclear. This study was undertaken to investigate the usage patterns of the Google Earth Engine platform and whether researchers in developing countries were making use of the opportunity. Analysis of published literature showed that a total of 300 journal papers were published between 2011 and June 2017 that used GEE in their research, spread across 158 journals. The highest number of papers were in the journal Remote Sensing, followed by Remote Sensing of Environment. There were also a number of papers in premium journals such as Nature and Science. The application areas were quite varied, ranging from forest and vegetation studies to medical fields such as malaria. Landsat was the most widely used dataset; it is the biggest component of the GEE data portal, with data from the first to the current Landsat series available for use and download. Examination of data also showed that the usage was dominated by institutions based in developed nations, with study sites mainly in developed nations. There were very few studies originating from institutions based in less developed nations and those that targeted less developed nations, particularly in the African continent.

Keywords: Google Earth Engine; web portal; satellite imagery; trends; earth observation

1. Introduction

The Google Earth Engine (GEE) is a web portal providing global time-series satellite imagery and vector data, cloud-based computing, and access to software and algorithms for processing such data [1]. The data repository is a collection of over 40 years of satellite imagery for the whole world, with many locations having two-week repeat data for the whole period, and a sizeable collection of daily and sub-daily data as well. The data available is from multiple satellites, such as the complete Landsat series; Moderate Resolution Imaging Spectrometer (MODIS); National Oceanographic and Atmospheric Administration Advanced very high resolution radiometer (NOAA AVHRR); Sentinel 1, 2, and 3; Advanced Land Observing Satellite (ALOS) etc. Table A1 gives a list of various satellite-based products, including raw and pre-processed bands, indices, composites, and elevation models that have worldwide coverage. It does not include other derived products, such as landcover and topographic features, that are available on the GEE platform. The table also does not include the datasets with the spatial coverage at national and regional extents. It also does not include most of the geophysical, demographic, and climate and weather data. The complete list can be obtained from the portal webpage (https://earthengine.google.com/datasets/).

While the initial setup included remote sensing data only, large amounts of vector, social, demographic, digital elevation models, and weather and climate data layers have now been added [2]. Most of the images have already been cleaned of cloud cover and have been mosaicked (by previous users) for quicker and easier processing; however, original imagery is available as well and the amount of original imagery far outweighs the amount of pre-build cloud-removed mosaics. In practice, images do not have the cloud pixels removed by default, and users have access to ancillary layers (e.g., Landsat Collection 1 Level-1 Quality Assessment Band) or algorithms (e.g., SimpleCloudScore, F-Mask) and decide when they use it in their scripts. All the raw imagery is available, along with cloud-cleared and mosaicked imagery. Some datasets have been pre-processed to convert raw digital numbers to top-of-the-atmosphere reflectance and even surface reflectance, rendering them suitable for further analysis without needing specialized software for solar and atmospheric corrections. Other ready-to-use computed products, such as Enhanced Vegetation Index (EVI) and Normalized Difference Vegetation Index (NDVI), are also available.

The programming interface allows users to create and run custom algorithms, and analysis is parallelized so that many processors are involved in any given computation, thus speeding up the process considerably. This enables global-scale analysis to be performed with considerable ease, as compared to desktop computing. One such example is the work by Hansen et al. [3] where the authors identified global-scale forest cover change between the years 2000 and 2012 using 654,178 Landsat 7 scenes (30 m spatial resolution), totaling 707 terabytes of data. The processing took 100 h on GEE, compared to approximately 1,000,000 h it would have taken on a standard desktop computer [4]. Images can be exported from Earth Engine in GeoTIFF or TFRecord format. GEE allows raw or processed images, map tiles, tables, and video to be exported from Earth Engine to a user's Google Drive account, to Google Cloud Storage or to a new Earth Engine asset. Google cloud storage is a fee-based service for which one needs set up a project, enables billing for the project, and creates a storage bucket. Users are free to upload other datasets and decide whether to share the data they have uploaded and scripts they have written with others or not.

This plethora of multi-temporal data, with local to global coverage, presents researchers with an unprecedented opportunity to undertake research with minimal cost and equipment. The cloud computing power of GEE enables the processing of petabytes of image data, combined with other vector data, within the cloud environment and removes the need to store, process, and analyze the large volumes of satellite data on an office computer. There is now a reduced need for computers with fast processing speeds and large storage capacities. Users do not have to entirely depend on specialist remote sensing software, such as Environment for Visualizing Images (ENVI) and Earth Resources Data Analysis System (ERDAS) Imagine; however, they may still be needed for special functions that are not available on GEE (such as object-based image analysis). Satellite images do not need to be downloaded, which is a major boon for regions with slow internet speeds, but internet connection is still needed to use GEE.

Such a development presents great opportunities and is also a great levelling field for researchers. The availability of data sets and innovative data processing algorithms provided by GEE or shared by other users of GEE should help to improve our capability to process earth observation data to support management decisions, irrespective of where we reside. A researcher living in Zimbabwe has relatively similar opportunities as one living in Australia. All that is needed is a basic desktop computer and internet connectivity. The capability to import and upload data on the Earth Engine's public data catalogue provides immense opportunities for data updating and sharing. Users can upload their own raster and vector data sets and even share scripts with other users for free, thereby promoting knowledge exchange. The data request forums and troubleshooting platforms makes data accessibility easier and provides fingertip assistance on processing. Given this background, the question that arises is whether researchers are making use of this opportunity, particularly those based in the less developed parts of the world? It is researchers in such environments who have been at a distinct disadvantage in utilizing earth observation data to its fullest capacity due to funding and infrastructure constraints [5]. This bridge has now been narrowed considerably, so one would expect researchers in such regions to take advantage of this noble innovation.

This research set out to investigate whether the availability of such large amounts of data at worldwide coverage, with free access (for research, education, and non-profit organizations) to data processing algorithms and cloud computing facilities, had led to increased research in less-developed nations, and whether researchers from these regions were embracing the opportunity. Research publications from 2010 (establishment of GEE) onwards were searched for using the keyword "Google Earth Engine" and all resulting publications were individually analyzed to record the origin of the principal author's affiliated institution, the origins of the affiliated institutions of all authors, the primary study site, the scale of the study, the subject area, datasets used, and the number of papers for each year of publication. The data was used to investigate patterns, authorship origins, and whether there was a general take-up of opportunities in the less developed nations.

2. Materials and Methods

Google Scholar and Web of Science were used to search for all articles with the words "Google Earth Engine" or "GEE" anywhere in the article, except in the references section. All such articles were downloaded into Endnote (Clarivate Analytics, 1500 Spring Garden Philadelphia, PA 19130, USA) and duplicates were then removed. Conference papers, books and book chapters, audio-visual material, newspaper articles, reports, thesis, websites, and abstracts from other sources were also discarded. A key reason for this was that the full versions of these were generally not available to enable extractions of all required data. Patents and review papers in journals were also removed. The remaining papers (journal articles) were manually screened to identify the subject areas, study sites, the scale of the study (global, continental, regional, country, or sub-country), number of authors, origin of the principal author's affiliated institution, the origins of the affiliated institutions of all authors, datasets used, and number of papers per year. The information about author's affiliated institutions was obtained from the contact address on the manuscript. Where multiple addresses were given for an author, the first address was used. It should be clarified that this research was not about the origin of authors but where they were based when the research was conducted. The subject areas were quite varied, so they were grouped into 16 broad categories. Data was analyzed in Excel and ArcGIS software (ESRI, 380 New York Street Redlands, CA 92373-810, USA).

3. Results

The initial search resulted in 785 articles and, after screening for duplicates, 485 articles remained. Of these, there was one audiovisual material, five books, 24 book sections, 66 conference proceedings, two films or broadcasts, one newspaper article, 11 reports, 20 theses, one web page, one generic material, 49 manuscripts where GEE was only mentioned in the references or was not in English or was a review paper, and 304 were journal articles. From these 304 journal articles that had actually used GEE in their research, four were either technical responses to other GEE related papers or only mentioned GEE as a graphics interface from which other relevant data could not be extracted. Thus 300 papers remained that were actual research papers published in journals between January 2010 and June 2017, which were then subjected to detailed analysis. The 300 journal papers were published across 158 different journals, many of which had a single paper. The majority of the papers were in *Remote Sensing* (32, IF: 3.41), *Remote Sensing of Environment* (19, IF: 6.46), *Science of Total Environment* (8, IF: 4.61), *PLoS ONE* (8, IF: 2.77), *IEEE Journal of Selected Topics in Applied Earth Observation* (6, IF: 2.78), *International Journal of Remote Sensing* (5, IF: 1.72), and *Remote Sensing Applications—Society and Environment* (5, IF: NA). *Applied Geography, Environmental Modelling and Software, Forest Ecology and Management, International Journal of Digital Earth, ISPRS Journal of Geoinformation, Science*, and *Malaria Journal* had four papers each.

3.1. Publication Trend

The trend of publication using GEE increased slowly in the first few years but had gained pace in the last three years. The first paper using GEE was published in 2011, and a total of nine papers were published that year. This increased to 109 in 2016 and 90 for the first half of 2017 (Table 1).

Table 1. Number of journal papers utilizing the GEE platform published each year since inception. Note that for 2017, the data is until 30 June.

Year	Number of Papers
2011	9
2012	10
2013	10
2014	25
2015	47
2016	109
2017	90

3.2. Application Regions

Fifty papers out of 300 (17%) covered the whole world as an application region, while there were 28 studies at a continental scale. Sixty-three studies were at the country scale and 139 at the sub-country scale. There were another 20 studies that were at variable scales or for extraterrestrial work. At the continental scale, most studies covered the American continent (seven for North America and another seven for South America), seven for Africa, three for Asia, three for Europe and one for Australia (Figure 1). The spatial distribution of the papers shows that at the country and sub-country level, the highest number of publications selected USA (60) as their application region, followed by China (24), Brazil (19), India (15), Indonesia (15), Australia (14), and Canada (11) (Figure 2, Table A2). A number of countries in the Middle East and Africa had no studies. Europe combined had 45 studies, with Italy having 7, Germany having 5, and U.K. having 4 (Appendix A Table A2).

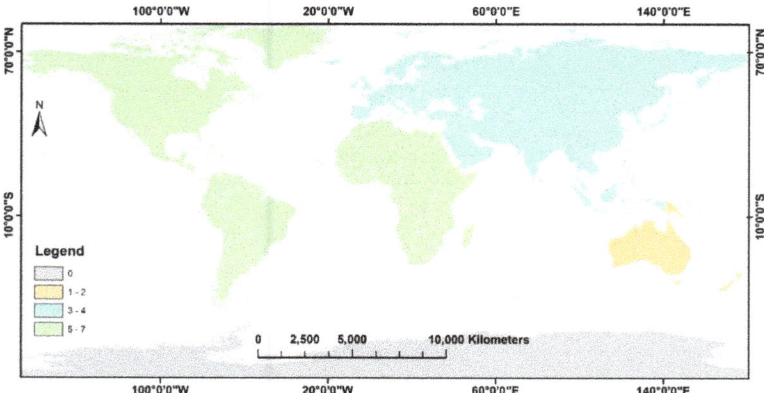

Figure 1. A breakdown of applications of GEE by continents. This figure includes those studies conducted at continental scales only. There were 50 such studies in total.

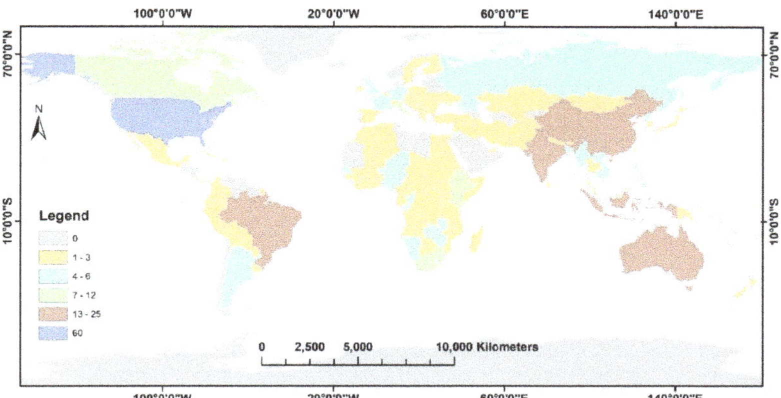

Figure 2. A breakdown of applications of GEE by country. If a study included several countries, then all those countries were included in the count each time. The countries in grey had no studies. This figure does not show the institution of the authors but the actual study sites.

3.3. Application Disciplines

There was a wide variety of application areas, ranging from agriculture, forestry, and ecology to economics and medicine (diseases). These were broadly placed into 16 categories, as shown in Figure 3. The highest number of studies were in the forest and vegetation category (17%); followed by 10% in land use and land cover studies; 8% in ecosystem and sustainability, wetland and hydrology, and in data manipulation; 7% in agriculture; 5% in mapping and change detection; 4% in both remote sensing applications and modelling and geoscience research; 3% each in cloud computing, soil, disease, climate science, and urban studies; and 2% in natural hazard and disaster studies. In addition, 11% of application disciplines were incorporated in the "others" category as their numbers were too small to create their own classes. These consisted of applications in areas such as economics, air pollution, virtual environments, air temperature, and archaeology. Therefore, overall, a large portion of the studies were in the natural resources mapping and management domains.

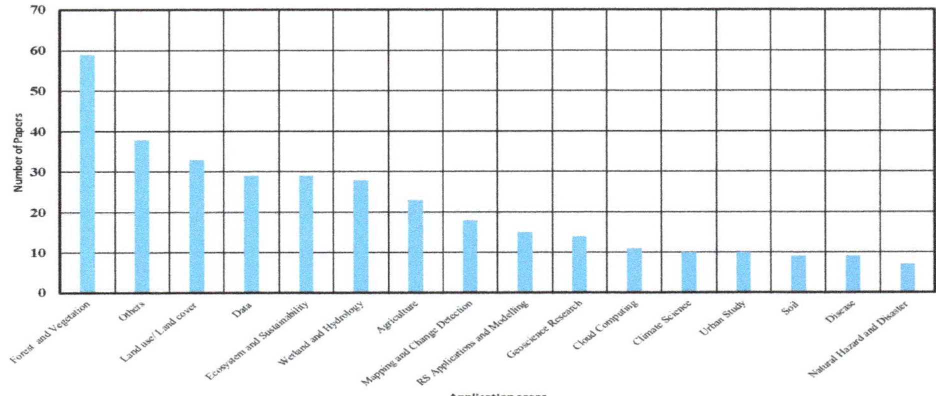

Figure 3. A broad categorization of application disciplines of GEE across the 300 papers surveyed in this research. There is always subjectivity involved in differentiating the "Forest and Vegetation" category and the "Landuse/Landcover" category. Together these two categories account for most of the GEE applications across the 300 papers surveyed.

3.4. Data Used in GEE Research

The majority of the work analyzed as part of this research used Landsat data (159) followed by MODIS (80). The other common datasets used were Google Earth (24), Satellite Probatoire d'Observation de la Terre (SPOT) (19), Sentinel (19), Shuttle Radar Topography Mission (SRTM) (18) and Advanced Land Observing Satellite (ALOS) Phased Array type L-band Synthetic Aperture Rada (PALSAR) (17) (Figure 4). Note that Google Earth data cannot be subjected to spectral analysis as with other remote sensing data, so it most probably was used for visualization or for manual identification of classes.

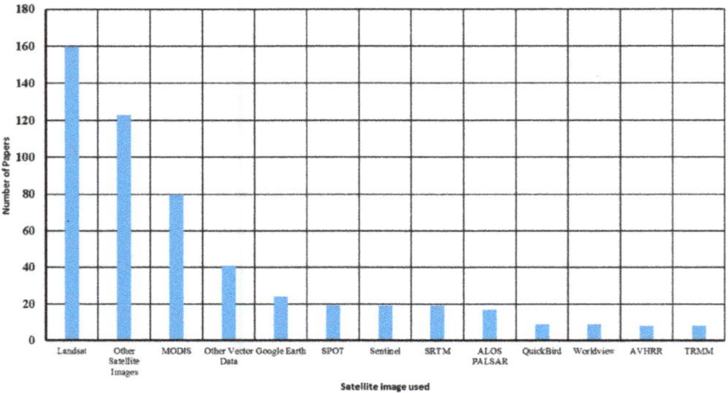

Figure 4. Satellite data source used in GEE research across the 300 papers surveyed. The "Other Satellite Imagery" category consisted of data such as IKONOS, Advanced Spaceborne Thermal Emission and Reflection Radiometer (ASTER), Medium Resolution Imaging Spectrometer (MERIS), Light Detection and ranging (LiDAR), Synthetic Aperture Radar (SAR), Proba, and Environmental Satellite (ENVISAT).

3.5. Authorship Patterns

Of the 300 papers that were analyzed, 137 had the primary author based in the United States, followed by 20 in Italy, 18 in Germany, 16 in China, and 15 in the U.K. (Figure 5). Very few papers were from researchers based in less developed countries, particularly Africa. When taking into consideration all authors of the 300 manuscripts (1447 authors in total), the majority (589, or 41%) were based in the United States, followed by Italy (120), Australia (86), Germany (79), China (68), and the U.K. (65). Again, there were no or very few authors from institutions from less developed countries contributing to these researches (Figure 6, Appendix A Table A3).

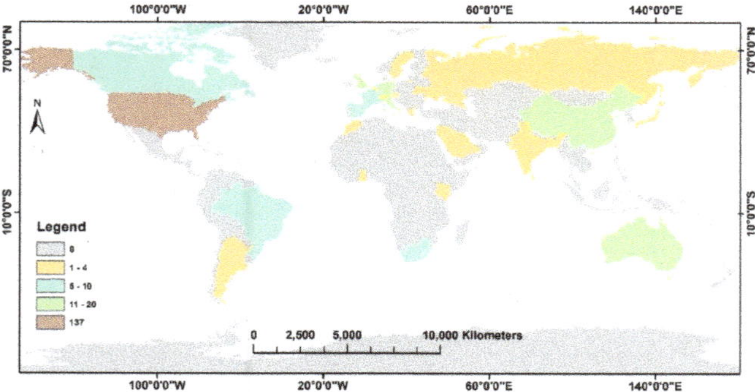

Figure 5. Distribution of the first author's affiliated institution (based on the address provided in the manuscript) for the 300 papers surveyed. Where multiple affiliations were provided by an author, only the first affiliation was considered.

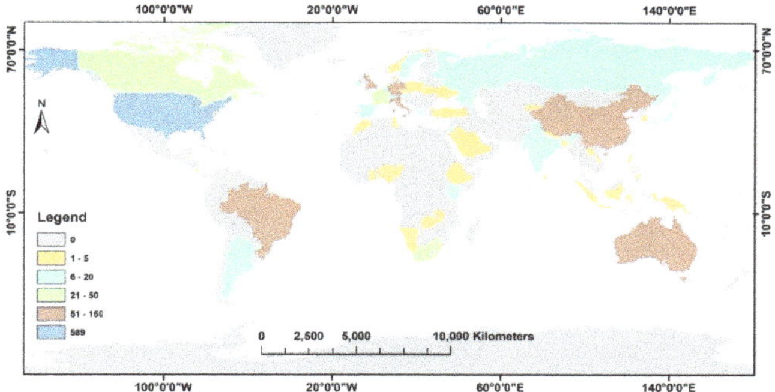

Figure 6. Distribution of all authors' affiliated institutions (based on the address provided in the manuscript) for the 300 papers surveyed. There were a total of 1447 authors across the 300 journal papers. Where multiple affiliations were provided by an author, only the first affiliation was considered.

4. Discussion

This paper has unpacked, among others, trends in use, areas of major use, major data sources being used, and authorship patterns of GEE. The results of the paper have shown that usage of GEE has been burgeoning over the years, with clear advantages of usage being shown in the number of publications. Researchers have highlighted the opportunities provided, including storage capacity, archival and processed data usage, as well as powerful processing capabilities engrained in the platform. A number of studies have taken advantage of easy accessibility to more than 40 years of free historical imagery and scientific data sets in order to develop large-scale and long-term monitoring applications [1].

The platform, through a dedicated high-performance computing infrastructure, provides computational capability for data preparation and allows systematic creation and training of many classifiers and algorithms [6–9]. For example, using machine learning algorithms, cloud computing allows for efficient testing of various base classifiers and combination, as well as training of the ensemble with a wide range of spectral, spatial, and temporal bands derived from very high spatial resolution satellite images [10].

This study has shown that there is variability in the application regions of GEE, with 17% applied at the global scale and the U.S.A. dominating at the country level. The 17% global application is a significant percentage, which is attributed to the processing as well as data storage capabilities of GEE [11]. To analyze data at a global scale requires more than 10 petabytes of data and efficient processing power, a strength engrained in GEE [11–13].

The authorship pattern also shows institutions from developed countries dominating the first author position with very few scholars from institutions based in less-developed counties. The wide application and dominant authorship in the U.S.A. and other developed countries clearly indicates the ease of accessibility to the technology as well as the capability to utilize the platform. There is also a wide use of data sets, such as Landsat, which are open source [13,14]. This is not a surprising result since an archive of such images are already available on the GEE platform and some are pre-processed (cloud-cleared, georeferenced, Top of atmosphere (TOA), and surface reflectance, etc.) for immediate application. This signals the importance of expanding the database to include higher-resolution datasets.

Variability in the application fields is also quite significant in this analysis. Most papers applied GEE in the field of vegetation and forest monitoring as well as landcover/use change mapping. This is commensurate with trends in remote sensing where most of the global research and journals target landcover and vegetation. There are few studies on disaster monitoring, disease, and soils. There could be an overlap between diseases category with vegetation and agriculture, where studies have been undertaken on disease infestation in vegetation and crops [15]. Natural disasters and geological applications require more publicity since the benefits of the platform are huge; however, it should be noted that natural disasters mapping needs to be processed in a more timely manner and the delay in the availability of images on GEE means that it is not an ideal platform for such an application. Having said that, however, users are still able to upload their own images on the platform and use GEEs massive processing power to deliver timely products.

While the development of GEE targeted poor countries as part of the REDD program for forest measurement and monitoring [16], results from this study have shown that institutions based in most developing countries are not using the facility effectively. Of serious concern is a lack of studies that have used GEE in Africa by Africa-based researchers. This could be attributed to a number of factors, which include data accessibility, technological skills to process the data, and opportunities for research. Due to the huge data sets involved, it is a technical and financial challenge to process the data using traditional local processing methods (e.g., image file download) and many researchers may not be aware that by using GEE, one does not have to download the datasets. A similar study on above ground biomass research showed that most satellite images, such as LiDAR and radar, are beyond the reach of many researchers in Africa due to the costs involved; thus, even with the availability of big data processing engines, the cost of data acquisition is a challenge, especially if such data is not available in the GEE repository [17]. The same pattern was also observed with crop health monitoring [18], where applications are limited to a few African countries, such as South Africa. In addition, most scientists in the application field do not have a background in programming. While the routines offered by GEE are relatively easier to follow, its application remains elusive until a proper graphical user interface (GUI), such as those found in ERDAS Imagine or ENVI, are fully developed. Therefore, the opportunity offered by the GEE platform falls short of being fully utilized in less-developed countries. However, GEE scripts can be made into web applications with a GUI and has the capability to increase usability and transferability. The development of Information Technology (IT) processing platforms, with user friendly GUI, should incorporate the size of the data, complexities, and analytical functions of the Big Data [19]. The study by Xiong et al. [20], undertaken to automate cropland mapping of continental Africa using GEE, shows the potential of the platforms application for the African continent.

The other problem is also a lack of awareness and poor data sharing networks in less developed countries. Of a few universities rapidly surveyed in South Africa through email questions to GIS departmental lecturers, very few apply GEE in teaching at undergraduate and post graduate levels, and neither is there widespread knowledge of its existence. In addition, the spatial data infrastructure in less developed countries is poorly networked for easy sharing of information [21,22]. This leads to disaggregated research silos with constricted cross-pollination of ideas and technology.

Overall, the Google Earth Engine goes a long way in providing realistic solutions to challenges of processing Big Data into usable information for addressing environmental concerns. It has been specifically designed to manage large volumes of data, a major bottleneck for researchers utilizing satellite images. GEE also has the advantage that many data layers are already available on the platform, with some already screened to remove cloud cover, converted to top-of-atmosphere reflectance, surface reflectance, and georeferenced. The platform also allows for the sharing of computer codes, so users do not have to be proficient in Javascript or Python coding, and there is a very active online community providing support. Given GEE's speed, flexibility, and accessibility, there is tremendous opportunity for the research community to make use of this platform for earth observation studies. GEE provides an answer and opportunity for researchers based in developing countries who often complain about data accessibility, funding for computer hardware and software, and general lack of resources. Hopefully, there is a bigger uptake of this opportunity by researchers in such regions and more applications are developed to better manage our rapidly diminishing resources.

With regard to limitations, Salager-Meyer [23] highlighted the disparities and inequities that exist in the world of scholarly publishing and the results seen in this paper are perhaps a result of those issues. It is seen from this research that in many developing countries, the trend of publishing research output using GEE in most disciplines drops behind that for the developed nations. The absence of citations or publications does not mean that the researchers from developing countries do not engage with scientific research or they do not consider the visibility of their work. Gibbs [24] stated that the share of mainstream journal articles in developing countries is insignificant even though it comprises of 24.1% of the world's scientists. There are many reasons for under-representation of publications from developing countries, such as financial restrictions [23], poor facilities (i.e., Internet access gap, infrastructure, inadequate laboratory equipment), lack of international collaboration, political legitimacy, limited technical support, and inadequate training [25–27]. Further, the language barrier on intercultural communication is the most frequent root cause for this problem as many researchers from developing countries do not speak English as their first language, which can pose a barrier for them to share their scientific findings [23]. Our study did not account for any of these issues.

It should also be noted that while this paper used the publication record as a proxy for "usage" of GEE, not everyone using GEE does it for scientific publication or is interested in sharing the findings with other academics. Many applications, such as the Global Forest Watch, are for operational use and are not captured in an analysis of publication data that is geared towards academia.

5. Conclusions

This research investigated the uptake and usage of the Google Earth Engine (GEE) platform, mainly in terms of the geographic location of users, the datasets used, and the broad fields of study. As one of the key goals of GEE is to provide a platform for planetary-scale geospatial analysis that is accessible for everyone, the central question of this manuscript is invaluable. Peer-reviewed literature was used for assessing the authorship patterns, the geographic scope of analysis, and major area of use. The results show that the use of GEE is dominated by developed countries, both in terms of user nationality (as given by institutional affiliation), and geographic application, while the applications of GEE in terms of subject matter are quite diverse. GEE provides substantial opportunities for earth observation and geospatial applications, and that it potentially eliminates some of the barriers, particularly in the developing world. However, this has yet to be fully realized, and opportunities exist to improve on this.

Overall, GEE has opened a new big data paradigm for storage and analysis of remotely sensed data at a scale that was not feasible using desktop processing machines. A more aggressive intervention approach could be taken to increase applications in developing countries. Similar to the European Space Agency (ESA) Thematic Exploitation Platforms (TEPs) initiated in 2014 (https://tep.eo.esa.int/about-tep), developing countries should also be involved in the Earth Observation Exploitation Platforms. The TEPS are an initiative by the ESA to create research and development interconnected themes around coastal, hydrology, urban, and food security, among others. Such initiatives provide platforms for collaboration and virtual work environments, including provision of access to earth observation data and tools, processors, and IT resources required using one coherent interface.

Author Contributions: L.K. conceptualized the idea, collected and analyzed data and wrote certain sections of the paper. O.M. conceptualized the idea, interpreted data, wrote certain sections and proof read the paper.

Funding: The paper was partly funded by the SARChI Chair in Land use Planning and Management.

Acknowledgments: The authors would like to thank Manoj Kumer Ghosh, Sadeeka Layomi Jayasinghe, Sushil Lamichhane and Cate MacGregor for help with database development and analysis.

Conflicts of Interest: "The authors declare no conflicts of interest".

Appendix A

Table A1. A summary of the main satellite imagery available in Google Earth Engine.

Image Collection	Description	Data Availability (Time)	Resolution (Meters)	Revisit Interval (Days)	Provider
Sentinel-1 SAR GRD	C-band Synthetic Aperture Radar Ground Range Detected, log scaling	3 October 2014–present	10	3	European Union/ESA/Copernicus
Sentinel-2 MSI	Multi Spectral Instrument, Level-1C	23 June 2015–present	10, 20, 60	5	European Union/ESA/Copernicus
Sentinel-3 OLCI EFR	Ocean and Land Color Instrument Earth Observation Full Resolution	18 October 2016–present	300	2	European Union/ESA/Copernicus
Landsat 1 MSS	Tier 1 and 2 (raw)	23 July 1972–7 January 1978	30, 60	16	USGS
Landsat 2 MSS	Tier 1 and 2 (raw)	22 January 1975–26 February 1982	30, 60	16	USGS
Landsat 3 MSS	Tier 1 and 2 (raw)	5 March 1978–31 March 1983	30, 60	16	USGS
Landsat 4 MSS	Tier 1 and 2 (raw)	16 July 1982–14 December 1993	30, 60	16	USGS
Landsat 4 TM	Tier 1 and 2 (raw, TOA reflectance, surface reflectance); 8 day, 32 day and annual composites (BAI, EVI, NDSI, NDVI, NDWI, Raw, TOA Reflectance), Annual greenest-pixel TOA Reflectance Composite	22 August 1982–14 December 1993	30	16	USGS
Landsat 5 MSS	Tier 1 and 2 (raw)	1 March 1984–31 January 2013	30, 60	16	USGS
Landsat 5 TM	Tier 1 and 2 (Raw, TOA reflectance, surface reflectance); 8 day, 32 day and annual composites (same as Landsat 4)	1 January 1984–5 May 2012	30	16	USGS
Landsat 7	Tier 1 and 2 (Real time, Raw, TOA reflectance, surface reflectance); 8 day, 32 day and annual composites (same as Landsat 4)	1 January 1999–present	15, 30	16	USGS
Landsat 8	Tier 1 and 2 (Real time, Raw, TOA reflectance, surface reflectance); 8 day, 32 day and annual composites (same as Landsat 4)	11 April 2013–present	15, 30	16	USGS
MODIS (Aqua and Terra)	Various bands, indices and composites	24 February 2000–present	250, 500, 1000	1	NASA LP DAAC at the USGS EROS Center
DMSP OLS	Global Radiance-Calibrated Nighttime Lights Version 4, Defense Meteorological Program Operational Linescan System	16 March 1996–July 2011	≈1 km (30 arc seconds)		NOAA
DMSP OLS	Nighttime Lights Time Series Version 4, Defense Meteorological Program Operational Linescan System	1 January 1992–1 January 2014	≈1 km (30 arc seconds)		NOAA
NOAA AVHRR	Various bands, indices and composites	24 June 1981–present	≈1.09 km (Different products at different resolutions)	1	NOAA

Table A1. *Cont.*

Image Collection	Description	Data Availability (Time)	Resolution (Meters)	Revisit Interval (Days)	Provider
ALOS/AVNIR-2 ORI	Orthorectified imagery from the Advanced Visible and Near Infrared Radiometer type 2 (AVNIR-2) sensor on-board the Advanced Land Observing Satellite (ALOS) "DAICHI".	26 April 2006–18 April 2011	10		JAXA Earth Observation Research Center
ALOS DSM	Global AW3D30		30 (1 arc second)		JAXA Earth Observation Research Center
SRTM	DEM 30m	11 February 2000–22 February 2000	30 (1 arc second)		NASA/USGS/JPL-Caltech
SRTM	DEM 90m version 4	11 February 2000–22 February 2000	90		NASA/CGIAR
ASTER	L1T Radiance	4 March 2000–present	15, 30, 90	5	NASA LP DAAC at the USGS EROS Center
ASTER Global Emissivity Dataset	This product includes the mean emissivity and standard deviation for all five ASTER thermal infrared bands, mean land surface temperature (LST) and standard deviation, a re-sampled ASTER GDEM, land-water mask, mean Normalized Difference Vegetation Index (NDVI) and standard deviation, and observation count.	1 January 2000–31 December 2008	100		NASA
TRMM 3B42	3-Hourly Precipitation Estimates	1 January 1998–31 May 2018	0.25 arc degrees		NASA GSFC
TRMM 3B43	Monthly Precipitation Estimates	1 January 1998–1 May 2018	0.25 arc degrees		NASA GSFC
GPM Global Precipitation Measurement v5	Data provided at 30 min cadence	12 March 2014–present	0.1 arc degrees		NASA PMM
GSMaP Operational	Data provided at hourly cadence	1 March 2014–present	0.1 arc degrees		JAXA Earth Observation Research Center
GSMaP Reanalysis	Data provided at hourly cadence	1 March 2000–12 March 2014	0.1 arc degrees		JAXA Earth Observation Research Center
CHIRPS Daily precipitation	Climate Hazards Group InfraRed Precipitation with Station Data (version 2.0 final)	1 January 1981–31 July 2018	0.05 arc degrees		UCSB/CHG
CHIRPS Pentad precipitation	Climate Hazards Group InfraRed Precipitation with Station Data (version 2.0 final)	1 January 1981–26 July 2018	0.05 arc degrees		UCSB/CHG
WorldClim V1	Climatological and Bio variables	1 January 1960–1 January 1991	30 arc seconds		University of California, Berkeley
TerraClimate	Monthly Climate and Climatic Water Balance for Global Terrestrial Surfaces, University of Idaho	1 January 1958–1 December 2017	2.5 arc minutes		1 January 1958–1 December 2017

Abbreviations: European Space Agency (ESA), United States Geological Survey (USGS), TOA (Top of Atmosphere), Burn Area Index (BAI), Enhanced Vegetation index (EVI), Normalised Difference Snow index (NDSI), Normalised Difference Vegetation index (NDVI), Normalised Difference Water Index (NDWI), Moderate Resolution Imaging Spectrometer (MODIS), National Oceanographic and Atmospheric Administration Advanced very high resolution radiometer (NOAA AVHRR), The Defense Meteorological Program (DMSP) Operational Line-Scan System (OLS), Shuttle Radar Topography Mission (SRTM), Advanced Spaceborne Thermal Emission and Reflection Radiometer (ASTER), The Tropical Rainfall Measuring Mission (TRMM), Goddard Space Flight Center (GSFC). The Japan Aerospace Exploration Agency (JAXA).

Table A2. Country-wise breakdown of application regions of GEE based on the 300 journal papers surveyed.

Study Area	No. of Papers	Study Area	No. of Papers	Study Area	No. of Papers	Study Area	No. of Papers
U.S.A.	60	Spain	3	South Sudan	2	Rwanda	1
China	24	Burkina Faso	3	Mauritania	2	Djibouti	1
Brazil	19	Nepal	3	Hong Kong	1	Mauritius	1
India	15	Thailand	3	Trinidad and Tobago	1	Seychelles	1
Indonesia	15	Greece	3	Botswana	1	Benin	1
Australia	14	Sierra Leone	4	Morocco	1	Ivory Coast	1
Canada	11	Peru	3	Sweden	1	Cape Verde	1
South Africa	10	Brunei	3	Uruguay	1	Gambia	1
Italy	7	Singapore	3	Panama	1	Guinea	2
Kenya	7	Tajikistan	3	Mongolia	1	Guinea-Bissau	1
Malaysia	7	Bolivia	2	Taiwan	1	Liberia	2
Vietnam	6	Mexico	2	East Timor	1	Togo	2
Myanmar	6	Poland	2	Finland	1	Albania	1
Russia	6	Turkey	2	Iraq	1	Austria	1
Bangladesh	5	Papua New Guinea	2	Iran	1	Bulgaria	1
Zambia	5	Ethiopia	3	Afghanistan	1	Croatia	1
Germany	5	Malawi	2	Tibet (China)	1	Czech Republic	1
Philippines	5	Japan	2	Scotland	1	Hungary	1
Nigeria	6	South Korea	2	New Zealand	1	Macedonia	1
Namibia	5	United Arab Emirates	2	Paraguay	1	Moldova	1
Jordan	4	Madagascar	3	Colombia	1	Montenegro	1
Argentina	4	Mozambique	3	Angola	1	Romania	1
Laos	4	Tanzania	3	French Guiana	1	Serbia	1
U.K.	4	Pakistan	2	Ecuador	1	Slovak Republic	1
Zimbabwe	4	Sri Lanka	2	Belize	1	Slovenia	1
Senegal	4	Bhutan	2	American Samoa	1	Switzerland	1
Swaziland	4	Kazakhstan	2	Vanuatu	1	Cyprus	1
North Korea	4	Gabon	3	Tonga	1	Lebanon	1
Cambodia	4	Syria	2	French Polynesia	1	Palestine	1
Niger	4	Puerto Rico	2	Cuba	1	Egypt	1
France	4	Bosnia and Herzegovina	2	Dominican Republic	1	Algeria	1
Israel	3	Ghana	3	Jamaica	1	Chad	1
Costa Rica	3	Somalia	2	Maldives	1	Cameroon	2
Ireland	3	Ukraine	2	Uzbekistan	1	Central African Republic	2
Democratic Republic of Congo	4	Haiti	2	Burundi	2	Cote d'Ivoire	1
Mali	3	Eritrea	2	Uganda	2	Equatorial Guinea	1

Table A3. Origin of affiliated institutions of all authors in the 300 GEE based research publications.

Country	No. of Authors	Country	No. of Authors	Country	No. of Authors	Country	No. of Authors
U.S.A.	589	Greece	13	Namibia	4	Philippines	2
Italy	120	Israel	13	Puerto Rico	4	Turkey	2
Australia	86	India	12	Saudi Arabia	4	Belize	1
Germany	79	Japan	12	Sri Lanka	4	Costa Rica	1
China	68	Singapore	9	Swaziland	4	Czech Republic	1
U.K.	65	Kenya	8	Ukraine	4	French Polynesia	1
Brazil	58	Russia	8	Benin	3	Kazakhstan	1
Netherland	46	Jordan	7	Denmark	3	Kuwait	1
Canada	37	Sweden	7	Hong Kong	3	Laos	1
France	28	Argentina	6	Indonesia	3	Nepal	1
South Africa	24	Ireland	6	Papua New Guinea	3	Nigeria	1
Belgium	16	Ghana	5	Trinidad and Tobago	3	Poland	1
Spain	16	Norway	5	Cyprus	2	South Korea	1
Switzerland	16	Zambia	5	Ethiopia	2	Tunisia	1
Austria	15	Bangladesh	4	Morocco	2		

References

1. Gorelick, N.; Hancher, M.; Dixon, M.; Ilyushchenko, S.; Thau, D.; Moore, R. Google Earth Engine: Planetary-scale geospatial analysis for everyone. *Remote. Sens. Environ.* **2017**, *202*, 18–27. [CrossRef]
2. Moore, R.; Parsons, E. Beyond SDI, bridging the power of cloud based computing resources to manage global environment issues. In Proceedings of the INSPIRE Conference, Edinburgh, UK, 27 June–1 July 2011.
3. Hansen, M.; Potapov, P.; Moore, R.; Hancher, M.; Turubanova, S.; Tyukavina, D.; Stehman, S.; Goetz, S.; Loveland, T.; Kommareddy, A. Observing the forest and the trees: The first high resolution global maps of forest cover change. *Science* **2013**, *342*, 850–853. [CrossRef] [PubMed]
4. Xiong, J. *Cloud Computing for Scientific Research*; Scientific Research Publishing Inc.: New York, NY, USA, 2018; p. 256.
5. Shoko, C.; Mutanga, O.; Dube, T. Progress in the remote sensing of C3 and C4 grass species aboveground biomass over time and space. *ISPRS J. Photogramm. Remote. Sens.* **2016**, *120*, 13–24. [CrossRef]
6. Gorelick, N. Google Earth Engine. In *EGU General Assembly Conference Abstracts*; American Geophysical Union: Vienna, Austria, 2013; p. 11997.
7. Hansen, C.H. Google Earth Engine as a Platform for Making Remote Sensing of Water Resources a Reality for Monitoring Inland Waters. Available online: https://www.researchgate.net/profile/Carly_Hansen/publication/277021226_Google_Earth_Engine_as_a_Platform_for_Making_Remote_Sensing_of_Water_Resources_a_Reality_for_Monitoring_Inland_Waters/links/555f8c2a08ae9963a118b3e2.pdf (accessed on 18 July 2018).
8. Huntington, J.L.; Hegewisch, K.C.; Daudert, B.; Morton, C.G.; Abatzoglou, J.T.; McEvoy, D.J.; Erickson, T. Climate Engine: Cloud Computing and Visualization of Climate and Remote Sensing Data for Advanced Natural Resource Monitoring and Process Understanding. *Bull. Am. Meteorol. Soc.* **2017**. [CrossRef]
9. Johnson, B.A.; Iizuka, K.; Bragais, M.A.; Endo, I.; Magcale-Macandog, D.B. Employing crowdsourced geographic data and multi-temporal/multi-sensor satellite imagery to monitor land cover change: A case study in an urbanizing region of the Philippines. *Comput. Environ. Urban Syst.* **2017**, *64*, 184–193. [CrossRef]
10. Aguilar, R.; Zurita-Milla, R.; Izquierdo-Verdiguier, E.; de By, R.A. A Cloud-Based Multi-Temporal Ensemble Classifier to Map Smallholder Farming Systems. *Remote. Sens.* **2018**, *10*, 729. [CrossRef]
11. Lemoine, G.; Léo, O. Crop mapping applications at scale: Using Google Earth Engine to enable global crop area and status monitoring using free and open data sources. In Proceedings of the 2015 IEEE International Geoscience and Remote Sensing Symposium (IGARSS), Milan, Italy, 26–31 July 2015; pp. 1496–1499.
12. Horowitz, F.G. MODIS Daily Land Surface Temperature Estimates in Google Earth Engine as an Aid in Geothermal Energy Siting. In Proceedings of the World Geothermal Congress 2015, Melbourne, Australia, 19–25 April 2015.
13. Huang, H.; Chen, Y.; Clinton, N.; Wang, J.; Wang, X.; Liu, C.; Gong, P.; Yang, J.; Bai, Y.; Zheng, Y. Mapping major land cover dynamics in Beijing using all Landsat images in Google Earth Engine. *Remote. Sens. Environ.* **2017**. [CrossRef]
14. Johansen, K.; Phinn, S.; Taylor, M. Mapping woody vegetation clearing in Queensland, Australia from landsat imagery using the Google Earth Engine. *Remote. Sens. Appl. Soc. Environ.* **2015**, *1*, 36–49. [CrossRef]
15. Abdel-Rahman, E.M.; Mutanga, O.; Adam, E.; Ismail, R. Detecting Sirex noctilio grey-attacked and lightning-struck pine trees using airborne hyperspectral data, random forest and support vector machines classifiers. *ISPRS J. Photogramm. Remote. Sens.* **2014**, *88*, 48–59. [CrossRef]
16. Moore, R.; Hansen, M. Google Earth Engine: A New Cloud-Computing Platform for Global-Scale Earth Observation Data and Analysis. Available online: http://adsabs.harvard.edu/abs/2011AGUFMIN43C..02M (accessed on 20 September 2019).
17. Dube, T.; Mutanga, O.; Ismail, R. Quantifying aboveground biomass in African environments: A review of the trade-offs between sensor estimation accuracy and costs. *Trop. Ecol.* **2016**, *57*, 393–405.
18. Mutanga, O.; Dube, T.; Omer, G. Remote Sensing of Crop Health for Food Security in Africa: Potentials and Constraints. *Remote. Sens. Appl. Soc. Environ.* **2017**, *8*, 231–238. [CrossRef]
19. Cossu, R.; Petitdidier, M.; Linford, J.; Badoux, V.; Fusco, L.; Gotab, B.; Hluchy, L.; Lecca, G.; Murgia, F.; Plevier, C. A roadmap for a dedicated Earth Science Grid platform. *Earth Sci. Inform.* **2010**, *3*, 135–148. [CrossRef]

20. Xiong, J.; Thenkabail, P.S.; Gumma, M.K.; Teluguntla, P.; Poehnelt, J.; Congalton, R.G.; Yadav, K.; Thau, D. Automated cropland mapping of continental Africa using Google Earth Engine cloud computing. *ISPRS J. Photogramm. Remote. Sens.* **2017**, *126*, 225–244. [CrossRef]
21. Du Plessis, H.; Van Niekerk, A. A new GISc framework and competency set for curricula development at South African universities. *S. Afr. J. Geomat.* **2014**, *3*, 1–12.
22. Maguire, D.J.; Longley, P.A. The emergence of geoportals and their role in spatial data infrastructures. *Comput. Environ. Urban Syst.* **2005**, *29*, 3–14. [CrossRef]
23. Salager-Meyer, F. Scientific publishing in developing countries: Challenges for the future. *J. Engl. Acad. Purp.* **2008**, *7*, 121–132. [CrossRef]
24. Gibbs, W.W. Lost science in the third world. *Sci. Am.* **1995**, *273*, 92–99. [CrossRef]
25. Galvez, A.; Maqueda, M.; Martinez-Bueno, M.; Valdivia, E. Scientific Publication Trends and the Developing World: What can the volume and authorship of scientific articles tell us about scientific progress in various regions? *Am. Sci.* **2000**, *88*, 526–533.
26. Krishna, V.; Waast, R.; Gaillard, J. *Scientific Communities in the Developing World*; Sage: Thousand Oaks, CA, USA, 1996.
27. Duque, R.B.; Ynalvez, M.; Sooryamoorthy, R.; Mbatia, P.; Dzorgbo, D.-B.S.; Shrum, W. Collaboration paradox: Scientific productivity, the Internet, and problems of research in developing areas. *Soc. Stud. Sci.* **2005**, *35*, 755–785. [CrossRef]

Article

Global Estimation of Biophysical Variables from Google Earth Engine Platform

Manuel Campos-Taberner [1,*], Álvaro Moreno-Martínez [2,3], Francisco Javier García-Haro [1], Gustau Camps-Valls [3], Nathaniel P. Robinson [2], Jens Kattge [4] and Steven W. Running [2]

1 Department of Earth Physics and Thermodynamics, Faculty of Physics, Universitat de València, Dr. Moliner 50, 46100 Burjassot, València, Spain; j.garcia.haro@uv.es

2 Numerical Terradynamic Simulation Group, College of Forestry & Conservation, University of Montana, Missoula, MT 59812, USA; alvaro.moreno@ntsg.umt.edu (Á.M.-M.); Nathaniel.Robinson@umontana.edu (N.P.R.); swr@ntsg.umt.edu (S.W.R.)

3 Image Processing Laboratory (IPL), Universitat de València, Catedrático José Beltrán 2, 46980 Paterna, València, Spain; gustau.camps@uv.es

4 Max-Planck-Institute for Biogeochemistry, Hans-Knöll Straβe 10, 07745 Jena, Germany; jkattge@bgc-jena.mpg.de

* Correspondence: manuel.campos@uv.es; Tel.: +34-963-543-256

Received: 6 June 2018; Accepted: 20 July 2018; Published: 24 July 2018

Abstract: This paper proposes a processing chain for the derivation of global Leaf Area Index (LAI), Fraction of Absorbed Photosynthetically Active Radiation (FAPAR), Fraction Vegetation Cover (FVC), and Canopy water content (CWC) maps from 15-years of MODIS data exploiting the capabilities of the Google Earth Engine (GEE) cloud platform. The retrieval chain is based on a hybrid method inverting the PROSAIL radiative transfer model (RTM) with Random forests (RF) regression. A major feature of this work is the implementation of a retrieval chain exploiting the GEE capabilities using global and climate data records (CDR) of both MODIS surface reflectance and LAI/FAPAR datasets allowing the global estimation of biophysical variables at unprecedented timeliness. We combine a massive global compilation of leaf trait measurements (TRY), which is the baseline for more realistic leaf parametrization for the considered RTM, with large amounts of remote sensing data ingested by GEE. Moreover, the proposed retrieval chain includes the estimation of both FVC and CWC, which are not operationally produced for the MODIS sensor. The derived global estimates are validated over the BELMANIP2.1 sites network by means of an inter-comparison with the MODIS LAI/FAPAR product available in GEE. Overall, the retrieval chain exhibits great consistency with the reference MODIS product (R^2 = 0.87, RMSE = 0.54 m^2/m^2 and ME = 0.03 m^2/m^2 in the case of LAI, and R^2 = 0.92, RMSE = 0.09 and ME = 0.05 in the case of FAPAR). The analysis of the results by land cover type shows the lowest correlations between our retrievals and the MODIS reference estimates (R^2 = 0.42 and R^2 = 0.41 for LAI and FAPAR, respectively) for evergreen broadleaf forests. These discrepancies could be attributed mainly to different product definitions according to the literature. The provided results proof that GEE is a suitable high performance processing tool for global biophysical variable retrieval for a wide range of applications.

Keywords: Google Earth Engine; LAI; FVC; FAPAR; CWC; plant traits; random forests; PROSAIL

1. Introduction

Earth vegetation plays an essential role in the study of global climate change influencing terrestrial CO_2 flux exchange and variability through plant respiration and photosynthesis [1,2]. Vegetation monitoring can be achieved through the evaluation of biophysical variables such as LAI (Leaf Area

Index), FVC (Fraction Vegetation Cover) and FAPAR (Fraction of Absorbed Photosynthetically Active Radiation) [3,4]. LAI accounts for the amount of green vegetation that absorbs or scatters solar radiation, FVC determines the partition between soil and vegetation contributions, while FAPAR is a vegetation health indicator related with ecosystems productivity. In addition, canopy water content (CWC) accounts for the amount of water content at canopy level, varies with vegetation water status, and is usually computed as the product of leaf water content (C_w) and LAI [5,6]. These essential variables can be estimated using remote sensing data and are key inputs in a wide range of ecological, meteorological and agricultural applications and models.

Biophysical variables can be derived from remote sensing data using statistical, physical and hybrid retrieval methods [7–9]. Statistical methods rely on models to relate spectral data with the biophysical variable of interest, usually through some form of regression. Statistical methods such as neural networks [10], random forests [11] or kernel methods [12], extract patterns and trends from a data set and approximate the underlying physical laws ruling the relationships between them from data. Physically-based retrieval methods are based on the physical knowledge describing the interactions between incoming radiation and vegetation through radiative transfer models (RTMs) [13,14]. In particular, the MODIS LAI/FAPAR product is based on a three-dimensional RTM which links surface spectral bi-directional reflectance factors (BRFs) to both canopy and soil spectral and structural parameters [15]. On the other hand, hybrid methods couple statistical with physically-based approaches inverting a database generated by an RTM [16–18]. For example, CYCLOPES global products [19] were derived inverting the PROSAIL radiative transfer model [20] using neural networks, while the global EUMETSAT LAI/FAPAR/FVC products are being produced inverting PROSAIL with multi-output Gaussian process regression from the EUMETSAT Polar System (EPS) [21]. The use of RTMs implies modeling leaf and canopy structural and biochemical parameters. The ranges and distribution of parameters used for running the simulations are usually based on field measurements that are very useful for simulating specific land covers [16,17]. Nevertheless, when the objective is to simulate a wide range of vegetation situations and land covers, ground data is often a limitation. In these scenarios, the scientific community uses distributions based on several experimental datasets such as the HAWAII, ANGERS, CALMIT-1/2 and LOPEX [22–25] which together embrace hundreds of observations [26]. Continuous update of ground measurements used for radiative transfer modeling are key in order to better constrain the RTM inversion process [27]. In this framework, the use of global plant traits database (TRY) [28] containing thousands of leaf data could alleviate this limitation.

From an operational standpoint, processing remote sensing data on an ongoing basis demands high storage capability and efficient computational power mainly when dealing with time series of long term global data sets. This situation also occurs because of the wide variety of free available remote sensing data disseminated by agencies such as the National Aeronautics and Space Administration (NASA) (e.g., MODIS), the United States Geological Survey (USGS) (e.g., Landsat), and the European Space Agency (ESA) (e.g., data from the Sentinel constellation) [29]. Recently, Google (Mountain View, Cal., USA) developed the Google Earth Engine (GEE) [30], a cloud computing platform specifically designed for geospatial analysis at the petabyte scale. The GEE data catalog is composed by widely used geospatial data sets. The catalog is continuously updated and data are ingested from different government-supported archives such as the Land Process Distributed Active Archive Center (LP DAAC), the USGS, and the ESA Copernicus Open Access Hub. The GEE data catalogue contains numerous remote sensing data sets such as top and bottom of atmosphere reflectance, as well as atmospheric and meteorological data. Data processing is performed in a parallel on Google's computational infrastructure, dramatically improving processing efficiency, and opens up excellent prospects especially for multitemporal and global studies that include vegetation, temperature, carbon exchange, and hydrological processes [31–35].

The present study proposes a generic retrieval chain for the production of global LAI, FAPAR, FVC and CWC estimates from 15 years of MODIS data (MCD43A4) on the GEE platform. The methodology is based on a hybrid method inverting a PROSAIL radiative transfer model database with random forests (RFs) regression. Major contributions of the presented work are:

- The development of a general methodology for global LAI/FAPAR estimation including FVC and CWC which are not provided by MODIS.
- The use of a global plant traits database (composed of thousands of data) for probability density function (PDF) estimation with copulas to be used for radiative transfer modeling leaf parameterization.
- The enforceability of biophysical parameter retrieval chain over GEE exploiting its capabilities to provide climate data records of global biophysical variables at computationally both affordable and efficient way.

Validation was performed by means of inter-comparison with the official MODIS LAI/FAPAR product available on GEE (MCD15A3H) over a network of globally distributed sites. The only process computed locally is the RTM simulation, while the inversion of the database, the derivation of the global maps, and the assessment of the retrievals have been performed into the GEE platform. Furthermore, the proposed methodology estimates both FVC and CWC variables which are not part of the official MODIS products, giving an added value to this work.

The remainder of the paper is structured as follows. Section 2 describes the data used in this work while Section 3 outlines the followed methodology. Section 4 exhibits the obtained results and the validation of the global estimates, and Section 6 discusses the main conclusions of this work.

2. Data Collection

2.1. MODIS Data

In this study, we used the MCD43A4 and the MCD15A3H MODIS products both available in GEE. Both the MCD (MODIS Combined Data) reflectance and LAI/FAPAR products are generated combining data from Terra and Aqua spacecrafts and are disseminated in a level-3 gridded data set. The MCD43A4 product provides a Bidirectional Reflectance Distribution Function (BRDF) from a nadir view in the 1–7 MODIS bands (i.e., red, near infrared (NIR), blue, green, short wave infrared-1 (SWIR-1), short wave infrared-2 (SWIR-2), and middle wave infrared (MWIR), see Table 1). MCD43A4 offers global surface reflectance data at 500 m spatial resolution with 8-day temporal frequency.

Table 1. Spectral specifications of the MODIS MCD43A4 product.

MCD43A4 Band	Wavelength (nm)
Band 1 (red)	620–670
Band 2 (NIR)	841–876
Band 3 (blue)	459–479
Band 4 (green)	545–565
Band 5 (SWIR-1)	1230–1250
Band 6 (SWIR-2)	1628–1652
Band 7 (MWIR)	2105–2155

On the other hand, GEE also offers access to MODIS derived LAI and FAPAR estimates through the MCD15A3H collection 6 product. The temporal frequency of the biophysical estimates is every four days, and the retrieval algorithm chooses the "best" pixel available from all the acquisitions of both MODIS sensors from within the 4-day period. The MCD15A3H main retrieval algorithm uses a look-up-table (LUT) approach simulated from a 3D RTM. Basically, this method searches for plausible values of LAI and FAPAR for a specific set of angles (solar and view), observed bidirectional reflectance factors at certain spectral bands, and biome types [15]. In addition, the MCD15A3H employs a back-up algorithm (when the main one fails) that uses empirical relationships between NDVI (Normalized Difference Vegetation Index) and the biophysical parameters. Similarly to MCD43A4, the pixel's spatial resolution is 500 m.

2.2. Global Plant Traits

The TRY database represents the biggest global effort to compile a massive global repository for plant trait data (6.9 million trait records for 148,000 plant taxa) at unprecedented spatial and climatological coverage [28,36]. So far, the TRY initiative has delivered to the scientific community around 390 million trait records which have resulted in more than 170 publications (https://www.try-db.org/). The applications of the database range from functional and community ecology, plant geography, species distribution, and vegetation models parameterizations [37–40].

We use a realistic representation of global leaf trait variability provided by the TRY to optimize a vegetation radiative transfer model (PROSAIL), commonly used by the remote sensing community [41]. Instead of using the common lookup tables available in the literature to parametrize the model [19,21,42], we exploit the potential of the TRY database to infer the distributions and correlations among some key leaf traits (leaf chlorophyll C_{ab}, leaf dry matter C_{dm} and water C_w contents) required by PROSAIL. Table 2 shows some basic information about the considered traits extracted from the TRY.

Table 2. Information about the TRY data used in this work.

Trait Name	Number of Samples	Number of Species
C_{ab}	19,222	941
C_{dm}	69,783	11,908
C_w	32,020	4802

3. Methodology

Physical approaches to retrieve biophysical variables rely on finding the best match between measured and simulated spectra. The solution can be achieved by means of numerical optimization or Monte Carlo approaches which are computationally expensive and do not guarantee the convergence to an optimal solution. Recently, new and more efficient algorithms relying on Machine Learning (ML) techniques have emerged and have become the preferred choice for most RTM inversion applications [16–19,21]. In this work, we have followed the latter hybrid approach, combining radiative transfer modeling and the parallelized machine learning RFs implementation available in GEE to retrieve the selected biophysical variables. Figure 1 shows a schema of the work flow.

3.1. Creation of Leaf Plant Traits' Distributions

Recent research has highlighted the importance of exploiting a priori knowledge to constrain solutions of the ill-posed inversion problem in RTMs [27,43]. In this work, we used the TRY database and the available literature to extract prior knowledge and improve our results. Despite using the biggest plant trait database available, the representation of trait observations in a spatial and climate context in the TRY is still limited, and it shows significant deviations among observed and modelled distributions [28]. Trait measurements in the TRY represent the variation of single leaf measurements because they are not abundance-weighted with respect to natural occurrence [28]. Trait distributions are biased due to the availability of samples, which vary significantly due to technical difficulties for sampling (e.g., very dense forests and remote areas) or the availability of funds to carry out expensive field measurement campaigns in the different parts of the globe. To overcome these issues, we have computed our leaf traits' univariate distribution functions by combining the plant trait database (TRY) with a global map of plant functional types (PFTs). The chosen global map of plant functional types was the official MODIS (MCD12Q1) land cover product [44], we used this product to compute global fractions of PFTs to weight more realistically species' occurrence for the selected traits (leaf chlorophyll, leaf dry matter, and leaf water contents).

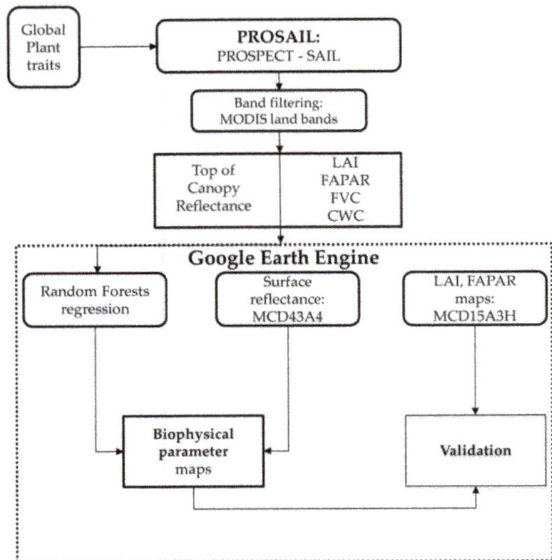

Figure 1. Work flow of the proposed retrieval chain over GEE.

The categorical information available in TRY allowed us to group leaf trait measurements in the common PFT definitions. These grouped data were then used to compute individual normalized histograms for each PFT, whereas the final leaf trait histogram was calculated as the weighted sum of each PFT normalized histogram according to the global PFT's spatial occurrence fractions. Repeating this process for all considered leaf traits, we obtained their final univariate distribution functions. These functions were inferred directly from the data by means of a non-parametric kernel density estimation (KDE, [45]). This approach allowed us to model leaf probability distributions without requiring any assumption regarding parametric families.

Some combinations of plant traits, like the ones considered in this work, exhibit significant correlations and tradeoffs as a result of different plant ecological strategies [46]. In order to capture these dependencies among traits, we created distributions that were also able to model correlated multivariate data by means of different copula functions. These functions separate marginal distributions from the dependency structure of a given multivariate distribution [47]. More precisely, we use a multivariate Gaussian copula function [48]. Using the calculated marginal univariate distributions for each trait and the Gaussian multivariate copula, computed from leaf measurements of the TRY database, we created a set of random training samples while preserving the correlation structure among them (see Figure 2).

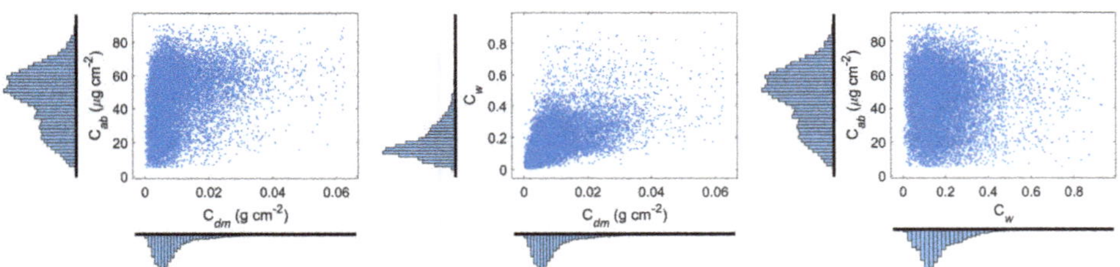

Figure 2. Constrained random samples of leaf chlorophyll (C_{ab}), leaf dry matter (C_{dm}), and leaf water (C_w) contents based upon prior knowledge of the TRY database, the MODIS land cover (MCD12Q1), kernel density estimators, and copulas.

3.2. Radiative Transfer Modeling

We used the PROSAIL RTM, which results from coupling the PROSPECT leaf optical model [49] with the SAIL canopy reflectance model [50]. Note that we used the PROSPECT-5B [26] for the coupling, which accounts for chlorophylls and carotenoids separately. PROSAIL was run in forward mode for building a database mimicking MODIS canopy reflectance. These data were then used for training the retrieval model assuming turbid medium canopies with randomly distributed leaves. PROSAIL simulates top of canopy bidirectional reflectance from 400 to 2500 nm with a 1 nm spectral resolution as a function of leaf biochemistry variables, canopy structure and background, as well as the sun-view geometry. Leaf optical properties are given by the mesophyll structural parameter (N), leaf chlorophyll (C_{ab}) and carotenoid (C_{ar}) contents, leaf brown pigment (C_{bp}) content, as well as leaf dry matter (C_m) and water (C_w) contents. The average leaf angle inclination (ALA), the LAI, and the hot-spot parameter (Hotspot) characterize the canopy structure. A multiplicative brightness parameter (β_s) was used to represent different background reflectance types [19]. The system's geometry was described by the solar zenith angle, the view zenith angle, and the relative azimuth angle between both angles, which in our case corresponded to illumination and observation zenith angles of 0°. Sub-pixel mixed conditions (i.e., spatial heterogeneity) were tackled assuming a linear spectral mixing model, which pixels are composed by a mixture of pure vegetation (vCover) and bare soil (1-vCover) fractions' [21].

The leaf variables were randomly generated following the calculated kernel density distributions from the available leaf traits measurements, whereas distributions of the canopy variables as well as the soil brightness parameter, were similar to those adopted in other global studies [19,21]. Brown pigments were intentionally set to zero in order to account only for photosynthetic elements of the canopy (see Table 3). In addition, with the aim of accounting for different sources of noise (e.g., atmospheric correction, BRDF normalization or radiometric calibration) a wavelength dependent white Gaussian noise was added to the reflectances of the PROSAIL simulations. Specifically, a Gaussian noise with $\sigma = 0.015$ was added in the blue, green, and red channels, $\sigma = 0.025$ in the NIR, and $\sigma = 0.03$ in the SWIR-1, SWIR-2 and MWIR.

Table 3. Distributions of the parameters within the PROSAIL RTM at leaf (PROSPECT-5B) and canopy (SAIL) levels. * KDE refers to kernel density estimation method, which does not provide any parameters being a non parametric model of the marginal distributions.

Parameter		Min	Max	Mode	Std	Type
Leaf	N	1.2	2.2	1.6	0.3	Gaussian
	C_{ab} ($\mu g \cdot cm^{-2}$)	-	-	-	-	KDE *
	C_{ar} ($\mu g \cdot cm^{-2}$)	0.6	16	5	7	Gaussian
	C_{dm} ($g \cdot cm^{-2}$)	-	-	-	-	KDE *
	C_w	-	-	-	-	KDE *
	C_{bp}	0	0	0	0	-
Canopy	LAI (m^2/m^2)	0	8	3.5	4	Gaussian
	ALA (°)	35	80	60	12	Gaussian
	Hotspot	0.1	0.5	0.2	0.2	Gaussian
	vCover	0.3	1	0.99	0.2	Truncated Gaussian
Soil	β_s	0.1	1	0.8	0.6	Gaussian

3.3. Random Forests Regression

The inversion of PROSAIL was done using standard regression. There is a wide variety of machine learning models for regression and function approximation. In this paper, we focus on the particular family of methods called random forests (RFs). An RF is essentially an ensemble method that constructs a multitude of decision trees (each of them trained with different subsets of features and examples), and yields the mean prediction of the individual trees [51]. RFs' classification and regression have been applied in different areas of concern in forest ecology, such as modelling the gradient of coniferous species [52], the occurrence of fire in Mediterranean regions [53], the classification of species or land cover type [54,55], and the analysis of the relative importance of the proposed drivers [55] or the selection of drivers [54,56,57]. The selection of RFs in our study is not incidental, and we capitalize on several useful properties. The main advantage of using RFs over other traditional machine learning algorithms like neural networks (NNETs) is that they can cope with high dimensional problems very easily thanks to their pruning strategy. In addition, unlike kernel machines (KMs), RFs are more computationally efficient. The RF strategy is very beneficial by alleviating the often reported overfitting problem of simple decision trees. Moreover, this paper training data set has been split into train and an independent test set that was only used for the assessment of the RFs. In addition, RFs excel in the presence of missing entries, heterogeneous variables, and can be easily parallelized to tackle large scale problems, which is especially relevant in the application described in this work. This way, we can exploit large datasets and run predictions within Google Earth Engine easily. In this work, we predicted the considered biophysical variables (LAI, FAPAR, FVC, and CWC) using the full set of MODIS land bands shown in Table 1.

4. Results and Validation

4.1. Random Forests Theoretical Performance

In this section, we evaluate the RFs' theoretical capabilities for LAI, FAPAR, FVC and CWC retrieval. The training database was composed of 14,700 cases of reflectances in the MODIS channels (Table 1) and the corresponding biophysical variables (i.e., LAI, FAPAR, FVC, and CWC) accounting for any combination of the PROSAIL parameters. We first trained RFs with 70% of the PROSAIL samples and then evaluated the estimation results over the remaining 30% of the samples (not in the training). Figure 3 shows the scatter plots of the RFs' estimates of every biophysical parameter over the unseen test set. High correlations (R^2 = 0.84, 0.89, 0.88, and 0.80 for LAI, FAPAR, FVC and CWC, respectively) low Root-Mean-Squared Errors (RMSE = 0.91 m^2/m^2, 0.08, 0.06, and 0.27 kg/m^2 for LAI, FAPAR, FVC and CWC, respectively) and practically no biases were found in all cases (see Figure 3).

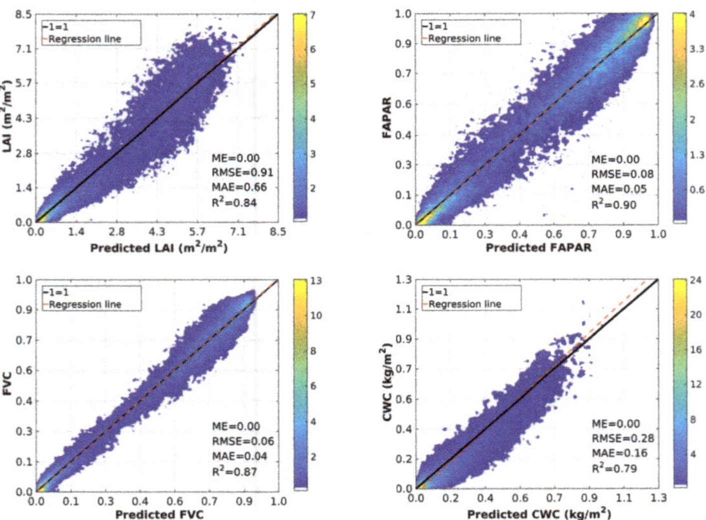

Figure 3. Theoretical performance of the Random forest regression over PROSAIL simulations of LAI, FAPAR, FVC and CWC. The colorbar indicates density of points in the scatter plots.

4.2. Obtained Estimates over GEE

After the RFs' regression assessment undertaken in the previous section, we ran the retrieval chain in GEE and obtained 15 years of global biophysical parameters. Here, we show the global mean values of LAI, FAPAR, FVC and CWC derived from 2010 to 2015 (Figure 4). The spatial distribution of retrieved parameters is expected, reaching the highest mean values close to the Equatorial zones (Central Africa forests and Amazon basin) followed by the Northern latitudes (e.g., boreal forests). In addition, Figure 5 shows the mean LAI and FAPAR values for the same period computed from the GEE MODIS reference product (MOD15A3H) freely distributed from the Land Processes Distributed Active Archive Center (LP DAAC) portal https://lpdaac.usgs.gov/.

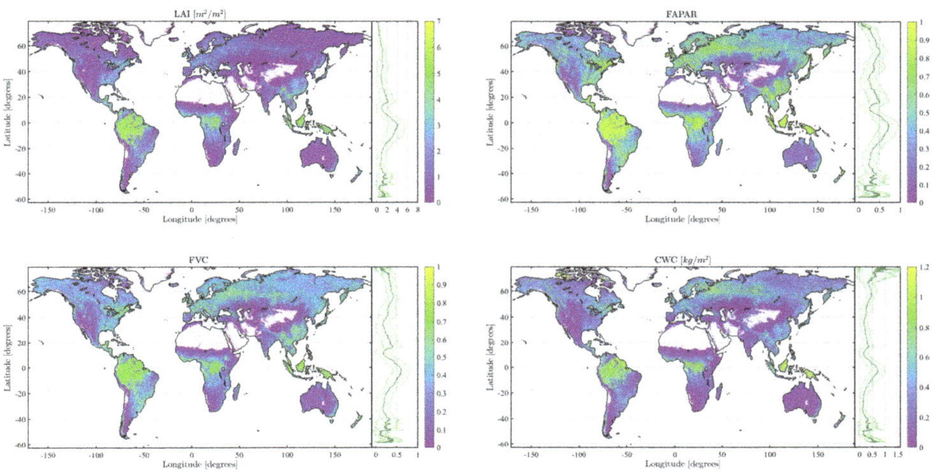

Figure 4. LAI, FAPAR, FVC, and CWC global maps and latitudinal transects corresponding to the mean values estimated by the proposed retrieval chain for the period 2010–2015.

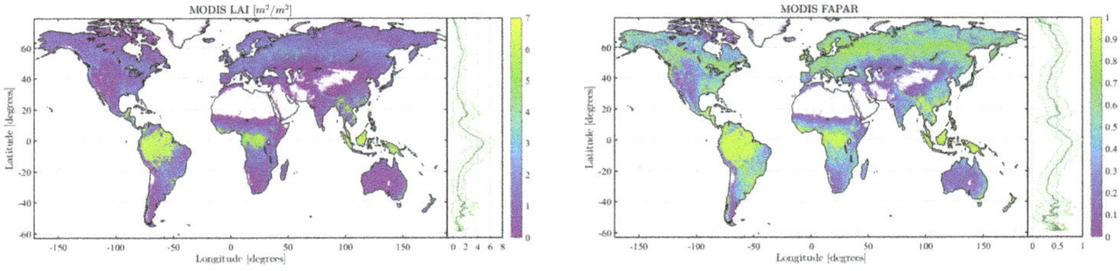

Figure 5. LAI and FAPAR global maps and latitudinal transects corresponding to the mean values of the GEE MODIS reference product (MOD15A3H) for the period 2010–2015.

4.3. Validation

Validation of LAI and FAPAR retrievals was undertaken by means of inter-comparison with the available LAI/FAPAR product (MCD15A3H) on GEE. The inter-comparison was conducted over a network of sites named BELMANIP-2.1 (Benchmark Land Multisite Analysis and Intercomparison of Products). These sites were especially selected for representing the global variability of vegetation , making them suitable for global intercomparison of land biophysical products [58]. BELMANIP-2.1

is an updated version of the original BELMANIP sites which includes 445 sites located in relatively homogeneous areas all over the globe (see Figure 6). The sites are aimed to be representative of the different planet biomes over an 10 × 10 km^2 area, mostly flat, and with minimum fractions of urban area and permanent water bodies.

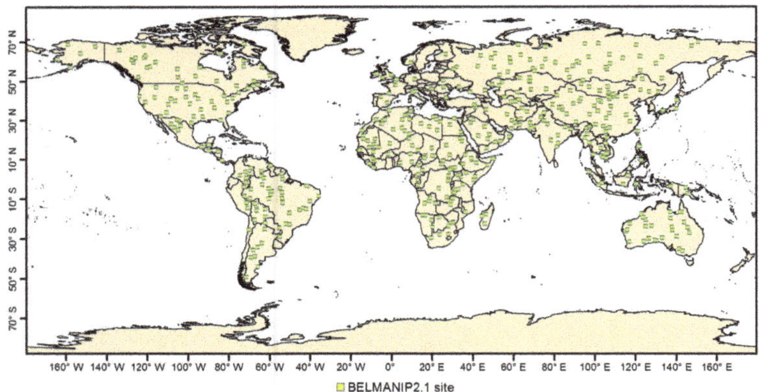

Figure 6. Sites location of the BELMANIP-2.1 network used for intercomparison of LAI and FAPAR retrievals and MOD15A3H LAI/FAPAR product.

We selected the MODIS pixels for every BELMANIP-2.1 location and then we computed the mean value of the MODIS valid pixels within a 1 km surrounding area. We also considered the contribution of partial boundary pixels by weighting their contribution to the mean according to their fractions included within the selected area. Non-valid pixels (clouds, cloud shadows) and low-quality pixels (back-up algorithm or fill values) were excluded according to the pixel-based quality flag in the MCD products. In addition, since the MCD15A3H and MCD43A4 differ in temporal frequency, only the coincident dates between them were selected for comparison. Due to the large amount of data available in GEE, we were able to select only high-quality MODIS pixels, resulting in ∼60,000 valid pixels from 2002–2017 for validation. Figure 7 shows per biome scatter plots between the estimates provided by the proposed retrieval chain and the reference MODIS LAI product over the BELMANIP2.1 sites from 2002 to 2012. Goodness of fit (R^2) ranging from 0.70 to 0.86 and low errors (RMSE) ranging from 0.23 to 0.57 m^2/m^2 are found between estimates in all biomes except for evergreen broadleaf forest, where R^2 = 0.42 and RMSE = 1.13 m^2/m^2 are reported.

Similarly, Figure 8 shows the obtained scatter plots for FAPAR. In this case, very good agreement (R^2 ranging from 0.89 to 0.92) and low errors (RMSE ranging from 0.06 to 0.08) are found between retrievals and the MODIS FAPAR product, over bare areas, shrublands, herbaceous, cultivated, and broadleaf deciduous forest biomes. For needle-leaf and evergreen broadleaf forests lower correlations (R^2 = 0.57 and 0.41) and higher errors (RMSE = 0.18 and 0.09) are obtained. It is worth mentioning that over bare areas, the MODIS FAPAR presents an unrealistic minimum value (∼0.05) through the entire period.

Figure 9 shows the LAI and FAPAR difference maps computed from the mean estimates (2010–2015) provided by the proposed retrieval chain and the mean reference MODIS LAI/FAPAR product. Mean LAI map revealed that most of the pixels fall within the range of ±0.5 m^2/m^2, which highlights the consistency between products. However, for high LAI values, there is an underestimation of the provided estimates over very dense canopies that may reach up to 1.4 m^2/m^2. In the case of FAPAR, there is a constant negative bias of ≈0.05 which is also noticeable in the scatter plots shown in Figure 8. This is related with a documented systematic overestimation of MODIS FAPAR retrievals [59–61], which is partly corrected by the proposed retrieval approach. The spatial consistency of LAI/FAPAR estimates was also compared over the African continent (Figure 10). The latitudinal transects provided by Figure 10 clearly show an underestimation of LAI retrievals in equatorial forests, having a better agreement for the remaining biomes.

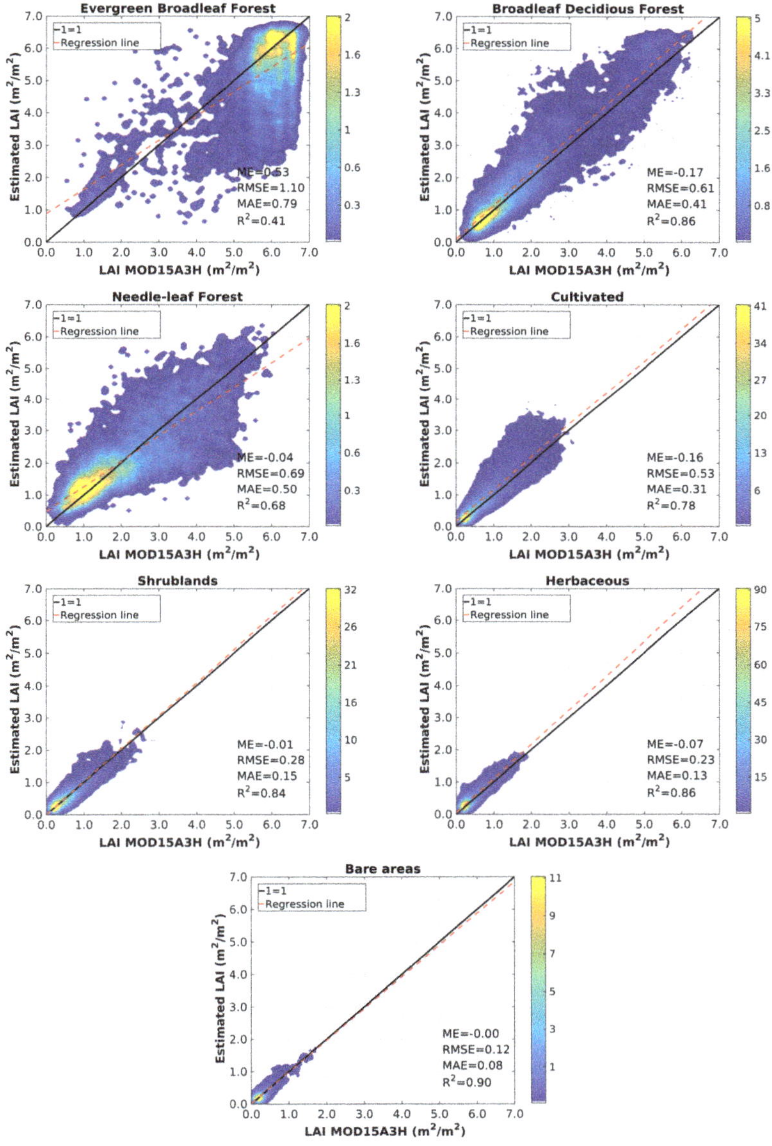

Figure 7. Biome-dependent scatter plots of the retrieved LAI over BELMANIP2.1 sites for the period 2002–2017. The colorbar indicates density of points in the scatter plots.

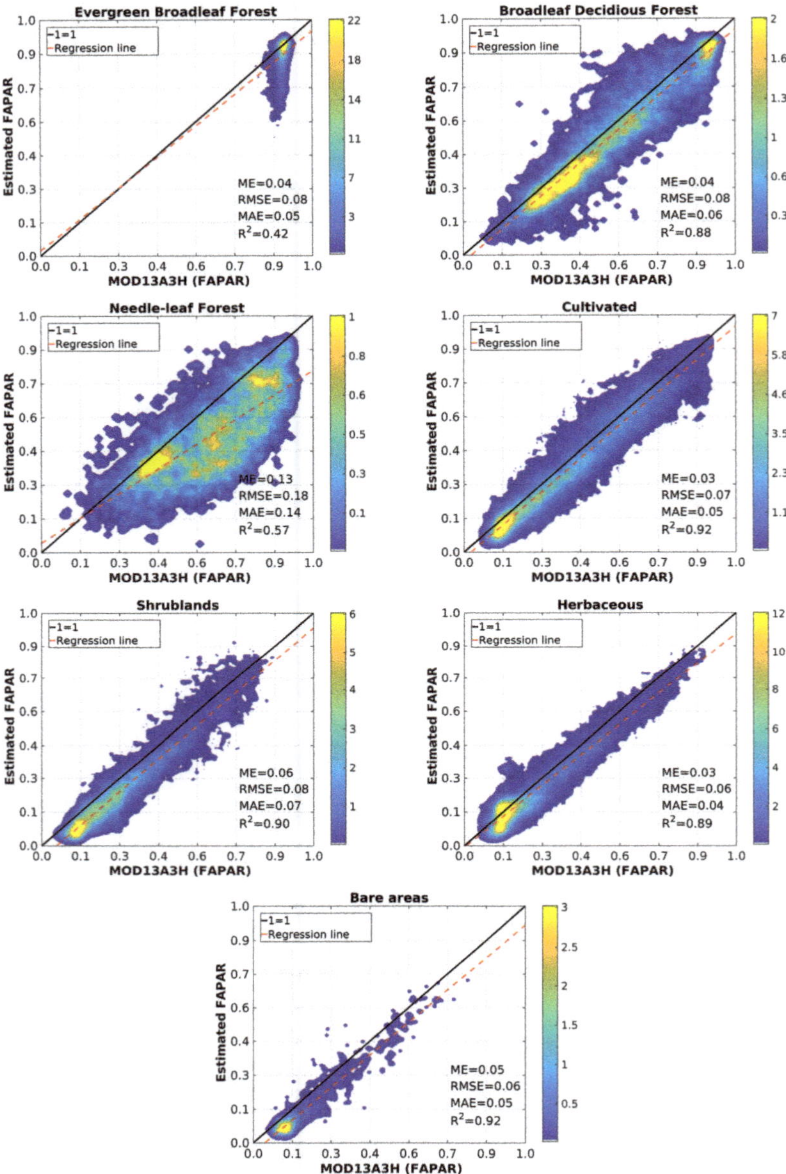

Figure 8. Biome-dependent scatter plots of the retrieved FAPAR over BELMANIP2.1 sites for the period 2002–2017. The colorbar indicates density of points in the scatter plots.

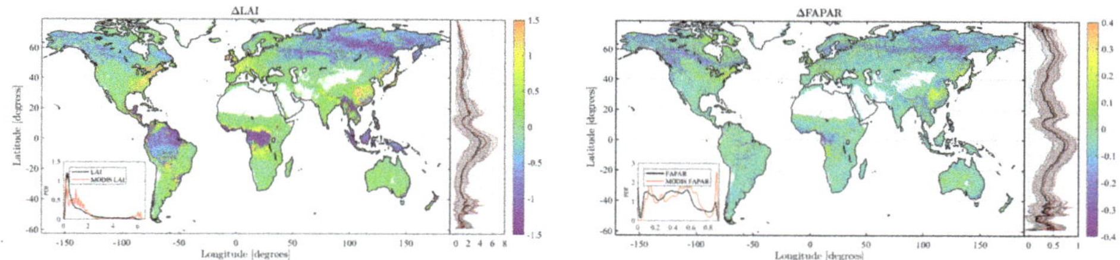

Figure 9. LAI and FAPAR global maps and latitudinal transects corresponding to the difference of mean values between derived estimates by the proposed retrieval chain and the GEE MODIS reference product for the period 2010–2015.

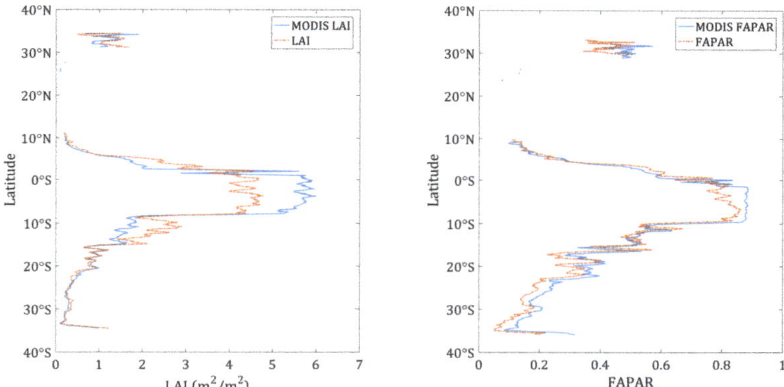

Figure 10. LAI (**left**) and FAPAR (**right**) latitudinal profiles over Africa (longitude 22°E) corresponding to the mean values of the GEE MODIS reference product and estimated by the proposed retrieval chain for the period 2010–2015.

5. Discussion

The usefulness of GEE for providing global land surface variables related to vegetation status was demonstrated in this work. GEE offers some major advantages mainly related to storage capacity and processing speed. Despite the variety of algorithms implemented in GEE, its capabilities are constrained by the number of state-of-the-art algorithms (in this case, regression-based) which are currently implemented in GEE. However, this limitation is being overcome by the increasing number of users developing algorithms that may be potentially implemented in GEE for a wide range of applications. The functions in GEE utilize several built-in parallelization and data distribution models to achieve high performance [30]. The RFs' implementation in GEE is not an exception to that. It allowed for the exploitation of large data sets and to obtain global estimates very efficiently. In GEE, the system handles and hides nearly every aspect of how a computation is managed, including resource allocation, parallelism, data distribution, and retries. These decisions are purely administrative; none of them can affect the result of a query, only the speed at which it is produced [30]. Under these circumstances, it is very difficult to give exact computation times because they vary in every run. As an example, in the present work, to compute the mean biophysical maps implied to process 230 (46 yearly images $\times$ 5 years) FAPAR images at 500 m spatial resolution ($\sim$440 million cells) and compute their annual mean, it took around 6 h.

The GEE data catalogue includes MODIS surface reflectance daily products, which can be advantageous to fully exploit the information contained in the reflectance signal of the surface. In this paper, we have preferred to use the normalized reflectance (MCD43A4) as an input. The BRDF normalization and temporal compositing steps assume: (1) perfectness of the linear kernel model inversion; (2) change of vegetation cover within temporal window is insignificant. These underlying assumptions are approximate but allow robust estimates of the BRDF kernel coefficients. A number of algorithms to retrieve satellite products (LAI, FAPAR and FVC) such as CYCLOPES (SPOT/VGT), Copernicus (SPOT/VGT and PROBA-V), and LSA-SAF (MSG and EPS) use as input top of canopy (TOC) normalized to a standard geometrical configuration [19,21,62]. This approach reduces considerably the requirements in terms of number of inputs and computational load. Despite the good speed potential of the proposed chain, the computational cost may be relevant when the aim is to generate time series of global products. In order to reduce input uncertainties, we used the quality flag provided by MCD43 products to filter non-valid pixels (persistent clouds and/or cloud shadows in the MCD43 composite) as well as for identifying zones with low-quality pixels. Because of the above-mentioned assumptions of the MCD43 product, further improvements of the proposed methodology will include the uncertainty propagation of the reflectance input data to our retrievals for operational use.

The comparison results between retrieved LAI/FAPAR and the reference MODIS product revealed good spatial consistency. However, there are differences in mean LAI values over dense forests (up to 1.4 m^2/m^2). The underestimation in high LAI values could be partly explained by two factors: (1) differences in the algorithms used to estimate the LAI; and (2) the use of distinct LAI definitions for each product estimates. Namely, the LAI retrieved by the proposed chain is based on the inversion of a RTM assuming the canopy as a turbid medium. This approximation provides estimates closer to an *effective* LAI (LAI_{eff}). In turn, the MODIS retrieval algorithm accounts for vegetation clumping at leaf and canopy scales through radiative transfer formulations, therefore estimated values should be closer to *actual* LAI (LAI_{actual}). The relationship between LAI_{eff} and LAI_{actual} is given by $LAI_{actual} \frac{LAI_{eff}}{\Omega}$ being Ω the cumpling index. Similar underestimation behaviour was found by other studies when comparing MODIS LAI products and LAI retrievals from RTM inversion [59,63]. Yan et al. [59] found RMSE = 0.66 m^2/m^2 and RMSE = 0.77 m^2/m^2 when comparing MODIS C6 LAI estimates with ground LAI_{actual} and LAI_{eff} measurements respectively, as well as larger uncertainties in high LAI values. Regarding FAPAR, an overall negative bias is found for all biomes. However, this bias could not be regarded as an issue in the estimations, since different studies have pointed out a systematic overestimation of MODIS retrievals in both C5 and C6 at low FAPAR values as a main drawback of the product [59–61]. For example, Xu et al. [64] assessed MODIS FAPAR through comparisons to ground measurements available from 2012–2016, obtaining a reasonable agreement (R^2 = 0.83, RMSE = 0.10) but with an overall overestimation tendency (bias = 0.08, scatters distributed within 0–0.2 difference). Similar results (R^2 = 0.74, RMSE = 0.15) were reported by Yan et al. [59] using globally distributed FAPAR measurements. The study evidenced a clear overestimation of FAPAR over sparsely-vegetated areas, as noted previously in other studies [60].

It is worth mentioning that neither the FVC nor the CWC products are available on GEE. Moreover, there is no global and reliable CWC product with which to compare the CWC estimates derived by the proposed retrieval chain. Regarding FVC, there are only a few global products that differ in retrieval approaches and spatiotemporal features. Since the main objective of the manuscript is to provide a generic biophysical retrieval chain, including the validation with the corresponding biophysical variables over GEE, the comparison of parameters not provided by GEE is out of the scope of the paper and could be addressed in future works.

6. Conclusions

This paper proposed a processing chain for the estimation of global biophysical variables (LAI, FAPAR, FVC, and CWC) from long-term (15-year) MODIS data in GEE. The approach takes the advantage of exploiting Earth observation data rapidly and efficiently through the GEE cloud storage and parallel computing capabilities. The retrieval methodology is based on a hybrid approach combining physically-based radiative transfer modelling (PROSAIL) and random forests regression.

The leaf parameter co-distributions employed during the radiative transfer modelling step were obtained by means of exploiting the TRY database. This allowed a better PROSAIL parametrization based on thousands of chlorophyll, water, and dry matter content ground measurements at leaf level. The increasing amount of available plant trait data in TRY (containing thousands of records) alleviates the need of a more realistic representation for some of the input parameters in radiative transfer models.

A validation exercise was undertaken over the BELMANIP2.1 network of sites by means of inter-comparison of the derived LAI and FAPAR with the MODIS reference LAI/FAPAR product available on GEE. The obtained results highlight the consistency of the estimates provided by the retrieval chain with the reference MODIS product. However, lower/poorer correlations were found for evergreen broadleaf forests when compared with the rest of biomes. These discrepancies could be mainly attributed to different retrieval approaches and variables definition, since derived LAI estimates are closer to LAI_{eff} rather than LAI_{actual} derived by the MOD15A3H product. In addition, derived FAPAR stands only for photosynthetic elements of the canopy while FAPAR provided by MODIS also accounts for non-photosynthetic elements. The proposed retrieval chain also derived globally both FVC and CWC variables which are not provided by any GEE dataset.

The results demonstrated the usefulness of GEE for global biophysical parameter retrieval and opened the door to user self-provisioning of leaf and canopy parameters in GEE for a wide range of applications including data assimilation and sensor fusion.

Supplementary Materials: A toy example of the code is available at https://code.earthengine.google.com/e3a2d589395e4118d97bae3e85d09106.

Author Contributions: All co-authors of this manuscript significantly contributed to all phases of the investigation. They contributed equally to the preparation, analysis, review and editing of this manuscript.

Funding: The research leading to these results was funded by the European Research Council under Consolidator Grant SEDAL ERC-2014-CoG 647423, the NASA Earth Observing System MODIS project (Grant NNX08AG87A), and supported by the LSA SAF CDOP3 project, and the Spanish Ministry of Economy and Competitiveness (MINECO) through the ESCENARIOS (CGL2016-75239-R) project.

Acknowledgments: The authors want to acknowledge the efforts of the TRY initiative on plant traits (http://www.try-db.org), hosted at the Max Planck Institute for Biogeochemistry, Jena, Germany.

Conflicts of Interest: The authors declare no conflict of interest.

References

1. Raich, J.W.; Schlesinger, W.H. The global carbon dioxide flux in soil respiration and its relationship to vegetation and climate. *Tellus B* **1992**, *44*, 81–99. [CrossRef]
2. Beer, C.; Reichstein, M.; Tomelleri, E.; Ciais, P.; Jung, M.; Carvalhais, N.; Rödenbeck, C.; Arain, M.A.; Baldocchi, D.; Bonan, G.B.; et al. Terrestrial Gross Carbon Dioxide Uptake: Global Distribution and Covariation with Climate. *Science* **2010**, *329*, 834–838. [CrossRef] [PubMed]
3. Huete, A.; Didan, K.; Miura, T.; Rodriguez, E.P.; Gao, X.; Ferreira, L.G. Overview of the radiometric and biophysical performance of the MODIS vegetation indices. *Remote Sens. Environ.* **2002**, *83*, 195–213. [CrossRef]
4. Fensholt, R. Earth observation of vegetation status in the Sahelian and Sudanian West Africa: Comparison of Terra MODIS and NOAA AVHRR satellite data. *Int. J. Remote Sens.* **2004**, *25*, 1641–1659. [CrossRef]
5. Clevers, J.G.P.W.; Kooistra, L.; Schaepman, M.E. Estimating canopy water content using hyperspectral remote sensing data. *Int. J. Appl. Earth Obs. Geoinf.* **2010**, *12*, 119–125. [CrossRef]
6. Yebra, D.P.; Chuvieco, E.; Riaño, D.; Zylstra, P.; Hunt, R.; Danson, F.M.; Qi, Y.; Jurdao, S. A global review of remote sensing of live fuel moisture content for fire danger assessment, moving towards operational products. *Remote Sens. Environ.* **2013**, *136*, 455–468. [CrossRef]

7. Wulder, M. Optical remote-sensing techniques for the assessment of forest inventory and biophysical parameters. *Prog. Phys. Geogr.* **1998**, *22*, 449–476. [CrossRef]
8. Zheng, G.; Monika, M. Retrieving leaf area index (LAI) using remote sensing: theories, methods and sensors. *Sensors* **1998**, *9*, 2719–2745. [CrossRef] [PubMed]
9. Verrelst, J.; Camps-Valls, G.; Muñoz-Marí, J.; Rivera, J.P.; Veroustraete, F.; Clevers, J.G.; Moreno, J. Optical remote sensing and the retrieval of terrestrial vegetation bio-geophysical properties—A review. *ISPRS J. Photogramm. Remote Sens.* **2015**, *108*, 273–290. [CrossRef]
10. Haykin, S. *Neural Networks—A Comprehensive Foundation*, 2nd ed.; Prentice Hall: Upper Saddle River, NJ, USA, 1999.
11. Breiman, L. Random forests. *Mach. Learn.* **2001**, *45*, 5–32. [CrossRef]
12. Camps-Valls, G.; Bruzzone, L. *Kernel Methods for Remote Sensing Data Analysis*; Wiley & Sons: Chichester, UK, 2009; p. 434, ISBN 978-0-470-72211-4.
13. Myneni, R.B.; Ramakrishna, R.; Nemani, R.; Running, S.W. Estimation of global leaf area index and absorbed PAR using radiative transfer models. *IEEE Trans. Geosci. Remote Sens.* **1997**, *35*, 1380–1393. [CrossRef]
14. Kimes, D.S.; Knyazikhin, Y.; Privette, J.L.; Abuelgasim, A.A.; Gao, F. Inversion methods for physically-based models. *Remote Sens. Rev.* **2000**, *18*, 381–439. [CrossRef]
15. Knyazikhin, Y.; Glassy, J.; Privette, J.L.; Tian, Y.; Lotsch, A.; Zhang, Y.; Wang, Y.; Morisette, J.T.; Votava, P.; Myneni, R.B.; et al. *MODIS Leaf Area Index (LAI) and Fraction of Photosynthetically Active Radiation Absorbed by Vegetation (FPAR) Product (MOD15) Algorithm Theoretical Basis Document*; NASA Goddard Space Flight Center: Greenbelt, MD, USA, 1999; Volume 20771.
16. Campos-Taberner, M.; García-Haro, F.J.; Camps-Valls, G.; Grau-Muedra, G.; Nutini, F.; Crema, A.; Boschetti, M. Multitemporal and multiresolution leaf area index retrieval for operational local rice crop monitoring. *Remote Sens. Environ.* **2016**, *187*, 102–118. [CrossRef]
17. Campos-Taberner, M.; García-Haro, F.J.; Camps-Valls, G.; Grau-Muedra, G.; Nutini, F.; Busetto, L.; Katsantonis, D.; Stavrakoudis, D.; Minakou, C.; Gatti, L.; et al. Exploitation of SAR and Optical Sentinel Data to Detect Rice Crop and Estimate Seasonal Dynamics of Leaf Area Index. *Remote Sens.* **2017**, *9*, 248. [CrossRef]
18. Svendsen, D.H.; Martino, L.; Campos-Taberner, M.; García-Haro, F.J.; Camps-Valls, G. Joint Gaussian Processes for Biophysical Parameter Retrieval. *IEEE Trans. Geosci. Remote Sens.* **2018**, *56*, 1718–1727. [CrossRef]
19. Baret, F.; Hagolle, O.; Geiger, B.; Bicheron, P.; Miras, B.; Huc, M.; Berthelot, B.; Niño, F.; Weiss, M.; Samain, O.; et al. LAI, fAPAR and fCover CYCLOPES global products derived from VEGETATION: Part 1: Principles of the algorithm. *Remote Sens. Environ.* **2007**, *110*, 275–286. [CrossRef]
20. Baret, F.; Jacquemoud, S.; Guyot, G.; Leprieur, C. Modeled analysis of the biophysical nature of spectral shifts and comparison with information content of broad bands. *Remote Sens. Environ.* **1992**, *41*, 133–142. [CrossRef]
21. García-Haro, F.J.; Campos-Taberner, M.; Muñoz-Marí, J.; Laparra, V.; Camacho, F.; Sánchez-Zapero, J.; Camps-Valls, G. Derivation of global vegetation biophysical parameters from EUMETSAT Polar System. *ISPRS J. Photogramm. Remote Sens.* **2018**, *139*, 57–74. [CrossRef]
22. Jacquemound, S.; Bidel, L.; Francois, C.; Pavan, G. ANGERS Leaf Optical Properties Database (2003). Data Set. Available online: http://ecosis.org (accessed on 5 June 2018).
23. Gitelson, A.A.; Merzlyak, M.N. Remote sensing of chlorophyll concentration in higher plant leaves. *Adv. Space Res.* **1998**, *22*, 689–692. [CrossRef]
24. Gitelson, A.A; Buschmann, C.; Lichtenthaler, H.K. Leaf chlorophyll fluorescence corrected for re-absorption by means of absorption and reflectance measurements. *J. Plant Physiol.* **1998**, *152*, 283–296. [CrossRef]
25. Hosgood, B.; Jacquemoud, S.; Andreoli, G.; Verdebout, J.; Pedrini, G.; Schmuck, G. *Leaf Optical Properties Experiment 93 (LOPEX93)*; European Commission—Joint Research Centre: Ispra, Italy, 1994; p. 20. Available online: https://data.ecosis.org/dataset/13aef0ce-dd6f-4b35-91d9-28932e506c41/resource/4029b5d3-2b84-46e3-8fd8-c801d86cf6f1/download/leaf-optical-properties-experiment-93-lopex93.pdf (accessed on 5 June 2018).
26. Feret, J.B.; François, C.; Asner, G.P.; Gitelson, A.A.; Martin, R.E.; Bidel, L.P.R.; Ustin, S.L.; Le Maire, G.; Jacquemoud, S. PROSPECT-4 and 5: Advances in the leaf optical properties model separating photosynthetic pigments. *Remote Sens. Environ.* **2008**, *112*, 3030–3043. [CrossRef]

27. Combal, B.; Baret, F.; Weiss, M.; Trubuil, A.; Mace, D.; Pragnere, A.; Myneni, R.B.; Knyazikhin, Y.; Wang, L. Retrieval of canopy biophysical variables from bidirectional reflectance using prior information to solve the ill-posed inverse problem. *Remote Sens. Environ.* **2002**, *84*, 1–15. [CrossRef]
28. Kattge, J.; Díaz, S.; Lavorel, S.; Prentice, I.C.; Leadley, P.; Bönisch, G.; Garnier, E.; Westoby, M.; Reich, P.B.; Wright, I.J.; et al. TRY—A global database of plant traits. *Glob. Chang. Biol.* **2011**, *17*, 2905–2935. [CrossRef]
29. Wulder, M.A.; Coops, N.C. Make Earth observations open access: Freely available satellite imagery will improve science and environmental-monitoring products. *Nature* **2014**, *513*, 30–32. [CrossRef] [PubMed]
30. Gorelick, N.; Hancher, M.; Dixon, M.; Ilyushchenko, S.; Thau, D.; Moore, R. Google Earth Engine: Planetary-scale geospatial analysis for everyone. *Remote Sens. Environ.* **2017**, *202*, 18–27. [CrossRef]
31. Robinson, N.P.; Allread, B.W.; Jones, M.O.; Moreno, A.; Kimball, J.S.; Naugle, D.E.; Erickson, T.A.; Richardson, A.D. A dynamic Landsat derived Normalized Difference Vegetation Index (NDVI) product for the Conterminous United States. *Remote Sens.* **2017**, *9*, 863. [CrossRef]
32. Attermeyer, K.; Flury, S.; Jayakumar, R.; Fiener, P.; Steger, K.; Arya, V.; Wilken, F.; Van Geldern, R.; Premke, K. Invasive floating macrophytes reduce greenhouse gas emissions from a small tropical lake. *Sci. Rep.* **2016**, *6*, 20424. [CrossRef] [PubMed]
33. Yu, M.; Gao, Q.; Gao, C.; Wang, C. Extent of night warming and spatially heterogeneous cloudiness differentiate temporal trend of greenness in mountainous tropics in the new century. *Sci. Rep.* **2017**, *7*, 41256. [CrossRef] [PubMed]
34. He, M.; Kimball, J.S.; Maneta, M.P.; Maxwell, B.D.; Moreno, A.; Begueria, S.; Wu, X. Regional Crop Gross Primary Productivity and Yield Estimation Using Fused Landsat-MODIS Data. *Remote Sens.* **2018**, *10*, 372. [CrossRef]
35. Kraaijenbrink, P.D.A.; Bierkens, M.F.P.; Lutz, A.F.; Immerzeel, W.W. Impact of a global temperature rise of 1.5 degrees Celsius on Asia's glaciers. *Nature* **2017**, *549*, 257–260. [CrossRef] [PubMed]
36. Reichstein, M.; Bahn, M.; Mahecha, M.D.; Kattge, J.; Baldocchi, D.D. Linking plant and ecosystem functional biogeography. *Proc. Natl. Acad. Sci. USA* **2014**, *111*, 13697–13702. [CrossRef] [PubMed]
37. Van Bodegom, P.M.; Douma, J.C.; Verheijen, L.M. A fully traits-based approach to modeling global vegetation distribution. *Proc. Natl. Acad. Sci. USA* **2014**, *111*, 13733–13738. [CrossRef] [PubMed]
38. Madani, N.; Kimball, J.S.; Ballantyne, A.P.; Affleck, D.L.R.; Bodegom, P.M.; Reich, P.B.; Kattge, J.; Sala, A.; Nazeri, M.; Jones, M.; et al. Future global productivity will be affected by plant trait response to climate. *Sci. Rep.* **2018**, *8*, 2870. [CrossRef] [PubMed]
39. Wirth, C.; Lichstein, J.W. The Imprint of Species Turnover on Old-Growth Forest Carbon Balances-Insights From a Trait-Based Model of Forest Dynamics. In *Old-Growth Forests*; Wirth, C., Heimann, M., Gleixner, G., Eds.; Springer: Jena, Germany, 2009; pp. 81–113, ISBN 978-3-540-92705-1.
40. Ziehn, T.; Kattge, J.; Knorr, W.; Scholze, M. Improving the predictability of global CO_2 assimilation rates under climate change. *Geophys. Res. Lett.* **2011**, *38*, 10. [CrossRef]
41. Berger, K.; Atzberger, C.; Danner, M.; D'Urso, G.; Mauser, W.; Vuolo, F.; Hank, T. Evaluation of the PROSAIL Model Capabilities for Future Hyperspectral Model Environments: A Review Study. *Remote Sens.* **2018**, *10*, 85. [CrossRef]
42. Bacour, C.; Baret, F.; Béal, D.; Weiss, M.; Pavageau, K. Neural network estimation of LAI, fAPAR, fCover and LAIxCab, from top of canopy MERIS reflectance data: Principles and validation. *Remote Sens. Environ.* **2014**, *105*, 313–325. [CrossRef]
43. Si, Y.; Schlerf, M.; Zurita-Milla, R.; Skidmore, A.; Wang, T. Mapping spatio-temporal variation of grassland quantity and quality using MERIS data and the PROSAIL model. *Remote Sens. Environ.* **2012**, *121*, 415–425. [CrossRef]
44. Friedl, M.A.; Sulla-Menashe, D.; Tan, B.; Schneider, A.; Ramankutty, N.; Sibley, A.; Huang, X. MODIS Collection 5 global land cover: Algorithm refinements and characterization of new datasets. *Remote Sens. Environ.* **2010**, *114*, 168–182. [CrossRef]
45. Parzen, E. On estimation of a probability density function and mode. *Ann. Math. Stat.* **1962**, *33*, 1065–1076. [CrossRef]
46. Reich, P.B. The world-wide 'fast–slow' plant economics spectrum: A traits manifesto. *J. Ecol.* **2014**, *102*, 275–301. [CrossRef]
47. Nelsen, R.B. *An Introduction to Copulass*, 2nd ed.; Springer Science & Business Media: New York, NY, USA, 2009; ISBN 978-0387-28659-4.

48. Žežula, I. On multivariate Gaussian copulas. *J. Stat. Plan. Inference* **2009**, *111*, 3942–3946. [CrossRef]
49. Jacquemoud, S.; Baret, F. PROSPECT: A model of leaf optical properties spectra. *Remote Sens. Environ.* **1990**, *34*, 75–191. [CrossRef]
50. Verhoef, W. Light scattering by leaf layers with application to canopy reflectance modeling: The SAIL model. *Remote Sens. Environ.* **1984**, *16*, 125–141. [CrossRef]
51. Breiman, L.; Friedman, J.H. Estimating Optimal Transformations for Multiple Regression and Correlation. *J. Am. Stat. Assoc.* **1985**, *391*, 1580–598.
52. Evans J.S.; Cushman, S.A. Gradient modeling of conifer species using random forests. *Landsc. Ecol.* **2009**, *24*, 673–683. [CrossRef]
53. Oliveira, S.; Oehler, F.; San-Miguel-Ayanz, J.; Camia, A.; Pereira, J.M.C. Modeling spatial patterns of fire occurrence in Mediterranean Europe using Multiple Regression and Random Forest. *For. Ecol. Manag.* **2012**, *275*, 117–129. [CrossRef]
54. Gislason, P.O.; Benediktsson, J.A.; Sveinsson, J.R. Random Forests for land cover classification. *Pattern Recognit. Lett.* **2006**, *27*, 294–300. [CrossRef]
55. Cutler, D.R.; Edwards, T.C.; Beard, K.H.; Cutler, A.; Hess, K.T.; Gibson, J.; Lawler, J.J. Random Forests for classification in ecology. *Ecology* **2007**, *88*, 2783–2792. [CrossRef] [PubMed]
56. Genuer, R.; Poggi, J.M.; Tuleau-Malot, C. Variable selection using random forests. *Pattern Recognit. Lett.* **2010**, *31*, 2225–2236. [CrossRef]
57. Jung, M.; Zscheischler, J. A Guided Hybrid Genetic Algorithm for Feature Selection with Expensive Cost Functions. *Procedia Comput. Sci.* **2013**, *18*, 2337–2346. [CrossRef]
58. Baret, F.; Morissette, J.T.; Fernandes, R.; Champeaux, J.L.; Myneni, R.B.; Chen, J.; Plummer, S; Weiss, M.; Bacour, C.; Garrigues, S.; et al. Evaluation of the representativeness of networks of sites for the global validation and intercomparison of land biophysical products: proposition of the CEOS-BELMANIP. *IEEE Trans. Geosci. Remote Sens.* **2006**, *44*, 1794–1803. [CrossRef]
59. Yan, K.; Park, T.; Yan, G.; Liu, Z.; Yang, B.; Chen, C.; Nemani, R.R.; Knyazikhin, Y.; Myneni, R.B. Evaluation of MODIS LAI/FPAR Product Collection 6. Part 2: Validation and Intercomparison. *Remote Sens.* **2016**, *8*, 460. [CrossRef]
60. Camacho, F.; Cernicharo, J.; Lacaze, R.; Baret, F.; Weiss, M. GEOV1: LAI, FAPAR essential climate variables and FCOVER global time series capitalizing over existing products. Part 2: Validation and intercomparison with reference products. *Remote Sens. Environ.* **2013**, *137*, 310–329. [CrossRef]
61. Nestola, E.; Sánchez-Zapero, J.; Latorre, C.; Mazzenga, F.; Matteucci, G.; Calfapietra, C.; Camacho, F. Validation of PROBA-V GEOV1 and MODIS C5 & C6 fAPAR Products in a Deciduous Beech Forest Site in Italy. *Remote Sens.* **2017**, *9*, 126.
62. Baret, F.; Weiss, M.; Lacaze, R.; Camacho, F.; Makhmara, H.; Pacholcyzk, P.; Smets, B. GEOV1: LAI and FAPAR essential climate variables and FCOVER global time series capitalizing over existing products. Part 1: Principles of development and production. *Remote Sens. Environ.* **2013**, *137*, 299–309. [CrossRef]
63. Campos-Taberner, M.; García-Haro, F.J.; Busetto, L.; Ranghetti, L.; Martínez, B.; Gilabert, M.A.; Camps-Valls, G.; Camacho, F.; Boschetti, M. A Critical Comparison of Remote Sensing Leaf Area Index Estimates over Rice-Cultivated Areas: From Sentinel-2 and Landsat-7/8 to MODIS, GEOV1 and EUMETSAT Polar System. *Remote Sens.* **2018**, *10*, 763. [CrossRef]
64. Xu, B.; Park, T.; Yan, K.; Chen, C.; Zeng, Y.; Song, W.; Yin, G.; Li, J.; Liu, Q.; Knyazikhin, Y.; et al. Analysis of Global LAI/FPAR Products from VIIRS and MODIS Sensors for Spatio-Temporal Consistency and Uncertainty from 2012–2016. *Forests* **2018**, *9*, 73. [CrossRef]

Article

An Operational Before-After-Control-Impact (BACI) Designed Platform for Vegetation Monitoring at Planetary Scale

Ate Poortinga [1,2,*], Nicholas Clinton [3], David Saah [1,4], Peter Cutter [1,2], Farrukh Chishtie [2,5], Kel N. Markert [6,7], Eric R. Anderson [6,7], Austin Troy [1,8], Mark Fenn [9], Lan Huong Tran [9], Brian Bean [9], Quyen Nguyen [2,5], Biplov Bhandari [2,5], Gary Johnson [1] and Peeranan Towashiraporn [2,5]

1. Spatial Informatics Group, LLC, 2529 Yolanda Ct., Pleasanton, CA 94566, USA; dssaah@usfca.edu (D.S.); pcutter@sig-gis.com (P.C.); austin@sig-gis.com (A.T.); gjohnson@sig-gis.com (G.J.)
2. SERVIR-Mekong, SM Tower, 24th Floor, 979/69 Paholyothin Road, Samsen Nai Phayathai, Bangkok 10400, Thailand; peeranan@adpc.net (P.T.); farrukh.chishtie@adpc.net (F.C.); nguyen.quyen@adpc.net (Q.N.); biplov.b@adpc.net (B.B.)
3. Google, Inc., 1600 Amphitheatre Parkway, Mountain View, CA 94043, USA; nclinton@google.com
4. Geospatial Analysis Lab, University of San Francisco, 2130 Fulton St., San Francisco, CA 94117, USA
5. Asian Disaster Preparedness Center, SM Tower, 24th Floor, 979/69 Paholyothin Road, Samsen Nai Phayathai, Bangkok 10400, Thailand
6. Earth System Science Center, The University of Alabama in Huntsville, 320 Sparkman Dr., Huntsville, AL 35805, USA; kel.markert@nasa.gov (K.N.M.); eric.anderson@nasa.gov (E.R.A.)
7. SERVIR Science Coordination Office, NASA Marshall Space Flight Center, 320 Sparkman Dr., Huntsville, AL 35805, USA
8. Department of Urban and Regional Planning, University of Colorado Denver, Campus Box 126, PO Box 173364, Denver, CO 80217-3364, USA
9. Winrock International, Vietnam Forests and Deltas program, 98 to Ngoc Van, Tay Ho, Hanoi 100803, Vietnam; markfenn@hotmail.com (M.F.); lan.huong.tran294@gmail.com (H.T.L.); BBean@winrock.org (B.B.)

* Correspondence: apoortinga@sig-gis.com or poortinga.ate@gmail.com

Received: 6 April 2018; Accepted: 13 May 2018; Published: 15 May 2018

Abstract: In this study, we develop a vegetation monitoring framework which is applicable at a planetary scale, and is based on the BACI (Before-After, Control-Impact) design. This approach utilizes Google Earth Engine, a state-of-the-art cloud computing platform. A web-based application for users named EcoDash was developed. EcoDash maps vegetation using Enhanced Vegetation Index (EVI) from Moderate Resolution Imaging Spectroradiometer (MODIS) products (the MOD13A1 and MYD13A1 collections) from both Terra and Aqua sensors from the years 2000 and 2002, respectively. to detect change in vegetation, we define an EVI baseline period, and then draw results at a planetary scale using the web-based application by measuring improvement or degradation in vegetation based on the user-defined baseline periods. We also used EcoDash to measure the impact of deforestation and mitigation efforts by the Vietnam Forests and Deltas (VFD) program for the Nghe An and Thanh Hoa provinces in Vietnam. Using the period before 2012 as a baseline, we found that as of March 2017, 86% of the geographical area within the VFD program shows improvement, compared to only a 24% improvement in forest cover for all of Vietnam. Overall, we show how using satellite imagery for monitoring vegetation in a cloud-computing environment could be a cost-effective and useful tool for land managers and other practitioners

Keywords: BACI; Enhanced Vegetation Index; Google Earth Engine; cloud-based geo-processing

1. Introduction

Forest ecosystems provide a wide range of benefits to humans [1–3] but remain under great pressure due to population growth and economic development. The protection of forests and their resources is important as local and distant human populations benefit directly from food, fuel, fiber and eco-tourism from healthy ecosystems. Functioning ecosystems also stabilize the climate, provide fresh water, control floods, and provide non-material benefits such as aesthetic views and recreational opportunities [4–8]. Deforestation and degradation are a major source of greenhouse gas emissions, while forest management and restoration programs can improve livelihoods, create jobs, and improve economic growth in local communities. They can also lead to healthier environments, functioning ecosystem services, and reduce global greenhouse gas emissions.

This latter issue, the protection of forest ecosystems and subsequent reduction of greenhouse gas emissions is an important item in the international environmental fora. REDD+ (Reducing Emissions from Deforestation and forest Degradation) is a major global initiative which, for example, aims to reduce land-use related emissions from developing countries. Payment for Ecosystem Services (PES) is another exemplar initiative, which creates voluntary agreements between individuals generating benefits from extracting forest resources, and those individuals negatively impacted by the deforestation [9]. The challenge in all of these initiatives is that developing countries often need extensive support to implement climate resilient strategies and protect their natural resources for future generations. Many international Non-Governmental Organizations (NGOs) offer generous support for the implementation of such strategies, but have strict guidelines on monitoring, evaluation and report on the impact of the measures which may be difficult for the host country to adhere to without specialized technical support.

A common method for evaluating the impact of environmental and ecological interventions is the BACI (Before-After, Control-Impact) method [10]. Figure 1 provides a schematic overview of the BACI framework. For the intervention area, the before and after variables of interest are measured. These are compared with the before and after measures of the same variables at a control site. The differences between the intervention and control sites determine the impact generated by the interventions [11,12]. Other studies have used BACI to study Marine protected areas [13], integrated marsh management [14] and ecosystem recovery [15].

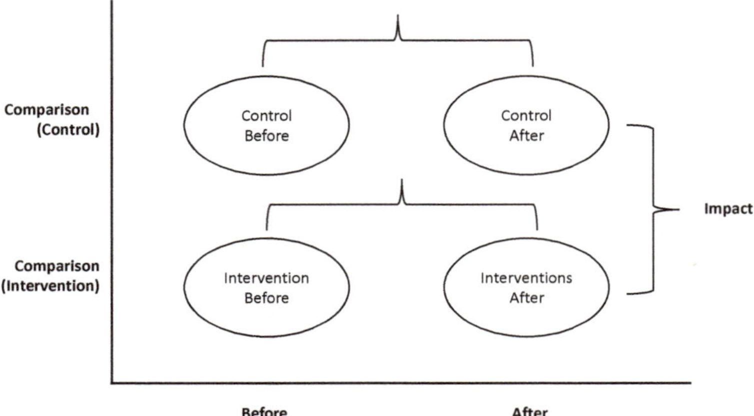

Figure 1. Schematic overview of the BACI (Before-After, Control-Impact) framework. For both the control and impact sites the before and after situations are evaluated. The difference between the two after situations defines the impact of the measures. Image was modified from [16].

Monitoring forest ecosystems is important but difficult due to their highly distinctive complex spatial and temporal patterns [17,18]. Conventional methods for forest evaluation include extensive field research on a wide range of biophysical parameters such as vegetation health, tree height, tree cover, species distributions, animal movement patterns, and many more. Important work is being conducted by the United Nations Food and Agricultural Organization (FAO) in their 5-yearly Global Forest Resources Assessments [19] where forest area and characteristics are identified. However, these approaches are expensive in terms of time and resources.

Recent international scientific developments have led to high resolution global satellite derived data products for assessing the state of vegetation and forest cover for the entire globe. These products have the resolution and coverage required for an adequate quantitative assessment of many environmental and ecological features and patterns. Together with recent advances in cloud-based remote sensing and geo-computational platforms, these technologies have led to greater open scientific data for use by policy makers and practitioners outside of academia. The hybridization and simplification of these technologies also allow scientists to provide policy-makers, international donors, NGOs, and other development partners with tailor-made products to monitor and value their ecosystems in near real-time, and requiring less advanced technical expertise than in the past.

In this paper we demonstrate a near real-time method for quantifying vegetation using cloud computing technology and remote-sensing data. We developed a novel custom indicator to monitor vegetation on a planetary scale, and then combined this with remote-sensing data for near real-time customized quantification of vegetation change. We demonstrate how new cloud-based geo-computational technology can be used to temporally, geographically and adaptively filter data collections, all while performing calculations on a global scale. Finally, we present a case study how these technologies can help policy makers, project managers and other non-experts to quantify vegetation for various purposes including monitoring and ecosystem valuation, and allow them to use the results for local economic and social progress in a developing nation.

2. Methods

We developed a framework to quantify and monitor vegetation using two common remote-sensing data products that incorporate the Before and After Control Impact (BACI) design [20]. This approach is used in ecology and environmental studies as an experimental design to evaluate the impact of an environment change on an ecosystem [21]. The framework is based on repeated measurements of several vegetation variables at various times over an observational period. Measurements were taken at a location which was known to be unaffected by vegetational change (control location) and at another location which potentially would be affected this same change (be explicit) (treatment location) for each timestep [20]. This approach is applicable for evaluating both natural and man-made changes to an ecosystem especially when it is not possible to randomly select treatment sites [22]. The framework is based on the Google Earth Engine cloud computing platform, which is a technology that is able to rapidly deliver information derived from remotely sensed imagery in near-real time.

2.1. Data

Vegetation conditions of a landscape were calculated from the Moderate Resolution Imaging Spectroradiometer (MODIS) Enhanced Vegetation Index (EVI) products (Table 1). The MODIS EVI products used in this study are provided every 16 days at 500 m spatial resolution as a gridded level-3 product. MYD13A1 and MOD13A1 are derived from MODIS Aqua and Terra satellites respectively and thus have a difference in temporal coverage. Both products contain 12 layers, including Normalized Difference Vegetation Index (NDVI), EVI, red reflectance, blue reflectance, Near Infrared (NIR) reflectance, view zenith, solar zenith, relative azimuth angle, Summary QA, detailed QA and day of the year.

The MODIS EVI products minimize canopy background variations and maintain sensitivity over dense vegetation conditions [23]. The blue band is used to remove residual atmosphere contamination caused by smoke and sub-pixel thin clouds. These products are computed from atmospherically corrected bi-directional surface reflectance that has been masked for water, clouds, heavy aerosols, and cloud shadows [24]. Many studies have been conducted which compare the relationship of MODIS EVI to biophysical conditions and Gross Primary Productivity of an ecosystem (e.g., [25–27]) making it a suitable remote sensing product for monitoring biophysical variables.

Table 1. MODIS products used to calculate the biophysical health of an area.

Product	Time Series	Temporal	Spatial	Sensor
MYD13A1	4 July 2002-present	16 days	500 m	MODIS Aqua
MOD13A1	18 February 2000-present	16 days	500 m	MODIS Terra

2.2. Vegetation Cover

To quantify changes in vegetation we adopted a climatological change approach. A user-defined baseline is calculated for a specified region and time period. The baseline defines the initial condition of the selected area. The baseline is calculated for pixels on a monthly timescale using all images in the baseline time-series. Equation (1) shows that the average monthly baseline (EVI_{B_m}) is calculated from the monthly EVI maps (EVI_{m_n}). The user specified study period is calculated from changes from the baseline, as shown in Equation (2), where EVI_{S_m} is the monthly averaged EVI map. Equation (3) is applied to calculate the cumulative sum at time t iteratively over the time-series.

$$EVI_{B_m} = \frac{1}{n}(EVI_{m_1} + EVI_{m_2} + \ldots + EVI_{m_n}) \quad (1)$$

$$\Delta EVI_{S_m} = EVI_{S_m} - EVI_{B_m} \quad (2)$$

$$EVI_t = \sum_{t=1}^{t} \Delta EVI_{S_m} \quad (3)$$

Both EVI products, namely, MYD13A1 and MOD13A1 are merged into one image collection. A time filter is then applied to create two image collections; one for the baseline period and one for the study period. Box 1 shows the JavaScript code to calculate the monthly EVI anomaly (Equations (1) and (2)). The monthly means of the baseline are calculated and subtracted from the monthly mean in the study period. The map function is used to apply the calculation to each month in the study period. These calculations are executed in parallel on the cloud-computing platform.

The calculation of the cumulative anomaly is computationally most expensive. First, a list with one image containing zeros is created (see box 2). Next, an image of the sorted (date) image collection of the anomaly is added to the last image in the newly created list. The iterate function is used to apply the function (box below) to each image in the collection. The iteration on a sorted image collection makes the calculation computational more intensive, as the results are dependent on results of the previous calculation.

2.3. Computational Platform

Recent technological advances have greatly enhanced computational capabilities and facilitated increased access to the public. In this regard, Google Earth Engine (GEE) is an online service that applies state-of-the-art cloud computing and storage frameworks to geospatial datasets. The archive contains a large catalog of earth observation data which enables the scientific community to perform calculations on large numbers of images in parallel. The capabilities of GEE as a platform which can deliver at a planetary scale are detailed in Gorelick, et al [28]. Various studies have been carried out using the GEE at a variety of scales for different purposes (see e.g., [29–31]).

Box 1. JavaScript code to calculate the monthly EVI anomaly.

```
// calculate the anomaly
var anomaly = study.map(img){

    // get the month of the map
    month = ee.Number.parse(ee.Date(img.get("system:time_start")).format("M"))

    // get the day in month
    day = ee.Number.parse(ee.Date(img.get("system:time_start")).format("d"))

    // select image in reference period
    referenceMaps = reference.filter(ee.Filter.calendarRange(month,month,"Month"))
    referenceMaps = referenceMaps.filter(ee.Filter.calendarRange(day,day,"day_of_month"))

    // get the mean of the reference and multiply with scaling factor
    referenceMean = ee.Image(referenceMaps.mean()).multiply(0.0001)

    // get date
    time = img.get('system:time_start')

    // multiply image by scaling factor
    study = img.multiply(0.0001)

    // subtract reference from image
    result = ee.Image(study.subtract(referenceMean).set('system:time_start',time))
    }
```

Box 2. JavaScript code to calculate the cumulative anomaly.

```
// Get the timestamp from the most recent image in the reference collection.
var time0 = monthlyMean.first().get('system:time_start');

// The first anomaly image in the list is just 0
var first = ee.List([ee.Image(0).set('system:time_start', time0)
                        .select([0], ['EVI'])]);

// This is a function to pass to Iterate().
// As anomaly images are computed and added to the list.
var accumulate = function(image, list) {

  // get(-1) the last image in the image collection
  var previous = ee.Image(ee.List(list).get(-1));

  // Add the current anomaly to make a new cumulative anomaly image.
  var added = image.add(previous)
              .set('system:time_start', image.get('system:time_start'));

  // Return the list with the cumulative anomaly inserted.
  return ee.List(list).add(added);
};

// Create an ImageCollection of cumulative anomaly images by iterating.
var cumulative =  ee.List(monthlyMean.iterate(accumulate, first));
```

The framework to request data, perform spatial calculations, and serve the information in a browser is shown in Figure 2. The front-end relies on Google App engine technology. Code developed from either the JavaScript or Python APIs are interpreted by the relevant client library (JavaScript or Python, respectively) and sent to Google as JSON request objects. Results are sent to either the Python command line or the web browser for display and/or further analysis. Spatial information is displayed with the Google Maps API and other information is sent to a console or the Google Visualization API.

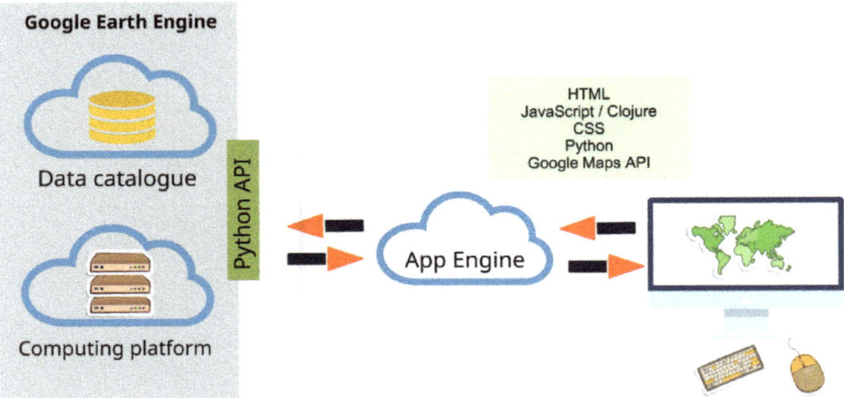

Figure 2. The infrastructure for spatial application development provided by Google. The Google Earth Engine consists of a cloud-based data catalogue and computing platform. The App Engine framework is used to host the Earth Engine application.

3. Results

To demonstrate the computational power of cloud-based geo-computational systems, we applied an algorithm on a planetary scale using countries as administrative boundaries. Our algorithm was applied to each country to investigate vegetation from 2015 onwards, using 2002-2015 as a baseline. We defined areas with a negative cumulative EVI anomaly as locations with vegetation loss, whereas positive numbers were associated with increased vegetation. The total area under stress can be seen in Figure 3a. It was found that countries in Africa, South America and South-East Asia have large areas with negative trends. Countries in Europe only have small areas with a negative trend, with an exception of Belarus and Ukraine. Similarly, we calculated areas which show a positive trend from the baseline on a country scale. It can be seen in Figure 3b that East Asia, Central Asia, Europe and North America have relatively large areas which show increased vegetation or greening. Also, countries such as Argentina, Paraguay, Uruguay and Australia show notable positive increase in vegetation. On the other hand, Russia, South East Asia and Africa show a low percentage of areas with positive trends.

Vegetation growth is a highly dynamic process in space and time. to estimate the net changes resulting in either growth or decline of each country, we used results of Figure 3a,b. The final result was obtained by calculating the difference between vegetation growth (positive trend) and vegetation decline (negative trend) over any given area. Negative numbers indicate a net negative trend whereas positive numbers indicates net greening. These results are shown in Figure 3c. It can be seen that tropical and sub-tropical regions show mostly negative trends. Also, most countries in Africa show negative numbers. Countries in Europe, Central and East Asia mostly have positive trends, which indicates an overall greening of their local environment. Whereas we have a baseline (before) and study period (after) defined no impact and control were defined.

The results in the previous figures are based on administrative country boundaries for the sake of simplicity. However, with the sort of geospatial technology we have used for this study, one can also draw or select custom geographies to investigate trends in cumulative EVI in relation to other geographies. As noted previously, the ultimate goal of this study was to create a user-friendly interface that would enable policy-makers, land use managers and other non-technical practitioners to use advanced monitoring and imaging techniques. Therefore, we developed an Ecological Monitoring Dashboard (global EcoDash; http://globalecodash.sig-gis.com/; the link and github repository are in Supplementary Materials), built on the Google App Engine framework, that communicates with

Google Earth Engine. Figure 4 shows the user interface of the EcoDash tool. Users can here define the baseline and analysis time periods, as well as define the geographies they wish to compare or investigate. Users then receive output that includes time series graphs and statistics on the change in bio-physical health for user-defined regions.

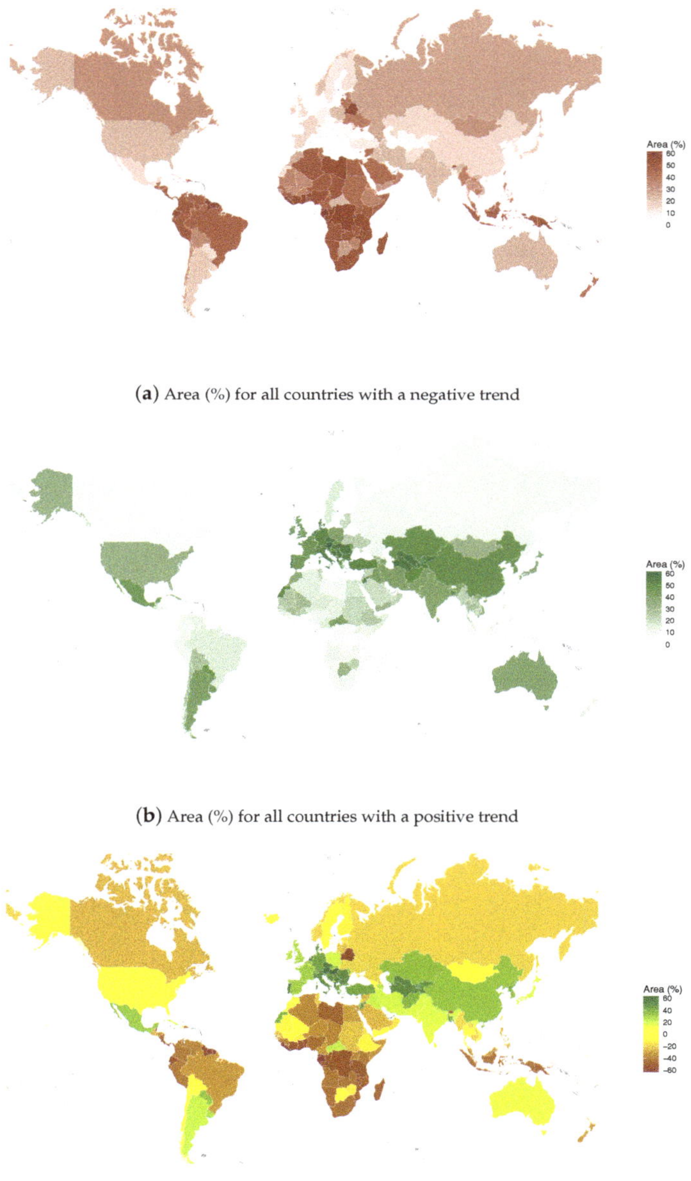

(**a**) Area (%) for all countries with a negative trend

(**b**) Area (%) for all countries with a positive trend

(**c**) Net result of area with increase-area with a decrease

Figure 3. Cumulative EVI anomaly on a country level.

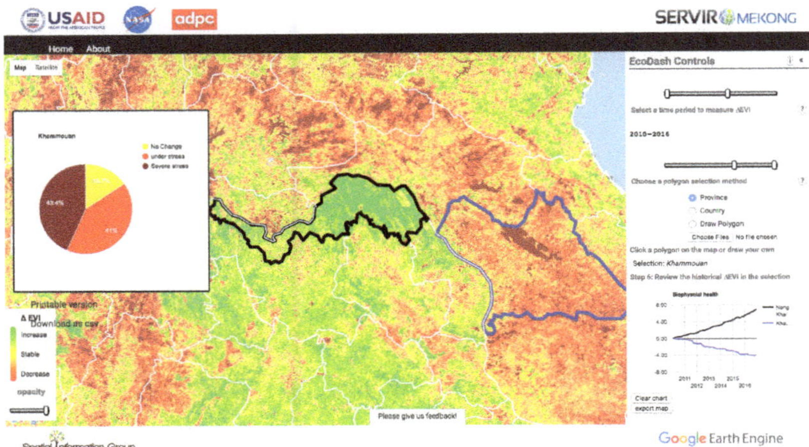

Figure 4. A screenshot of the Ecological Monitoring Dashboard (EcoDash) tool developed at SERVIR-Mekong.

For the final step of this study, EcoDash was applied to demonstrate its usability in a developing country for land monitoring, with the USAID-funded Vietnam Forests and Deltas (VFD) program. Vietnam's forests remain under development pressure and their deforestation and degradation are a source of emissions, while improved management and restoration programs offers opportunities to sequester carbon and leverage funding to further support management and livelihoods development. The VFD development program is focused on promoting practices which restore degraded landscapes and promote green growth in the forestry and agricultural sectors. The component to support adoption of land use practices and improved and sustainable forest management that slow, stop and reverse emissions from deforestation and degradation of forests and other landscapes, and can leverage mitigation finance opportunities which was started in October 2012 in the Nghe An and Thanh Hoa provinces (Figure 5). Project impacts should include improved biophysical conditions for the intervention areas, however, no baseline data were available. Then, we used EcoDash to measure and compare EVI indices in order to estimate the impact of the VFD program. We used the period before 2012 as the baseline, and we used the Nghe An and Thanh Hoa provinces as impact areas, and the remainder of Vietnam as control areas.

Figure 6 shows the cumulative EVI anomaly for the period 2011–2017 using the previous period as baseline. The green line shows the net change for the whole country (control-after) and the red line for the intervention area (impact -after). The blue line shows the difference between the control and impact topographies/areas. A rapid improvement in vegetation growth can be seen at the onset of the project in the impact areas, whereas negative vegetation growth was found for the control sites. Figure 6 clearly shows that the impact area under the VFD program experienced increased vegetation growth over the course of the project, as compared to the rest of the provinces in Vietnam, which were under similar environmental conditions such as soil and climate. As of March 2017, 86% of the net area in the intervention zones show increased vegetation cover while the overall increase in vegetation for the rest of Vietnam was only 24%. The 2015 drought event can also be seen in Figure 6.

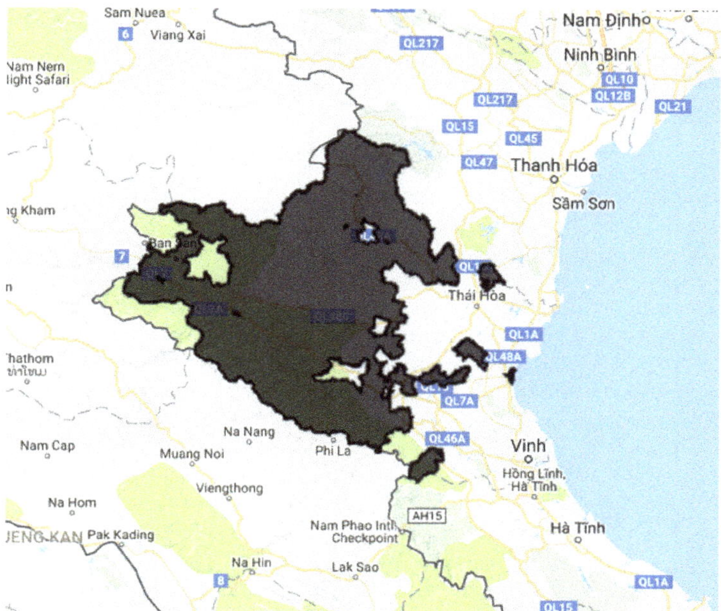

Figure 5. Location of the intervention area in Nghe An and Thanh Hoa, Vietnam.

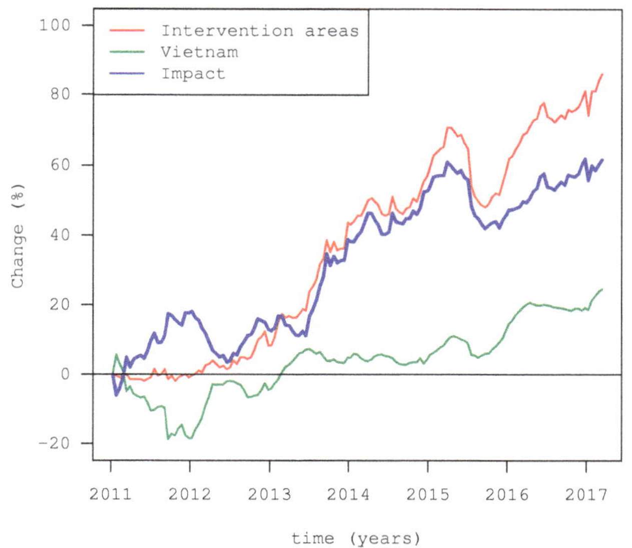

Figure 6. Cumulative EVI anomaly for the VFD intervention areas and Vietnam.

4. Discussion

The EcoDash platform for vegetation monitoring at planetary scale enables users to measure the relative change of vegetation in an impact area and compare it with a control area over a period of time. This allows end users to quickly and easily assess impacts of land management projects and enables better future planning based on the effectiveness of the various interventions. EcoDash has limitations that it does not link vegetation dynamics to the biological, climatological, or anthropological drivers behind vegetation change. Ultimately, land use managers and practitioners need data about drivers in order to appropriately plan for remediation, however, that is beyond the scope of this study. Additionally, natural systems are interconnected by feedback between systems [32] and mapping or measuring such feedback was also beyond the scope of the current study. However, in this study we did attempt to distinguish natural environmental drivers of vegetative change from anthropogenic drivers, by including a user-defined baseline for vegetation cover in a similar climatological and geographical region. The relationship between vegetation growth and ecological feedback loops can be inferred by the users on a case-by-case basis as each ecosystem/region will have varying drivers and feedbacks.

EcoDash and its underlying platform are also limited by the relatively coarse spatial resolution of the products. The spatial resolution for this study was 500 m square image segments. This sort of resolution is cheaper and easier to produce than finer resolution images/senses, and is adequate for end users who want to measure large scale temporal change quickly and cost-effectively. However, it must be noted that many projects related to landscape protection or payment for ecosystem services may require a higher spatial resolution. The Landsat satellite can provide higher spatial resolution products, but at the cost of offering a lower temporal resolution. Users could misinterpret the significance of vegetation change by blindly comparing control and impact areas that are not biophysically similar. Others have begun to addressed such misinterpretations from econometric and geospatial perspectives, but further investment is required to globally scale their methods and to transition data processing to the cloud for global access [33]. The new Sentinel-2 satellite also offers higher spatial resolution data. However, Sentinel also tends to provide lower temporal resolution data; therefore, even using data from both of these systems together means that there may be data scarcity in regions with high cloud cover, as clouds impede a clear view on the vegetation.

All results in this study were computed from satellite data. Therefore, these methods are highly suitable for areas where on-the-ground data is scarce due to financial limitations and/or inaccessible terrains. The method described here is also transparent, repeatable and suitable for near real-time monitoring. Practitioners using such methods need to be aware that satellite measurements can be affected by atmospheric conditions, errors related to the angle of the sun or sensor characteristics. Therefore, field validation of results will should always be implemented, where possible, in order to corroborate and refine results, allowing practitioners to present a more comprehensive picture of ecological change in their study areas. Finally, it should be noted that the current computational framework is highly effective for pixel-to-pixel calculations, but less suitable in situations where pixels in a single map have dependencies.

5. Conclusions

Forest ecosystems are vital for the overall well-being of our planet, as they provide habitat that maintains biodiversity, they sequester carbon, and they contribute to clean air and water in local communities, and worldwide. Monitoring changes in vegetation and forest cover is a necessary task for conscientious land managers in the wake of extensive deforestation, urban growth and other land use change. Monitoring changes in vegetative cover is also important in the context of various international initiatives, such as REDD+, which allow developing countries to finance other areas of economic growth in exchange for land preservation and the concomitant carbon sequestration and reduction in GHG emissions. Using an interface like EcoDash, previously difficult-to-access earth observations can now be leveraged by non-technical end users using cloud-based computing platforms such as the Google Earth Engine, which provides free access and ease-of-use across a vast diversity of users.

In this study, we demonstrated the practical application of this BACI technical framework and EcoDash user interface both across the globe, and more specifically for the case of the VFD program for the Nghe An and Thanh Hoa provinces in Vietnam.

This framework is therefore usable across a planetary scale, and with EcoDash applied on top of this framework, it is cost-effective and simple to use. This makes it an ideal tool for land use managers, conservationists, and development organizations worldwide who need inexpensive and quick methods to assess progress and results of environmental interventions. Ideally, future technologies will be able to provide higher resolution imagery and sensing so that this framework can be used in applications that require more fine scale data. Additionally, future sensing and computing technologies will also be able to help practitioners determine drivers of change, as well as feedback loops between ecosystems. However, this is an extremely complex topic that will not be easy to solve in the near term. Regardless, the availability of earth observations and cloud computing platforms are ushering in a new era of ways in which environmental impacts and interventions can be cheaply and quickly monitored across vast areas, which should be a boon to development professionals, land managers, urban planners, and other similar practitioners worldwide.

Supplementary Materials: The online global EcoDash tool is available at: http://globalecodash.sig-gis.com/; source code for the online global EcoDash application is available at: https://github.com/servir-mekong/ecodash/tree/global_ecodash/.

Author Contributions: D.S., N.C., M.F. and A.P. designed the methodology; A.P., N.C., D.S., P.C., F.C. and K.N.M. performed the data analysis; E.R.A., A.T., H.T.L., B.B., Q.N., B.B., G.J. and P.T. contributed to the methodology, analysis tools, website and paper.

Acknowledgments: Support for this work was provided through the joint US Agency for International Development (USAID) and National Aeronautics and Space Administration (NASA) initiative SERVIR, partially through the NASA Applied Sciences Capacity Building Program, NASA Cooperative Agreement NNM11AA01A.

Conflicts of Interest: The authors declare no conflict of interest.

References

1. Farber, S.; Costanza, R.; Childers, D.L.; Erickson, J.; Gross, K.; Grove, M.; Hopkinson, C.S.; Kahn, J.; Pincetl, S.; Troy, A. Linking ecology and economics for ecosystem management. *Bioscience* **2006**, *56*, 121–133. [CrossRef]
2. Wilson, M.A.; Troy, A.; Costanza, R. The economic geography of ecosystem goods and services. In *Cultural Landscapes and Land Use*; Springer: Heidelberg/Berlin, Germany, 2004; pp. 69–94.
3. Simons, G.; Poortinga, A.; Bastiaanssen, W.; Saah, D.; Troy, D.; Hunink, J.; de Klerk, M.; Rutten, M.; Cutter, P.; Rebelo, L.M.; et al. *On Spatially Distributed Hydrological Ecosystem Services: Bridging the Quantitative Information Gap using Remote Sensing and Hydrological Models*; FutureWater: Wageningen, The Netherlands, 2017; 45p. [CrossRef]
4. Costanza, R.; d'Arge, R.; De Groot, R.; Farber, S.; Grasso, M.; Hannon, B.; Limburg, K.; Naeem, S.; O'neill, R.V.; Paruelo, J. The value of the world's ecosystem services and natural capital. *Ecol. Econ.* **1998**, *25*, 3–16. [CrossRef]
5. Poortinga, A.; Bastiaanssen, W.; Simons, G.; Saah, D.; Senay, G.; Fenn, M.; Bean, B.; Kadyszewski, J. A Self-Calibrating Runoff and Streamflow Remote Sensing Model for Ungauged Basins Using Open-Access Earth Observation Data. *Remote Sens.* **2017**, *9*, 86. [CrossRef]
6. Bui, Y.T.; Orange, D.; Visser, S.; Hoanh, C.T.; Laissus, M.; Poortinga, A.; Tran, D.T.; Stroosnijder, L. Lumped surface and sub-surface runoff for erosion modeling within a small hilly watershed in northern Vietnam. *Hydrol. Process.* **2014**, *28*, 2961–2974. [CrossRef]
7. Wallace, K.J. Classification of ecosystem services: Problems and solutions. *Biol. Conserv.* **2007**, *139*, 235–246. [CrossRef]
8. Markert, K.N.; Griffin, R.E.; Limaye, A.S.; McNider, R.T. Spatial Modeling of Land Cover/Land Use Change and Its Effects on Hydrology Within the Lower Mekong Basin. In *Land Atmospheric Research Applications in Asia*; Vadrevu, K.P., Ohara, T., Justice, C., Eds.; Springer: Heidelberg/Berlin, Germany, 2018; pp. 667–698.
9. Suhardiman, D.; Wichelns, D.; Lestrelin, G.; Hoanh, C.T. Payments for ecosystem services in Vietnam: Market-based incentives or state control of resources? *Ecosyst. Serv.* **2013**, *6*, 64–71. [CrossRef]
10. Smith, E.P.; Orvos, D.R.; Cairns, J., Jr. Impact assessment using the before-after-control-impact (BACI) model: Concerns and comments. *Can. J. Fish. Aquat. Sci.* **1993**, *50*, 627–637. [CrossRef]

11. Conquest, L.L. Analysis and interpretation of ecological field data using BACI designs: Discussion. *J. Agric. Biol. Environ. Stat.* **2000**, *5*, 293–296. [CrossRef]
12. Underwood, A. Beyond BACI: The detection of environmental impacts on populations in the real, but variable, world. *J. Exp. Mar. Biol. Ecol.* **1992**, *161*, 145–178. [CrossRef]
13. Moland, E.; Olsen, E.M.; Knutsen, H.; Garrigou, P.; Espeland, S.H.; Kleiven, A.R.; André, C.; Knutsen, J.A. Lobster and cod benefit from small-scale northern marine protected areas: Inference from an empirical before–after control-impact study. *Proc. R. Soc. B R. Soc.* **2013**, *80*, 20122679. [CrossRef] [PubMed]
14. Rochlin, I.; Iwanejko, T.; Dempsey, M.E.; Ninivaggi, D.V. Geostatistical evaluation of integrated marsh management impact on mosquito vectors using before-after-control-impact (BACI) design. *Int. J. Health Geogr.* **2009**, *8*, 35. [CrossRef] [PubMed]
15. Hawkins, S.; Gibbs, P.; Pope, N.; Burt, G.; Chesman, B.; Bray, S.; Proud, S.; Spence, S.; Southward, A.; Langston, W. Recovery of polluted ecosystems: The case for long-term studies. *Mar. Environ. Res.* **2002**, *54*, 215–222. [CrossRef]
16. Center for International Forestry Research. *REDD Subnational Initiatives*; Center for International Forestry Research: Bogor, Indonesia, 2018.
17. Stürck, J.; Poortinga, A.; Verburg, P.H. Mapping ecosystem services: The supply and demand of flood regulation services in Europe. *Ecol. Ind.* **2014**, *38*, 198–211. [CrossRef]
18. Troy, A.; Wilson, M.A. Mapping ecosystem services: Practical challenges and opportunities in linking GIS and value transfer. *Ecol. Econ.* **2006**, *60*, 435–449. [CrossRef]
19. Keenan, R.J.; Reams, G.A.; Achard, F.; de Freitas, J.V.; Grainger, A.; Lindquist, E. Dynamics of global forest area: Results from the FAO Global Forest Resources Assessment 2015. *For. Ecol. Manag.* **2015**, *352*, 9–20. [CrossRef]
20. Upton, G.; Cook, I. *A Dictionary of Statistics 3e*; Oxford University Press: Oxford, UK, 2014.
21. Carpenter, S.R.; Frost, T.M.; Heisey, D.; Kratz, T.K. Randomized intervention analysis and the interpretation of whole-ecosystem experiments. *Ecology* **1989**, *70*, 1142–1152. [CrossRef]
22. Conner, M.M.; Saunders, W.C.; Bouwes, N.; Jordan, C. Evaluating impacts using a BACI design, ratios, and a Bayesian approach with a focus on restoration. *Environ. Monit. Assess.* **2016**, *188*, 555. [CrossRef] [PubMed]
23. Huete, A.; Didan, K.; Miura, T.; Rodriguez, E.P.; Gao, X.; Ferreira, L. Overview of the radiometric and biophysical performance of the MODIS vegetation indices. *Remote Sens. Environ.* **2002**, *83*, 195–213. [CrossRef]
24. Didan, K.; Huete, A. *MODIS Vegetation Index Product Series Collection 5 Change Summary*; TBRS Lab, The University of Arizona: Tucson, AZ, USA, 2006.
25. Rahman, A.F.; Cordova, V.D.; Gamon, J.A.; Schmid, H.P.; Sims, D.A. Potential of MODIS ocean bands for estimating CO2 flux from terrestrial vegetation: A novel approach. *Geophys. Res. Lett.* **2004**, *31*, L10503, doi:10.1029/2004GL019778. [CrossRef]
26. Xiao, X.; Hollinger, D.; Aber, J.; Gholtz, M.; Davidson, E.A.; Zhang, Q.; Moore, B., III. Satellite-based modeling of gross primary production in an evergreen needleleaf forest. *Remote Sens. Environ.* **2004**, *89*, 519–534. [CrossRef]
27. Xiao, X.; Zhang, Q.; Braswell, B.; Urbanski, S.; Boles, S.; Wofsy, S.; Moore, B.; Ojima, D. Modeling gross primary production of temperate deciduous broadleaf forest using satellite images and climate data. *Remote Sens. Environ.* **2004**, *91*, 256–270. [CrossRef]
28. Gorelick, N.; Hancher, M.; Dixon, M.; Ilyushchenko, S.; Thau, D.; Moore, R. Google Earth Engine: Planetary-scale geospatial analysis for everyone. *Remote Sens. Environ.* **2017**, *202*, 18–27. [CrossRef]
29. Klein, T.; Nilsson, M.; Persson, A.; Håkansson, B. From Open Data to Open Analyses—New Opportunities for Environmental Applications? *Environments* **2017**, *4*, 32. [CrossRef]
30. Chen, B.; Xiao, X.; Li, X.; Pan, L.; Doughty, R.; Ma, J.; Dong, J.; Qin, Y.; Zhao, B.; Wu, Z. A mangrove forest map of China in 2015: Analysis of time series Landsat 7/8 and Sentinel-1A imagery in Google Earth Engine cloud computing platform. *ISPRS J. Photogramm. Remote Sens.* **2017**, *131*, 104–120. [CrossRef]
31. Deines, J.M.; Kendall, A.D.; Hyndman, D.W. Annual Irrigation Dynamics in the US Northern High Plains Derived from Landsat Satellite Data. *Geophys. Res. Lett.* **2017**, *44*, 9350–9360. [CrossRef]

32. Bennett, E.M.; Peterson, G.D.; Gordon, L.J. Understanding relationships among multiple ecosystem services. *Ecol. Lett.* **2009**, *12*, 1394–1404, doi:10.1111/j.1461-0248.2009.01387.x. [CrossRef] [PubMed]

33. Blackman, A. *Ex-Post Evaluation of Forest Conservation Policies Using Remote Sensing Data: An Introduction and Practical Guide;* Technical Report; Inter-American Development Bank: Washington, DC, USA, 2012.

Article

Mapping Vegetation and Land Use Types in Fanjingshan National Nature Reserve Using Google Earth Engine

Yu Hsin Tsai [1,*], Douglas Stow [1], Hsiang Ling Chen [2,3], Rebecca Lewison [2], Li An [1] and Lei Shi [4]

1 Department of Geography, San Diego State University, San Diego, CA 92182-4493, USA; stow@sdsu.edu (D.S.); lan@sdsu.edu (L.A.)
2 Department of Biology, San Diego State University, San Diego, CA 92182-4493, USA; hsiangling@dragon.nchu.edu.tw (H.L.C.); RLEWISON@sdsu.edu (R.L.)
3 Department of Forestry, National Chung Hsing University, Taichung City 402, Taiwan
4 Fanjingshan National Nature Reserve Administration, Jiangkou County, Guizhou 554400, China; slwy893433@163.com
* Correspondence: cindyxtsai@gmail.com

Received: 2 May 2018; Accepted: 11 June 2018; Published: 12 June 2018

Abstract: Fanjingshan National Nature Reserve (FNNR) is a biodiversity hotspot in China that is part of a larger, multi-use landscape where farming, grazing, tourism, and other human activities occur. The steep terrain and persistent cloud cover pose challenges to robust vegetation and land use mapping. Our objective is to develop satellite image classification techniques that can reliably map forest cover and land use while minimizing the cloud and terrain issues, and provide the basis for long-term monitoring. Multi-seasonal Landsat image composites and elevation ancillary layers effectively minimize the persistent cloud cover and terrain issues. Spectral vegetation index (SVI) products and shade/illumination normalization approaches yield significantly higher mapping accuracies, compared to non-normalized spectral bands. Advanced machine learning image classification routines are implemented through the cloud-based Google Earth Engine platform. Optimal classifier parameters (e.g., number of trees and number of features for random forest classifiers) were achieved by using tuning techniques. Accuracy assessment results indicate consistent and effective overall classification (i.e., above 70% mapping accuracies) can be achieved using multi-temporal SVI composites with simple illumination normalization and elevation ancillary data, despite the fact limited training and reference data are available. This efficient and open-access image analysis workflow provides a reliable methodology to remotely monitor forest cover and land use in FNNR and other mountainous forested, cloud prevalent areas.

Keywords: Landsat; Google Earth Engine; protected area; forest and land use mapping; machine learning classification; China; temporal compositing

---

1. Introduction

Despite their protected status, nature reserves can be strongly influenced by adjacent or overlapping anthropogenic activities [1]. Given this sensitivity, accurately mapping vegetation community and land use types is important to maintain the integrity of reserve habitat and biodiversity. Fanjingshan National Nature Reserve (FNNR), a national forest reserve in Guizhou province, China, has been identified as one of the 25 global biodiversity hotspots [2] with over 100 endemic species. However, human activities such as farming, grazing, tourism, and related development frequently occur from the 21,000 people living within or near the Reserve [3].

To protect ecosystem services (limiting soil erosion and runoff) and FNNR biodiversity, Chinese government agencies have implemented payment for ecosystem services (PES) policies to promote afforestation, reduce logging, and limit farming on high sloping lands surrounding the Reserves [4–6]. Such PES programs include the National Forest Conservation Program (NFCP) in 1998, which seeks to ban logging and promote afforestation to restore forests through incentives paid to forest enterprises or users. One year later, China started another large PES program, the Grain-To-Green Program (GTGP). This program aims to reduce soil erosion and increase vegetation cover through tree planting in steep farmland areas (>15° slope in northwestern China, and 25° in southwestern China; [7,8]). This context makes monitoring and mapping forest vegetation and land use types an essential element of such programs. At FNNR, these two programs have been implemented for over 16 years, yet quantitative, large scale data about PES effectiveness remains scarce. The reserve management would benefit from mapping and monitoring forest composition and cover in a reliable and extensive manner. While the most feasible and efficient means for such mapping and monitoring is through satellite remote sensing, the persistent cloud cover and steep terrain associated with the FNNR region pose a great challenge to forest mapping with optical or microwave remote sensing approaches.

Landsat satellite imagery has several characteristics that can support long-term mapping and monitoring of vegetation and land cover changes. Landsat systems provide regular image collection at 30 m spatial resolution with a potential frequency of every 16 days and a freely available image archive dating to the early- to mid-1980s [9]. More stable and reliable land change analyses with multi-temporal Landsat data are enabled when digital numbers are converted to surface reflectance values. The conversion to surface reflectance accounts for some atmospheric and solar illumination effects and ensures multi-date images are more comparable [10,11]. Landsat surface reflectance products are processed through algorithms of the Landsat Ecosystem Disturbance Adaptive Processing System (LEDAPS; [12]) for Landsat 4, 5, and 7, and Landsat Surface Reflectance Code (LaSRC) for Landsat 8 imagery.

Spectral vegetation indices (SVIs) are commonly derived from multispectral images to characterize vegetation. Normalized indices such as simple ratio, normalized difference vegetation index (NDVI), and enhanced vegetation index (EVI) can partially suppress illumination, terrain, and soil reflectance influences in the image data to more reliably monitor vegetation [13]. For example, EVI was found to be resilient to residual atmospheric effects in a recent study [14]. Modified soil adjusted vegetation index (MSAVI) and EVI contain a soil adjustment factor that minimizes the soil background while increasing the range of vegetation signal [13]. MSAVI demonstrated the most linear relationship when regressed with biomass in the Amazonian region among the commonly utilized SVIs [15]. However, in mountainous regions, the soil adjustment factor was found to cause EVI to be more sensitive to topographic effects when comparing to NDVI [16].

Compositing or exploiting dense layer stacks of multi-temporal images have been demonstrated to improve forest and land use type mapping accuracy [17]. Images can be composited to form a multi-layer time series stack to map forest cover, whether they are from the same or different seasons. By compositing multiple images, clouds, and other missing data that occur in single images can be ignored, and seasonal phenology signals may be exploited [18]. Potapov [19] found that Landsat composites of the growing season ensured anomaly-free pixels, and were effective in mapping boreal forest cover and change. Forest types were successfully mapped with multi-date Landsat images in New Hampshire [20] and Wisconsin [21], with accuracies of 74% and 83%, respectively. Including ancillary data during the classification has also been found to improve classification accuracy [9,22,23]. In a steep mountainous study area, Dorren et al. mapped forest types using Landsat Thematic Mapper data and a digital elevation model (DEM) layer to account for variable illumination effects, with a 73% overall accuracy [24].

Machine learning type classifiers may require larger amounts of training data [25], but higher mapping accuracy can also be achieved than conventional classifiers [26]. A variety of machine learning image classification methods have been used to map vegetation type and land use, such as artificial neural networks (NN), support vector machine (SVM), decision tree (i.e., CART), and random forest classifiers. SVM classifiers assign pixels to classes by maximizing class separability from the training data, and labels pixels according to their nearest class in feature space [27,28]. Decision tree classifiers [29] apply a multi-stage binary decision making system to classify images. At each stage, pixels are divided according to the binary classification rule. Groups of pixels can be further divided based on tree growing and pruning parameters, until optimal classification is achieved. Decision tree models can be sensitive to small changes to the training data and parameters [30]. Random forest classifiers construct a multitude of decision trees that are sampled independently during training, typically improving classification results over a single decision tree model [31,32]. These machine learning classifiers require various input parameters, which can be optimized through cross-validation. In a previous study on the classification of natural vegetation in the Mediterranean region, Sluiter and Pebesma [22] utilized HyMap, ASTER optical bands, and Landsat 7 images and found that machine learning classifiers yielded up to 75% accuracy and outperformed conventional statistical-based classifiers. Johansen et al. [33] mapped woody vegetation in Australia using Landsat 5 and 7 images, and concluded that CART and random forest classifiers produced highly accurate vegetation change maps. A study on crop and land cover mapping in Ukraine compared different machine learning image classifiers, with the highest map accuracy (~75%) achieved with CART [32]. With a random forest classifier and MODIS data, Parente et al. [34] achieved almost 80% accuracy when mapping pastureland in Brazil. Other studies conclude that random forest classifiers yield higher classification accuracies, require less model training time, and are less sensitive to training sample qualities compared to SVM and NN classifiers [35,36].

Cloud-computing resources enable efficient image processing on otherwise computational intensive tasks, such as with classification of large volumes of image data, and particularly when using advanced machine learning algorithm. Google Earth Engine (https://earthengine.google.com/) is a cloud-based platform for geospatial analysis [37] that is open-access and free of charge for research, education, and non-profit purposes. The platform requires a simple online application and a Google user account to access. With a JavaScript code editor platform, Earth Engine provides a massive imagery data collection (including almost the entire Landsat archive and associated surface reflectance products) that can be retrieved directly, allowing users to interactively test and develop algorithms and preview results in real time. Earth Engine also provides various pixel-based supervised and unsupervised classifiers, including machine learning type algorithms, for mapping implementation. Google Earth Engine was utilized by Hansen et al. [38] to generate global forest cover change products. Over 650 thousand Landsat 7 scenes were incorporated, and the processes took just a few days. Other studies have also demonstrated the ease of incorporating various sources of imagery data and automating image classification routines for crop and vegetation mapping using Earth Engine [32–34].

The objective of this study is to develop and test advanced image classification techniques on the cloud-based platform Google Earth Engine for mapping vegetation and land use types in the FNNR region, and analyze their spatial distributions. A secondary objective is to determine if multi-temporal composites, SVIs, and digital elevation data enable more accurate forest and land cover mapping results in this cloud prone and complex terrain study area. Tree-based machine learning image classifiers—decision tree and random forest classifiers—are applied to multi-temporal Landsat data to generate vegetation and land cover maps. Cloud-free multi-seasonal image composites consisting of spectral vegetation indices (SVI) and ancillary data are tested for effectiveness in vegetation type mapping. Terrain shading normalization approaches are implemented to improve the mapping results in the mountainous study area. Vegetation type maps are assessed for accuracy by comparison with reference data collected through field assessment with sampling plots. The vegetation and land use mapping workflow, which includes cloud-based image processing approaches, provides a reliable method to remotely map forest and land use composition in FNNR.

2. Study Area and Materials

Fanjingshan National Nature Reserve (FNNR, 27.92°N, 108.70°E), as shown in Figure 1, is roughly 419 km^2 in size. FNNR was established in 1978 as a protected area, and was included in the UNESCO Man and Biosphere Protection network in 1986. The mountainous terrain displays a vertical elevation difference of over 2000 m, and is located in the humid subtropical climate zone. FNNR is also referred to as "ecosystem kingdom", because of its diverse microclimate that creates habitat for over 6000 different types of plants and animals, and over 100 endemic species. The vegetation communities of the FNNR region are complex and normally mixed, and almost no single-species cover type exists [39]. Based on the dominant species, we generalized the vegetation communities into five common types for this study: deciduous, evergreen broadleaf, mixed deciduous and evergreen, bamboo, and conifer. Non-reserve land use types, namely built and terraced agriculture, tend to be located along the periphery of the forest reserve.

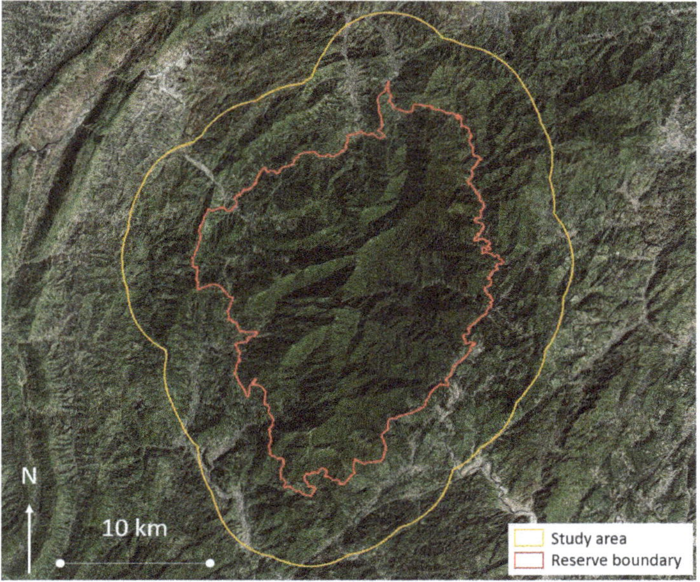

Figure 1. Study area map. Fanjingshan National Nature Reserve boundary is outlined in red; a 6-km buffer based on the reserve boundary shown in orange defines the mapping area for this study. Base imagery source: Esri, DigitalGlobe, GeoEye, Earthstar Geographics, CNES/Airbus DS, USDA, USGS, AeroGRID, IGN, and the GIS User Community.

Landsat 5 Thematic Mapper (TM) and Landsat 8 Operational Land imager (OLI) surface reflectance images of FNNR (located within Worldwide Reference System 2 path 126, row 41) on the USGS Earth Explorer website (EarthExplorer, http://earthexplorer.usgs.gov) were reviewed, and selected dates were retrieved from the Google Earth Engine image library for image processing and analysis. Landsat 7 Enhanced Thematic Mapper Plus (ETM+) images were not utilized for this study due to the Scan Line Corrector failure since 2003. Based on cloud cover and image availability for coverage throughout a vegetation growing season, two study years were selected for this study, 2011 and 2016. Images were also visually inspected to ensure the quality of the analysis. Table 1 provides information on specific image dates, sensors, and number of images used.

The FNNR study area is extremely mountainous and cloud-prone. We evaluated the cloud cover statistics with the C version of Function of Mask products (CFmask; [40]) provided with the Landsat

LEDAPS/LaSRC processed data sets. Landsat imagery for both study periods contains persistent cloud cover. For the circa 2011 period, portions of the study area have up to ten out of 13 available images that were covered by clouds. A minimum of eight image dates out of 17 images for the circa 2016 period have high amounts of cloud cover. For areas such as the highest ridge and peak located at above 2560 m elevation, there was only one image during the latter study period that provided cloud-free coverage. As Table 1 shows, two mostly cloud-free August images were selected to analyze single-date classification accuracies for 2011 and 2016. Two types of multi-temporal image stacks were also generated for the study periods: cloud-free layerstacks and seasonal composites. The compositing method is described in Section 3.1.

Table 1. Image dates and number of images associated with the study periods for this study.

Study Period & Sensor	Image Dates		
	Single Summer Date	Cloud-Free Layerstack	Seasonal Composite
Circa 2011 Landsat 5	16 August 2011	1 November 2010 28 May 2011 16 August 2011	2010–2011 13 images
Circa 2016 Landsat 8	29 August 2016	14 October 2015 29 August 2016 28 May 2017	2015–2016 17 images

Ground reference data on vegetation composition within the portions of the FNNR were collected during Fall 2012, Spring 2013, Spring 2015, Fall 2015, and Spring 2016. Relatively homogeneous 20×20 m^2 and 30×30 m^2 areas were chosen as survey plots, on the ground, based on accessibility and visual inspection. Survey plot locations were selected to ensure sampling within the five common vegetation types in FNNR—deciduous, evergreen broadleaf, mixed deciduous and evergreen, bamboo, and conifer. At every plot, we determined the dominant vegetation community type based on species cover through a rapid assessment process similar to the California Native Plant Society's Rapid Assessment Protocol [41]. We also collected digital photographs and vegetation structure information. All the survey plot locations were recorded with a global navigation satellite system receiver.

The survey locations were recorded often under the dense vegetation cover in the mountainous terrain in the study area. These factors led to difficulty in collecting more reference data, and led to a higher degree of uncertainty in positional accuracy. Prior to utilizing the survey data points, we improved the reliability of the reference dataset by cross validating with an unpublished vegetation community map. The vegetation community map was created through a collaborative project between Global Environmental Facility (GEF) and the FNNR management office in 2007. This map was generated through the combination of forest inventory field surveys and visual interpretation of a Landsat image. The map depicted 37 dominant overstory species for the reserve and was rendered to the five common FNNR vegetation community types (i.e., deciduous, evergreen, mixed deciduous and evergreen, bamboo, and conifer). While we use the map as additional reference data, the accuracy of the map has not been determined and we have observed errors when comparing the map to high spatial resolution satellite imagery. The locations and vegetation types of the survey data samples were cross-validated and co-located with the GEF 2007 vegetation map. When field assessment and map data did not agree, vegetation type classes were determined using other reference sources, such as PES locations derived from household surveys, high spatial resolution Google Earth images (namely 2013 and 2017 Pleiades), and ArcGIS Basemap imagery (pan-sharpened QuickBird, GeoEye, and WorldView-2 images from 2004 to 2012).

3. Methods

The majority of the image processing and analysis for this study was implemented through Google Earth Engine. The methods include image normalization for illumination effects (i.e., shade), generating multi-seasonal image stacks, tuning machine learning classifier parameters, generating classification maps, and assessing accuracies of vegetation/land use products.

3.1. Multi-Temporal Image Stacks

Two types of multi-temporal image stacks were generated: cloud-free layerstacks, and seasonal composites. The cloud-free layerstack image input was formed by stacking the most cloud-free Spring and Fall images available within two consecutive years of the selected Summer image. Images were combined to form a three-date stack. To minimize cloud pixels, seasonal image composites were also generated. For the seasonal composites, all Landsat images that were captured during the years of 2010–2011, and 2015–2016 were utilized. In order for the composites to preserve seasonal vegetation signals, we split the images into Spring, Summer, and Fall season groups. For each seasonal group, the mean value between all available images was calculated. Lastly, the three season layers were combined (i.e., layerstacked) to form the seasonal composites.

3.2. Classification Feature Input

Due to the extreme elevation range and steep slopes in the FNNR region, a reflectance normalization process was applied to Landsat images by dividing each reflectance band by the reflectance sum of all bands [42]. Spectral band derivatives were utilized to further minimize terrain illumination effects and maximize vegetation signature differences. Six SVI derived from the reflectance normalized Landsat spectral bands, along with elevation, slope, and aspect layers derived from a SRTM DEM were used as feature inputs for vegetation and land use classification. The slope layer was calculated in degrees, ranging from 0 to 90°. The aspect layer had a value range from 0 to 360°, and was transformed by taking the trigonometric sine values of aspect to avoid circular data values [43]. Sine values of aspect represents the degree of east-facing slopes, as the values range from 1 (i.e., east-facing) to −1 (west-facing). Clouds, cloud shadow, and water bodies were masked using the CFmask products.

NDVI, normalized difference blue and red (NDBR), normalized difference green and red (NDGR), normalized difference shortwave infrared and near infrared (NDII), MSAVI, and spectral variability vegetation index (SVVI) were derived from the Landsat data as defined below. MSAVI is calculated as Equation (1) [13]:

$$\mathrm{MSAVI} = \frac{2\rho_{\mathrm{NIR}} + 1 - \sqrt{(2\rho_{\mathrm{NIR}} + 1)^2 - 8(\rho_{\mathrm{NIR}} - \rho_{\mathrm{red}})}}{2} \quad (1)$$

NDVI is calculated as Equation (2) [44]:

$$\mathrm{NDVI} = \frac{\rho_{\mathrm{NIR}} - \rho_{\mathrm{red}}}{\rho_{\mathrm{NIR}} + \rho_{\mathrm{red}}} \quad (2)$$

where ρ_{NIR} and ρ_{red} in Equations (1) and (2) represent the near infrared and red reflectance values for a given pixel. The other three normalized difference indices: NDBR, NDGR, and NDII, were calculated as the form of NDVI in Equation (2), only with blue and red bands for NDBR, green and red bands for NDGR, and infrared bands (NIR and SWIR) for NDII. SVVI is calculated as the difference between the standard deviation (*SD*) of all Landsat bands (excluding thermal) and *SD* of all three infrared bands, as displayed in Equation (3) [45]:

$$\mathrm{SVVI} = SD(\rho_{all\ bands}) - SD(\rho_{NIR\ and\ SWIR}) \quad (3)$$

3.3. Classifiers

Two pixel-based, supervised machine learning type image classifiers were implemented and tested: decision tree (DT) and random forest (RF). To train and test the image classifiers, we utilized the forest composition survey data and manually digitized agriculture and built sample areas based on high spatial resolution satellite imagery. We derived a total of 109 samples for image classifier training and testing purposes. Of the 109 samples, 34 represented mixed, 12 broadleaf, and eight deciduous vegetation, 16 conifer, 10 bamboo, six bare land, 11 agriculture, and 12 built land uses. These samples were stratified by image illumination to account for the drastic spectral difference between illuminated and shaded slopes [46]. We organized the samples in a Google Fusion Table and retrieved in Google Earth Engine. The corresponding input image values for the 109 samples were extracted at the Landsat image pixel level.

Cross-validation and grid search techniques were implemented to optimize classifier parameters and ensure model fitting. The 109 samples were randomly selected and split into two parts (i.e., cross-validation): 2/3 for training and 1/3 for testing. Samples were selected by each class to maintain the class proportion. Different combinations of classifier parameters were systematically tested (i.e., grid search) with the training samples. The trained models were then evaluated using the reserved 1/3 testing samples for the estimated model performance. The parameter combination that yielded the highest testing accuracy was used as the optimal classifier parameter. Final vegetation and land use maps were produced based on the optimal classifier parameters utilizing the entire set of data samples. The derived maps portray the five vegetation types (bamboo, conifer, deciduous, evergreen broadleaf, and mixed deciduous and evergreen) and three land use types: agriculture, bare soil, and built.

3.4. Accuracy Assessment

The classification products were evaluated for mapping accuracy using an independent set of accuracy data points. We had difficulties discerning certain forest community types. Thus, the seven-class mapping scheme was generalized into four classes—built, agriculture, forest, and bamboo/conifer vegetation to create a second, more generalized map as part of the accuracy assessment. Conifer and bamboo were grouped into a single class, while the forest class contained deciduous, evergreen, and mixed deciduous and evergreen. A total of 128 points (32 points per class) for the study area were generated using a distance-restricted random method (i.e., points to be at least five Landsat pixels apart) and manual editing. Points were overlaid on the Planet imagery captured on July 2017 [47] and manually labeled as forest, bamboo/conifer, agriculture, or built class. The labeled reference points were compared to the corresponding classified pixels, and the percent of agreement was recorded in an accuracy table.

To examine the forest community type classification accuracy, the Landsat-derived classification products for 2011 were compared to the 2007 GEF vegetation map. A spatial correspondence matrix was generated for each product to quantify the site-specific and areal coverage similarities and differences between the classification maps and the GEF map. Only the 2011 classification maps were evaluated for they correspond in time better with the GEF map.

Classification products from the same time period that were derived using different inputs and classifiers were also compared to each other to evaluate differences in classifiers and how they represented the vegetation and land use of the study area. The most reliable classification approach was determined based on mapping accuracies, and the map comparison and evaluation results between the GEF map and the circa 2011 classification products.

4. Results

Figure 2 shows the circa 2016 classification map using RF classifier with the seasonal composite image inputs. The reserve is mostly classified as mixed evergreen and deciduous type (displayed in light green color in Figure 2). Evergreen broadleaf cover (displayed in yellow) has a distinct distribution along the river and stream channels that originate from the reserve, in addition to the concentration on the eastern and southern side of the study area. Deciduous cover type (displayed in brown) is concentrated along the high elevation ridge in the middle of the reserve, as well as the two clusters found on the south end. As for the bamboo and conifer vegetation types, more bamboo cover (displayed in red) is mapped on the eastern side of the reserve, and conifer (displayed in dark green) were found more concentrated to the west and north. The road network that surrounds the reserve is depicted as part of the built class. Mapped in close proximity to built areas are agriculture land, bamboo, and conifer cover types. They are found distributed towards the periphery and outside of the reserve boundary.

Key image classifier parameters were tuned for optimizing the classification accuracies. Optimal parameters were identified for each study period and image input type, and they can be found in Table 2. For the DT classifier, the minimum number of training points required to create a terminal node was the parameter tuned. Parameters tuned for the RF classifier were number of trees to create, and the number of variables per split.

Table 2. The optimal image classifier parameters derived through the grid search tuning. The parameters listed here yielded the highest testing accuracies and were utilized for the final map classifications. Parameters were tuned for decision tree (DT) and random forest (RF) classifiers.

Study Period	Classifier & Parameters		Classification Parameters		
			Single Summer Date	Cloud-Free Layerstack	Seasonal Composite
Circa 2011	DT	# of leaf	7	1	2
	RF	# of trees; features	80; 8	133; 2	109; 9
Circa 2016	DT	# of leaf	2	1	1
	RF	# of trees; features	17; 3	11; 5	85; 14

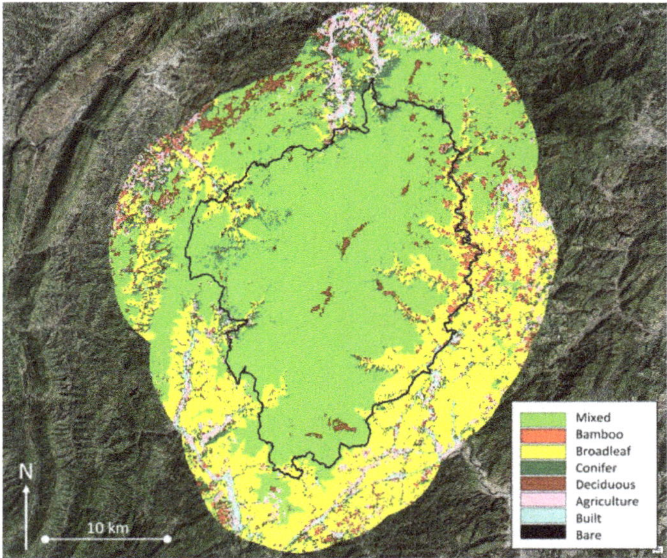

Figure 2. Vegetation type and land use classification map for circa 2016. The map was derived from the seasonal composite image input with the RF classifier.

The generalized four-class (i.e., forest, agriculture, built, and bamboo/conifer vegetation) classification map products yielded moderate overall accuracies, particularly with RF classifier with multi-date image inputs. Table 3 shows the generalized map classification accuracies for different study periods and image inputs. Of the classifiers tested, the RF classifier consistently yielded higher accuracies compared to the DT classifier (as shown in Table 3) regardless of study periods or input types. The RF classifier with the circa 2011 cloud-free layerstack image input yielded the highest accuracy value at 77%. Table 3 also shows the multi-image stack approaches yielded higher classification accuracies compared to the single-date input except one instance. On average, the multi-image stack accuracies were 2 to 11% higher. The RF classifier yielded 72 and 74% average accuracies with the multi-image input, while the single date input produced an average accuracy of 65%.

Table 3. The classification accuracy on the generalized, four-class land cover maps for different study periods and image input types. The average accuracy values are calculated for each input type and image classifier.

Study Period & Classifiers		Generalized Map Classification Accuracy		
		Single Summer Date	Cloud-Free Layerstack	Seasonal Composite
Circa 2011	DT	0.61	0.66	0.67
	RF	0.62	0.77	0.70
Circa 2016	DT	0.64	0.63	0.66
	RF	0.67	0.70	0.73
Average	DT	0.63	0.65	0.67
	RF	0.65	0.74	0.72

Table 4 shows the accuracy assessment results for the circa 2016 period using RF classifier on various methods. Based on classification accuracies and visual inspection, classification products generated using RF classifier with seasonal composite image inputs yielded the most stable and consistent high accuracy maps. Specifically, the most reliable method (in terms of accuracy and consistency) as seen in Table 4 is utilizing SVIs derived from shade and illumination normalized data in conjunction with a DEM layer as the classification input. This most reliable method yielded 73% mapping accuracy, and is significantly higher compared to using spectral bands (accuracy = 55%). Incorporating elevation information from SRTM DEM as part of the classification input also substantially improved the classification accuracy. RF classification without the DEM layer yielded 66% accuracy, seven percent lower compared to the optimal method. Classification accuracy was slightly lower when using image inputs normalized for shade and illumination, compared to using the non-normalized input products (the excluding normalization method in Table 4). However, without the normalization procedure, the classification result portrayed a substantial amount of conifer misclassification due to the extreme terrain shading in the study area.

Table 4. Mapping accuracies for circa 2016 using random forest classifier on various seasonal composite image inputs. The x symbols mark the techniques applied, and the mapping accuracy is listed for each method.

Method	Technique				Map Accuracy
	SVIs	SRTM DEM	Shade Stratification	Illumination Normalization	
Most Reliable Method	x	x	x	x	73%
Spectral Input		x	x	x	55%
Excluding DEM	x		x	x	66%
Excluding Normalization	x	x			79%

Table 5 shows the accuracy assessment and confusion matrix for the circa 2016 classification product that was generated with RF classifier using the seasonal composite image input. The forest class is likely overclassified, suggested by 97% producer's accuracy with a lower, 58% user's accuracy as Table 5 shows. Agriculture land and bamboo/conifer vegetation were the sources of confusion with the forest class. On the other hand, the built class yielded 100% user's accuracy with a lower producer's accuracy (69%), which indicate under-classification. Most of the confusion for the built class is with agriculture land, as small villages and fallow or emergent croplands exhibit similar image signatures. Agriculture activities in the study area are mostly sparse and low stature plantations, which exposes a lot of bare soil. The high reflectance spectral signature of exposed soil and fallow fields could be the source of misclassification between agriculture and built.

The spatial correspondence products generated comparing the circa 2011 classification map (generated using RF classifier with the seasonal composite image inputs) to the GEF reference map indicated 53% of the reserve area was classified as the same vegetation community types as the reference map. Mixed deciduous and evergreen, evergreen broadleaf, and conifer types were the three classes that showed highest mapping agreement. Most of the reserve core and eastern side of the reserve indicated high agreement, mostly consisting of mixed deciduous and evergreen, evergreen broadleaf, and bamboo types. The north-western portion of FNNR also showed high mapping agreement, and consisting mostly of conifer cover. The Landsat classification maps portrayed the south-western side of the reserve mostly as mixed deciduous and evergreen community types, which made up 30% of the reserve. The same area on the reference map is identified as more heterogeneous, distinguished communities of dominant evergreen broadleaf, deciduous, and conifer. Two ridges to the south of the reserve are mapped as deciduous type cover surrounded by mixed vegetation. The deciduous community makes up roughly 2% of the reserve area. This area is correctly mapped with the Landsat-derived classification products, while the GEF map portrays it as mixed deciduous and evergreen type.

Table 5. Accuracy assessment results for the c. 2016 classification product generated with the seasonal composite image input and RF classifier. These values were derived using the final accuracy assessment data on the four-class generalized map. Gray cells indicate agreement.

c. 2016 Classified Class	2016 Reference Class				User's Accuracy
	Forest	Agriculture	Built	Bamboo/Conifer	
Forest	31	9	2	11	58%
Agriculture	0	19	8	0	70%
Built	0	0	22	0	100%
Bamboo/Conifer	1	4	0	21	81%
Producer's Accuracy	97%	59%	69%	66%	Overall Accuracy 73%

5. Discussion

Our cloud-based, multi-temporal composite classification approach of satellite-based land cover data overcame challenges associated with persistent cloud cover and terrain shading effects in the FNNR region. Our results suggest Google Earth Engine is efficient and effective in accessing pre-processed satellite imagery, implementing machine learning type image classifiers, and generating classification products for FNNR. The entire workflow described in this study on average takes less than 30 min to complete. The open-access platform and the procedures described in this study enable reserve managers to monitor protected areas in an effective manner without having to purchase or download data and software. The scripting also allows users to streamline complex image processing workflow, and execute processes with minimum intervention. Collaborating and sharing scripts is also efficient with the online platform. Like most commercial image processing software, Google Earth Engine provides online documentation and tutorials as user support. It also has a discussion forum where users can post questions and share their knowledge. Earth Engine requires no specific hardware setup like most commercial image processing software. However, it does require stable internet connection which might not always be available.

Based on the classification accuracies and visual inspection, we determined that classification products generated with the RF classifier using seasonal composite image input yielded the most stable, consistent, and accurate maps for the study area. Our accuracy assessment results were comparable with many studies mapping mixed forest types in mountainous terrain [20,22,24]. Higher mapping accuracy would likely be achieved with larger and more certain training datasets [25,33]. Shade and illumination normalization techniques were helpful in minimizing the terrain shading effects and greatly decreased the misclassification of conifer cover. Incorporating elevation and its derived products in addition to SVI layers were also found to improve the classification accuracy and mapping quality significantly. The scaled sine values of the aspect data, which measures east-ness, was found to increase the map accuracy. Likely due to most of the mountain ridges in FNNR being north-south oriented and slopes are facing east-west, the scaled aspect layer using sine function produced higher accuracy than the aspect layer scaled by cosine values (which measures north-ness).

The cloud cover issue for FNNR is effectively minimized with the multi-temporal seasonal composite approach, and the RF image classifiers. Cloud cover is prevalent in most of the Landsat images utilized in this study. The CFmask products were used to exclude cloud pixels prior to image classification, and the pixels were treated as no data values. Utilizing all available data within each season maximizes the amount of cloud-free observations, thus reducing misclassification of no data pixels. Both the single summer date and cloud-free layerstack classification approaches yielded map products with apparent and substantial misclassification due to no data pixels originated from cloud cover. The DT classifier also produced map products with mislabeled no data pixels. In those instances, pixels were commonly mislabeled as bare, agriculture, or built classes. The RF image classifier tested in this study were able to consistently minimize the effects of clouds and the derived no data pixels.

The Landsat-based vegetation type classifications for the core and eastern portion of the FNNR reserve were generally similar to those portrayed in the GEF reference map. The majority of the disagreement occurred at the forest community type level, particularly mixed deciduous and evergreen class. With relatively heavy anthropogenic activities in the western portion of the reserve, it was documented that the pristine, primary forest cover has degraded to mixed primary and secondary forest type, particularly in the lower elevation [39]. This could explain the mapping differences between our classification products and the GEF reference map. A combination of subjective survey work and limited training samples are also likely why the mixed type was not further discerned into evergreen broadleaf or deciduous as in the GEF map. The GEF mapping incorporated forest inventory survey knowledge and was susceptible to labeling biases. Our survey efforts were constrained by the steep terrain and access restrictions from the Reserve Administration Office. There were only a dozen training samples collected in this portion of the reserve, nine of which were labeled as mixed through fieldwork, and only two were recorded as broadleaf type. Also, the field-based vegetation rapid assessment procedures are subjective and uncertain due to differences in seasonality and surveyors.

We encountered a few challenges in this study, mostly pertaining to uncertainties between the Landsat-derived maps and the reference data. The reference data samples that were utilized during the classifier training and testing phases were limited in quantity, and involved positional uncertainty. The bagging method as part of the RF classifier [29] likely improved the small training data limitation in this study. Another major challenge we encountered was the mismatch in time between the available cloud-free Landsat images and the reference data. The GEF reference map was produced four years earlier than the 2011 study period. The field survey (conducted between 2012 and 2015) and the high spatial resolution reference images retrieved from Google Earth and Planet (captured in 2013, 2016, and 2017) are minimally a year apart from the two study periods. This posted difficulty in analyzing the classification products in conjunction with the available reference dataset.

6. Conclusions

Frequent anthropogenic disturbances at biodiversity hotspots can degrade ecosystems and ecosystem function. This study demonstrated an effective approach to mapping vegetation cover and land use utilizing cloud-based image processing tools, even with persistent cloud cover and extreme terrain and illumination effects. The use of freely available Landsat imagery and the Earth Engine image analysis tools ensure that FNNR managers have the resources needed to continue to monitor forest cover and land use changes. Although future studies will need to continue to improve classification accuracy, particularly for the bamboo/conifer and agriculture classes where mapping errors were higher, this method can be used to evaluate impacts of afforestation policy and identify areas of ongoing human disturbance. With the generalized, four-class maps from multiple dates, land transitions of interest could be identified. For example, areas that were mapped as agriculture before 2001 (prior to PES implementation) and transitioned to bamboo/conifer at a more recent image date could be mapped as locations of PES implementation. Our image classification techniques will generate reliable information with regard to forest dynamics (especially in cloud prevalent forested areas like FNNR), which is of great importance not only for assessment of PES efficacy, but also for the long-term monitoring and assessment of generic environmental changes or conservation efforts.

Author Contributions: Conceptualization, Y.H.T. and D.S.; Methodology, Y.H.T. and D.S.; Software, Y.H.T.; Validation, Y.H.T., H.L.C., and L.S.; Formal Analysis, Y.H.T.; Investigation, Y.H.T.; Resources, H.L.C. and L.S.; Data Curation, Y.H.T. and H.L.C.; Writing-Original Draft Preparation, Y.H.T.; Writing-Review & Editing, D.S., H.L.C., R.L., and L.A.; Visualization, Y.H.T.; Supervision, D.S.; Project Administration, L.A.; Funding Acquisition, L.A.

Funding: This research was funded by the National Science Foundation under the Dynamics of Coupled Natural and Human Systems program [Grant DEB-1212183] and by Long Gen Ying Travel Grant with the support of Arthur Getis.

Acknowledgments: This research benefited from San Diego State University for providing graduate assistant support. The anonymous reviews also provided constructive comments and helpful suggestions.

Conflicts of Interest: The authors declare no conflict of interest.

References

1. Liu, J.; Linderman, M.; Ouyang, Z.; An, L.; Yang, J.; Zhang, H. Ecological degradation in protected areas: The case of Wolong Nature Reserve for giant pandas. *Science* **2001**, *292*, 98–101. [CrossRef] [PubMed]
2. Myers, N.; Mittermeier, R.A.; Mittermeier, C.G.; Da Fonseca, G.A.; Kent, J. Biodiversity hotspots for conservation priorities. *Nature* **2000**, *403*, 853–858. [CrossRef] [PubMed]
3. Wandersee, S.M.; An, L.; López-Carr, D.; Yang, Y. Perception and decisions in modeling coupled human and natural systems: A case study from Fanjingshan National Nature Reserve, China. *Ecol. Model.* **2012**, *229*, 37–49. [CrossRef]
4. Uchida, E.; Xu, J.; Rozelle, S. Grain for green: Cost-effectiveness and sustainability of China's conservation set-aside program. *Land Econ.* **2005**, *81*, 247–264. [CrossRef]
5. Liu, J.; Li, S.; Ouyang, Z.; Tam, C.; Chen, X. Ecological and socioeconomic effects of China's policies for ecosystem services. *Proc. Natl. Acad. Sci. USA* **2008**, *105*, 9477–9482. [CrossRef] [PubMed]
6. Liu, J.; Yang, W. Integrated assessments of payments for ecosystem services programs. *Proc. Natl. Acad. Sci. USA* **2013**, *110*, 16297–16298. [CrossRef] [PubMed]
7. Bennett, M.T. China's sloping land conversion program: Institutional innovation or business as usual? *Ecol. Econ.* **2008**, *65*, 699–711. [CrossRef]
8. Liu, J.G.; Diamond, J. China's environment in a globalizing world. *Nature* **2005**, *435*, 1179–1186. [CrossRef] [PubMed]
9. Xie, Y.; Sha, Z.; Yu, M. Remote sensing imagery in vegetation mapping: A review. *J. Plant Ecol.* **2008**, *1*, 9–23. [CrossRef]
10. Hall, F.; Strebel, D.; Nickeson, J.; Goetz, S. Radiometric rectification: Toward a common radiometric response among multidate, multisensor images. *Remote Sens. Environ.* **1991**, *35*, 11–27. [CrossRef]
11. Moran, M.S.; Jackson, R.D.; Slater, P.N.; Teillet, P.M. Evaluation of simplified procedures for retrieval of land surface reflectance factors from satellite sensor output. *Remote Sens. Environ.* **1992**, *41*, 169–184. [CrossRef]
12. Masek, J.G.; Vermote, E.F.; Saleous, N.E.; Wolfe, R.; Hall, F.G.; Huemmrich, K.F.; Gao, F.; Kutler, J.; Lim, T.K. A Landsat surface reflectance dataset for North America, 1990–2000. *IEEE Geosci. Remote Sens. Lett.* **2006**, *3*, 68–72. [CrossRef]
13. Qi, J.; Chehbouni, A.; Huete, A.R.; Kerr, Y.H.; Sorooshian, S. A modified soil adjusted vegetation index. *Remote Sens. Environ.* **1994**, *48*, 119–126. [CrossRef]
14. Davies, K.P.; Murphy, R.J.; Bruce, E. Detecting historical changes to vegetation in a Cambodian protected area using the Landsat TM and ETM+ sensors. *Remote Sens. Environ.* **2016**, *187*, 332–344. [CrossRef]
15. Wang, C.; Qi, J.; Cochrane, M. Assessment of tropical forest degradation with canopy fractional cover from Landsat ETM+ and IKONOS imagery. *Earth Interact.* **2005**, *9*, 1–18. [CrossRef]
16. Matsushita, B.; Yang, W.; Chen, J.; Onda, Y.; Qiu, G. Sensitivity of the enhanced vegetation index (EVI) and normalized difference vegetation index (NDVI) to topographic effects: A case study in high-density cypress forest. *Sensors* **2007**, *7*, 2636–2651. [CrossRef] [PubMed]
17. Lu, D.; Weng, Q. A survey of image classification methods and techniques for improving classification performance. *Int. J. Remote Sens.* **2007**, *28*, 823–870. [CrossRef]

18. Franco-Lopez, H.; Ek, A.R.; Bauer, M.E. Estimation and mapping of forest stand density, volume, and cover type using the k-nearest neighbors method. *Remote Sens. Environ.* **2001**, *77*, 251–274. [CrossRef]
19. Potapov, P.; Turubanova, S.; Hansen, M.C. Regional-scale boreal forest cover and change mapping using Landsat data composites for European Russia. *Remote Sens. Environ.* **2011**, *115*, 548–561. [CrossRef]
20. Schriever, J.R.; Congalton, R.G. Evaluating Seasonal Variability as an Aid to Cover-Type Mapping from Landsat Thematic Mapper Data in the Northwest. *Photogramm. Eng. Remote Sens.* **1995**, *61*, 321–327.
21. Wolter, P.T.; Mladenoff, D.J.; Host, G.E.; Crow, T.R. Improved Forest Classification in the Northern Lake States Using Multi-Temporal Landsat Imagery. *Photogramm. Eng. Remote Sens.* **1995**, *61*, 1129–1143.
22. Sluiter, R.; Pebesma, E.J. Comparing techniques for vegetation classification using multi-and hyperspectral images and ancillary environmental data. *Int. J. Remote Sens.* **2010**, *31*, 6143–6161. [CrossRef]
23. Domaç, A.; Süzen, M.L. Integration of environmental variables with satellite images in regional scale vegetation classification. *Int. J. Remote Sens.* **2006**, *27*, 1329–1350. [CrossRef]
24. Dorren, L.K.; Maier, B.; Seijmonsbergen, A.C. Improved Landsat-based forest mapping in steep mountainous terrain using object-based classification. *For. Ecol. Manag.* **2003**, *183*, 31–46. [CrossRef]
25. Kotsiantis, S.B. Supervised machine learning: A review of classification techniques. *Informatica* **2007**, *31*, 249–268.
26. Rodriguez-Galiano, V.F.; Ghimire, B.; Rogan, J.; Chica-Olmo, M.; Rigol-Sanchez, J.P. An assessment of the effectiveness of a random forest classifier for land-cover classification. *Photogramm. Eng. Remote Sens.* **2012**, *67*, 93–104. [CrossRef]
27. Boser, B.E.; Guyon, I.; Vapnik, V. A training algorithm for optimal margin classifiers. In Proceedings of the Fifth Annual Workshop on Computational Learning Theory, Pittsburgh, PA, USA, 27–29 July 1992; ACM Press: New York, NY, USA, 1992; pp. 144–152.
28. Mountrakis, G.; Im, J.; Ogole, C. Support vector machines in remote sensing: A review. *Photogramm. Eng. Remote Sens.* **2011**, *66*, 247–259. [CrossRef]
29. Breiman, L.; Friedman, J.; Stone, C.J.; Olshen, R.A. *Classification and regression trees*; Routledge: New York, NY, USA, 1984; ISBN 9781351460491.
30. Bishop, C.M. *Pattern Recognition and Machine Learning*; Springer: New York, NY, USA, 2006; ISBN 8132209060.
31. Breiman, L. Random forests. *Mach. Learn.* **2001**, *45*, 5–32. [CrossRef]
32. Shelestov, A.; Lavreniuk, M.; Kussul, N.; Novikov, A.; Skakun, S. Exploring Google Earth Engine platform for big data processing: Classification of multi-temporal satellite imagery for crop mapping. *Front. Earth Sci.* **2017**, *5*, 17. [CrossRef]
33. Johansen, K.; Phinn, S.; Taylor, M. Mapping woody vegetation clearing in Queensland, Australia from Landsat imagery using the Google Earth Engine. *Remote Sens. Appl. Soc. Environ.* **2015**, *1*, 36–49. [CrossRef]
34. Parente, L.; Ferreira, L. Assessing the Spatial and Occupation Dynamics of the Brazilian Pasturelands Based on the Automated Classification of MODIS Images from 2000 to 2016. *Remote Sens.* **2018**, *10*, 606. [CrossRef]
35. Pal, M. Random forest classifier for remote sensing classification. *Int. J. Remote Sens.* **2005**, *26*, 217–222. [CrossRef]
36. Belgiu, M.; Drăguţ, L. Random forest in remote sensing: A review of applications and future directions. *ISPRS J. Photogramm. Remote Sens.* **2016**, *114*, 24–31. [CrossRef]
37. Gorelick, N.; Hancher, M.; Dixon, M.; Ilyushchenko, S.; Thau, D.; Moore, R. Google Earth Engine: Planetary-scale geospatial analysis for everyone. *Remote Sens. Environ.* **2017**, *202*, 18–27. [CrossRef]
38. Hansen, M.C.; Potapov, P.V.; Moore, R.; Hancher, M.; Turubanova, S.; Tyukavina, A.; Thau, D.; Stehman, S.V.; Goetz, S.J.; Loveland, T.R.; et al. High-resolution global maps of 21st-century forest cover change. *Science* **2013**, *342*, 850–853. [CrossRef] [PubMed]
39. Zhou, Z. (Ed.) Department of Forestry of Guizhou Province; Fanjingshan National Nature Reserve Administration Office. In *Research on the Fanjing Mountain*; Guizhou People's Publishing House: Guizhou, China, 1990; ISBN 7-221-01372-1.
40. Foga, S.; Scaramuzza, P.L.; Guo, S.; Zhu, Z.; Dilley, R.D.; Beckmann, T.; Schmidt, G.L.; Dwyer, J.L.; Hughes, M.J.; Laue, B. Cloud detection algorithm comparison and validation for operational Landsat data products. *Remote Sens. Environ.* **2017**, *194*, 379–390. [CrossRef]
41. California Native Plant Society (CNPS) Vegetation Committee Rapid Assessment (RA) Protocol. Available online: http://www.cnps.org/cnps/vegetation/pdf/protocol-combined-2016.pdf (accessed on 11 June 2018).

42. Wu, C. Normalized spectral mixture analysis for monitoring urban composition using ETM+ imagery. *Remote Sens. Environ.* **2004**, *93*, 480–492. [CrossRef]
43. Xu, J.; Zhang, X.; Zhang, Z.; Zheng, G.; Ruan, X.; Zhu, J.; Xi, B. Multi-scale analysis on wintering habitat selection of Reeves's pheasant (*Syrmaticus reevesii*) in Dongzhai National Nature Reserve, Henan Province, China. *Acta Ecol. Sin.* **2006**, *26*, 2061–2067. [CrossRef]
44. Carlson, T.N.; Ripley, D.A. On the relation between NDVI, fractional vegetation cover, and leaf area index. *Remote Sens. Environ.* **1997**, *62*, 241–252. [CrossRef]
45. Coulter, L.L.; Stow, D.A.; Tsai, Y.; Ibanez, N.; Shih, H.C.; Kerr, A.; Benza, M.; Weeks, J.R.; Mensah, F. Classification and assessment of land cover and land use change in southern Ghana using dense stacks of Landsat 7 ETM+ imagery. *Remote Sens. Environ.* **2016**, *184*, 396–409. [CrossRef]
46. Tsai, Y.; Stow, D.; Shi, L.; Lewison, R.; An, L. Quantifying canopy fractional cover and change in Fanjingshan National Nature Reserve, China using multi-temporal Landsat imagery. *Remote Sens. Lett.* **2016**, *7*, 671–680. [CrossRef]
47. Planet Team. Planet Application Program Interface: In Space for Life on Earth. San Francisco, CA. Available online: https://www.planet.com (accessed on 11 June 2018).

Technical Note

A Dynamic Landsat Derived Normalized Difference Vegetation Index (NDVI) Product for the Conterminous United States

Nathaniel P. Robinson [1,2,*], Brady W. Allred [1,2], Matthew O. Jones [1,2], Alvaro Moreno [2], John S. Kimball [1,2], David E. Naugle [1], Tyler A. Erickson [3] and Andrew D. Richardson [4,5]

1 W.A. Franke College of Forestry and Conservation, University of Montana, Missoula, MT 59812, USA; allredbw@gmail.com (B.W.A.); matt.jones@ntsg.umt.edu (M.O.J.); johnk@ntsg.umt.edu (J.S.K.); david.naugle@umontana.edu (D.E.N.)

2 Numerical Terradynamic Simulation Group, University of Montana, Missoula, MT 59812, USA; alvaro.moreno@ntsg.umt.edu

3 Google, Inc., Mountain View, CA 94043, USA; tylere@google.com

4 School of Informatics, Computing and Cyber Systems, Northern Arizona University, Flagstaff, AZ 86011, USA; Andrew.richardson@nau.edu

5 Center for Ecosystem Science and Society, Northern Arizona University, Flagstaff, AZ 86011, USA

* Correspondence: Nathaniel.Robinson@umontana.edu; Tel.: +1-406-243-5521

Academic Editors: Prasad Thenkabail, Lalit Kumar and Onisimo Mutanga
Received: 13 July 2017; Accepted: 18 August 2017; Published: 21 August 2017

Abstract: Satellite derived vegetation indices (VIs) are broadly used in ecological research, ecosystem modeling, and land surface monitoring. The Normalized Difference Vegetation Index (NDVI), perhaps the most utilized VI, has countless applications across ecology, forestry, agriculture, wildlife, biodiversity, and other disciplines. Calculating satellite derived NDVI is not always straight-forward, however, as satellite remote sensing datasets are inherently noisy due to cloud and atmospheric contamination, data processing failures, and instrument malfunction. Readily available NDVI products that account for these complexities are generally at coarse resolution; high resolution NDVI datasets are not conveniently accessible and developing them often presents numerous technical and methodological challenges. We address this deficiency by producing a Landsat derived, high resolution (30 m), long-term (30+ years) NDVI dataset for the conterminous United States. We use Google Earth Engine, a planetary-scale cloud-based geospatial analysis platform, for processing the Landsat data and distributing the final dataset. We use a climatology driven approach to fill missing data and validate the dataset with established remote sensing products at multiple scales. We provide access to the composites through a simple web application, allowing users to customize key parameters appropriate for their application, question, and region of interest.

Keywords: Google Earth Engine; NDVI; vegetation index; Landsat; remote sensing; phenology; surface reflectance

---

1. Introduction

The Normalized Difference Vegetation Index (NDVI) is arguably the most widely implemented remote sensing spectral index for monitoring Earth's land surface. Since the earliest report of use in 1973 [1,2], the term NDVI is found in nearly 121,000 scientific articles, conference papers, and books (Google Scholar). The index capitalizes on the optical properties of the cellular structure of leaves; the photosynthetic pigments (chlorophyll, associated light-harvesting pigments, and accessory pigments) efficiently absorb radiation in the visible range of the spectrum (to power photosynthesis) and reflect radiation in the near-infrared (NIR) range. The simple formula of NDVI and its direct relationship to vegetation photosynthetic capacity is a proxy for a wide range of essential vegetation characteristics and functions (e.g., fraction of photosynthetic radiation absorbed by the canopy, leaf area, canopy "greenness", gross primary productivity) with countless applications in agriculture, forestry, ecology, biodiversity, habitat modeling, species migrations, land surface phenology, earth system processes (nutrient cycling, net primary productivity, evapotranspiration), and even economic, social, and medical sciences.

Satellite remote sensing (SRS) allows for the calculation of NDVI globally at a range of temporal intervals and spatial resolutions dependent on sensor characteristics and the satellite orbit, with a common inverse relationship between temporal and spatial resolutions. The Landsat Mission, with its first sensor launched in 1972, is the only uninterrupted long-term (>30 years) high-resolution remote sensing dataset that can provide a continuous historic NDVI record globally. The Landsat record at 30-m resolution is ideally suited for local or regional scale time-series applications, particularly with the recent release of higher-level surface reflectance products from Landsat sensors 5 ETM, 7 ETM+, and 8 OLI from 1984 to present. Utilizing these products across scenes and through time, however, is not without complications [3], particularly for users without GIS and Remote Sensing training and resources. To create consistent mosaics or long-term time series, users must account for data record gaps, radiometric differences across sensors [4], scene overlaps, malfunctions (e.g., the Landsat 7 scan line corrector malfunction), and inherent noise (due to clouds, atmospheric contamination, missing auxiliary data, etc.). As the region of interest and temporal extent increases, data volume and compute processing needs present significant barriers to many users without access to high performance computing facilities or the necessary skills to manipulate such data. These limitations often prevent the implementation of such a dataset in ecological studies, conservation monitoring efforts, or teaching exercises despite the clear value of its application.

The rise of high performance computing clusters, public access to supercomputing facilities and cloud computing and storage removes many of the computational barriers associated with Landsat data. The ability to create user friendly applications that interacts with these computing services eliminates additional barriers associated with data manipulation and enables users with minimal technical coding skills to access and process data. We capitalize on the abilities of high performance computing resources and web-based software to provide a Landsat derived conterminous U.S. (CONUS), 30-m resolution, NDVI product (Figure 1). We use Landsat 5 ETM, 7 ETM+, and 8 OLI sensors, with a user specified climatology (historic NDVI value limited by a user-defined time-period) for temporal smoothing, and Google Earth Engine (a cloud-based geospatial platform for planetary-scale data analysis) for rapid data processing and visualization [5], to produce 16 day NDVI composites from 1984 to 2016. We validate the NDVI product by comparing against other established remote sensing products across multiple spatial scales. The resulting NDVI record enables greater use of Landsat data in answering crucial ecological questions across broad spatio-temporal scales at a higher level of spatial detail than possible with other currently available NDVI products. While Landsat composite products exist (e.g., the Web Enabled Landsat Data product [6] and the ability to create simple mean/median/max composites) our product improves upon these with the novel gap-filling and smoothing approaches (Figure 2). Additionally, we make the composites available through a dynamic web application, allowing users to customize key parameters to produce NDVI composites more suited to specific regions or ecological questions.

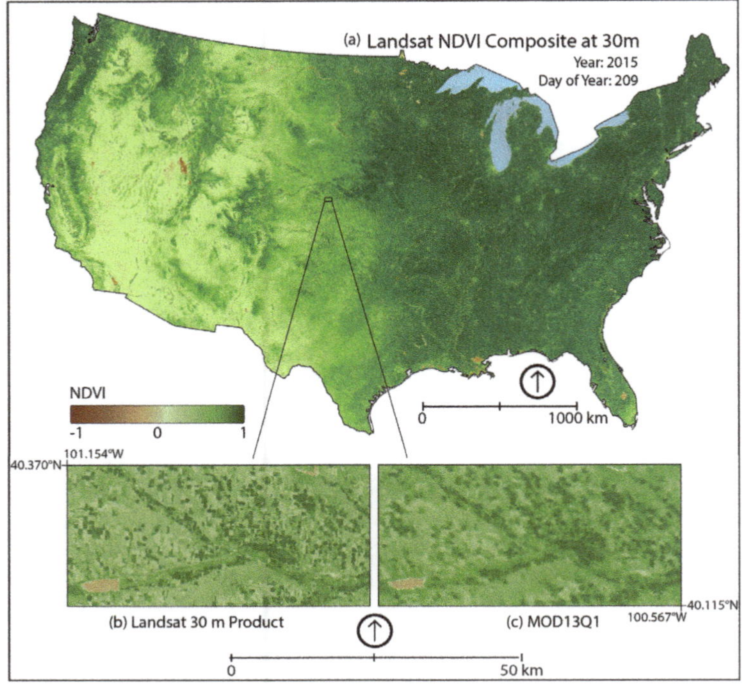

Figure 1. (**a**) A 30 m continuous CONUS Landsat NDVI composite for 28 July 2015. Our methods produce broad scale composites with minimal gaps in data and reduce the effect of scene edges and Local scale comparison of (**b**) Landsat NDVI at 30 m and (**c**) MODIS MOD13Q1 at 250 m from the same composite period. The Landsat product provides added spatial detail important in measuring certain ecological processes.

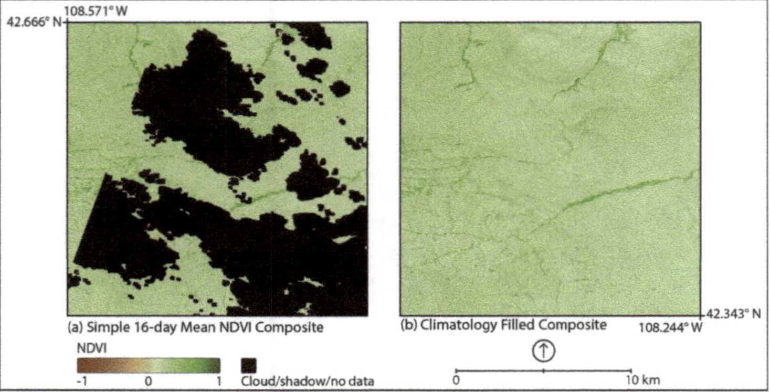

Figure 2. (**a**) A simple 16-day mean NDVI composite from 28 July to 12 August 2015 created from Landsat 7 and 8 sensors. The composite contains missing data due to cloud cover and scene edges are apparent due to differing acquisition dates. (**b**) A 16-day climatology (5-year) gap filled composite for the same time and location. The climatology is user defined in order to produce an appropriate composite for the question being asked.

2. Materials and Methods

2.1. Data

We use the surface reflectance (SR) products from Landsat 5 ETM, 7 ETM+, and 8 OLI sensors to create NDVI composites. The Landsat satellites have near-polar orbits with a repeat overpass every 16 days; throughout the Landsat missions; however, two satellites have often operated simultaneously (Figure 3) in asynchrony, creating an eight-day return overpass for a given area.

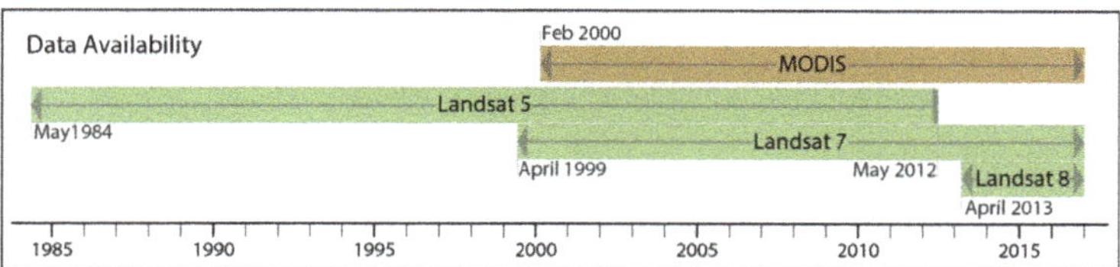

Figure 3. A timeline showing the data availability for Landsat NDVI, based upon Landsat surface reflectance products and MOD13Q1. The extended Landsat record provides a longer continuous record of high resolution NDVI.

Furthermore, adjacent orbits of a single sensor spatially overlap from 7% at the equator to 68.7% at 70° latitude [7]. During a single 16-day period there may be as many as four independent views for a given point. Our compositing method (Figure 4) capitalizes on the operation of multiple sensors and views to maximize the potential of retrieving an NDVI observation every 16 days.

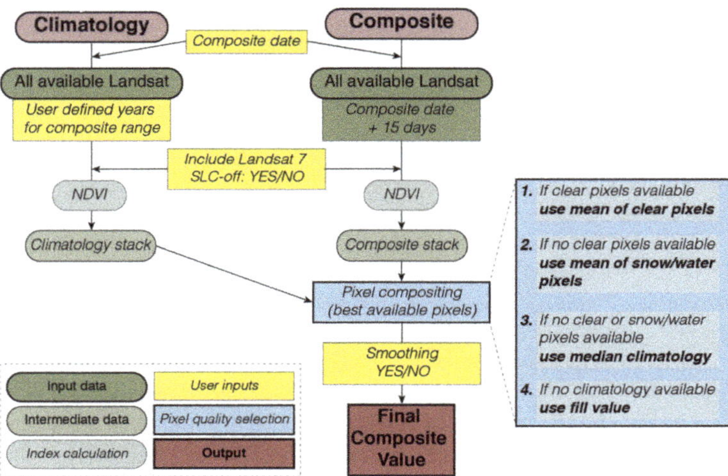

Figure 4. A flow chart demonstrating the NDVI compositing process, in which the best available pixels from all available Landsat sensors are selected and combined to produce the final NDVI composite value.

The Landsat SR products [8,9] correct for atmospheric and illumination/viewing geometry effects, and are the highest level of image processing available for Landsat data. Although some images are not processed due to missing auxiliary data, the use of SR is generally more appropriate for measuring and monitoring vegetation at the land surface [10,11]. Landsat Surface reflectance products also contain useful pixel data quality flag information indicating clear, water, snow, cloud or shadow conditions, as determined by the CFMask algorithm [12]. We employ this information to select the best available data within each composite period.

2.2. Compositing

To produce a pixel-wise 16-day composite (date of composite plus subsequent 15 days), all available Landsat surface reflectance images (from 5 ETM, 7 ETM+, and 8 OLI) are processed. Landsat scenes are resampled bilinearly to a Geographic Coordinate System WGS84 grid of approximately 30 m (1/5000 degrees) resolution. NDVI is calculated as:

$$NDVI = \frac{(\rho NIR - \rho RED)}{(\rho NIR + \rho RED)} \quad (1)$$

where ρNIR is surface reflectance in the near infrared band (band 4—Landsat 5, 7; band 5—Landsat 8) and ρRED is surface reflectance in the red band (band 3—Landsat 5, 7; band 4—Landsat 8). To account for sensor differences, we adjusted landsat NDVI values from Landsat 5 ETM and 7 ETM+ to match Landsat 8 OLI using a simple linear transformation: [13].

$$NDVI_{L8} = 0.0235 + 0.9723 \times NDVI_{L5,7} \quad (2)$$

Additionally, Landsat 5 scenes often contain abnormalities along scene edges, resulting in both missing data and erroneously high NDVI values. These pixels are removed by buffering 450 m inwards from the image mask (Figure S1). The buffer size was determined from visual inspection of a subset of Landsat 5 scenes, ensuring removal of all the erroneous pixels without losing substantial amounts of valid data. To ensure the best available data for each composite, a pixels are selected and used based on their quality flag. First, all pixels flagged as clear during a 16-day period are selected and the mean NDVI calculated. If no 'clear' pixels are available, the mean NDVI value of all 'water' and 'snow' pixels is used. If there are still no available pixels, (i.e., all pixels within the 16-day period are flagged as cloud or shadow, or no surface reflectance images are available) the pixel is filled with a climatology. The climatology is calculated as the median NDVI of 'clear', 'water' and 'snow' pixels over the same 16-day period from previous years, with the user specifying the number of years. The median climatology is used to minimize the effects abnormally wet or dry years within the climatology record. In rare instances when no climatology is available (i.e., all pixels within the set climatology length are flagged as cloud or shadow), the composite is filled with a no-data value.

2.3. Smoothing

As NDVI is a proxy for vegetation greenness, it is expected to follow a relatively smooth and continuous temporal profile. Outside of disturbance or land cover change events, a sudden drop in NDVI is likely due to atmospheric contamination or a quality issue not identified in the Landsat surface reflectance product [14,15]. To account for these anomalous declines, we employ a smoothing method, similar to iterative Interpolation for Data Reconstruction (IDR) [16]. If a composite NDVI value is less than the mean of the previous and following time step composites by a threshold of 0.1, it is replaced by that mean value. While Julien and Sobrino suggest iteratively smoothing until convergence is reached, we only smooth once as multiple runs significantly increases computational time at large scales. Invocation of the smoothing algorithm by the user is optional.

2.4. Quality

A quality band is provided to specify the attributes of the raw data used to calculate each pixel's composite value. The quality band indicates if a composite value was calculated from clear pixels; water or snow pixels; or if the climatology was used. The quality band also indicates if a composite value is the result of smoothing. Table 1 shows the range of quality band values and descriptions.

Table 1. NDVI quality band values and descriptions.

Pixel Value	Description
10	Clear not smoothed
11	Clear and smoothed
20	Snow or water not smoothed
21	Snow or water smoothed
30	Climatology not smoothed
31	Climatology smoothed

2.5. Product Creation and Distribution

Landsat derived NDVI is available through a simplified web-interface (Figure 5, http://ndvi.ntsg.umt.edu/) that utilizes Google Earth Engine. Users define a region of interest, select a time period, the length of the climatology used for gap filling (2, 5, 10, 15, 20, 25, or 30 years), inclusion of Landsat 7 ETM+ SLC-off data, and whether to apply the smoothing algorithm. The customized NDVI composite is then produced (as a GeoTIFF) as requested based on the user defined parameters.

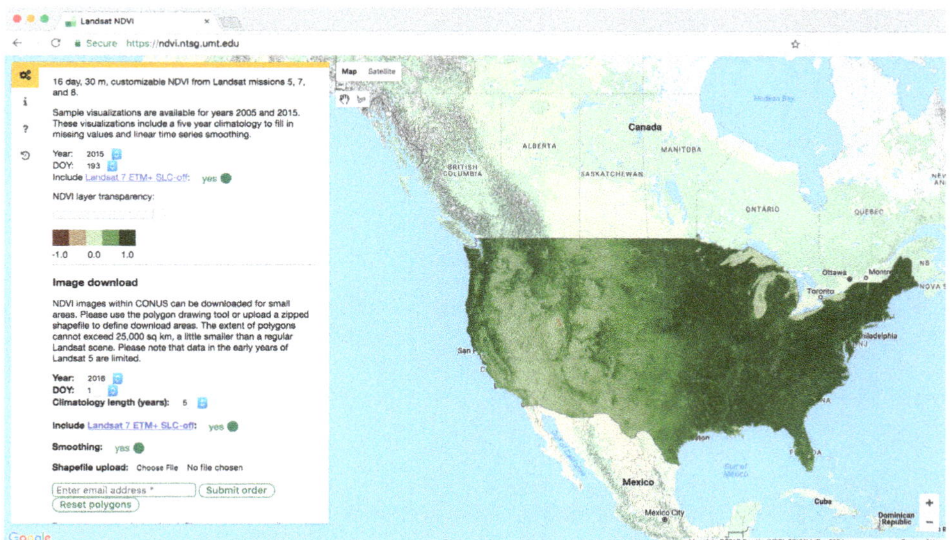

Figure 5. A screen shot of the NDVI web application (https://ndvi.ntsg.umt.edu). To download a composite, users set their desired parameters in the left panel. The region of interest can either be an uploaded shapefile or a polygon drawn directly on the map. The composite is processed on the fly and users are notified via email when it is ready to download.

2.6. NDVI Comparisons Across Spatial Scales

We compare the Landsat derived NDVI record to independently derived finer and coarser resolution data, including: the green chromatic coordinate from in situ phenology camera (phenocam) observations and the Moderate Resolution Imaging Spectroradiometer 16-day 250 m NDVI product (MOD13Q1). We use Landsat derived NDVI composites with a five-year climatology for gap filling and employ the IDR smoothing algorithm for the validation comparisons.

2.6.1. Phenology Cameras

The PhenoCam Network provides automated, sub-daily, near-surface remote sensing of canopy phenology through digital repeat photography [17]. The images are continuous in time and robust to variation in illumination condition, with minimal influence from clouds or atmospheric effects, particularly when calculating vegetation indices [18]. Numerous studies [18–21] have demonstrated that the green chromatic coordinate (GCC; [22]); can be used to identify phenology phases and monitor canopy development, with strong correlations to NDVI time series. The GCC is calculated as:

$$GCC = \frac{DN_g}{(DN_r + DN_g + DN_b)} \quad (3)$$

where *DN* is the digital number recorded by the camera and r, g, and b denote red, green, and blue channels respectively. PhenoCam Network sites within CONUS that had at least four years of continuous imagery were selected for analysis; resulting in 43 sites that include agriculture/crops, shrublands, grasslands, deciduous broadleaf forests, and evergreen needleleaf forests (Figure S2). We use the daily GCC90 data provided by the PhenoCam Network, which represents the daily 90th percentile of the GCC during daylight hours. A 16-day mean is calculated from the daily GCC90, using the same 16-day period as the Landsat NDVI product. The corresponding Landsat NDVI time series is extracted over each PhenoCam site, followed by calculation of Pearson correlation coefficients.

Within each image field of view (FOV), a predefined region of interest (ROI) is used to calculate the GCC, isolating the plant functional type (PFT) of interest. Depending on the FOV, more than one ROI can be defined, providing two independent time series of different PFTs. Four of the 43 sites contained two ROIs and we compare both ROIs at these sites to the single broader scale (30 m) Landsat NDVI time series.

The comparison of two independent vegetation indices derived from sensors with different bandwidths, fields of view, and viewing geometries is not without issue [23]. The GCC is more sensitive to leaf pigmentation than NDVI [24] and the Landsat pixel may not capture the camera FOV or may be smaller than the FOV. However, the PhenoCam data provides the only multi-year, high spatial and temporal resolution standardized product comparable to the 30 m land surface phenology signal. The correlations provide an assessment of the Landsat NDVI composites seasonal response to vegetation conditions either within or in close proximity to the camera FOV.

2.6.2. MOD13Q1

The MODIS VI products (MOD13) are designed to provide consistent spatiotemporal observations of vegetation conditions, have been continually produced since 2001 [25], and employed in at least 1700 peer-reviewed research articles (Google Scholar). The MOD13Q1 product has a 16-day NDVI composite with an approximate spatial resolution of 250 m. Like the Landsat NDVI product, the MOD13Q1 16-day composite period includes the composite date and 15 ensuing days. MOD13Q1 composites are created using a constrained-view angle, maximum value composite technique, and the MODIS surface reflectance product [26].

We compare the Landsat derived NDVI to the MOD13Q1 NDVI from 2000 to 2016. Time series of both products are extracted for a set of points across the CONUS domain (Figure S2) using a stratified random sample across land cover classes. Points are only selected within areas of homogenous land cover at the MODIS resolution, determined using the National Land Cover Dataset (NLCD) for 2001, 2006, and 2011 [27–29]. Within these homogenous regions, up to 50 random points are created, using Google Earth Engine's random point function, for 12 major land cover classes across the domain (evergreen forest, deciduous forest, mixed forest, shrubland, grassland, pasture/hay, herbaceous wetland, wooded wetland, barren, developed-open space, and developed-low intensity). For certain land cover classes, less than 50 random points in homogeneous pixels are available, resulting in a total sample size of 356 points across the domain. To match resolutions, the Landsat NDVI was degraded to the MODIS 250 m resolution where the mean Landsat NDVI value was calculated within the extent of each MODIS pixel. The time series for both products were extracted, disregarding any null values, resulting in 131,973 paired observations. The Pearson correlation coefficients (r-value), mean bias, mean absolute bias (MAB), and root mean square error (RMSE) are calculated for the entire series and each location separately.

3. Results

3.1. Phenology Cameras Results

The phenocam correlation analysis (Table S1) resulted in 36 of the 47 ROIs exhibiting r-values greater than 0.70, and just three ROIs with r-values less than 0.30 (all ROIs: mean r-value = 0.72; range: −0.35–0.92; $p < 0.01$ for all cases). The high and significant correlations demonstrate that the 16-day Landsat composites do well in capturing the seasonal greenness patterns exhibited by the phenocam GCC90. The sites with the three lowest correlations provide good examples where the resulting NDVI values and their comparison to other data products requires careful interpretation. One site (drippingsprings; $r = 0.22$) presents a mismatch between the vegetation in the extent of the Landsat pixel and the ROI of the phenocam image. The phenocam ROI delineates a single deciduous broadleaf tree canopy in a narrow ravine, while the extent of the Landsat pixel includes other riparian zone species and shrubs above the ravine. Another low correlation site (oregonMP; $r = -0.24$) is from an evergreen needleleaf forest in Oregon. Examination of the quality band indicates this site is often obscured by clouds and snow in the winter months, resulting in a spurious NDVI time series with poorly defined seasonality, while the GCC90 time series provides a well-defined seasonal signal. The site with the lowest correlation (sedgwick SH; $r = -0.35$) contained two ROIs and is discussed below.

Three of the four sites with two ROIs displayed strong correlations both between ROI's ($0.81 < r < 0.94$) and versus the Landsat NDVI ($0.72 < r < 0.88$). Therefore, even though the two ROIs within a site delineated separate PFTs, the PFTs displayed a common seasonality. The fourth site with one grass ROI and one shrub ROI, located on the Sedgwick Reserve in southern California, displayed contrasting results: Shrub vs. Grass ROI, r-value = −0.20; Shrub ROI vs. Landsat NDVI, r-value = −0.35; Grass ROI vs. Landsat NDVI, r-value = 0.75. Examination of the time series revealed that the Shrub ROI was out of phase with the Grass ROI, with a seasonal lag of approximately three months, resulting in negative correlations when compared to the grassland dominated NDVI signal.

The low correlation sites highlight two important considerations that must be accounted for when comparing satellite and ground-level observations. First, vegetation indices from satellite data represent integrated measures of the vegetation at the pixel scale often confounding comparisons to canopy scale indices, such as those derived from phenology cameras particularly over heterogeneous landscapes [30,31]. Second, phenology camera FOVs will vary from site to site, and in some cases an ROI may be beyond the extent of the satellite pixel that contains the camera, particularly when implementing high resolution (30 m) data.

3.2. MOD13Q1 Results

We found high correlations between the Landsat NDVI product and coarser MOD13Q1 observations (Table 2, Figure 6), with an overall r-value of 0.94. When disaggregated by the Landsat product quality flag these data show a higher correlation for clear pixels (r-value = 0.97), slightly lower correlation for climatology filled pixels (r-value = 0.88) and still lower correlation for snow/water pixels (r-value = 0.70).

Table 2. Mean bias, mean absolute bias (MAB), root mean square error (RMSE), and r-values for all the MOD13Q1 and Landsat NDVI sample points combined. Each statistic is calculated for all pixels and each quality flag separately.

Statistic	All Pixels	Clear Pixels	Snow/Water Pixels	Climatology Pixels
Mean Bias	−0.03	−0.03	−0.01	−0.02
MAB	0.06	0.05	0.10	0.09
RMSE	0.10	0.08	0.15	0.14
Pearson's r	0.94	0.97	0.71	0.88

When disaggregated to individual points, 258 of the 356 points (72%) exhibit r-values greater than 0.70, while 24 points had correlations lower than 0.30 (all points: mean r-value = 0.74; range: 0.01–0.97). The generally favorable results demonstrate that the 16-day Landsat NDVI composites track the greenness trends captured by the MOD13Q1 product. The relationship breaks down at some sites, especially within certain land cover classes (Figure 6).

The poorest performing land cover classes, with r-values less than 0.70, represent barren, evergreen needleleaf forest, and herbaceous wetland (mean r-values: 0.41, 0.57, and 0.64, respectively) land cover conditions. NDVI over barren land may be highly variable due to the high saturation of background soils affecting the sensors differently. The low mean correlations in evergreen forest is largely due to a few influential outliers. Many of these sites are located in the northwest. Similar to the oregonMP PhenoCam site, the time series are often contaminated with clouds and snow, and exhibit little NDVI seasonality. Temporal profiles of the Landsat NDVI and MOD13Q1 product (Figure 7), for a selection of points representing the major land cover classes across CONUS (Figure S2), demonstrate the strong correlation between the two products.

The profiles are particularly analogous during the growing season. It is mainly during the winter months where the profiles tend to diverge, as the Landsat composites are more likely contaminated with cloud and/or snow cover, with lower signal-to-noise. Additionally, in heterogeneous landscapes, the 30 m Landsat NDVI product better reflects the spatial variability of the underlying land cover (Figures 1 and 8).

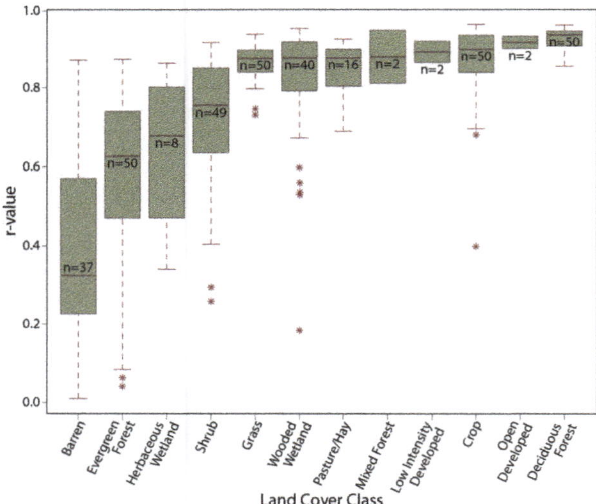

Figure 6. The distribution of Pearson correlation coefficients between MOD13Q1 NDVI and Landsat NDVI for each land cover class. * represent suspected outliers (observations that fall outside the upper or lower quartiles plus or minus 1.5 times the interquartile distance).

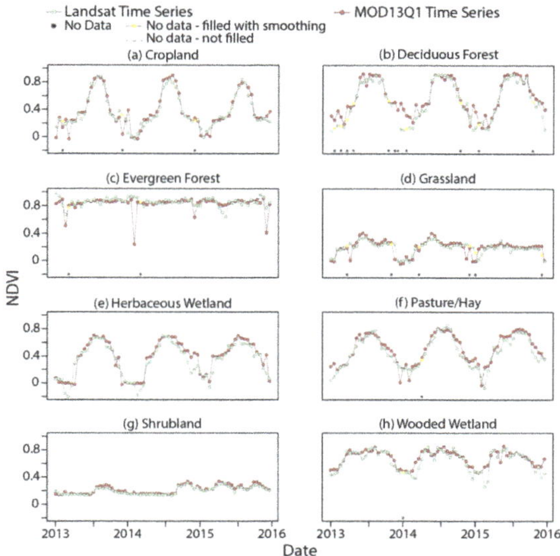

Figure 7. Time series of 30 m Landsat NDVI and 250 m MOD13Q1 NDVI time series from 2013 to 2015, separated by land cover class. After April 2013, the Landsat NDVI time series include data from both Landsat 7 and 8, while before April 2013 they included just Landsat 7 data. Each time series is from a single point, within a homogenous area (i.e., pixels where both Landsat and MOD13Q1 represent the same land cover), sampled at a location indicative of the major land cover classes.

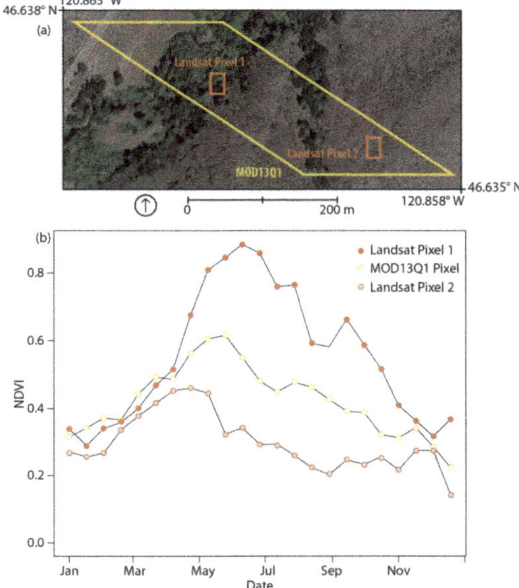

Figure 8. (**a**) Pixel locations in central Washington, USA. Landsat derived NDVI can provide increased detail in heterogeneous landscapes. The difference in pixel shape is due to native projections being transformed to a common projection. (**b**) Chart for 2015 of a Landsat derived NDVI and MOD13Q1 NDVI time series.

4. Discussion

The first-ever 16-day continuous and customizable Landsat derived NDVI composites produced here (30 m resolution for CONUS; 1984–2016) overcome many of the previous barriers of working with Landsat imagery (e.g., obtaining current or historical images; managing overlapping scenes; image storage and processing; etc.), permitting ecologists to focus time and effort on specific questions rather than data/imagery manipulation. The composites are well correlated with other observational benchmarks, including in situ phenocam observations of local vegetation conditions and coarser satellite observations from MODIS (MOD13Q1), demonstrating product capabilities for tracking greenness trends from local to regional extents. Fine spatial resolution products such as these, with a longer historical record (Figure 3), open the door to numerous analytical possibilities and applications, ranging from change detection (Figure S3) to conservation monitoring to ecosystem assessment [32–34]. The ability to customize the NDVI composite, per user specification, grants the use of a priori knowledge of the region to obtain the most suitable composite for the question at hand, producing an application ready product without the need for post-processing.

As with all remotely sensed products, the scope of Landsat derived NDVI has limitations, and is best suited for local or regional applications, where incomplete data are minimized due to a smaller spatial extent. Due to the infrequent return time of Landsat observations, data may be limited during the 16-day compositing period; cloudy pixels or the lack of surface reflectance images will reduce the overall data available for the composite. Additionally, due to the orbital paths of the Worldwide Reference System 2, a composite may be created from multiple scenes obtained from different dates within the 16-day period (e.g., different scenes that intersect an area of interest, but are acquired at the beginning and end of the 16-day period). If data are incomplete (e.g., cloudy pixels, scan line corrector errors of Landsat 7 ETM+, etc.) within these scenes, it is possible that two adjacent pixels can represent two different acquisition dates; if no data for the period are available then a climatology is used for gap filling, further distancing the dates used in the composite. Frequency of gap filling that occurs varies both geographically and seasonally, and is more likely when only a single Landsat sensor is operational. Furthermore, gap filling with climatology may produce anomalies, particularly during unusually wet or dry years, yielding systematically low or high values, respectively. These caveats may result in visual artifacts in areas with incomplete data or along scene edges.

The real power of emerging big data, cloud and web-based applications, and technologies (e.g., Google Earth Engine, GeoTrellis, GeoMesa, Apache Spark, etc.) is our new-found ability to create customizable geospatial products. Publicly available applications may be built upon these technologies, ultimately allowing users greater flexibility to provide input data, set spatial or temporal restrictions, modify parameters of algorithms, or perform on the fly testing and validation before final analysis. Such capabilities change the paradigm of static geospatial products to dynamic geospatial products, where the output is dependent upon the user's knowledge of both the system and the question. Although this requires products to be generated as needed, it provides the ability to create a much more appropriate product for any given system and question. The Landsat NDVI product and its associated web application (http://ndvi.ntsg.umt.edu/) provide a glimpse into this reality of dynamic geospatial products.

5. Conclusions

The present work introduces a unique approach to creating and disseminating high resolution spatially and temporally continuous Landsat derived NDVI. Our motivation is to remove the barriers of these datasets to further conservation and ecological research. Sixteen-day composites are created by selecting the best available pixels during each 16-day composite period from all available Landsat sensors. Missing values, due to unprocessed scenes, atmospheric contamination, or sensor malfunction are gap filled with a user-defined climatology. The resulting NDVI time series is then smoothed to approximate natural vegetative phenology. We validate the NDVI dataset using established remote sensing products at multiple scales, demonstrating the effectiveness of our approach. We provide open access to the dataset through a simple web application (http://ndvi.ntsg.umt.edu/) enabling ecologists, land managers, conservationists, and others–who may not have the compute processing capacity or technical skills–to process massive amounts of remote sensing data. This process is simplified with Google Earth Engine, an advanced planetary-scale cloud-based geospatial processing platform, which processes and distributes the product. Each 16-day composite for CONUS requires processing of at least 2700 individual Landsat scenes (more if the climatology is used for gap filling). The web application permits on-the-fly processing with customizable parameters, eliminating the need to store large amounts of data. Although we limit this study to CONUS, the framework can be expanded beyond CONUS where Landsat surface reflectance data are available and to include other useful vegetation indices (e.g., EVI, SAVI), and can be updated to accommodate updates or reorganization of the Landsat archive (e.g., Collection 1) or be modified to utilize other satellite remote sensing datasets.

Supplementary Materials: The following are available online at www.mdpi.com/2072-4292/9/8/863/s1, Figure S1: Landsat 5 Edge Removal, Figure S2: Map of PhenoCam Sites and MOD13Q1 comparison points, Figure S3: A visual example of long term land use change using the NDVI product, Table S1: PhenoCam Correlation Analysis, Acknowledgements S1: PhenoCam Site Acknowledgments.

Acknowledgments: We thank the Google Earth Engine developers for their support and technical advice. This work was funded through a Google Earth Engine research award and by the NRCS Wildlife Conservation Effects Assessment Project and Sage Grouse Initiative. The development of PhenoCam has been supported by the Northeastern States Research Cooperative, NSF's Macrosystems Biology program (award EF-1065029 and EF-1702697), DOE's Regional and Global Climate Modeling program (award DE-SC0016011), and the US National Park Service Inventory and Monitoring Program and the USA National Phenology Network (grant number G10AP00129 from the United States Geological Survey). We thank Koen Hufkens and Tom Milliman for their contributions to producing the PhenoCam data. We thank the PhenoCam site collaborators and funding sources (listed in the Supplementary Materials) for their support of the PhenoCam project.

Author Contributions: Nathaniel P. Robinson, Brady W. Allred, Matthew O. Jones, and Alvaro Moreno conceived and designed the project. Nathaniel P. Robinson and Brady W. Allred, with assistance from Tyler A. Erickson, wrote the Google Earth Engine code. Andrew D. Richardson provided PhenoCam data. All authors contributed to the writing and review of the manuscript.

Conflicts of Interest: The authors declare no conflict of interest.

References

1. Rouse, J.W., Jr.; Haas, R.H.; Schell, J.A.; Deering, D.W. Monitoring vegetation systems in the Great Plains with ERTS. In Proceedings of the Third Earth Resources Technology Satellite-1 Symposium, Washington, DC, USA, 10–14 December 1973.
2. Tucker, C.J.; Miller, L.D.; Pearson, R.L. Measurement of the combined effect of green biomass, chlorophyll, and leaf water on canopy spectroreflectance of the shortgrass prairie. *Remote Sens. Earth Resour.* **1973**, *1973*, 2.
3. Wijedasa, L.S.; Sloan, S.; Michelakis, D.G.; Clements, G.R. Overcoming limitations with Landsat imagery for mapping of peat swamp forests in Sundaland. *Remote Sens.* **2012**, *4*, 2595–2618. [CrossRef]
4. She, X.; Zhang, L.; Cen, Y.; Wu, T.; Huang, C.; Baig, M.H.A. Comparison of the continuity of vegetation indices derived from Landsat 8 OLI and Landsat 7 ETM+ data among different vegetation types. *Remote Sens.* **2015**, *7*, 13485–13506. [CrossRef]
5. Gorelick, N.; Hancher, M.; Dixon, M.; Ilyushchenko, S.; Thau, D.; Moore, R. Google Earth Engine: Planetary-scale geospatial analysis for everyone. *Remote Sens. Environ.* **2016**. [CrossRef]
6. Roy, D.P.; Ju, J.; Kline, K.; Scaramuzza, P.L.; Kovalskyy, V.; Hansen, M.C.; Loveland, T.R.; Vermote, E.F.; Zhang, C. Web-enabled Landsat Data (WELD): Landsat ETM+ Composited Mosaics of the Conterminous United States. *Remote Sens. Environ.* **2010**, *114*, 35–49. [CrossRef]
7. Pekel, J.-F.; Cottam, A.; Gorelick, N.; Belward, A.S. High-resolution mapping of global surface water and its long-term changes. *Nature* **2016**, *540*, 418–422. [CrossRef] [PubMed]
8. Masek, J.G.; Vermote, E.F.; Saleous, N.E.; Wolfe, R.; Hall, F.G.; Huemmrich, K.F.; Gao, F.; Kutler, J.; Lim, T.-K. A Landsat surface reflectance dataset for North America, 1990–2000. *IEEE Geosci. Remote Sens. Lett.* **2006**, *3*, 68–72. [CrossRef]
9. Vermote, E.; Justice, C.; Claverie, M.; Franch, B. Preliminary analysis of the performance of the Landsat 8/OLI land surface reflectance product. *Remote Sens. Environ.* **2016**, *185*, 46–56. [CrossRef]

10. Feng, M.; Huang, C.; Channan, S.; Vermote, E.F.; Masek, J.G.; Townshend, J.R. Quality assessment of Landsat surface reflectance products using MODIS data. *Comput. Geosci.* **2012**, *38*, 9–22. [CrossRef]
11. Song, C.; Woodcock, C.E.; Seto, K.C.; Lenney, M.P.; Macomber, S.A. Classification and change detection using Landsat TM data: When and how to correct atmospheric effects? *Remote Sens. Environ.* **2001**, *75*, 230–244. [CrossRef]
12. Foga, S.; Scaramuzza, P.L.; Guo, S.; Zhu, Z.; Dilley, R.D., Jr.; Beckmann, T.; Schmidt, G.L.; Dwyer, J.L.; Joseph Hughes, M.; Laue, B. Cloud detection algorithm comparison and validation for operational Landsat data products. *Remote Sens. Environ.* **2017**, *194*, 379–390. [CrossRef]
13. Roy, D.P.; Kovalskyy, V.; Zhang, H.K.; Vermote, E.F.; Yan, L.; Kumar, S.S.; Egorov, A. Characterization of Landsat-7 to Landsat-8 reflective wavelength and normalized difference vegetation index continuity. *Remote Sens. Environ.* **2016**, *185*, 57–70. [CrossRef]
14. Bradley, B.A.; Jacob, R.W.; Hermance, J.F.; Mustard, J.F. A curve fitting procedure to derive inter-annual phenologies from time series of noisy satellite NDVI data. *Remote Sens. Environ.* **2007**, *106*, 137–145. [CrossRef]
15. Reed, B.C.; Brown, J.F.; VanderZee, D.; Loveland, T.R.; Merchant, J.W.; Ohlen, D.O. Measuring phenological variability from satellite imagery. *J. Veg. Sci.* **1994**, *5*, 703–714. [CrossRef]
16. Julien, Y.; Sobrino, J.A. Comparison of cloud-reconstruction methods for time series of composite NDVI data. *Remote Sens. Environ.* **2010**, *114*, 618–625. [CrossRef]
17. Richardson, A.D.; Braswell, B.H.; Hollinger, D.Y.; Jenkins, J.P.; Ollinger, S.V. Near-surface remote sensing of spatial and temporal variation in canopy phenology. *Ecol. Appl.* **2009**, *19*, 1417–1428. [CrossRef] [PubMed]
18. Sonnentag, O.; Hufkens, K.; Teshera-Sterne, C.; Young, A.M.; Friedl, M.; Braswell, B.H.; Milliman, T.; O'Keefe, J.; Richardson, A.D. Digital repeat photography for phenological research in forest ecosystems. *Agric. For. Meteorol.* **2012**, *152*, 159–177. [CrossRef]
19. Richardson, A.D.; Jenkins, J.P.; Braswell, B.H.; Hollinger, D.Y.; Ollinger, S.V.; Smith, M.-L. Use of digital webcam images to track spring green-up in a deciduous broadleaf forest. *Oecologia* **2007**, *152*, 323–334. [CrossRef] [PubMed]
20. Ahrends, H.E.; Etzold, S.; Kutsch, W.L.; Stoeckli, R.; Bruegger, R.; Jeanneret, F.; Wanner, H.; Buchmann, N.; Eugster, W. Tree phenology and carbon dioxide fluxes: Use of digital photography for process-based interpretation at the ecosystem scale. *Clim. Res.* **2009**, *39*, 261–274. [CrossRef]
21. Zhao, J.; Zhang, Y.; Tan, Z.; Song, Q.; Liang, N.; Yu, L.; Zhao, J. Using digital cameras for comparative phenological monitoring in an evergreen broad-leaved forest and a seasonal rain forest. *Ecol. Inform.* **2012**, *10*, 65–72. [CrossRef]
22. Toomey, M.; Friedl, M.A.; Frolking, S.; Hufkens, K.; Klosterman, S.; Sonnentag, O.; Baldocchi, D.D.; Bernacchi, C.J.; Biraud, S.C.; Bohrer, G.; et al. Greenness indices from digital cameras predict the timing and seasonal dynamics of canopy-scale photosynthesis. *Ecol. Appl.* **2015**, *25*, 99–115. [CrossRef] [PubMed]
23. Petach, A.R.; Toomey, M.; Aubrecht, D.M.; Richardson, A.D. Monitoring vegetation phenology using an infrared-enabled security camera. *Agric. For. Meteorol.* **2014**, *195–196*, 143–151. [CrossRef]
24. Keenan, T.F.; Darby, B.; Felts, E.; Sonnetag, O.; Friedl, M.A.; Hufkens, K.; O'Keefe, J.; Klosterman, S.; Munger, J.W.; Toomey, M.; et al. Tracking forest phenology and seasonal physiology using digital repeat photography: A critical assessment. *Ecol. Appl.* **2014**, *24*, 1478–1489. [CrossRef]
25. Solano, R.; Didan, K.; Jacobson, A.; Huete, A. *MODIS Vegetation Index User's Guide (MOD13 Series)*; Version 2.0; Vegetation Index and Phenology Lab, the University of Arizona: Tucson, AZ, USA, 2010; pp. 1–38.
26. Didan, K.; Munoz, A.B.; Solano, R.; Huete, A. *MODIS Vegetation Index User's Guide (MOD13 Series)*; Version 3.0; University of Arizona: Tucson, AZ, USA, 2015.
27. Homer, C.; Dewitz, J.; Fry, J.; Coan, M.; Hossain, N.; Larson, C.; Herold, N.; McKerrow, A.; VanDriel, J.N.; Wickham, J. Completion of the 2001 national land cover database for the conterminous United States. *Photogramm. Eng. Remote Sens.* **2007**, *73*, 337–341.
28. Fry, J.A.; Xian, G.; Jin, S.; Dewitz, J.A.; Homer, C.G.; Limin, Y.; Barnes, C.A.; Herold, N.D.; Wickham, J.D. Completion of the 2006 National Land Cover Database for the conterminous United States. *Photogramm. Eng. Remote Sens.* **2011**, *77*, 858–864.

29. Homer, C.G.; Dewitz, J.A.; Yang, L.; Jin, S.; Danielson, P.; Xian, G.; Coulston, J.; Herold, N.D.; Wickham, J.D.; Megown, K. Completion of the 2011 National Land Cover Database for the conterminous United States-Representing a decade of land cover change information. *Photogramm. Eng. Remote Sens.* **2015**, *81*, 345–354.
30. Hufkens, K.; Friedl, M.; Sonnentag, O.; Braswell, B.H.; Milliman, T.; Richardson, A.D. Linking near-surface and satellite remote sensing measurements of deciduous broadleaf forest phenology. *Remote Sens. Environ.* **2012**, *117*, 307–321. [CrossRef]
31. Klosterman, S.T.; Hufkens, K.; Gray, J.M.; Melaas, E.; Sonnentag, O.; Lavine, I.; Mitchell, L.; Norman, R.; Friedl, M.A.; Richardson, A.D. Evaluating remote sensing of deciduous forest phenology at multiple spatial scales using PhenoCam imagery. *Biogeosciences* **2014**, *11*, 4305–4320. [CrossRef]
32. Hansen, M.C.; Loveland, T.R. A review of large area monitoring of land cover change using Landsat data. *Remote Sens. Environ.* **2012**, *122*, 66–74. [CrossRef]
33. Jensen, J.R.; Rutchey, K.; Koch, M.S.; Narumalani, S. Inland wetland change detection in the Everglades Water Conservation Area 2A using a time series of normalized remotely sensed data. *Photogramm. Eng. Remote Sens.* **1995**, *61*, 199–209.
34. Nouvellon, Y.; Moran, M.S.; Seen, D.L.; Bryant, R.; Rambal, S.; Ni, W.; Bégué, A.; Chehbouni, A.; Emmerich, W.E.; Heilman, P.; et al. Coupling a grassland ecosystem model with Landsat imagery for a 10-year simulation of carbon and water budgets. *Remote Sens. Environ.* **2001**, *78*, 131–149. [CrossRef]

Article

High Spatial Resolution Visual Band Imagery Outperforms Medium Resolution Spectral Imagery for Ecosystem Assessment in the Semi-Arid Brazilian Sertão

Ran Goldblatt, Alexis Rivera Ballesteros and Jennifer Burney *

School of Global Policy and Strategy, University of California, San Diego, San Diego, CA 92093, USA; rgoldblatt@ucsd.edu (R.G.); alexis.rivera.b@gmail.com (A.R.B.)
* Correspondence: jburney@ucsd.edu; Tel.: +1-858-534-4149

Received: 17 October 2017; Accepted: 10 December 2017; Published: 20 December 2017

Abstract: Semi-arid ecosystems play a key role in global agricultural production, seasonal carbon cycle dynamics, and longer-run climate change. Because semi-arid landscapes are heterogeneous and often sparsely vegetated, repeated and large-scale ecosystem assessments of these regions have to date been impossible. Here, we assess the potential of high-spatial resolution visible band imagery for semi-arid ecosystem mapping. We use WorldView satellite imagery at 0.3–0.5 m resolution to develop a reference data set of nearly 10,000 labeled examples of three classes—trees, shrubs/grasses, and bare land—across 1000 km^2 of the semi-arid Sertão region of northeast Brazil. Using Google Earth Engine, we show that classification with low-spectral but high-spatial resolution input (WorldView) outperforms classification with the full spectral information available from Landsat 30 m resolution imagery as input. Classification with high spatial resolution input improves detection of sparse vegetation and distinction between trees and seasonal shrubs and grasses, two features which are lost at coarser spatial (but higher spectral) resolution input. Our total tree cover estimates for the study area disagree with recent estimates using other methods that may underestimate treecover because they confuse trees with seasonal vegetation (shrubs and grasses). This distinction is important for monitoring seasonal and long-run carbon cycle and ecosystem health. Our results suggest that newer remote sensing products that promise high frequency global coverage at high spatial but lower spectral resolution may offer new possibilities for direct monitoring of the world's semi-arid ecosystems, and we provide methods that could be scaled to do so.

Keywords: remote sensing; semi-arid; ecosystem assessment; land use change; image classification; seasonal vegetation; carbon cycle; Google Earth Engine

---

1. Introduction

Approximately 2 billion people live in the semi-arid regions of the world—agro-ecological zones characterized by low-to-medium rainfall (~600 mm per year) typically confined to one rainy season [1]. These ecosystems are critical to planetary health for several reasons. First, more than 600 million people in the semi-arid tropics are smallholder farmers and pastoralists, living on a few hectares of land, at most. These populations, who are among the world's poorest, depend primarily on rain-fed agriculture for their livelihoods (for cultivation of cereals, forage for livestock, and more). Even small changes in precipitation in these regions may lead to changes in vegetation, which ultimately have important consequences for human welfare, subsequent land cover and land use change—including deforestation (for example, when farmers must expand their land footprint to meet basic needs), and climate change. Second, in steady-state, vegetation in the world's semi-arid regions is a main driver of the seasonal carbon cycle. Moreover, because these regions often have a larger percentage of precipitable water from vegetation, vegetation and rainfall (and, by extension, human livelihoods) are more tightly coupled than in other ecosystems, and in a potential positive feedback loop [2–4]. Over recent decades, semi-arid zones have seen—and are expected to see in the future—larger climate signals compared to the global mean, in part due to this feedback [5–7]. As such, the world's semi-arid regions sit at the nexus of climate impacts, adaptation, food security, and economic development. Monitoring vegetation and biomass dynamics in these regions is critical for a sustainable development.

Ecosystem assessment—or the mapping of the landcover and services provided by a given ecosystem—has traditionally been difficult and costly even locally, and impossible at a global scale or at high temporal resolution. Though remotely sensed data have offered new possibilities for mapping landcover and landuse, until very recently, satellite instruments—especially those that are publicly available—prioritized spectral information (higher radiometric sensitivity) over increased spatial and temporal resolution, in part due to data requirements and computational limitations. Satellite data have been critical to monitoring the world's dense forests [8], but has been limited in its applicability for monitoring more sparsely-vegetated and heterogeneous landscapes, where key features (e.g., individual trees) are too small to be identified as individual objects by the satellite's sensors. In the world's drylands, for example, coarse spatial resolution data has limited our ability to understand the meaning of changes in overall 'greenness' [9]; at the same time, higher spatial resolution data may not fully leverage meaningful spectral information about the reflectance of the land cover and is often difficult to scale [10]. Existing and proposed devices, however, like small satellite flocks and coordinated unmanned aerial vehicles (UAV) flights, promise frequent coverage at high spatial but lower spectral resolution (typically the visible red, green, and blue (RGB) bands, and possibly RGB together with near infrared (RGB + NIR)). With the reduced cost of computing, including through analysis on cloud-based platforms like Google Earth Engine, there are new possibilities for monitoring the world's semi-arid ecosystems on a global seasonal scale.

In light of both the importance of semi-arid ecosystems and the increased availability of high-resolution data and computational capabilities, we seek to answer three related questions here: (1) How well do classifiers based on high-spatial/low-spectral (HSLS) resolution imagery as input perform in classifying the land cover of semi-arid ecosystems, compared to classifiers based on more conventional medium-spatial- resolution input traditionally used for long-run monitoring, like Landsat? (2) Given the importance of the seasonal vegetation cycle in semi-arid regions to both human livelihoods and climate, does HSLS resolution imagery offer a potential source for seasonal carbon cycle information? Finally, (3) can HSLS resolution imagery be used effectively for ecosystem assessment in the heterogeneous and often sparsely vegetated landscapes of the semi-arid tropics? How do large-scale estimates of land cover based on these methods compare with other existing methods?

We evaluate the potential of high-resolution satellite imagery for ecosystem assessment in the case study of the Sertão, a semi-arid biome of northeastern Brazil. In the Sertão, we identified ten 100 km^2 regions representative of the regional landscape heterogeneity. In each region, we performed a 3-class classification into three types of land cover—trees, shrubs, and bare land—using WorldView-2, WorldView-3 and Landsat-8 satellites imagery (representing high- and medium- spatial resolution imagery—or 0.3–0.5 m and 30 m, respectively). These three coarse classes represent the types of landcover that an ecosystem assessment would want to track—bare land as indicative of degraded habitat, shrubs as a broad category of non-tree and largely seasonal vegetation, and trees. We then compared the accuracy of classification with the two sets of imagery as input to the classifiers (WorldView and Landsat), along with several combinations between the sets of inputs. We also tested performance in 2-class classification tasks, as well as in classification across seasons. Our paper proceeds as follows: Section 2 describes the satellite imagery used, construction of the reference data sets, and the methods we used for classification in Google Earth Engine and for cross-validation; Section 3 describes the results, also in the context of seasonal variations; in Section 4, we offer a concluding discussion; we compare our whole-area estimates to existing estimates based on other methods and provide a discussion of the potential to scale our methods to a global, seasonal ecosystem assessment.

2. Materials and Methods

2.1. Overview

Various machine-learning approaches (e.g., supervised, unsupervised and semi-supervised) can be combined with satellite imagery for ecosystem land cover classification, but accurate classification of distinct types of biomes remains challenging due to several constraints. First, there is a scarcity of reference data to facilitate training of classifiers or for validation of existing maps. Second, distinguishing between types of biomes and assessing their health, requires "good enough" satellite imagery. The definition of "good enough"—i.e., the optimal spatial and spectral characteristics of the imagery necessary for classification—remains unspecified. For example, very high-resolution imagery provided by private commercial companies is expensive, and prohibitively so at the global scale. On the other hand, publicly available global-scale satellite imagery is generally characterized by a lower spatial resolution (but by a higher spectral resolution compared to the former). Third, mapping ecosystems on global scales requires overcoming computational constraints (i.e., storage and analysis) which, until recently, with the emergence of cloud based computational platforms such as Google Earth Engine, have limited the scale of the analyzed regions.

In this study, we performed supervised pixel-based image classification of ecosystems in the semi-arid Sertão using an extensive dataset of almost 10,000 examples (points) that were hand-labeled as either trees (within the tree canopy), shrubs (all non-tree vegetation), or land (bare ground). We utilized both medium spatial-resolution imagery (Landsat 8; hereafter referred to as Landsat) and 3-band (RGB) high spatial-resolution imagery (WorldView-2 and WorldView-3; hereafter referred to as WV2 and WV3, respectively). Although the spatial resolution of Landsat is lower than of WorldView (30 m vs. 0.3–0.5 m, respectively), Landsat imagery contains a much wider range spectral information compared to the visible WV imagery (8 vs. 3 spectral bands, respectively), which can be used to calculate other composite spectral indices [11]. Most important, for the purposes of this study, is that the spectrum of 3 of Landsat bands (Bands 2–4 are BGR) overlaps very closely with the spectra of the RGB WorldView bands. This allows us to assess the trade-offs between spatial and spectral resolution in the input data that is used for classification. We perform classification using Landsat and WorldView as separate inputs to the classifiers and using the two inputs combined (where each WorldView pixel also includes Landsat's band properties). The process is shown visually in Figure 1 and described in detail below.

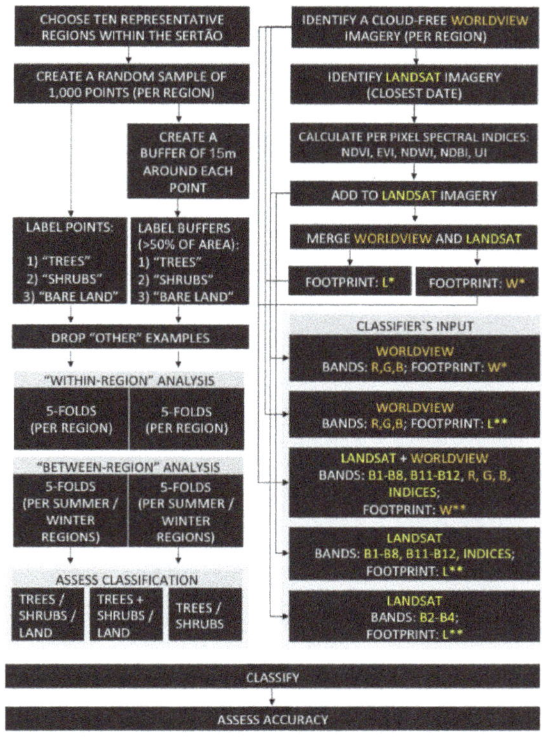

Figure 1. Methods flow chart for labeling and classification.

2.2. Study Area and Region Selection

The Sertão is a semi-arid region in northeast Brazil (Figure 2a). The region, part of the Brazilian Highlands, is characterized by low density vegetation, primarily shrubs and sparse forests. Through a research agreement, we were given access to a subset of WorldView 3-band imagery via the DigitalGlobe Basemap suite (My DigitalGlobe, DigitalGlobe, Inc., Westminster, CO, USA) that included intermittent coverage of parts of the Sertão starting in 2015. We first selected a subset of cloud-free scenes from the available imagery, which narrowed the selection set considerably. Second, to account for the potential seasonal variation known to be present in semi-arid regions, we split our test regions across two seasons—September and March. We did not have access to repeat images of any Sertão locations in the high-resolution imagery spanning two seasons within two calendar years—which would have been ideal in terms of experimental design—and were thus restricted to choosing images from the same year and different locations across seasons. This left us with approximately fifteen 10 km × 10 km regions, from which we selected ten final images (one image per region) spanning the largest area and with the most heterogeneity. These regions represent together a 1000 km^2 area, or a size equivalent to a smaller Brazilian municipality (Figure 2a,b).

Figure 2. Study Region and labeling examples. (**a**) the semi-arid Sertão lies in northeast Brazil (grey region). We sampled ten 100 km^2 regions, constrained by availability of WorldView imagery. The zoomed inset shows the ten regions. Four (red) had imagery available from March, and the other six (blue) had imagery available from September; (**b**) a 1 km × 1 km area from one of the 10 regions, showing landscape heterogeneity. We labeled (and then classified) examples as either trees (meant to capture more permanent vegetation), shrubs (meant to capture smaller, seasonal vegetation), and bare land (no vegetation); (**c**) in each region, 1000 examples were labeled according to (i) the class of the corresponding high-resolution WorldView pixel of the exact location (red point), and (ii) the majority class of all WorldView pixels within a 30 m diameter buffer (cyan ring). The examples shown here were labeled as bare land (top), shrub (middle), and tree (bottom) in both point and buffer labeling schemas. Grey stippled points in the map are the sampling regions from (**a**), drawn from [10], and the dashed bounding box is the area used for comparison to their treecover estimates.

2.3. Development of a Reference Training Set

In each region, we randomly sampled 1000 locations (points) to hand-label as reference examples. We overlaid these points on the WorldView imagery and labeled each location as trees, shrubs, or land using two image interpretation approaches: (i) in the first, we labeled examples according to the characteristics of the land cover in the WorldView pixel directly corresponding to the location (referred to as point dataset). In this version, for example, a point is labeled as tree only if it is located precisely over a tree; (ii) in the second approach, we labeled each location according to the majority land cover type in the WorldView imagery within a buffer of 30 m (radius-15 m) around each point (referred to as buffered dataset). In this version, a point is labeled as tree only if more than 50% of the buffer around it is covered with trees. We excluded examples that did not correspond with any of the three classes (i.e., polygons covered with different types of land cover). We pursued this dual labeling strategy to better compare the two different spatial resolutions. Since a Landsat pixel is much larger than a WorldView pixel, directly transferring a label from a fine resolution pixel to a coarse resolution pixel might be misleading (a Landsat pixel will often contain mixed classes). Indeed, the two labeling approaches result in a different distribution of labeled examples: in the point dataset, more examples were labeled as shrubs than trees (51% vs. 28%, respectively), while, in the buffered dataset, fewer examples were labeled as shrubs than trees (40% vs. 46%, respectively) (Table 1 presents the distribution of the labeled examples). Twenty-five examples that did not correspond with any of the three classes (e.g., bodies of water or built structure) were not included in our reference dataset. Examples were labeled by three research assistants who were trained extensively and supervised by the research team. Each labeled example is a point, which is associated with exactly one pixel. The classification is thus performed with 9975 labeled pixels (Landsat or WorldView), where each pixel includes inputs (band and indices values) and a response value (a class).

Table 1. Distribution of labeled examples by class and tile, for both point and buffer labeling systems (expressed in percentages).

Region	Point Classification			Buffer Classification		
	Land	Shrub	Tree	Land	Shrub	Tree
1	5.91	43.04	51.05	4.1	29.73	66.17
2	14.93	27.25	57.82	10.72	23.55	65.73
3	7.71	49.95	42.34	3.6	18.52	77.88
4	8.83	39.92	51.25	5.12	18.96	75.93
5	15.32	48.75	35.94	9.11	24.52	66.37
6	54.21	44.79	1.00	45.69	44.59	9.72
7	37.17	45.49	17.33	33.17	35.17	31.66
8	30.38	62.47	7.14	12.68	70.72	16.6
9	33.67	58.89	7.44	12.96	65.43	21.61
10	6.11	85.67	8.22	1.7	67.94	30.36
Average	21.42	50.62	27.95	13.89	39.91	46.20
Standard Error	5.22	4.97	6.93	4.53	6.61	8.40

2.4. Classification and Classifier Performance

We performed supervised pixel-based image classification in Google Earth Engine (GEE). GEE leverages cloud-computational services for planetary-scale analysis and consists of petabytes of geospatial and tabular data, including a full archive of Landsat scenes, together with a JavaScript, Python based API (GEE API), and algorithms for supervised image classification [12]. GEE has been previously used for various research applications, including mapping population [13,14], urban areas [15] and forest cover [8].

We used both Landsat and WorldView imagery as input for classification. We matched the selected WorldView regions to the co-located Landsat imagery (8-day TOA reflectance composites) nearest in date, and trimmed Landsat imagery to the same boundaries. We used Landsat 8 Top-of-Atmosphere (TOA) reflectance orthorectified scenes, with calibration coefficients extracted from the corresponding image metadata. The Landsat scenes were from the L1T level product (precision terrain-corrected), converted to TOA reflectance and rescaled to 8-bits for visualization purposes. These pre-processed calibrated scenes are available in GEE. It should be noted that, although TOA reflectance accounts for planetary variations between acquisition dates (e.g., the sun's azimuth and elevation), many images remain contaminated with haze, clouds, and cloud shadows, which may limit their effective utilization [16], especially for agriculture and ecological applications.

For each Landsat pixel, we used 11 spectral band designations, in a spatial resolution of between 15 m (band 8) to 30 m (bands 1–7, 9–11) and we calculated five spectral indices as additional inputs to the classifier. These indices are commonly used to identify water (Normalized Difference Water Index (NDWI)), built up and bare land areas (Normalized Difference Built-up Index (NDBI); Urban Index (UI)), and vegetation (Normalized Difference Vegetation Index (NDVI); Enhanced Vegetation Index (EVI)). We thus used a total of 15 features per pixel (11 spectral bands and 4 spectral indices). We added these nonlinear indices as additional inputs to the classifier to improve identification of vegetation and bodies of water (NDVI, NDWI and EVI) and other types of land cover (NDBI and UI). Although the latter two indices were originally designed to capture built-up and urban areas, they are also sensitive to the spectral characteristics of bare land [17], which are relatively similar to those of urban areas [18].

For each WorldView pixel, we used 3 spectral band designations (blue, green, and red) in a spatial resolution of 0.5 m (WV2) and 0.3 m (WV3) (bands for WorldView and Landsat are described in Table A2). It should be noted that, while WV2 collects imagery at eight multispectral bands (plus Panchromatic) and WV3 collects imagery at eight multispectral bands, eight-band short-wave infrared (SWIR) and 12 CAVIS (plus Panchromatic), we were only granted access to three of WorldView's imagery bands (the red, green and blue bands). WV2 and WV3 3-band pan sharpened natural color imagery was downloaded from

DigitalGlobe Basemap (My DigitalGlobe). The imagery included single-strip ortho with off-nadir angle of 0–30 degrees (the full metadata of the downloaded scenes is presented in Appendix A, Table A1). The WorldView imagery we analyzed was processed by DigitalGlobe. Though temporal analysis with this type of data should be performed with caution (while accounting, for example, for radiometric and atmospheric variations), it is sufficient for the purpose of our study: to evaluate classification performance in one moment of time with different spatial and spectral resolution input data (note that we do not include spectral indices that are based on the WorldView input). Our objective here was not to perform change detection or to classify the land cover in different moments in time. Instead, we performed a basic per-pixel supervised classification procedure, where the characteristics of the feature space (e.g., calibrated or not, DN or reflectance values) are similar in training and classification.

We performed classification with 5 different combinations of inputs to the classifier: using (I) Landsat bands alone (herafter LS8); (II) using WorldView bands alone (hereafter WV); (III) using WorldView bands, rescaled (averaged) to Landsat's spatial resolution (30 m) (hereafter WV30); (IV) using the subset of 3 Landsat bands, corresponding to WorldView's visible bands (Red, Green, and Blue) (hereafter LS8RGB); and, finally: (V) using WorldView bands combined with Landsat bands, at the spatial resolution and footprint of WorldView (hereafter WVLS8). The spatial and spectral information for each classifier is summarized in Table A2.

In each case, we used a Random Forest (RF) classifier with 20 decision trees. Random Forests are tree-based classifiers that include k decision trees (k predictors). When classifying an example, its variables (in this case, spectral bands and/or spectral indices) are run through each of the k tree predictors, and the k predictions are averaged to get a less noisy prediction (by voting on the most popular class). The learning process of the forest involves some level of randomness. Each tree is trained over an independently random sample of examples from the training set and each node's binary question in a tree is selected from a randomly sampled subset of the input variables. We used RF because previous studies found that the performance of RF is superior to other classifiers [15], especially when applied to large-scale high dimensional data [19]. Random Forests are computationally lighter than other tree ensemble methods [20,21] and can effectively incorporate many covariates with a minimum of tuning and supervision [22], RF often achieve high accuracy rates when classifying hyperspectral, multispectral, and multisource data [23]. In this study, we set the number of trees in the Random Forest classifier to 20. Previous studies have shown mixed results as for the optimal number of trees in the decision tree, ranging from 10 trees [24] to 100 trees [20]. According to [15], although the performance of Random Forest improves as the number of trees increases, this pattern only holds up to 10 trees. Performance remains nearly the same with 50 and with 100 decision trees. Similarly, Du et al. (2015) evaluated the effect of the number of trees (10 to 200) on the performance of RF classifier and showed that the number of trees does not substantially influence the classification accuracy [25]. Because our intention here was not to identify an optimal number of trees in the RF but to compare classification with different types of input, we chose to use a RF classifier with 20 trees and did not imply that this is the optimal number of trees.

In general, automatic classification of land use/land cover (LULC) from satellite imagery can be conducted at the level of a pixel (pixel-based), an object (object-based) or in a combined method. Pixel-based methods for LULC classification rely on the spectral information contained in individual pixels and have been extensively used for mapping LULC, including for change detection analysis [26]. Object-based methods rely on the characteristics of groups of pixels, where pixels are segmented into groups according to similarity and the classification is done per object, rather than per pixel. While several studies suggest that object-based classifiers outperform pixel-based classifiers in LULC classification tasks [27–30], other studies suggest that pixel-based and object-based classifiers perform similarly when utilizing common machine-learning algorithms [31,32]. In addition, object-based classification requires significantly more computational power than pixel-based classification and there is no universally accepted method to determine an optimal scale level for image segmentation [28], especially when analyzing large-scale and geographically diverse regions. Thus, object-based classification is typically conducted when the unit of analysis is relatively small, such as a city [28,30], or a region of a country [27,30–32], as we do in this study. Here, we map land cover at the level of the pixel.

We evaluated the performance of the classifiers in both 3-class classifications, utilizing the full set of examples: trees, shrubs, and land, and 2-class classification, where we classified (I) land versus all vegetation (trees + shrubs) and (II) trees versus shrubs (excluding land). Given a 2-class classification problem (into class C and D), the classifiers predict per-pixel probability of class membership, (posterior probability (p), in the range between 0 and 1) representing the probability that a pixel X belongs to class C (the probability the pixel belongs to class D is 1-p). A pixel X is classified as belonging to class C, if the probability it belongs to class C exceeds a given threshold. An important question in any machine learning task is what is the optimal posterior probability threshold above which a pixel is declared to belong to the 'positive' class. To address this question, we also assess performance of 2-class probability-mode classification, where a pixel is declared as positive if the posterior probability exceeds either 0.1, 0.2, 0.5 or 0.8. We evaluate the effect of these posterior probability thresholds on the performance of the classifiers (i.e., the True Positive Rate (TPR) and True Negative Rate (TNR) measures). Although previous studies suggest methods for probability estimates for multi-class classification (i.e., classification into more than two classes), for example, by combining the prediction of multiple or all available pairs of classes [33,34], here we only perform a 2-class probability classification, classifying land versus all vegetation and trees versus shrubs. According to this approach, a pixel is declared as positive (i.e., vegetation or trees, respectively) if the probability it belongs to these classes exceeds a threshold (0.1, 0.2, 0.5 or 0.8). Below this threshold, the pixel is classified as negative (land or shrubs, respectively).

We performed classification of trees versus shrubs only with the examples that were labeled as either one of these two classes. This classification procedure predicts the class of each pixel in the classified universe as either trees or shrubs (although the scenes, by their nature, also include other types of land cover such as bare land). Obviously, mapping this prediction will be misleading, as this is a 2-class classification. In this procedure, each pixel in the universe is classified as one of two classes (trees or shrubs) and the prediction is evaluated against the examples labeled as trees and shrubs in the reference data.

For the 3-class classification, each example (pixel) was classified as one of three classes (trees, shrubs, land), without treating the examples as 'positive' or 'negative'. Given a trained classifier consisting of n trees (in this case 20 trees), each new instance (pixel) is run across all the trees grown in the forest. Each tree predicts the probability an instance belongs to each one of the classes (in this case, one of 3 classes) and votes for the instance's predicted class. Then, the votes from all trees are combined and the class for which maximum votes are counted is declared as the predicted class of the instance [35,36].

The distinction between classes are all meaningful in an ecological sense, given the landscape heterogeneity and highly varying vegetation patterns of semi-arid regions like the Sertão, and the most appropriate choice of the classified classes depends on the application. For example, a 3-class classification would be most relevant for total ecosystem landcover assessment, whereas land-versus-vegetation would be most relevant for assessing degradation or recovery, and trees-versus-shrubs would be most relevant for assessing seasonal variations in vegetation.

The performance (accuracy) of a classifier refers to the probability that it will correctly classify a random set of examples [37]; the data used to train the classifier must thus be distinct from the data that is used to assess its performance. Labeled (or reference) data is typically divided into training and test sets (a validation set may also be used to "tune" the classifier's parameters). Different data splitting heuristics can be used to assure a separation between the training and test sets [37], including the holdout method, in which the data is divided into two mutually exclusive subsets: a training set and a test/holdout set; bootstrapping, in which the dataset is sampled uniformly from the data, with replacement; and cross-validation (also known as k-fold cross-validation), in which the data are divided into k subsets, and the classification is performed k times, rotating through each 'fold' as a test set. By averaging across the different draws, cross-validation gives a less biased estimate of classifier performance, [38] along with a variance estimation of the classification error [39,40]), but is less computationally intensive than (typically hundreds) of bootstrap re-samples.

Here, we adopted a *k*-fold cross validation procedure. We used 5-fold cross-validation in all experiments (i.e., 5 folds, or groups), with random examples allocated to folds stratified by labeled class (i.e., each fold had the same proportion of each class as the full region). We performed classification and accuracy assessment both within and between regions: in "within-region" analysis, we divided the examples in a region into 5 folds. For each experiment, we used all examples in 4 folds for training, and assessed the classification of the examples in the remaining fold. In "between-region" analysis, we used all the examples in 6 or 4 regions (the summer or the winter images). The latter procedure is designed to evaluate the spatial generalization of the classifiers. Thus, we repeat the 5-fold cross validation twice: with all the examples in regions 1–4 (March 2016) and with all the examples in regions 5–10 (September 2015).

We assess classifier performance using three estimated quantities from 2- and 3- class confusion matrices (e.g., as in [41] for 3-class): (1) True Positive Rate (TPR), or sensitivity, which refers to the percentage of positive examples classified correctly as positive (i.e., positive label = classification as positive); (2) True Negative Rate (TNR), or specificity, which refers to the percentage of negative examples (or "not positive") that were classified correctly as negative (or as "not positive") (e.g., if trees are positive and land and shrubs are negative, TNR refers to the percentage of land and shrubs examples that were correctly classified as not trees). Although reporting TPR and TNR is standard practice in a 2-class classification problem, we note that it is especially important to interpret these quantities appropriately when dealing with imbalanced datasets. In our setting, the weight of each class (i.e., the number of per-class examples, Table 1) depends on the coverage of each class in a region.

For each experiment, we calculated mean values and standard errors of TPR and TNR across all 5 folds. When assessing classifier performance by class, standard errors (95% confidence interval) for each metric were calculated across folds based on the number of examples per class per fold.

2.5. Experiment Summary and Ecosystem-Scale Assessments

In summary, we conducted 170 total experiments, with a 5-fold cross-validation: (1) within-region: 5 band/resolution combinations × 10 regions × 3 classifications; and (2) between-region: 5 band/resolution combinations × 4 season combinations. After assessing classifier performance using 5-fold cross-validation (with the exception of cross-season, where we used all the labeled examples from one season to classify the samples in the region sensed in the other season), we used each classifier to do an out-of-sample classification of all pixels in the 1000 km^2 region, and compared total area fractions across classifiers both internally and to existing estimates from the literature.

3. Results

3.1. Within-Region Classifier Performance

Performance metrics (TPR and TNR) of 3-class classification (land, shrubs, trees) by class for each of the 5 spatial/spectral resolution combinations are shown in Figure 3. TNR is higher than the TPR for classification with all inputs, and TNR confidence intervals are on average smaller than those for TPR: this is in large part due to the fact that, as noted above, TNR in a 3-class classification contains confusion between the two negative classes and is thus artificially inflated. The classifiers perform better, in general, when using the high spatial-resolution WorldView Data. Classification with WorldView by itself (WV) and with WorldView + Landsat 8 (WVLS8) (both at 0.3–0.5 m resolution), outperforms classification with Landsat 8 (LS8 and LS8RGB) and with WorldView at a 30 m resolution (WV30); we also observe that classification with WV is not significantly different from classification with WVLS8, indicated by all performance metrics. This suggests that, in this context, the addition of Landsat 8 spectral bands does not add much power to the high spatial-resolution RGB data.

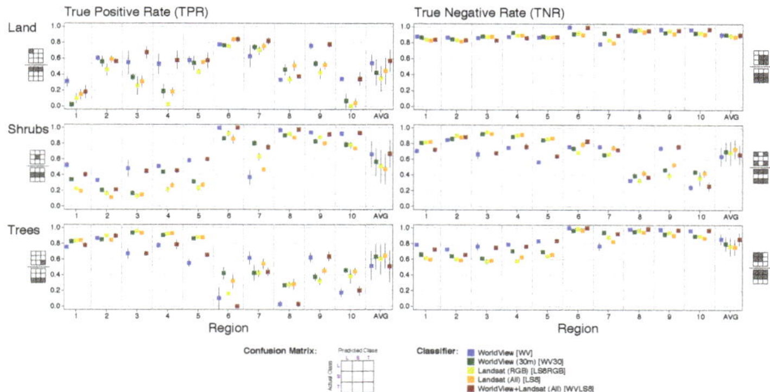

Figure 3. Performance metrics by class for five different combinations of spatial resolution and spectral bands from WorldView and Landsat. (**left column**) True Positive Rate (TPR) and (**right column**) True Negative Rate (TNR) are shown for land (**top row**), shrubs (**middle row**), and trees (**bottom row**). Metrics are provided for each of the ten study regions (Figure 2), along with the average across regions for each metric and classifier. (For each tile, error bars are the 95% Confidence Interval (CI) from the folds; per class are defined according to confusion matrix for a 3-class classification (representations modeled after [41]), shown at the bottom of the figure. We note that TNR is different from the 2-class classification case because it gets confused between the two negative classes. For this reason, we show TPR and TNR separately, and do not calculate the average between TPR and TNR (often referred to as balanced accuracy rate, or BAR). (A table version of these results can be found in the Appendix C, Table A7.)

On average, classification with high spatial-resolution inputs outperforms classification with coarser-resolution inputs in differentiating between bare land and shrubs, and is less likely to mis-classify bare land and shrubs as trees. In addition, we find variation in the performance of the classifiers across the study area. In particular regions 6 and 9 stand out for the accurate classification of land and shrubs, as shown in Table 1; this is likely a function of the major predominance of these two classes in those regions (there are simply more examples labeled as trees and shrubs).

The results suggest that classification with WV (high spatial-resolution but only three bands) outperforms classification with the full spectral range of the coarser spatial resolution Landsat (LS8). These results are not enough to tease out the relative importance of Landsat's non-RGB bands (spectral degrees of freedom) versus WorldView's pixel size (spatial degrees of freedom). For this reason, we performed an additional experiment designed to conceptually link these two different data sources with two 'intermediate' inputs—WV30 (WorldView, aggregated to 30 m pixel size) and LS8RGB (Landsat, using only RGB bands). Conceptually, classification with these two inputs should perform similarly. Indeed, this is the case, although performance is not identical, with WV30 slightly outperforming LS8RGB across the board. We then interpret the differences between classification with WV and WV30 as the impact of decreased spatial resolution with a constant spectral resolution (RGB bands). Similarly, we interpret the differences between classification with LS8RGB and with the full spectral information of Landsat (LS8) as the effect of increased spectral resolution, while spatial resolution is held fixed. Comparing within a given region (e.g., region 9), we see bigger improvement between classification with WV30 and WV than between classification with LS8RGB and LS8. These results indicate that the factor of 100 difference in pixel edge length out-weighs the additional (factor of >10 increase) in spectral degrees of freedom.

Assessment of 2-class classification accuracy with different posterior probability thresholds (Figure 4) suggests that classification with higher-resolution inputs to the classifier (WV and WVLS8) shows more sensitivity to the threshold than classification with lower spatial resolution inputs (LS8RGB,

LS8 and WV30). Namely, both TPR and TNR drop more sharply as the posterior probability decreases when classification is performed with WV and WVLS8 than when classification is performed with lower-resolution inputs. This implies that, when classification is performed with high-resolution imagery, it is 'easier' for the classifiers to predict an example as positive (i.e., many of the pixels have a high posterior probability value). On the other hand, higher sensitivity to the threshold indicates that the confidence is lower (i.e., if we set the threshold to a low value, we miss positive pixels). An optimal low posterior probability threshold means that it is 'harder' for the classifier to predict a positive example than to predict a negative example. A low threshold is appropriate when we don't want to weight keeping true positive examples more than having excess predicted positive examples (even if the probability they are true is low). Conversely, when the posterior probability threshold is set to a higher value, fewer examples are predicted correctly as positive (we 'miss' those with a lower probability value) but also have a much lower false-positive rate.

All of the classifiers perform well in distinguishing land from vegetation (shrubs and trees considered together), though classification with a higher spatial resolution input (WV and WVLS8) shows more sensitivity to the posterior probability threshold. As shown in Figure 4, at a low posterior probability threshold of 0.1, the 2-class classifications of land vs. vegetation and trees vs. shrubs have very high TPR and TNR values, suggesting that the separation of the two types of vegetation is a limiting factor for classifier performance. Indeed, classification of land vs. vegetation is significantly more accurate than classification of trees vs. shrubs, especially when the posterior probability threshold is high. Classification of land vs. vegetation with the higher resolution input (WV and WVLS8) performs worse than classification with the 30 m classifiers (LS8), especially at higher posterior probability thresholds. This is the only analysis we performed where increased spectral information performed better than increased spatial information. However, as discussed below, this is likely an artifact of averaging across regions (and therefore seasons).

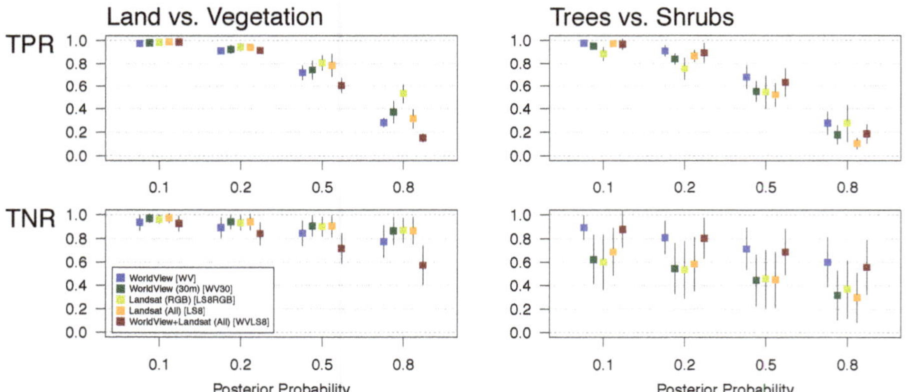

Figure 4. True Positive Rate (TPR) and True Negative Rate (TNR) across regions for (2-class) probability mode classification, with 95% CI calculated across tiles, grouped by classifier and posterior probability threshold.

The variation in the performance of the classifiers across regions also is meaningful. Performance metrics for images from March (regions 1–4) have less variation both within and across regions than for images from September (regions 5–10), especially for classification with high spatial resolution inputs (WV, WVLS8) and for classification of vegetation (trees, shrubs). March is the rainy season for the region, as vegetation starts to green up; while September is near the beginning of the dry season, where cumulative effects of the previous season's precipitation should be visible. We would thus naturally expect that classifiers show less variation in performance for differentiating between vegetation classes in March than September, and less overall variation between seasons for classification of land.

3.2. Between-Region Classifier Performance

We next assess the spatial generalization of the classifiers and the extent to which reference data collected from one region can be used to classify other regions. As shown in Figure 5, when using reference points from one region to classify another within the same season, classifiers perform better in March than the average (Figure 3), driven by a larger training data size (number of labeled examples). They perform closer to average or slightly worse in the images sensed in September. The performance difference between the two seasons can again be understood in terms of vegetation differences in monsoonal ecosystems—heterogeneity in rainfall-driven seasonal vegetation would be expected to peak after the rainy season has ended, early in the dry season.

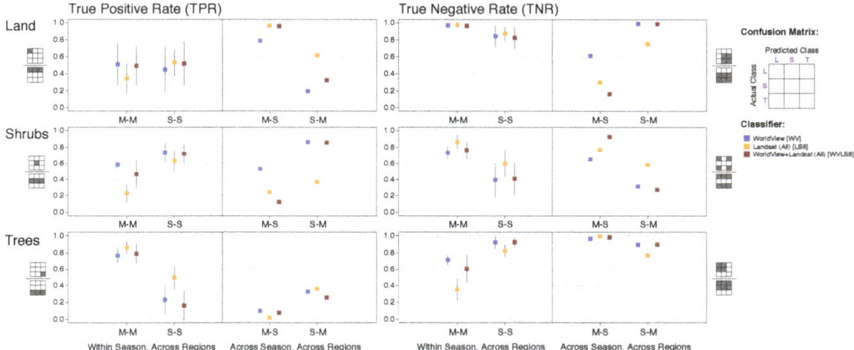

Figure 5. Spatial (cross-region) and seasonal cross-validation: TPR and TNR are shown for (**top row**) land, (**middle row**) shrubs, and (**bottom row**) trees. In each plot, the points to the left of the center line show the results from within-season classification, with 95% CI calculated across folds; the points to the right of the center line show across-season results. The month label for the across-season results indicates the training set; i.e., "M-M" means that the labeled examples from regions imaged in March were used to classify the other March regions; "S" indicates September. There are no error bars for the cross-season experiment as, for this, all of the data from one season was used to classify the other.

We also run the classification using the labeled examples from March to classify the samples of the region sensed in September and vice versa (shown on the right sides of plots in Figure 5). Performance is worse for classification with all combinations of inputs, and for all classes. While not surprising, these results confirm that seasonality and type of land cover matters when using one region to classify another: the ecosystem changes across seasons, affecting potentially both the reflectance of different classes (discussed below) and the stratification of the training data set (which may not match the test set).

3.3. Seasonality

We did not have access to co-located imagery from different seasons within the same year (the ideal experiment), so we instead chose regions from within the same calendar year, but different seasons, to examine the extent of meaningful variation in the spectral reflectance of the land cover. Though observing seasonal variation is rather intuitive and expected, this analysis is important for explaining differences between and within seasons in classification accuracy. To do this, we examine the reflectance profile of the three investigated land cover types (trees, shrubs, and land) in March (regions 1–4) and in September (regions 5–10). We define the reflectance profile for each type of land cover as the 8-bits average reflectance value of all pixels labeled as a given land cover type, per Landsat 8 and WorldView band. Note that Landsat and WorldView satellites collect data in a different dynamic range. WV2 and WV3 collect data in a dynamic range of 11 and 14 bits, respectively, while Landsat collects data over a 12-bit dynamic range (the data products are delivered in 16-bit unsigned integer format). The down scaling procedure to 8-bit values (which we use for classification) results in different reflectance values between the satellites. As expected, results show seasonal variations in the spectral profile of the three land cover types, which is observed in both Landsat 8 and WorldView (Figure 6, Tables A3 and A4). In September (the dry season), the reflectance of land and trees is significantly ($p < 0.05$) higher than in March (the rainy season) in all Landsat bands, besides the near-infrared wavelength (B5). Shrubs, however, are characterized by a relatively similar reflectance in Landsat's blue and green bands (B2 and B3, respectively) in March and in September, but a seasonal difference in the red band (drier or 'browner' vegetation in September).

Consistent with the high reflectance of live vegetation in the near infrared wavelength, the reflectance of the three land cover types is significantly ($p = 0.000$) higher in the near infrared wavelength (B5) in March than in September. Seasonal variation in the reflectance of trees and shrubs is also observed in WorldView imagery. In September, the reflectance of trees and shrubs are significantly ($p = 0.000$) higher than in March, in all bands. Moreover, the results show variations in the reflectance of the land cover also within season. As illustrated in Figure 6 (the reflectance profile of the land cover in Landsat and WorldView), the reflectance of shrubs is higher than the reflectance of trees in all Landsat and WorldView bands, in March and in September. However, the difference between the reflectance of trees and shrubs is more significant in March than in September. The difference between the mean reflectance of trees and shrubs is significantly ($p < 0.01$) larger in March than in September in all bands besides Landsat's thermal bands (bands B10 and B11) (Table A5). In other words, the difference between the reflectance of trees and shrubs is significantly larger in the rainy (or "greener") season than in the dry season.

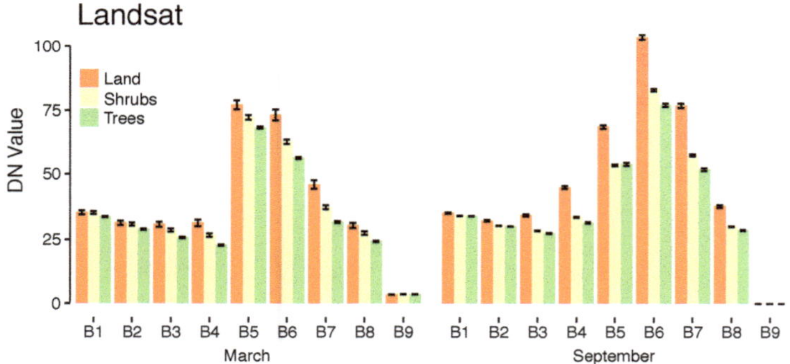

Figure 6. Spectral profile (mean values and 95% CI) of Landsat and WorldView bands, per class, in March and September (8-bit Digital Numbers (DNs)). Values are shown in table form in the Appendix B (Tables A3–A6).

The variations in the reflectance of the land cover (within and between seasons) is also expressed by the examined L8-based spectral indices (Table A6). Trees and shrubs are characterized by significantly ($p < 0.05$) different NDVI, NDWI, NDBI and UI values, both in the dry and in the rainy seasons. In both seasons, the NDVI value of trees in significantly ($p = 0.000$) higher than the NDVI value of shrubs. Moreover, the average NDVI value of the three land cover types is significantly higher in March than in September (Figure 7 presents the distribution of NDVI and NDWI values for trees and shrubs pixels, in March and in September). Similarly, the average NDWI of trees is significantly lower than of shrubs, and significantly lower in March than in September ($p = 0.000$, for both). This result is consistent with the higher reflection of live vegetation in the near infrared wavelength than in the green wavelength, especially in the rainy season. On the other hand, the average NDBI and UI values of all land cover types are significantly ($p = 0.000$) higher in September than in March (higher values of these indices indicate presence of bare land and soil, and less vegetation).

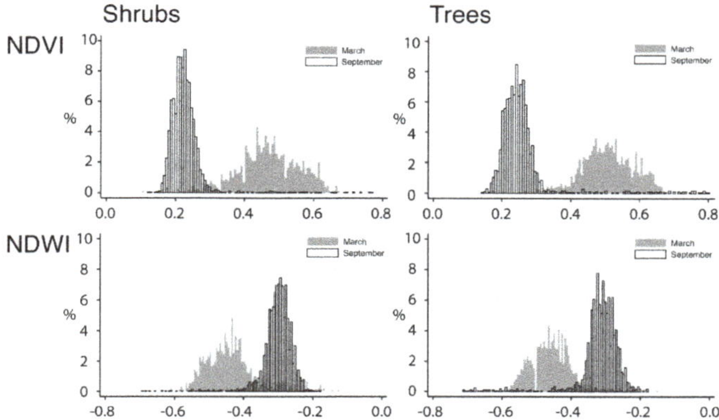

Figure 7. Histograms of Landsat 8 NDVI and NDWI values for trees and shrubs, in March and September.

Taken together, the variations in spectral reflectance—even in the visible RGB bands—indicate that the cross-season performance differences and the poor performance of training sets on imagery from the opposite season (Figure 5) are likely to be driven by actual changes in the reflectance of different land cover types between wet and dry seasons. This has important implications for ecosystem-scale assessments based on satellite imagery.

3.4. Full Area Estimates

To test the application potential of high-resolution classifiers over large areas, we used the labeled examples from each of the ten regions as training sets to then classify each region (using three classes), a total area of 1000 km^2. Figure 8 shows two representative samples of the classification in subsets of two of the ten regions. When classification is performed with high spatial resolution input, the distinction between features in the heterogeneous semi-arid landscape is clear, and contrasts profoundly with the coarse (by comparison) Landsat-based classifier. Table A8 shows the percentage of each 100 km^2 region covered by each land cover class. The totals derived from the WV and LS8 classifiers vary widely, with the greatest differences between trees and shrubs. This further supports what the 2-class performance metrics (Figure 4) suggested: namely that classification with high spatial resolution data outperforms in distinguishing between these two vegetation types.

Figure 8. Full ecosystem landcover classification examples using 3-class classifiers with WorldView and Landsat data. The top row shows a 1 km × 1 km section of one of the ten test regions (the same area shown in Figure 1); the bottom row shows a 380 m × 380 m region (to show greater detail). The first column shows the WorldView imagery in RGB color; the other columns show the classifier results for WorldView and Landsat 8 data, separated by class. In each case, the lighter color shows the Landsat pixel classification overlaid on the Worldview Imagery, and the darker color shows the WorldView classification. WorldView imagery in columns 2–4 is shown in grayscale for clarity. Full classification results of the 10 areas (1000 km^2) can be found at https://goo.gl/8fEj4r (see supplementary materials).

The WV classification estimates that 20% of the area is bare land, while 27% is tree/tree canopy and 53% is seasonal vegetation (shrubs); this is very close, as would be expected, to the distribution of our reference data in the original random selection of points. The full area estimates for classification with LS8 (15%/38%/47% for land, shrubs, and trees, respectively) is also closely aligned with the reference data that was labeled according to the buffer approach. However, Figures 2 and 8 show the fundamental issue with classification based on a coarse spatial resolution input: because the spatial scale is much larger than the scale of variation in land cover, the classification overestimates land cover for each class in some areas, and underestimates it in others. While the overall totals or averages may not differ dramatically from the finer-scale classifier, performance is clearly worse. In particular, bare land and seasonal vegetation often exist in narrow or winding tracks and sparse patterns, not within full pixels, in the Sertão, so it is fundamentally underestimated when classification is performed with coarser-resolution inputs; even a small fraction of trees within one 30 m resolution pixel may affect the spectral characteristics of the pixel, and lead to over-classification of trees. This problem of 'mixed-class pixels'—when two or more distinct classes may fall within a single pixel—has been widely addressed in the literature. The extent to which this occurs is dependent on the spatial resolution of the sensor and the spatial variability of the observed surface [42]. In this example of different types of vegetation, the mixed pixel problem appears to play a dominant role in the accuracy of the classification.

Lastly, we compare our full-area estimates to the estimates of a recent global dryland forest assessment by Bastin et al. [10], who use similar high-resolution imagery but a sampling and manual image interpretation approach to estimate tree canopy cover across the world's drylands (that is, they sample regions, sub-sample pixels within those regions, and manually label them; they do not run any classification of the full area). We use their data and match local estimates to our full-region estimates as shown in Figure 9. Our classification with high spatial resolution inputs (WV and WVLS8) estimate a higher percentage of tree cover than Bastin et al. (WV: Mean 26.9%, Standard Deviation (Std.) 24.3%;

Bastin: Mean 20.5%, Std. 34.0%), with the spatially-coarser classifier (LS8) overestimating treecover (LS8: Mean 47.2%, Std. 32.5%), largely due to the 'mixed pixel' problem described above and illustrated in Figure 8 and Table A8.

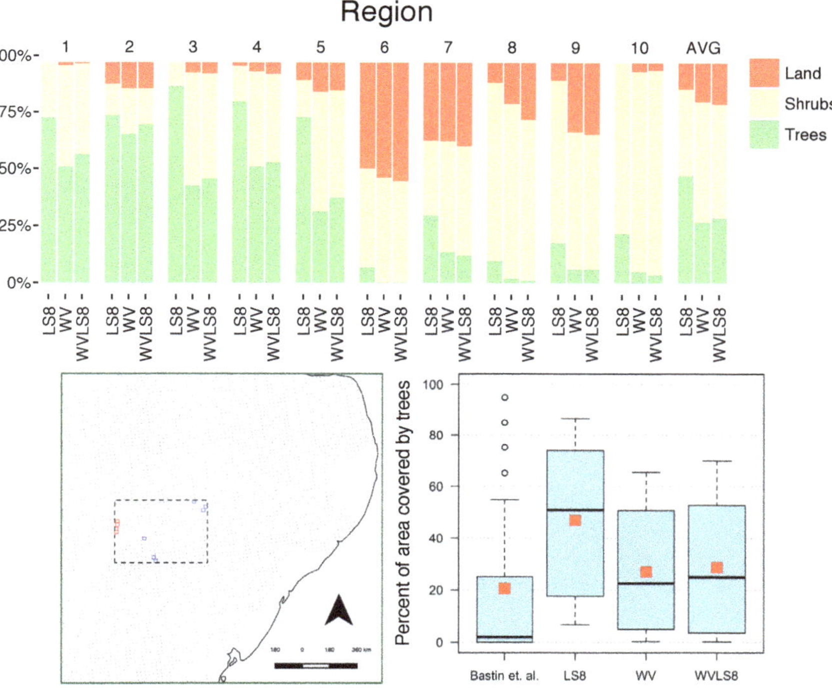

Figure 9. (**top panel**) Full area estimates, showing percentage of each region covered by each class, per classifier. Note: for full classification of each 1000 km^2 region we used the labeled examples from within that region as a training set. Corresponding values shown in Table A8. (**bottom panels**) Comparison of results for our pixel-based method using high spatial resolution data to Bastin et al.'s results [10] using a sampling and image interpretation approach. Box and whisker plots show the percentile distributions; red squares correspond to mean values. Our high resolution classifiers (WV and WVLS8) estimate a much larger average fraction of tree cover, when comparing our 1000 km^2 area (10 sample regions) to Bastin et al.'s sampled regions from the surrounding area (black dashed box). The black dashed box includes 8584 0.5 ha sampling areas, or just under 43 km^2. The pure Landsat (LS8) classifier over-estimates treecover, due to a relative inability to distinguish trees from seasonal shrubs and grasses. (A table version of the full area estimates can be found in Appendix C, Table A8.)

4. Discussion

Here, we show that pixel-based image classification with high spatial resolution satellite imagery and with only three bands—RGB (WV)—outperforms classification using inputs with a higher spectral but lower spatial resolution satellite imagery (LS8) for classification of non-built-up landcover in the semi-arid landscapes of the Brazilian Sertão. Our results are to some extent intuitive in that they show that high spatial resolution input gives us much more detailed information on land-cover patterns than medium or coarse spatial resolution input (even with higher spectral resolution). This study complements an emerging literature on why spatial resolution is considered one of the most significant emerging factors for accurate complex urban land cover mapping (e.g., [43]) and is especially essential when mapping heterogeneous and fragmented landscapes [44] as found in the semi-arid Brazilian Sertão.

We demonstrate that high spatial resolution imagery helps to mitigate the problem identified in previous literature as 'mixed class' pixels (e.g., [45]). In particular, we show that classification with high spatial resolution data as input to the classifier improves the ability to distinguish between more permanent and seasonal features in the landscape (trees versus shrubs). The distinction between trees and shrubs is critical for ecosystem assessment in dryland regions because while tree cover in semi-arid regions is an important driver of overall landscape processes and carbon sequestration, seasonal variation in shrub and grass cover is a principal driver of the interannual variability in the carbon cycle [5,46]. Our results thus complement recent studies showing that using high spatial resolution imagery can increase the accuracy of above-ground biomass estimation [47,48].

Although we highlight here that increased spatial resolution improves classifier performance, we do not discount the importance of spectral information, and, in fact, show that, even in the visible bands, spectral differences across seasons are meaningful in semi-arid regions. Although this study was limited by the WorldView imagery made available to us, use of the full spectral information available from WorldView (i.e., high spatial and high spectral resolution) would be extremely valuable for this type of assessment.

Our results confirm that seasonality and type of land cover matters when using one region to classify another: the ecosystem changes across seasons, affecting potentially both the spectral reflectance of different classes and the stratification of the reference training data set (which may not match the test set). However, we find that the seasonal difference in all three classes (trees, shrubs, and land) does not seem to be bigger in the RGB bands compared to other bands (at least according to the *t*-stat measures), implying that the high spatial-resolution RGB data is sufficient for detection of seasonal differences in the distribution of vegetation. We nevertheless emphasize that larger-scale implementations of this kind of classification for ecosystem assessments need to be undertaken with seasonality in mind. This is not new information for those already familiar with or working in semi-arid regions, but it is a distinct departure from the way treecover assessments are done in more dense forests.

Many previous studies utilize publicly available data for land cover mapping, including Landsat and Sentinel (although Sentinel-2 now offers spectral data at 10 m resolution, this pixel size is still too large to allow identification of many of the features that exists in semi-arid landscapes). Here, we compare classification with Landsat-8 (LS8) and with high-resolution data (WV) and demonstrate why, as described in [49], higher-resolution data is preferable to lower-resolution data when a landscape has small features and/or fine-scale variation in LULC, when a large portion of LULC change patches are smaller than the pixel size at lower-resolutions, and when high accuracy is necessary to inform decisions.

Our full area estimates of treecover in the Sertão are higher than, but not profoundly different from, the most recent attempt to quantify treecover in the world's drylands [10]. That effort used a sampling and image interpretation approach (but not classification) in an effort to deal both with 'mixed pixel' problems and seasonality issues. Both Bastin et al.'s approach and ours here require a human-in-the loop to either label examples (this study) or to do the classification itself (Bastin et al.). One benefit of our approach is that it is scalable to classify large-scale regions, as opposed to providing a sample-based estimate of the total quantity of interest. We have also shown that, within season, reference data can be used successfully across larger swaths with no reduction in performance. Finally, we classify the three main landcover types (bare land, trees, seasonal vegetation), as opposed to simply one.

These three features of supervised pixel-based image classification approach would be especially important for applications like monitoring land degradation at finer spatial scales (e.g., farm level seasonal changes) or estimating compliance with (for example) Brazil's Forest Code, which stipulates that 20% of all private landholdings outside the Amazon be preserved in native habitat [50], comprised of a wide range of vegetation types. Existing Landsat-based forest cover data sets (e.g., [8]) are able to assess changes in dense forest cover in the Amazon (where the Forest Code preservation requirement is 80%), but here we show that basic supervised image classification using high-resolution imagery in Google Earth Engine could be used to fully assess the state of compliance in the Sertão.

We acknowledge that at present very real financial obstacles exist to using private high-resolution data like that from WorldView for assessments of large-scale regions. However, as more and more high-resolution remotely sensed data become available to the research community, our study provides the best existing example of how they could be leveraged, in conjunction with the processing power of platforms like GEE, to conduct novel, and critical, large-scale ecosystem assessments of the world's semi-arid regions. Furthermore, in this study, we relied on WorldView orthorectified pan-sharpened 3-band imagery (using off the shelf imagery as delivered by DigitalGlobe). Extension to our work should utilize WorldView multispectral imagery to fully understand the contribution of the spatial and spectral dimension of the input data. In addition, understanding temporal dynamics in land cover patterns, including changes in vegetation land cover, requires radiometrically calibrated spectral data that was not available to us in this study.

5. Conclusions

This study shows that WorldView RGB imagery in a spatial resolution of 0.3–0.5 m can be utilized for high-quality and accurate classification of three types of natural landcover in the semi-arid Brazilian Sertão (land, trees and shrubs), outperforming classification with 30 m multi-spectral Landsat-8 imagery as input. We show that this superior performance classifying bare land, trees, and shrubs (here all non-tree vegetation) stems from two features of the high spatial resolution imagery. First, WorldView imagery, even with fewer spectral bands (RGB), allows for identifying small objects, such as individual trees and shrubs, that are captured in Landsat imagery as part of 'mixed-class' pixels; i.e., features smaller than the spatial resolution of the spectral Landsat imagery. Second, even with only the visible bands (RGB), classification with WorldView imagery as input can help to distinguish between trees and other seasonal vegetation with higher accuracy than with Landsat. This holds across the rainy and the dry seasons, although we show that, for the Sertão (and likely other semi-arid regions), performance is poor across seasons—a critical note for researchers using pixel-based classification in these agro-ecological zones.

By combining high spatial resolution imagery with the computation power of Google Earth Engine, and prototyping our classification on a 1000 km^2 area, we demonstrate the first viable methodology for conducting large-scale, high-resolution, and repeated ecosystem assessments of the world's semi-arid regions. Our full-area estimates of treecover differ from estimates based on both Landsat (which overestimate trees) and high-resolution sub-sampling and image-interpretation approaches, which appear to underestimate treecover, perhaps due to sparse sampling in heterogeneous areas. This implies that WorldView imagery, combined with relatively simple methods for supervised image classification in Google Earth Engine, can be utilized for totally new and perhaps counterintuitive land cover statistics of global semi-arid regions and the world's drylands.

Semi-arid regions, as home to hundreds of millions of smallholder farmers, are hotspots for food security and climate adaptation. Additionally, both longer-run trends in ecosystem degradation and land use changes, as well as interannual variability in seasonal vegetation, are key drivers of the global carbon cycle. Seasonal, global, and high-resolution assessments of the world's semi-arid regions based on the methodology described here would thus constitute a tremendously valuable new data source to understand earth system dynamics from scales small—e.g., the drivers of land use change at the farm scale—to large—e.g., land-atmosphere coupling globally.

Supplementary Materials: Classification results for the test region can be found at: http://goo.gl/8fEj4r and the Google Fusion Table containing the full set of labeled examples for the test region at: https://fusiontables.google.com/DataSource?docid=1duaYsQhbBpNV5mZeemCjweg0bZdAGTS_doBIkj3b.

Acknowledgments: This work was supported by the Big Pixel Initiative at UC San Diego (bigpixel.ucsd.edu), San Diego, CA, USA, including support for open access publication. The Digital Globe Foundation provided access to the WorldView imagery analyzed here.

Author Contributions: J.B. conceived the study; R.G. and J.B. designed the experiments; A.R.B. performed the experiments; R.G., A.R.B., and J.B. analyzed the data and wrote the paper.

Conflicts of Interest: The authors declare no conflict of interest.

Abbreviations

The following abbreviations are used in this manuscript:

WV	WorldView (Digital Globe Satellite)
LS8	Landsat 8
RF	Random Forest
GEE	Google Earth Engine
RGB	Red, Green, Blue
IR	Infra red
CI	Confidence Interval

Appendix A. Additional Imagery and Band Information

Here, we provide more detailed metadata on the imagery used from WorldView-2 and WorldView-3 (Table A1) and a summary of the band values for Landsat 8 and WorldView 2&3 used in the classifications (Table A2).

Table A1. Metadata for WorldView imagery used (all scenes are Pan Sharpened Natural Color). Regions correspond to the areas in Figure 2.

Region	Source	Resolution	Date (D/M/Y)	Cloud Cover	Product Type	Off Nadir Angle	Sun Elevation	Sun Azimuth	RMSE Accuracy
1	WV03	30 cm	1/3/16	3.31%	PSNC	12.0392°	67.0268°	86.4171°	3.914
2	WV03	30 cm	1/3/16	3.31%	PSNC	12.0392°	67.0268°	86.4171°	3.914
3	WV03	30 cm	1/3/16	3.31%	PSNC	12.0392°	67.0268°	86.4171°	3.914
4	WV03	30 cm	1/3/16	3.31%	PSNC	12.0392°	67.0268°	86.4171°	3.914
5	WV02	50 cm	13/9/15	0.00%	PSNC	28.7381°	65.2489°	58.3622°	3.914
6	WV02	50 cm	24/9/15	0.00%	PSNC	11.9266°	67.2595°	65.8200°	3.914
7	WV02	50 cm	24/9/15	0.00%	PSNC	11.9266°	67.2595°	65.8200°	3.914
8	WV02	50 cm	11/9/15	0.89%	PSNC	24.6518°	60.8081°	63.9840°	3.914
9	WV02	50 cm	19/9/15	0.18%	PSNC	14.7205°	64.2304°	67.6808°	3.914
10	WV02	50 cm	19/9/15	0.18%	PSNC	14.7205°	64.2304°	67.6808°	3.914

Table A2. Description of spectral bands from Landsat-8 and WorldView-2/WorldView-3 used in classification.

	Spectral Band	Wavelength (micrometers)	Resolution (meters)
	Landsat 8		
B1	Band 1—Ultra blue	0.43–0.45	30
B2	Band 2—Blue	0.45–0.51	30
B3	Band 3—Green	0.53–0.59	30
B4	Band 4—Đ Red	0.64–0.67	30
B5	Band 5—Near Infrared (NIR)	0.85–0.88	30
B6	Band 6—SWIR 1	1.57–1.65	30
B7	Band 7—SWIR 2	2.11–2.29	30
B8	Band 8—Panchromatic	0.50–0.68	15
B9	Band 9—Cirrus	1.36–1.38	30

Table A2. *Cont.*

	Spectral Band	Wavelength (micrometers)	Resolution (meters)
B10	Band 10—Thermal Infrared (TIRS) 1	10.60–11.19	100 (resampled to 30)
B11	Band 11—Thermal Infrared (TIRS) 2	11.50–12.51	100 (resampled to 30)
NDVI	(B5 − B4)/(B5 + B4)		30
NDWI	(B3 − B5)/(B3 + B5)		30
NDBI	(B6 − B5)/(B6 + B5)		30
EVI	2.5 * ((B5/B4)/(B5 + 6 * B4 − 7.5 * B2 + 1)		30
UI	(B7 − B5)/(B7 + B5)		30
	WorldView 2/3		
B	Blue	0.45–0.51	0.5/0.3
G	Green	0.51–0.58	0.5/0.3
R	Red	0.63–0.69	0.5/0.3

Appendix B. Spectral Band Values

Here, we provide data from Figures 6 and 7 in table format.

Table A3. Statistics for Landsat 8 reflectance values of training data set, per class and season (p-values shown for 95% confidence interval).

	Band:	B1	B2	B3	B4	B5	B6	B7	B8	B10	B11
LAND		Regions 1–4 (March) (n = 235)									
	Mean	35.30	31.37	30.86	31.37	77.09	73.19	45.96	30.47	193.95	190.82
	Std. Err.	0.38	0.42	0.46	0.60	0.84	1.05	0.83	0.49	0.14	0.12
		Regions 5–10 (September) (n = 1150)									
	Mean	35.19	32.30	34.35	44.86	68.62	103.52	76.75	37.76	212.79	210.55
	Std. Err.	0.11	0.14	0.20	0.27	0.33	0.47	0.44	0.22	0.12	0.11
		t-tests of equal means (March/September)									
	(t-stat)	−0.30	2.11	6.88	20.43	−9.38	26.38	32.79	13.54	104.22	117.89
	(p-value)	0.77	0.036	0.000	0.000	0.000	0.000	0.000	0.000	0.000	0.000
TREES		Regions 1–4 (March) (n = 2852)									
	Mean	33.72	28.99	25.79	22.79	68.32	56.28	31.71	24.32	193.13	190.12
	Std. Err.	0.12	0.13	0.14	0.16	0.20	0.21	0.17	0.14	0.03	0.03
		Regions 5–10 (September) (n = 1759)									
	Mean	33.98	30.09	27.46	31.48	53.82	76.92	51.76	28.60	208.05	206.34
	Std. Err.	0.05	0.07	0.10	0.15	0.30	0.31	0.27	0.11	0.08	0.07
		t-tests of equal means (March/September)									
	(t-stat)	1.97	7.55	10.01	39.89	−40.38	55.23	63.43	23.32	181.03	204.88
	(p-value)	0.048	0.000	0.000	0.000	0.000	0.000	0.000	0.000	0.000	0.000
SHRUBS		Regions 1–4 (March) (n = 906)									
	Mean	35.28	30.93	28.62	26.66	72.28	62.80	37.34	27.49	193.27	190.24
	Std. Err.	0.28	0.31	0.32	0.36	0.44	0.45	0.37	0.33	0.07	0.06
		Regions 5–10 (September) (n = 3073)									
	Mean	34.09	30.35	28.47	33.59	53.42	82.77	57.29	30.04	208.43	206.37
	Std. Err.	0.04	0.05	0.08	0.12	0.18	0.24	0.21	0.09	0.06	0.06
		t-tests of equal means (March/September)									
	(t-stat)	1.97	−1.87	−0.47	18.41	−39.64	39.62	47.12	7.44	166.37	185.38
	(p-value)	0.048	0.062	0.635	0.000	0.000	0.000	0.000	0.002	0.000	0.000

Table A4. Statistics for WorldView reflectance values of training data set, per class and season (p-values shown for 95% confidence interval).

	Band:	**B1**	**B2**	**B3**
LAND	Regions 1–4 (March) (n = 373)			
	Mean	144.27	135.65	132.74
	Std. Err.	1.60	1.64	1.66
	Regions 5–10 (September) (n = 1763)			
	Mean	178.32	148.44	123.47
	Std. Err.	0.95	1.00	0.94
	t-tests of equal means (Mar/Sept)			
	(t-stat)	18.33	6.67	−4.88
	(p-value)	0.000	0.000	0.000
TREES	Regions 1–4 (March) (n = 2021)			
	Mean	82.97	84.02	88.94
	Std. Err.	0.40	0.38	0.37
	Regions 5–10 (September) (n = 769)			
	Mean	110.81	100.90	93.67
	Std. Err.	1.05	0.84	0.83
	t-tests of equal means (Mar/Sept)			
	(t-stat)	24.82	18.23	5.20
	(p-value)	0.000	0.000	0.000
SHRUBS	Regions 1–4 (March) (n = 1599)			
	Mean	102.83	100.61	102.40
	Std. Err.	0.58	0.56	0.54
	Regions 5–10 (September) (n = 3450)			
	Mean	134.67	121.42	110.11
	Std. Err.	0.58	0.54	0.45
	t-tests of equal means (Mar/Sept)			
	(t-stat)	38.96	26.88	10.98
	(p-value)	0.000	0.000	0.000

Table A5. Two-sample t-test with unequal variances: comparison of mean reflectance (Landsat 8 and WorldView) of trees and shrubs.

	Landsat								
	B2			B3			B4		
	Diff.	Std. Err.	$Pr(\|T\| > \|t\|)$	Diff.	Std. Err.	$Pr(\|T\| > \|t\|)$	Diff.	Std. Err.	$Pr(\|T\| > \|t\|)$
March	−1.941	0.335	0.000	−2.832	0.345	0.000	−3.871	0.389	0.000
September	−0.256	0.080	0.001	−1.004	0.122	0.000	−2.110	0.195	0.000
	B5			B6			B7		
	Diff.	Std. Err.	$Pr(\|T\| > \|t\|)$	Diff.	Std. Err.	$Pr(\|T\| > \|t\|)$	Diff.	Std. Err.	$Pr(\|T\| > \|t\|)$
March	−3.956	0.484	0.000	−6.515	0.491	0.000	−5.636	0.407	0.000
September	0.398	0.347	0.252	−5.849	0.391	0.000	−5.537	0.337	0.000
	B8			B9			B10		
	Diff.	Std. Err.	$Pr(\|T\| > \|t\|)$	Diff.	Std. Err.	$Pr(\|T\| > \|t\|)$	Diff.	Std. Err.	$Pr(\|T\| > \|t\|)$
March	−3.166	0.361	0.000	−0.143	0.075	0.058	−0.120	0.070	0.088
September	−1.443	0.146	0.000	−0.384	0.097	0.000	−0.031	0.094	0.743

Table A5. *Cont.*

	WorldView								
	B1			B2			B3		
	Diff.	Std. Err.	Pr(|T| > |t|)	Diff.	Std. Err.	Pr(|T| > |t|)	Diff.	Std. Err.	Pr(|T| > |t|)
March	−19.86	0.704	0.000	−16.59	0.676	0.000	−13.46	0.651	0.000
September	−23.86	1.197	0.000	−20.51	1.001	0.000	−16.45	0.946	0.000

Table A6. Statistics for NDVI, EVI, NDWI, NDBI and UI spectral indices (calculated with Landsat 8), per season, with comparison for trees and shrubs.

		NDVI		EVI		NDWI		NDBI		UI	
		March	Sep.	March	Sep.	March	Sep.	March	Sep.	March	Sep.
Land	Mean	0.425	0.212	4.770	1.014	−0.427	−0.334	−0.032	0.202	−0.262	0.0052
	Std. Err.	0.005	0.002	2.280	0.124	0.004	0.002	0.005	0.003	0.007	0.003
Trees	Mean	0.504	0.257	−5.930	0.861	−0.453	−0.316	−0.099	0.179	−0.372	−0.017
	Std. Err.	0.002	0.002	0.681	0.374	0.001	0.002	0.001	0.003	0.002	0.003
Shrubs	Mean	0.468	0.228	−1.053	0.941	−0.435	−0.301	−0.073	0.217	−0.326	0.033
	Std. Err.	0.003	0.000	1.721	0.212	0.003	0.000	0.002	0.002	0.003	0.002
		t-tests of equal means (unequal variances) for trees and shrubs (within season)									
t-test *	(*t*-stat)	10.431	12.855	−2.630	−0.190	−6.210	−8.680	−9.940	−13.032	−13.273	−15.369
	(*p*-value)	0.000	0.000	0.010	0.850	0.000	0.000	0.000	0.000	0.000	0.000

Appendix C. Table Versions of Results

Results are presented graphically in the main text (Figure 9); here, we include table versions of our findings for completeness.

Table A7. Main classifier performance results (data from Figure 3).

Region	Class	Source	TPR	sd (TPR)	TNR	sd (TNR)	Region	Class	Source	TPR	sd (TPR)	TNR	sd (TNR)
1	LAND	LS8	0.145	0.107	0.825	0.007	6	LAND	LS8	0.834	0.048	0.896	0.032
1	LAND	LS8RGB	0.095	0.104	0.835	0.006	6	LAND	LS8RGB	0.744	0.035	0.917	0.012
1	LAND	WV	0.309	0.110	0.876	0.022	6	LAND	WV	0.773	0.023	0.995	0.008
1	LAND	WV30	0.020	0.045	0.865	0.006	6	LAND	WV30	0.763	0.045	0.911	0.030
1	LAND	WVLS8	0.181	0.134	0.837	0.018	6	LAND	WVLS8	0.839	0.057	0.992	0.012
1	SHRUBS	LS8	0.190	0.016	0.819	0.012	6	SHRUBS	LS8	0.854	0.094	0.789	0.055
1	SHRUBS	LS8RGB	0.221	0.025	0.812	0.024	6	SHRUBS	LS8RGB	0.924	0.054	0.684	0.037
1	SHRUBS	WV	0.521	0.071	0.707	0.043	6	SHRUBS	WV	0.995	0.011	0.762	0.027
1	SHRUBS	WV30	0.339	0.023	0.809	0.028	6	SHRUBS	WV30	0.863	0.052	0.741	0.059
1	SHRUBS	WVLS8	0.404	0.054	0.723	0.049	6	SHRUBS	WVLS8	1	0	0.831	0.056
1	TREES	LS8	0.835	0.015	0.596	0.019	6	TREES	LS8	0.319	0.154	0.963	0.022
1	TREES	LS8RGB	0.831	0.037	0.610	0.023	6	TREES	LS8RGB	0.155	0.038	0.983	0.012
1	TREES	WV	0.752	0.042	0.784	0.036	6	TREES	WV	0.100	0.224	0.999	0.002
1	TREES	WV30	0.826	0.031	0.660	0.013	6	TREES	WV30	0.419	0.129	0.963	0.018
1	TREES	WVLS8	0.778	0.049	0.727	0.029	6	TREES	WVLS8	0	0	0.999	0.002
2	LAND	LS8	0.589	0.067	0.815	0.009	7	LAND	LS8	0.750	0.071	0.795	0.033
2	LAND	LS8RGB	0.456	0.118	0.831	0.015	7	LAND	LS8RGB	0.698	0.085	0.831	0.022
2	LAND	WV	0.604	0.047	0.866	0.020	7	LAND	WV	0.627	0.219	0.786	0.033
2	LAND	WV30	0.559	0.109	0.844	0.009	7	LAND	WV30	0.740	0.062	0.919	0.010
2	LAND	WVLS8	0.564	0.025	0.832	0.010	7	LAND	WVLS8	0.820	0.066	0.895	0.023
2	SHRUBS	LS8	0.112	0.024	0.878	0.028	7	SHRUBS	LS8	0.466	0.049	0.743	0.056
2	SHRUBS	LS8RGB	0.162	0.048	0.899	0.011	7	SHRUBS	LS8RGB	0.632	0.083	0.663	0.036
2	SHRUBS	WV	0.331	0.046	0.847	0.018	7	SHRUBS	WV	0.368	0.151	0.762	0.045
2	SHRUBS	WV30	0.206	0.050	0.861	0.026	7	SHRUBS	WV30	0.802	0.055	0.659	0.016
2	SHRUBS	WVLS8	0.213	0.040	0.883	0.049	7	SHRUBS	WVLS8	0.751	0.060	0.724	0.031
2	TREES	LS8	0.842	0.024	0.584	0.014	7	TREES	LS8	0.538	0.101	0.825	0.022
2	TREES	LS8RGB	0.895	0.022	0.597	0.020	7	TREES	LS8RGB	0.421	0.073	0.880	0.022
2	TREES	WV	0.866	0.022	0.726	0.029	7	TREES	WV	0.614	0.135	0.768	0.087
2	TREES	WV30	0.851	0.039	0.641	0.015	7	TREES	WV30	0.423	0.161	0.941	0.013
2	TREES	WVLS8	0.896	0.052	0.657	0.018	7	TREES	WVLS8	0.438	0.084	0.927	0.019

Table A7. *Cont.*

Region	Class	Source	TPR	sd (TPR)	TNR	sd (TNR)	Region	Class	Source	TPR	sd (TPR)	TNR	sd (TNR)
3	LAND	LS8	0.306	0.119	0.875	0.009	8	LAND	LS8	0.508	0.103	0.934	0.020
3	LAND	LS8RGB	0.256	0.188	0.867	0.014	8	LAND	LS8RGB	0.334	0.077	0.962	0.005
3	LAND	WV	0.550	0.235	0.860	0.024	8	LAND	WV	0.331	0.042	0.960	0.018
3	LAND	WV30	0.362	0.064	0.880	0.012	8	LAND	WV30	0.460	0.110	0.952	0.017
3	LAND	WVLS8	0.673	0.121	0.830	0.017	8	LAND	WVLS8	0.374	0.027	0.955	0.011
3	SHRUBS	LS8	0.147	0.035	0.921	0.016	8	SHRUBS	LS8	0.871	0.038	0.427	0.059
3	SHRUBS	LS8RGB	0.126	0.051	0.943	0.026	8	SHRUBS	LS8RGB	0.920	0.013	0.327	0.053
3	SHRUBS	WV	0.480	0.171	0.665	0.092	8	SHRUBS	WV	0.968	0.029	0.332	0.029
3	SHRUBS	WV30	0.166	0.042	0.921	0.021	8	SHRUBS	WV30	0.901	0.028	0.392	0.059
3	SHRUBS	WVLS8	0.448	0.053	0.680	0.027	8	SHRUBS	WVLS8	0.976	0.017	0.373	0.015
3	TREES	LS8	0.932	0.012	0.583	0.030	8	TREES	LS8	0.279	0.108	0.943	0.019
3	TREES	LS8RGB	0.951	0.028	0.569	0.044	8	TREES	LS8RGB	0.273	0.061	0.962	0.010
3	TREES	WV	0.670	0.099	0.763	0.081	8	TREES	WV	0.029	0.064	0.982	0.009
3	TREES	WV30	0.936	0.024	0.611	0.016	8	TREES	WV30	0.267	0.025	0.952	0.008
3	TREES	WVLS8	0.671	0.048	0.750	0.024	8	TREES	WVLS8	0.029	0.064	0.988	0.007
4	LAND	LS8	0.179	0.083	0.891	0.008	9	LAND	LS8	0.515	0.079	0.913	0.024
4	LAND	LS8RGB	0.020	0.045	0.886	0.015	9	LAND	LS8RGB	0.415	0.114	0.947	0.013
4	LAND	WV	0.531	0.201	0.874	0.012	9	LAND	WV	0.758	0.066	0.968	0.010
4	LAND	WV30	0.188	0.148	0.927	0.004	9	LAND	WV30	0.534	0.063	0.927	0.021
4	LAND	WVLS8	0.576	0.184	0.860	0.014	9	LAND	WVLS8	0.777	0.055	0.958	0.016
4	SHRUBS	LS8	0.265	0.054	0.912	0.012	9	SHRUBS	LS8	0.798	0.048	0.530	0.052
4	SHRUBS	LS8RGB	0.214	0.058	0.901	0.018	9	SHRUBS	LS8RGB	0.875	0.024	0.392	0.073
4	SHRUBS	WV	0.511	0.047	0.748	0.039	9	SHRUBS	WV	0.939	0.018	0.746	0.049
4	SHRUBS	WV30	0.437	0.037	0.887	0.027	9	SHRUBS	WV30	0.827	0.048	0.468	0.038
4	SHRUBS	WVLS8	0.454	0.054	0.761	0.067	9	SHRUBS	WVLS8	0.916	0.028	0.764	0.035
4	TREES	LS8	0.925	0.013	0.626	0.033	9	TREES	LS8	0.451	0.067	0.903	0.025
4	TREES	LS8RGB	0.917	0.023	0.573	0.037	9	TREES	LS8RGB	0.316	0.078	0.937	0.012
4	TREES	WV	0.775	0.058	0.790	0.020	9	TREES	WV	0.622	0.099	0.979	0.011
4	TREES	WV30	0.906	0.025	0.704	0.049	9	TREES	WV30	0.372	0.079	0.920	0.019
4	TREES	WVLS8	0.787	0.092	0.765	0.024	9	TREES	WVLS8	0.634	0.093	0.972	0.005
5	LAND	LS8	0.550	0.041	0.869	0.016	10	LAND	LS8	0.040	0.089	0.905	0.007
5	LAND	LS8RGB	0.429	0.046	0.861	0.017	10	LAND	LS8RGB	0	0	0.921	0.023
5	LAND	WV	0.574	0.080	0.869	0.021	10	LAND	WV	0.344	0.023	0.966	0.009
5	LAND	WV30	0.540	0.143	0.879	0.010	10	LAND	WV30	0.067	0.149	0.922	0.023
5	LAND	WVLS8	0.574	0.159	0.873	0.013	10	LAND	WVLS8	0.344	0.134	0.971	0.012
5	SHRUBS	LS8	0.270	0.052	0.867	0.027	10	SHRUBS	LS8	0.736	0.018	0.425	0.073
5	SHRUBS	LS8RGB	0.231	0.046	0.854	0.014	10	SHRUBS	LS8RGB	0.778	0.063	0.360	0.137
5	SHRUBS	WV	0.584	0.055	0.566	0.041	10	SHRUBS	WV	0.925	0.019	0.246	0.062
5	SHRUBS	WV30	0.314	0.047	0.847	0.030	10	SHRUBS	WV30	0.785	0.060	0.441	0.040
5	SHRUBS	WVLS8	0.599	0.042	0.643	0.042	10	SHRUBS	WVLS8	0.936	0.026	0.265	0.091
5	TREES	LS8	0.873	0.020	0.659	0.042	10	TREES	LS8	0.442	0.071	0.870	0.011
5	TREES	LS8RGB	0.876	0.020	0.637	0.017	10	TREES	LS8RGB	0.379	0.142	0.892	0.031
5	TREES	WV	0.546	0.071	0.831	0.028	10	TREES	WV	0.175	0.090	0.965	0.009
5	TREES	WV30	0.861	0.029	0.696	0.036	10	TREES	WV30	0.459	0.043	0.895	0.029
5	TREES	WVLS8	0.655	0.045	0.838	0.021	10	TREES	WVLS8	0.203	0.103	0.970	0.012

Table A8. Full area estimates. Percentage of each region covered by each class, by classifier. Note: for full classification of each 1000 km^2 region, we used the labeled examples from within that region as a training set. (Percentages may not add perfectly to 100 due to rounding.)

		1	2	3	4	5	6	7	8	9	10	Average
WV	L	4.2	14.5	7.4	7.0	15.9	53.8	37.8	21.1	33.7	6.7	20.2
	S	44.6	20.2	49.9	42.1	52.6	45.9	48.7	76.8	60.2	88.2	52.9
	T	51.1	65.3	42.7	51.0	31.5	0.3	13.5	2.1	6.1	5.1	26.9
LS8	L	2.3	12.5	2.6	4.5	10.8	49.8	37.5	11.7	10.8	2.6	14.5
	S	25.1	13.7	10.7	15.6	16.2	43.3	32.7	78.6	71.6	75.7	38.3
	T	72.6	73.8	86.7	79.9	73.0	6.9	29.7	9.7	17.7	21.7	47.2
WVLS8	L	3.6	14.5	7.8	8.0	15.2	55.4	40.0	28.2	35.0	6.3	21.4
	S	40.1	15.8	46.4	39.0	47.3	44.4	47.9	70.6	58.9	90.1	50.1
	T	56.4	69.7	45.8	53.0	37.5	0.3	12.1	1.2	6.2	3.7	28.6

References

1. Reynolds, J.F.; Smith, D.M.S.; Lambin, E.F.; Turner, B.; Mortimore, M.; Batterbury, S.P.; Downing, T.E.; Dowlatabadi, H.; Fernández, R.J.; Herrick, J.E.; et al. Global desertification: Building a science for dryland development. *Science* **2007**, *316*, 847–851.

2. Koster, R.D.; Dirmeyer, P.A.; Guo, Z.; Bonan, G.; Chan, E.; Cox, P.; Gordon, C.; Kanae, S.; Kowalczyk, E.; Lawrence, D.; et al. Regions of strong coupling between soil moisture and precipitation. *Science* **2004**, *305*, 1138–1140.
3. Oyama, M.D.; Nobre, C.A. Climatic consequences of a large-scale desertification in northeast Brazil: A GCM simulation study. *J. Clim.* **2004**, *17*, 3203–3213.
4. Seddon, A.; Macias-Fauria, M.; Long, P.; Benz, D.; Willis, K. Sensitivity of global terrestrial ecosystems to climate variability. *Nature* **2016**, *531*, 229–232.
5. Poulter, B.; Frank, D.; Ciais, P.; Myneni, R.B.; Andela, N.; Bi, J.; Broquet, G.; Canadell, J.G.; Chevallier, F.; Liu, Y.Y.; et al. Contribution of semi-arid ecosystems to interannual variability of the global carbon cycle. *Nature* **2014**, *509*, 600–603.
6. Oyama, M.D.; Nobre, C.A. A new climate-vegetation equilibrium state for tropical South America. *Geophys. Res. Lett.* **2003**, *30*, doi:10.1029/2003GL018600.
7. Huang, J.; Yu, H.; Guan, X.; Wang, G.; Guo, R. Accelerated dryland expansion under climate change. *Nat. Clim. Chang.* **2016**, *6*, 166–171.
8. Hansen, M.C.; Potapov, P.V.; Moore, R.; Hancher, M.; Turubanova, S.A.; Tyukavina, A.; Thau, D.; Stehman, S.V.; Goetz, S.J.; Loveland, T.R.; et al. High-resolution global maps of 21st-century forest cover change. *Science* **2013**, *342*, 850–853.
9. Fensholt, R.; Langanke, T.; Rasmussen, K.; Reenberg, A.; Prince, S.D.; Tucker, C.; Scholes, R.J.; Le, Q.B.; Bondeau, A.; Eastman, R.; et al. Greenness in semi-arid areas across the globe 1981–2007—An Earth Observing Satellite based analysis of trends and drivers. *Remote Sens. Environ.* **2012**, *121*, 144–158.
10. Bastin, J.F.; Berrahmouni, N.; Grainger, A.; Maniatis, D.; Mollicone, D.; Moore, R.; Patriarca, C.; Picard, N.; Sparrow, B.; Abraham, E.M.; et al. The extent of forest in dryland biomes. *Science* **2017**, *356*, 635–638.
11. Barsi, J.A.; Lee, K.; Kvaran, G.; Markham, B.L.; Pedelty, J.A. The spectral response of the Landsat-8 operational land imager. *Remote Sens.* **2014**, *6*, 10232–10251.
12. Gorelick, N.; Hancher, M.; Dixon, M.; Ilyushchenko, S.; Thau, D.; Moore, R. Google Earth Engine: Planetary-scale geospatial analysis for everyone. *Remote Sens. Environ.* **2017**, *202*, 18–27, doi:10.1016/j.rse.2017.06.031.
13. Patel, N.N.; Angiuli, E.; Gamba, P.; Gaughan, A.; Lisini, G.; Stevens, F.R.; Tatem, A.J.; Trianni, G. Multitemporal settlement and population mapping from Landsat using Google Earth Engine. *Int. J. Appl. Earth Obs. Geoinf.* **2015**, *35 Pt B*, 199–208.
14. Trianni, G.; Lisini, G.; Angiuli, E.; Moreno, E.A.; Dondi, P.; Gaggia, A.; Gamba, P. Scaling up to national/regional urban extent mapping using Landsat data. *IEEE J. Sel. Top. Appl. Earth Obs. Remote Sens.* **2015**, *8*, 3710–3719.
15. Goldblatt, R.; You, W.; Hanson, G.; Khandelwal, A.K. Detecting the boundaries of urban areas in India: A dataset for pixel-based image classification in Google Earth Engine. *Remote Sens.* **2016**, *8*, 634.
16. Liang, S.; Fang, H.; Chen, M. Atmospheric correction of Landsat ETM+ land surface imagery. I. Methods. *IEEE Trans. Geosci. Remote Sens.* **2001**, *39*, 2490–2498.
17. Li, H.; Wang, C.; Zhong, C.; Su, A.; Xiong, C.; Wang, J.; Liu, J. Mapping urban bare land automatically from Landsat imagery with a simple index. *Remote Sens.* **2017**, *9*, 249.
18. He, C.; Shi, P.; Xie, D.; Zhao, Y. Improving the normalized difference built-up index to map urban built-up areas using a semiautomatic segmentation approach. *Remote Sens. Lett.* **2010**, *1*, 213–221.
19. Gislason, P.O.; Benediktsson, J.A.; Sveinsson, J.R. Random Forests for land cover classification. *Pattern Recognit. Lett.* **2006**, *27*, 294–300.
20. Rodriguez-Galiano, V.F.; Ghimire, B.; Rogan, J.; Chica-Olmo, M.; Rigol-Sanchez, J.P. An assessment of the effectiveness of a random forest classifier for land-cover classification. *ISPRS J. Photogramm. Remote Sens.* **2012**, *67*, 93–104.
21. Jean, N.; Burke, M.; Xie, M.; Davis, W.M.; Lobell, D.B.; Ermon, S. Combining satellite imagery and machine learning to predict poverty. *Science* **2016**, *353*, 790–794.
22. Stevens, F.R.; Gaughan, A.E.; Linard, C.; Tatem, A.J. Disaggregating Census Data for Population Mapping Using Random Forests with remotely-sensed and ancillary data. *PLoS ONE* **2015**, *10*, e0107042.
23. Guan, H.; Li, J.; Chapman, M.; Deng, F.; Ji, Z.; Yang, X. Integration of orthoimagery and lidar data for object-based urban thematic mapping using random forests. *Int. J. Remote Sens.* **2013**, *34*, 5166–5186.

24. Zhang, H.; Zhang, Y.; Lin, H. Urban land cover mapping using random forest combined with optical and SAR data. In Proceedings of the 2012 IEEE International Geoscience and Remote Sensing Symposium, Munich, Germany, 22–27 July 2012; pp. 6809–6812.
25. Du, P.; Samat, A.; Waske, B.; Liu, S.; Li, Z. Random forest and rotation forest for fully polarized SAR image classification using polarimetric and spatial features. *ISPRS J. Photogramm. Remote Sens.* **2015**, *105*, 38–53.
26. Aguirre-Gutiérrez, J.; Seijmonsbergen, A.C.; Duivenvoorden, J.F. Optimizing land cover classification accuracy for change detection, a combined pixel-based and object-based approach in a mountainous area in Mexico. *Appl. Geogr.* **2012**, *34*, 29–37.
27. Whiteside, T.; Ahmad, W. A comparison of object-oriented and pixel-based classification methods for mapping land cover in northern Australia. In *Proceedings of the SSC2005 Spatial Intelligence, Innovation and Praxis: The National Biennial Conference of the Spatial Sciences Institute, September 2005*; Spatial Sciences Institute: Melbourne, Australia, 2005; pp. 1225–1231.
28. Myint, S.W.; Gober, P.; Brazel, A.; Grossman-Clarke, S.; Weng, Q. Per-pixel vs. object-based classification of urban land cover extraction using high spatial resolution imagery. *Remote Sens. Environ.* **2011**, *115*, 1145–1161.
29. Whiteside, T.G.; Boggs, G.S.; Maier, S.W. Comparing object-based and pixel-based classifications for mapping savannas. *Int. J. Appl. Earth Obs. Geoinf.* **2011**, *13*, 884–893.
30. Bhaskaran, S.; Paramananda, S.; Ramnarayan, M. Per-pixel and object-oriented classification methods for mapping urban features using Ikonos satellite data. *Appl. Geogr.* **2010**, *30*, 650–665.
31. Duro, D.C.; Franklin, S.E.; Dubé, M.G. A comparison of pixel-based and object-based image analysis with selected machine learning algorithms for the classification of agricultural landscapes using SPOT-5 HRG imagery. *Remote Sens. Environ.* **2012**, *118*, 259–272.
32. Dingle Robertson, L.; King, D.J. Comparison of pixel-and object-based classification in land cover change mapping. *Int. J. Remote Sens.* **2011**, *32*, 1505–1529.
33. Wu, T.F.; Lin, C.J.; Weng, R.C. Probability estimates for multi-class classification by pairwise coupling. *J. Mach. Learn. Res.* **2004**, *5*, 975–1005.
34. Zadrozny, B.; Elkan, C. Transforming classifier scores into accurate multiclass probability estimates. In Proceedings of the Eighth ACM SIGKDD International Conference on Knowledge Discovery and Data Mining, Edmonton, AB, Canada, 23–26 July 2002; pp. 694–699.
35. Pal, M. Random forest classifier for remote sensing classification. *Int. J. Remote Sens.* **2005**, *26*, 217–222.
36. Kulkarni, V.Y.; Sinha, P.K. Random forest classifiers: A survey and future research directions. *Int. J. Adv. Comput.* **2013**, *36*, 1144–1153.
37. Kohavi, R. A study of cross-validation and bootstrap for accuracy estimation and model selection. In Proceedings of the 14th International Joint Conference on Artificial Intelligence, Montreal, QC, Canada, 20–25 August 1995; pp. 1137–1143.
38. Rodriguez, J.D.; Perez, A.; Lozano, J.A. Sensitivity analysis of k-Fold cross validation in prediction error estimation. *IEEE Trans. Pattern Anal. Mach. Intell.* **2010**, *32*, 569–575.
39. Salzberg, S.L. On comparing classifiers: Pitfalls to avoid and a recommended approach. *Data Min. Knowl. Discov.* **1997**, *1*, 317–328.
40. Arlot, S.; Celisse, A. A survey of cross-validation procedures for model selection. *Stat. Surv.* **2010**, *4*, 40–79.
41. Beleites, C.; Salzer, R.; Sergo, V. Validation of soft classification models using partial class memberships: An extended concept of sensitivity & co. applied to grading of astrocytoma tissues. *Chemom. Intell. Lab. Syst.* **2013**, *122*, 12–22.
42. Barnsley, M. Digital remotely-sensed data and their characteristics. *Geogr. Inf. Syst.* **1999**, *1*, 451–466.
43. Momeni, R.; Aplin, P.; Boyd, D.S. Mapping complex urban land cover from spaceborne imagery: The influence of spatial resolution, spectral band set and classification approach. *Remote Sens.* **2016**, *8*, 88.
44. Chen, D.; Stow, D.; Gong, P. Examining the effect of spatial resolution and texture window size on classification accuracy: An urban environment case. *Int. J. Remote Sens.* **2004**, *25*, 2177–2192.
45. Suwanprasit, C.; Srichai, N. Impacts of spatial resolution on land cover classification. *Proc. Asia-Pac. Adv. Netw.* **2012**, *33*, 39–47.
46. Ahlström, A.; Raupach, M.R.; Schurgers, G.; Smith, B.; Arneth, A.; Jung, M.; Reichstein, M.; Canadell, J.G.; Friedlingstein, P.; Jain, A.K.; et al. The dominant role of semi-arid ecosystems in the trend and variability of the land CO_2 sink. *Science* **2015**, *348*, 895–899.

47. Lu, D.; Hetrick, S.; Moran, E. Land cover classification in a complex urban-rural landscape with QuickBird imagery. *Photogramm. Eng. Remote Sens.* **2010**, *76*, 1159–1168.
48. Meng, B.; Ge, J.; Liang, T.; Yang, S.; Gao, J.; Feng, Q.; Cui, X.; Huang, X.; Xie, H. Evaluation of remote sensing inversion error for the Above-Ground Biomass of Alpine Meadow Grassland based on multi-source satellite data. *Remote Sens.* **2017**, *9*, 372.
49. Fisher, J.R.; Acosta, E.A.; Dennedy-Frank, P.J.; Kroeger, T.; Boucher, T.M. Impact of satellite imagery spatial resolution on land use classification accuracy and modeled water quality. *Remote Sens. Ecol. Conserv.* **2017**, doi:10.1002/rse2.61.
50. Soares-Filho, B.; Rajão, R.; Macedo, M.; Carneiro, A.; Costa, W.; Coe, M.; Rodrigues, H.; Alencar, A. Cracking Brazil's forest code. *Science* **2014**, *344*, 363–364.

Article

Assessing the Spatial and Occupation Dynamics of the Brazilian Pasturelands Based on the Automated Classification of MODIS Images from 2000 to 2016

Leandro Parente * and Laerte Ferreira

Image Processing and GIS Laboratory (LAPIG), Federal University of Goiás (UFG), Goiânia GO 74001-970, Brazil; laerte@ufg.br

* Correspondence: leal.parente@gmail.com; Tel.: +55-62-3521-1360

Received: 18 March 2018; Accepted: 4 April 2018; Published: 14 April 2018

Abstract: The pasturelands areas of Brazil constitute an important asset for the country, as the main food source for the world's largest commercial herd, representing the largest stock of open land in the country, occupying ~21% of the national territory. Understanding the spatio-temporal dynamics of these areas is of fundamental importance for the goal of promoting improved territorial governance, emission mitigation and productivity gains. To this effect, this study mapped, through objective criteria and automatic classification methods (Random Forest) applied to MODIS (Moderate Resolution Imaging Spectroradiometer) images, the totality of the Brazilian pastures between 2000 and 2016. Based on 90 spectro-temporal metrics derived from the Red, NIR and SWIR1 bands and distinct vegetation indices, distributed between dry and wet seasons, a total of 17 pasture maps with an approximate overall accuracy of 80% were produced with cloud-computing (Google Earth Engine). During this period, the pasture area varied from ~152 (2000) to ~179 (2016) million hectares. This expansion pattern was consistent with the bovine herd variation and mostly occurred in the Amazon, which increased its total pasture area by ~15 million hectares between 2000 and 2005, while the Cerrado, Caatinga and Pantanal biomes showed an increase of ~8 million hectares in this same period. The Atlantic Forest was the only biome in which there was a retraction of pasture areas throughout this series. In general, the results of this study suggest the existence of two relevant moments for the Brazilian pasture land uses. The first, strongly supported by the opening of new grazing areas, prevailed between 2000 and 2005 and mostly occurred in the Deforestation Arc and in the Matopiba regions. From 2006 on, the total pasture area in Brazil showed a trend towards stabilization, indicating a slight intensification of livestock activity in recent years.

Keywords: MODIS; Random Forest; pasture mapping; Brazilian pasturelands dynamics

---

1. Introduction

Historically, food production was driven by population expansion, consumption and increased per capita income, which gradually raised the global food demand [1]. In this context, Brazil has tremendous importance for food production, as it is the largest world beef exporter [2] and is responsible for 35% of the world soybean exportation, considering grain and derived products [3].

This increase in the Brazilian agricultural production occurred by the conversion of natural ecosystems into planted pastures, mainly altering the Amazon [4] and Cerrado [5] biomes and through the soybean expansion over natural ecosystems and planted pastures [6,7]. Recent works have revealed a reduction in this extensification process, as converted areas are being intensified to produce both commodities [8,9]. Beyond its economic importance, this production has great potential in the mitigation of greenhouse gas emissions [10,11], although only a small fraction of this potential can be achieved at a viable economic cost [12].

Considering the dynamics of the Brazilian agricultural land use, the pasturelands are an important asset for the country, occupying ~21% of its territory [13]; it can be used as both a land reserve [14] and as food for the herds containing ~218 million cattle [15]. Despite previous works having mapped these areas throughout the Brazilian territory [13,16], the absence of recurrent maps and the methodological differences between these initiatives make a temporal analysis of the Brazilian pasturelands difficult; its dynamics have important territorial, economic and environmental implications [17,18].

Faced with the challenge of systematic land cover and land use mapping on a large scale, the images provided by the MODIS sensor (Moderate Resolution Imaging Spectroradiometer) [19], (e.g., [20]), combined with new classification algorithms (e.g., [21]), made possible highly accurate and recurring representations of the Earth's surface (e.g., [22,23]). Specifically, this work uses MODIS data and a novel approach to produce annual pasture area maps for the entire Brazil. In addition, and based on our classification results, census data and socioeconomic statistics we conduct an analysis of the territorial dynamics of the pasture areas in the last 17 years, taking into account different aspects and characteristics of the Brazilian livestock.

2. Data and Methods

The pastureland mapping method, shown in Figure 1, was based on MODIS data, product MOD13Q1 Collection-6 [24,25] and on the supervised classification of spectral-temporal metrics. This approach, which is able to use several observations over one year to capture per-pixel seasonal responses, processed all of the images obtained over the Brazilian territory between 2000 and 2016, considering only the best observations (i.e., without cloud and cloud shadowing contamination), according to the MODIS pixel reliability [26].

The classification was performed on a yearly basis, considering a feature space with 90 spectral-temporal metrics, equally distributed between the dry and wet seasons (e.g., maximum red reflectance value in the wet season). These seasons were defined, pixel by pixel, through a percentile analysis of all NDVI values for one year, whereas all observations of the first quartile ($\leq$25th percentile NDVI) were associated with the dry season and the rest of the observations (>25th percentile NDVI) were associated with the wet season. For each season, a minimum, maximum, amplitude, median and standard deviation were calculated considering the Red, NIR (near infrared) and SWIR2 (short wavelength infrared) band reflectance values and the spectral indices NDVI (normalized difference vegetation index), EVI2 (enhanced vegetation index) [27] and AFRI (aerosol free vegetation index) [28].

The classification approach used the Random Forest, an algorithm that considers several statistical decision trees to choose, based on a majority voting, a final class [29]. The training dataset, with the classes "pasture" and "not-pasture," was obtained from a Brazilian pasture map with per-pixel probability information based on Landsat 8 data obtained in 2015 [13], which was resampled to 250 m using the mode criterion. Considering the difficulty with sampling throughout the Brazilian territory, this training dataset was automatically produced using random points with larger ($\geq$60%) and smaller probabilities ($\leq$40%), according to the classes "pasture" and "not-pasture," respectively.

Due to the absence of reference pasture maps on other dates, the first classification was performed with training samples from 2015, assuming a stability scenario (i.e., without major land-cover and land-use changes). This scenario, even if hypothetical, allowed the annual production of preliminary pasture maps, with a per-pixel probability information and consequently the use of a single and consistent automatic sampling approach for the entire analysis period. Then, for each year and region of interest, a second classification was performed, considering the samples derived from the initial classification, using the same probabilities thresholds, that is, pasture ($\geq$60%) and not-pasture ($\leq$40%).

The training of a single classification model capable to generalize the Brazilian pastureland identification in the temporal and spatial dimensions is quite improbable, since these areas are very susceptible to climatic intra- and inter-annual variations [30] and present different biophysical and management characteristics throughout the territory [31,32]. To handle this, the current study chose a geographical and temporal stratification approach, responsible for training models capable of capturing the regional characteristics and variability in a specific time window. The geographical stratification, based on the useful limits of the Landsat WRS-2 (World Reference System) tiles, avoided the use of a model, for instance, specific for the Pantanal pasturelands, to classify a region in the Atlantic Forest (Figure 1). Likewise, the temporal stratification considered a calendar year window, avoiding the use of a model, trained in 2010, for instance, to classify the same region in 2000. Thus, in total, 12,920 classification models were trained (i.e., 380 WRS-2 geographical regions, analyzed over 17 years in two classification phases), using a total of 2700 point samples for each model (comprising 500 trees), that is, 300 points extracted from the region of interest and eight adjacent regions (Figure 1).

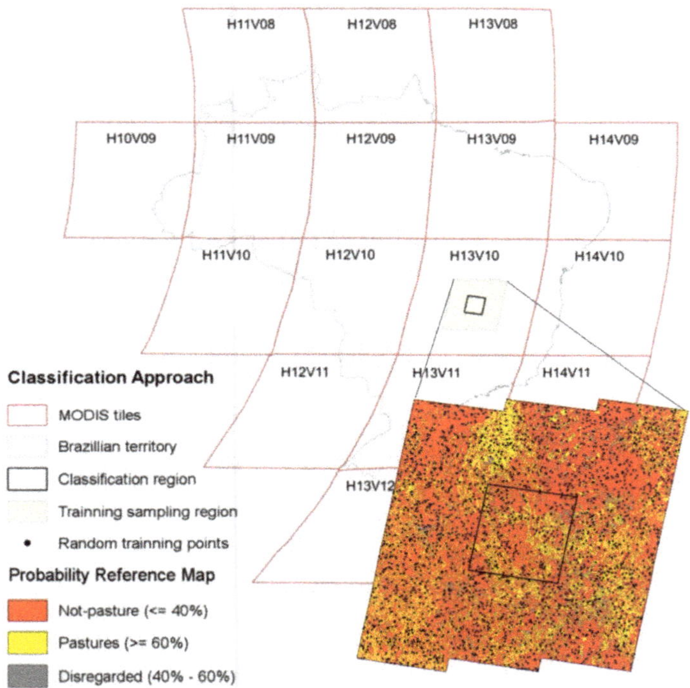

Figure 1. The stratified classification approach for mapping the Brazilian pasture areas, considering 16 MODIS tiles and the respective limits of 380 Landsat World Reference System (WRS)-2 regions. For each classification region, a total of 2700 training points (from the central WRS-2 path and row and eight neighboring regions), were randomly selected over a reference probabilistic map [13], considering both pasture (probabilities $\geq$60%) and not-pasture areas (probabilities $\leq$40%).

All steps described above were performed on Google Earth Engine, a cloud-computing platform capable of processing several remote-sensing analyses on a set of data and public images [33], using the python programing language (the scripts created within the scope of this work are available at: https://www.lapig.iesa.ufg.br/drive/index.php/s/NRP1VZ6kTyhh04i). The classification results, with the pasture class per-pixel probability, were exported to Google Drive. These were downloaded to a local workstation and combined to produce the annual pastureland maps for Brazil since 2000. The final maps of the pasture areas considered only the pixels with probability greater than or equal to 80%, used a majority rule filter to remove possible classification noises and discarded any pixels located in fully protected [34], urban [34] and permanent water areas [35]. All post-classification steps were performed with the GDAL library [36].

The pasture mapping results were evaluated using 5000 random points, equally distributed into pasture and not-pasture areas, conservatively assuming that the minimum mapping accuracy was 50% and that the accuracy assessment error was 1% within a 95% confidence interval [37]. These points were visually inspected by five trained interpreters, who analyzed, for each point, 34 Landsat images acquired between 2000 and 2016 (i.e., two images per year, considering the dry and wet seasons), classifying each point according to 10 land-cover and land-use classes (i.e., pasture, crop agriculture, planted forest, native vegetation, mixed use, water bodies, urban area, mining and others). This assessment was conducted in the Temporal Visual Inspection Tool, which also considers the respective MOD13Q1 NDVI time series and high-resolution Google Earth images [38]. Using these points, the overall accuracy (which is a robust indicator for assessing the quality of binary classifications, that is, pasture and not-pasture) and the omission and commission errors were estimated for the pasture class, for each one of the 17 maps produced for Brazil.

The flowchart in Figure 2 depicts the main datasets and strategies used for the annual mapping of the Brazilian pasturelands, between 2000 and 2016.

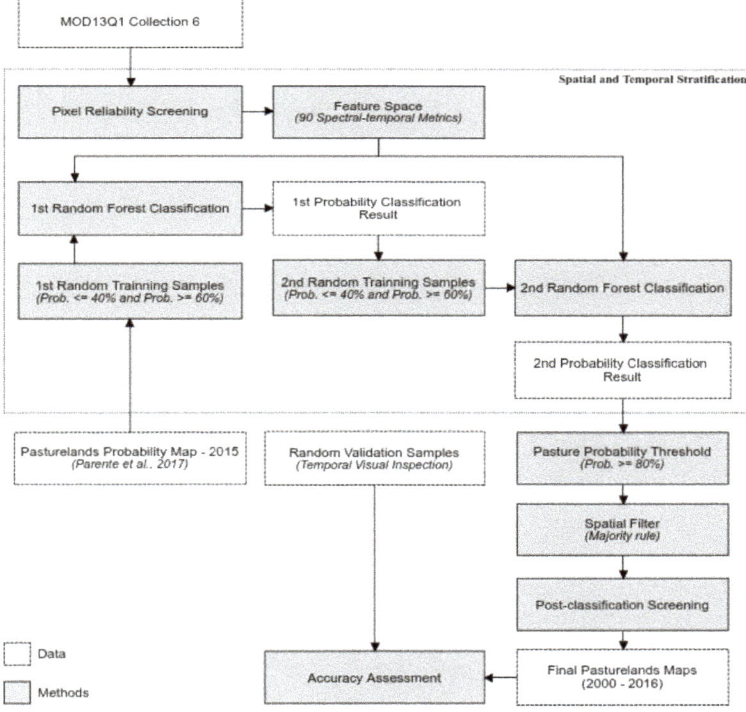

Figure 2. Methodological approaches used for mapping the Brazilian pasturelands between 2000 and 2016 based on MODIS data. The geographic stratification was based on the useful limits of the Landsat WRS-2 tiles, while the temporal stratification considered a calendar year window.

3. Results

The set of 17 Brazilian pasture maps (from 2000 to 2016), produced from the automated classification of MODIS images (Figure 3c), presented a relative convergence with existing mappings, for the years 2002 (Figure 3a) [16] and 2015 (Figure 3b) [13], based on Landsat images. In this study, the pasture area mapped in 2002 and 2015 was ~169 and ~176 million hectares, respectively, while the

Landsat-based mappings show, for the same years, ~149 and ~179 million hectares of pastures. In addition to these relatively close area values, the mapped areas have a very similar distribution pattern, which indicates that the proposed method is producing, in general, coherent and spatially consistent results, even though it is based on moderate spatial resolution data.

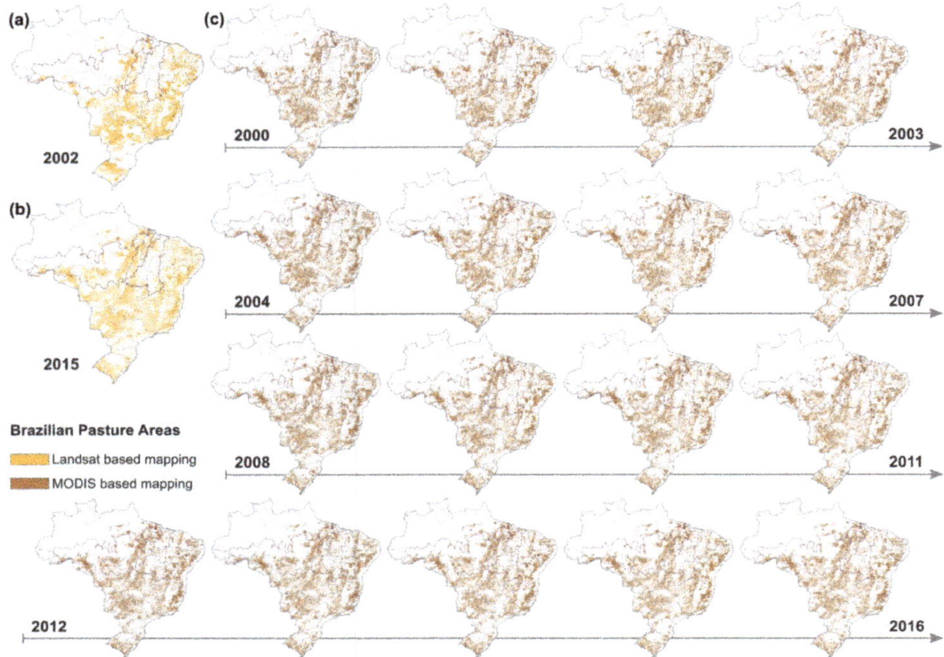

Figure 3. (**a**) The Brazilian pasture area as mapped by the segmentation and visual inspection of Landsat 7 images, with reference in 2002 [16]; (**b**) The Brazilian pasture area as mapped by the automated classification of Landsat 8 Images, with reference in 2015 [13]; (**c**) The Brazilian pasturelands time series, according to the automated classification of MODIS images obtained between 2000 and 2016.

In fact, for the entire time series, the pasture maps showed an overall accuracy of ~80% (Figure 4). In all years, the pasture class had a user accuracy (related to the commission error) greater than the producer accuracy, indicating that most of the mapping errors are related to pastures areas which have been omitted. According to this analysis, 2000 was the year with the largest mapping errors, while the highest time series accuracies were observed in 2010.

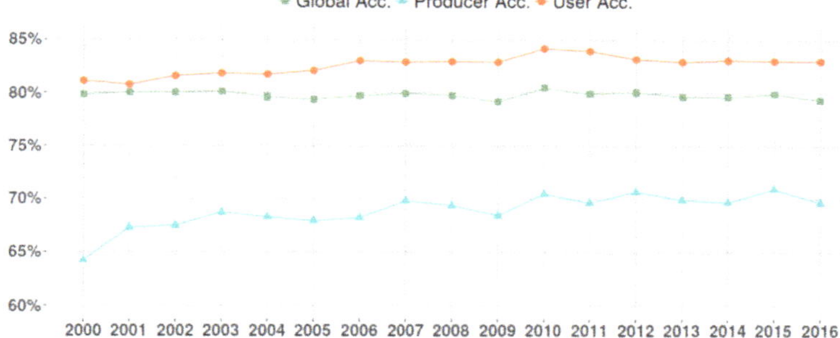

Figure 4. The accuracy assessment of the Brazilian pastureland mapping, considering 5000 random points, was inspected by five interpreters who considered 34 Landsat overpasses (two images per year), the NDVI-MOD13Q1 time series and high-resolution Google Earth images.

Our mapping revealed an increase of ~25 million hectares in the country's pasture area between 2000 and 2016; overall, ~80% of this increase occurred in the first six years of the analyzed period (Figures 3c and 5a). Most of this expansion occurred in the Amazon, increasing the biome pasture area by ~15 million hectares between 2000 and 2005, while in the Cerrado, Caatinga and Pantanal, the combined increased area, for the same period, was 8 million hectares (Figure 5b). The Atlantic Forest was the only biome that showed a retraction in pasture area throughout the analyzed time period, while the Pampa, a biome with a predominance of native grasslands, presented a relative stability over the years.

A temporal analysis of the Brazilian bovine herd, in animal units (AUs), also reveals an expressive increase in the first six years of the same period, starting at 125 million AU in 2000 and reaching 152 million AU in 2005 (Figure 5a). The animal unit calculation—1 AU is equivalent to 450 kg of live animal weight—considered the cattle herd composition, produced by the Agricultural Census for the year of 2006 only [39]—and which was assumed constant for other years—and the absolute number of cattle heads, estimated on a yearly basis by the Municipal Livestock Research [15]. The ratio between the AU and the pasture area showed a variation of 0.1 AU/hec in the Brazilian bovine stocking rate (over the last 17 years), which reached its lowest value in 2001 with 0.8 AU/hec and its peak in 2016, with 0.9 AU/hec. These results suggest the prevalence of an extensification process in the first years of the series, followed by a slight intensification of pasture areas in recent years (Figure 6).

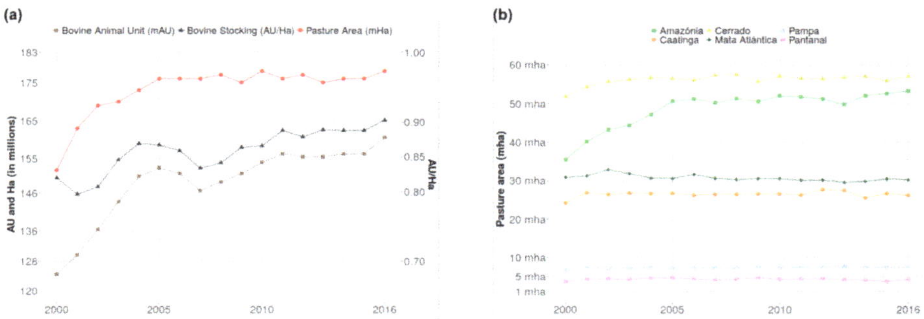

Figure 5. (**a**) Pasture area in Brazil, bovine animal unit (considering the composition and size of the herd) and Brazilian stocking rate time series; (**b**) Time series of the pasture areas in the six Brazilian biomes.

According to these mapping results, Pará was the state with the greatest pasture area expansion, equivalent to 8.2 million hectares (i.e., 6.61% of its territory), followed by Mato Grosso, Bahia, Maranhão and Rondônia states, with 4.6, 2.7, 2.4 and 2.2 million hectares of new areas, respectively (Figure 6). Considering these states, the municipalities that are responsible for this process are located in agriculture frontiers, such as the Deforestation Arc in the Cerrado and Amazon borders [4] and the Matopiba, in the Cerrado biome [40]; moreover, most of this extensification process occurred until 2005.

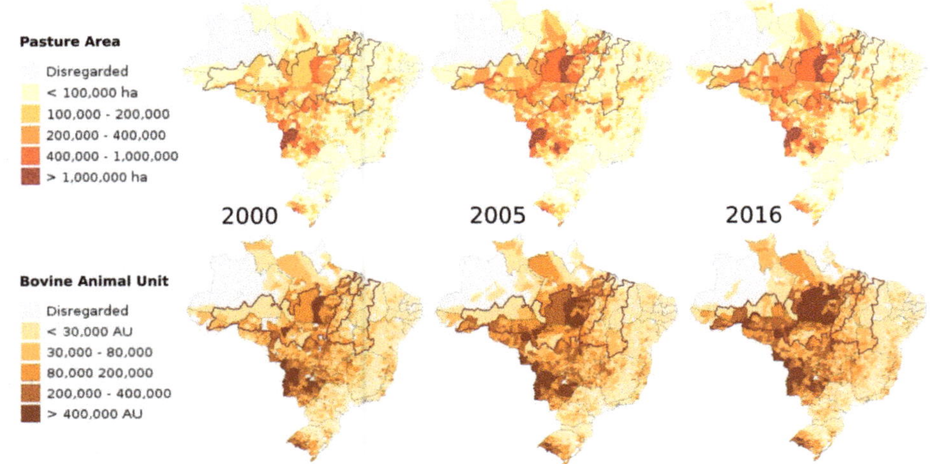

Figure 6. Geographic distribution of the mapped pasture and bovine animal unit, at the municipal basis, depicting the increase of both until the year 2005 and a subsequent intensification process, with stabilization of the pasture area and an increase in the number of cattle.

Within the same period, the only states in which there was a decrease in the absolute values of pasture areas were São Paulo, Paraná, Paraíba and the Federal District, being numerically inexpressive in the last two states. The São Paulo state reduced pasture areas by almost 2.4 million hectares (i.e., 9.62% of its territory), while a contraction of 0.4 million hectares was observed in Paraná. This analysis, at the municipal basis, reveals that none of the five municipalities with the greatest pasture area reduction—Rosário do Oeste (MT), Três Lagoas (MS), Rio Verde (GO), Sorriso (MT) and Nobres (MT)—are located in the São Paulo and Paraná states. This result suggests the existence of a spatial dynamic for the Brazilian pasture, intra- and inter-state, which displaces the land use of these areas to other municipalities according to the production needs and agricultural activity of the country.

4. Discussion

The relative stability of the accuracy values (Figure 4), in particular the overall accuracy, indicates that the proposed method is capable of producing comparable maps, which are spatially and temporally consistent. Nevertheless, these maps are subject to commission and omission errors. In part, the commission errors were minimized by the use of filters and masks, such as the water mask, which was used to remove pixels with high sediment concentration, wrongly classified as pasture (and mostly located in rivers of the Amazon biome). On the other hand, the more significant omission errors are related to the spatial resolution of the MODIS pixels (~6.25 hectares), which is not very compatible with the detection of small pasture fragments, usually on slope regions, especially located in east of the Serra do Espinhaço (Minas Gerais) and in the Serra Gaúcha (Rio Grande do Sul).

Despite these spatial resolution limitations, the MODIS MOD13Q1 data presented a high availability of observations free of clouds and cloud shadows (Figure 7), which provided denser time series for the generation of spectral-temporal metrics, enabling the appropriate capture of seasonal variations in pasture areas in the classification models. Considering the entire Brazilian territory, between 2000 and 2016, there was an average of ~18 MODIS good observations per pixel, a value which is relatively close to its full temporal resolution (i.e., 23 observations).

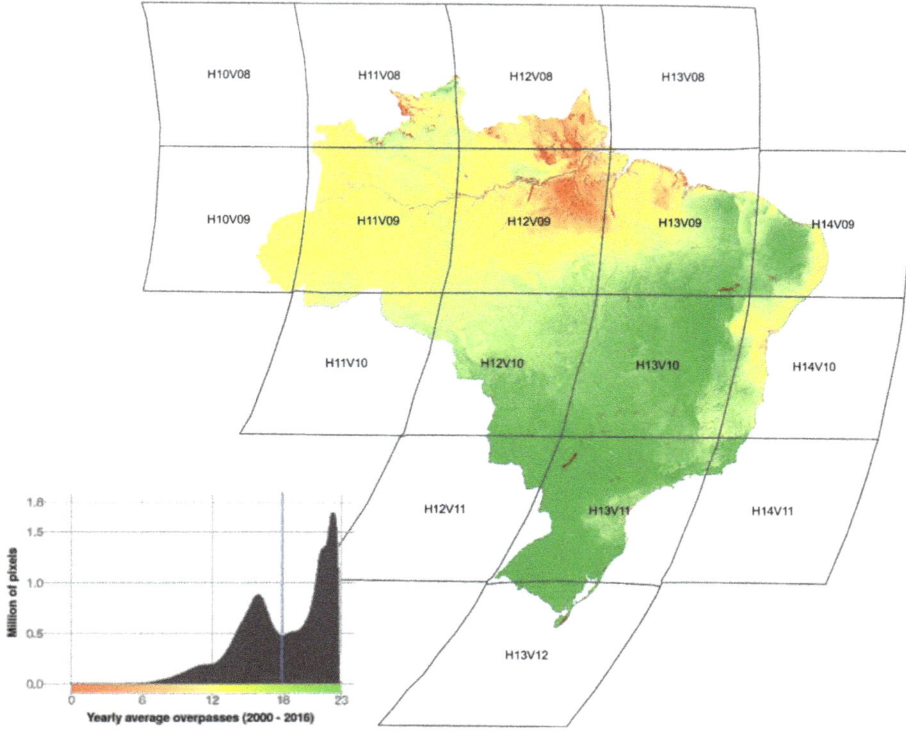

Figure 7. Annual average of MODIS observations considering only the best quality pixels (i.e., values 0 and 1, according to the pixel reliability image) in the period of 2000 and 2016. The blue line represents the average value of MODIS observations for the entire Brazilian territory during the same period.

In general, the results produced by this study suggest the existence of two relevant moments regarding the occupation of the Brazilian pasture areas. The first, strongly supported by the opening of new grazing areas, prevailed between 2000 and 2005 and mostly occurred in the regions of the Deforestation Arc and Matopiba. From 2006 on, the total pasture area in Brazil tended to stabilize and a correlation with bovine herd data suggested a mild intensification process, which enabled production increases, in animal units, on the same portion of land which was already opened. This hypothesis is reinforced by the ability of the pasture areas to model, on a municipal basis, the bovine animal unit values (Figure 8). Despite the good correlations between very distinct datasets (i.e., remote sensing and census data) and complementary periods, this analysis disregards the spatial dynamics of pasture areas. The process of converting older pastures to crop areas may have occurred, for instance, in the Matopiba municipalities, forcing the opening of new grazing areas for livestock, without any substantial changes in the region's pasture area balance. The spatial distribution of areas can vary at the intra- and inter-municipal level, according to market demand [41], infrastructure aspects [42] and land competition among the various productive sectors of society [43].

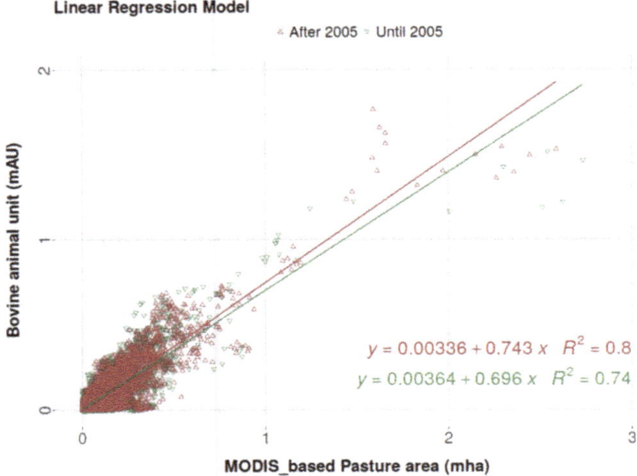

Figure 8. Linear regressions between pasture areas (based on MODIS data classification) and bovine animal units for all municipalities in Brazil, in two complementary periods (2000 to 2005 and 2006 to 2016).

Looking for a better understanding of these spatial dynamics, a linear regression was adjusted between the (MODIS based) pasture area and the cattle herd (considering the last 17 years) for each Brazilian municipality, which allowed a correlation spatialization between these variables, through the resulting goodness-of-fit (R2). This approach also considered the slope of both variables, calculated by their linear regressions against time. The combination of these several linear models revealed a set of municipalities relevant to the livestock activity, in which the expansion and retraction of pasturelands is historically related to the municipality's bovine animal unit (Figure 9). This analysis exposed a strong migration trend of the Brazilian livestock from center-south to the north, in a relatively short period—less than two decades—although the infrastructure—federal roads network [44] and slaughtering houses [45]—has remained in the center-south portion of the country. A probable consequence of this migration was the increase in beef production cost due to the transportation of the animals to slaughterhouses in the termination phase, possibly offset by the lower land's price in the identified livestock expansion zone [46,47].

All of the "top-5" pastureland increasing municipalities—São Félix do Xingu (PA), Altamira (PA), Novo Repartimento (PA), Porto Velho (RO) and Novo Progresso (PA)—are in the expansion zone, mostly concentrated in the Deforestation Arc and in the Matopiba, which indicates a pasture expansion over native vegetation areas of the Cerrado and Amazon in the last 17 years. Several studies have analyzed the environmental impacts and socioeconomic benefits resulting from the conversion of native areas to agricultural areas [48–50]. However, in the context of recent Brazilian livestock dynamics, the benefits appear to be less expressive than the impacts, since in 2010, several municipalities located in the expansion zone presented HDI (human development Index—[51]) levels smaller than, or close to, the general levels of the retraction zone in 2000 (Figure 10). Even with HDI increases of these municipalities in the same period, the livestock migration to the north shifted the cattle ranchers away from consumer markets, both internal and external, consolidating them in a portion of the territory with a lack of infrastructure and inducing various environmental impacts [52,53].

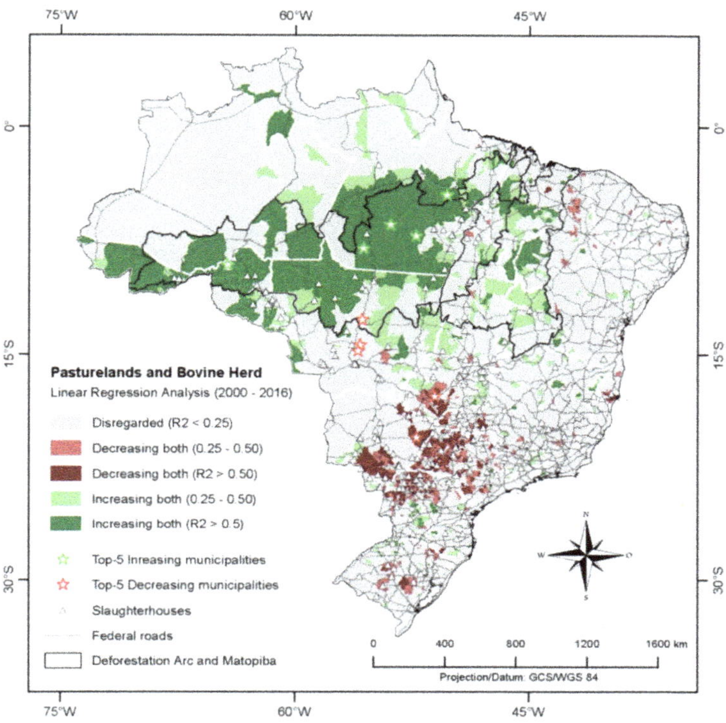

Figure 9. The expansion (green colors) and retraction (red colors) zones of the Brazilian livestock activity, produced through several linear regression models between the MODIS based pasture areas and the cattle animal unit at the municipality level (available at: http://maps.lapig.iesa.ufg.br/?layers=pa_br_correlacao_pastagem_ua_250_na_lapig).

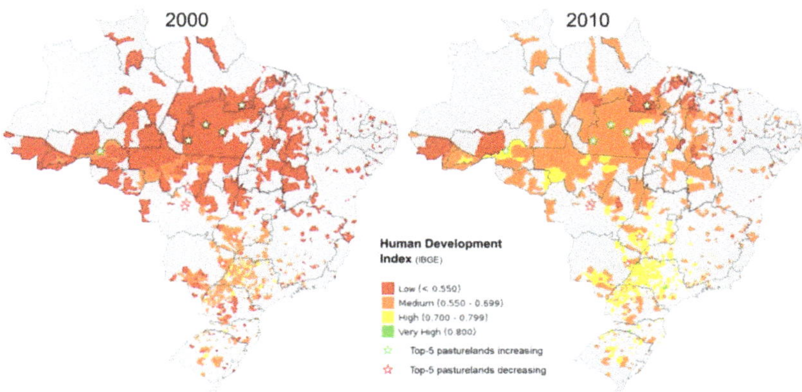

Figure 10. 2000 and 2001 Human Development Index (HDI) values of the municipalities located in the expansion and retraction zones of the Brazilian livestock activity.

One of the probable causes of this process was the pressure for already opened lands in the central south of Brazil, due to the production demand of other commodities, such as soybeans, sugarcane and wood. The replacement of livestock activity in this region increased the HDI from medium to high. In the north of Paraná and in the west of São Paulo, the main driver of this process was the rapid expansion of sugarcane cultivation [54], as well as in part of southeastern Goiás [55]. The decrease in livestock activity and the increase in the HDI of Três Lagoas-MS—the Brazilian municipality that lost the second most pasture area in the analyzed period—is justified by its consolidation as a relevant silviculture pole and by the installation of two large cellulose plants [56]. In the southeastern of Goiás, livestock activity was also replaced by grain cultivation [57], which justifies the presence of Rio Verde-GO—the Brazilian municipality that lost the third most pasture area in the analyzed period—in the identified retraction zone.

The retraction of pastures in the rest of the "top-5" pastureland decreasing municipalities, located in Mato Grosso, is driven by the grain cultivation expansion in the state, mostly soybean [58]. Interestingly, only Sorriso-MT had a decrease, even from 2010, of the bovine herd. The increasing trend of animal units, observed in Rosário do Oeste-MT and Nobres-MT, may indicate a livestock intensification close to these municipalities. The reduction of pastures and the availability of machinery and agricultural products, consolidated by the cultivation of grains, may have forced an intensification of the beef production in the region, through more efficient [59] or rotating [60] productive systems.

5. Concluding Remarks

This study used a unique automated methodological approach to produce, based on the Random Forest classification of MODIS images, a temporal series of 17 pasture maps for the entire Brazil (2000–2016). With an overall accuracy of ~80% and consistent spatial-temporal trends, the mapped area varied from ~152 million hectares in 2000, to ~178 million hectares in 2016, a pattern that is consistent with the cattle herd increase and prevalence of production growth through the conversion of new areas.

Specifically, the analysis of these maps revealed that there was a strong expansion of pasture areas in the first five years of the series, mostly in the Cerrado and in the Amazon. From 2005 on, the area balance stabilized and the livestock activity showed a slight intensification process, increasing its bovine stocking rate by 0.1 AU/hec in the entire analysis period. The high correlation of the cattle herd with the pasture areas, on a municipal basis, revealed a migration trend of livestock activity from the center-south to the north of the country, probably caused by the production needs of other Brazilian commodities, which demand greater infrastructure and are more economically attractive.

Despite the recent growth of the Brazilian livestock—currently, the country is the largest beef exporter in the world—the dynamic revealed by this study shows a displacement of the cattle ranchers from consumer markets, both internal and external and their consolidation in municipalities with a lack of infrastructure, at a high environmental cost. In this respect, the country demands public policies capable of improving the infrastructure and mitigate new environmental impacts in the north region. Public policies such as the ABC Plan (low carbon agriculture) can encourage sustainable management practices in regions with extensification history, like the Pará state, avoiding the opening of new pasture areas to meet the future beef production demand. On the other hand, productive intensification policies can focus on regions which already have an agricultural production infrastructure, such as the center of Mato Grosso, thereby reducing its implementation time. The effectiveness of these and other related public policies has great potential to meet the climate commitments signed by Brazil in the COP 21 and to promote the better utilization of the Brazilian territory.

Acknowledgments: This work, part of the MapBiomas initiative (http://mapbiomas.org), was supported by the Gordon and Betty Moore Foundation, The Nature Conservancy (TNC), the State of Goiás Research Foundation (FAPEG) and the Brazilian Research Council (CNPq).

Author Contributions: L.P. and L.F. conceived the idea of the study; L.P. processed the data and, together with L.F., analyzed the results; L.P. and L.F. contributed to the discussions, writing, and revision of the manuscript.

Conflicts of Interest: The authors declare no conflict of interest.

References

1. Godfray, H.C.J.; Beddington, J.R.; Crute, I.R.; Haddad, L.; Lawrence, D.; Muir, J.F.; Pretty, J.; Robinson, S.; Thomas, S.M.; Toulmin, C. Food security: The challenge of feeding 9 billion people. *Science* **2010**, *327*, 812–818. [CrossRef] [PubMed]
2. CNA Brasil Pode Se Tornar o Maior Produtor de Carne Bovina do Mundo. Available online: http://www.cnabrasil.org.br/noticias/brasil-pode-se-tornar-o-maior-produtor-de-carne-bovina-do-mundo (accessed on 15 January 2018).
3. Westcott, P.; Contact, E. *USDA Agricultural Projections to 2025 Interagency Agricultural Projections Committee*; USDA Long-term Projections; US Department of Agriculture: Washington, DC, USA, 2016.
4. Tyukavina, A.; Hansen, M.C.; Potapov, P.V.; Stehman, S.V.; Smith-Rodriguez, K.; Okpa, C.; Aguilar, R. Types and rates of forest disturbance in Brazilian Legal Amazon, 2000–2013. *Sci. Adv.* **2017**, *3*, e1601047. [CrossRef] [PubMed]
5. Rocha, G.; Ferreira, L.; Ferreira, N.; Ferreira, M. Detecção de desmatamentos no bioma cerrado entre 2002 e 2009: Padrões, tendências e impactos. *Rev. Bras. Cartogr.* **2011**, *63*, 341–349.
6. Macedo, M.N.; DeFries, R.S.; Morton, D.C.; Stickler, C.M.; Galford, G.L.; Shimabukuro, Y.E. Decoupling of deforestation and soy production in the southern Amazon during the late 2000s. *Proc. Natl. Acad. Sci. USA* **2012**, *109*, 1341–1346. [CrossRef] [PubMed]
7. Barretto, A.G.O.P.; Berndes, G.; Sparovek, G.; Wirsenius, S. Agricultural intensification in Brazil and its effects on land-use patterns: An analysis of the 1975–2006 period. *Glob. Chang. Biol.* **2013**, *19*, 1804–1815. [CrossRef] [PubMed]
8. Dias, L.C.P.; Pimenta, F.M.; Santos, A.B.; Costa, M.H.; Ladle, R.J. Patterns of land use, extensification, and intensification of Brazilian agriculture. *Glob. Chang. Biol.* **2016**, *22*, 2887–2903. [CrossRef] [PubMed]
9. De Oliveira, J.C.; Trabaquini, K.; Epiphanio, J.C.N.; Formaggio, A.R.; Galvão, L.S.; Adami, M. Analysis of agricultural intensification in a basin with remote sensing data. *GIScience Remote Sens.* **2014**, *51*, 253–268. [CrossRef]
10. Bustamante, M.M.C.; Nobre, C.A.; Smeraldi, R.; Aguiar, A.P.D.; Barioni, L.G.; Ferreira, L.G.; Longo, K.; May, P.; Pinto, A.S.; Ometto, J.P.H.B. Estimating greenhouse gas emissions from cattle raising in Brazil. *Clim. Chang.* **2012**, *115*, 559–577. [CrossRef]
11. Castanheira, É.G.; Freire, F. Greenhouse gas assessment of soybean production: Implications of land use change and different cultivation systems. *J. Clean. Prod.* **2013**, *54*, 49–60. [CrossRef]
12. Herrero, M.; Henderson, B.; Havlík, P.; Thornton, P.K.; Conant, R.T.; Smith, P.; Wirsenius, S.; Hristov, A.N.; Gerber, P.; Gill, M.; et al. Greenhouse gas mitigation potentials in the livestock sector. *Nat. Clim. Chang.* **2016**, *6*, 452–461. [CrossRef]
13. Parente, L.; Ferreira, L.; Faria, A.; Nogueira, S.; Araújo, F.; Teixeira, L.; Hagen, S. Monitoring the brazilian pasturelands: A new mapping approach based on the landsat 8 spectral and temporal domains. *Int. J. Appl. Earth Obs. Geoinform.* **2017**. [CrossRef]
14. Lambin, E.F.; Gibbs, H.K.; Ferreira, L.; Grau, R.; Mayaux, P.; Meyfroidt, P.; Morton, D.C.; Rudel, T.K.; Gasparri, I.; Munger, J. Estimating the world's potentially available cropland using a bottom-up approach. *Glob. Environ. Chang.* **2013**, *23*, 892–901. [CrossRef]
15. IBGE Pesquisa Pecuária Municipal. Available online: https://sidra.ibge.gov.br/pesquisa/ppm/quadros/brasil/2016 (accessed on 10 February 2018).
16. MMA. *Projeto de Conservação e Utilização Sustentável da Diversidade Biológica Brasileira: Relatório de Atividades*; Ministério do Meio Ambiente: Brasília, Brazil, 2002.
17. Lapola, D.M.; Martinelli, L.A.; Peres, C.A.; Ometto, J.P.H.B.; Ferreira, M.E.; Nobre, C.A.; Aguiar, A.P.D.; Bustamante, M.M.C.; Cardoso, M.F.; Costa, M.H.; et al. Pervasive transition of the Brazilian land-use system. *Nat. Clim. Chang.* **2014**, *4*, 27–35. [CrossRef]
18. Phalan, B.; Green, R.E.; Dicks, L.V.; Dotta, G.; Feniuk, C.; Lamb, A.; Strassburg, B.B.N.; Williams, D.R.; zu Ermgassen, E.K.H.J.; Balmford, A. CONSERVATION ECOLOGY. How can higher-yield farming help to spare nature? *Science* **2016**, *351*, 450–451. [CrossRef] [PubMed]

19. Justice, C.O.; Vermote, E.; Townshend, J.R.G.; Defries, R.; Roy, D.P.; Hall, D.K.; Salomonson, V.V.; Privette, J.L.; Riggs, G.; Strahler, A.; et al. The Moderate Resolution Imaging Spectroradiometer (MODIS): Land remote sensing for global change research. *IEEE Trans. Geosci. Remote Sens.* **1998**, *36*, 1228–1249. [CrossRef]
20. Friedl, M.; McIver, D.; Hodges, J.C.; Zhang, X.; Muchoney, D.; Strahler, A.; Woodcock, C.; Gopal, S.; Schneider, A.; Cooper, A.; et al. Global land cover mapping from MODIS: Algorithms and early results. *Remote Sens. Environ.* **2002**, *83*, 287–302. [CrossRef]
21. Maus, V.; Câmara, G.; Cartaxo, R.; Sanchez, A.; Ramos, F.M.; Ribeiro, G.Q. A Time-Weighted Dynamic Time Warping method for land use and land cover mapping. *IEEE J. Sel. Top. Appl. Earth Obs. Remote Sens.* **2015**, *20*, 1–10. [CrossRef]
22. Townshend, J.R.; Justice, C.O. Towards operational monitoring of terrestrial systems by moderate-resolution remote sensing. *Remote Sens. Environ.* **2002**, *83*, 351–359. [CrossRef]
23. Friedl, M.A.; Sulla-Menashe, D.; Tan, B.; Schneider, A.; Ramankutty, N.; Sibley, A.; Huang, X. MODIS Collection 5 global land cover: Algorithm refinements and characterization of new datasets. *Remote Sens. Environ.* **2010**, *114*, 168–182. [CrossRef]
24. Huete, A.; Didan, K.; Miura, T.; Rodriguez, E.; Gao, X.; Ferreira, L. Overview of the radiometric and biophysical performance of the MODIS vegetation indices. *Remote Sens. Environ.* **2002**, *83*, 195–213. [CrossRef]
25. Peng, D.; Zhang, X.; Zhang, B.; Liu, L.; Liu, X.; Huete, A.R.; Huang, W.; Wang, S.; Luo, S.; Zhang, X.; et al. Scaling effects on spring phenology detections from MODIS data at multiple spatial resolutions over the contiguous United States. *ISPRS J. Photogramm. Remote Sens.* **2017**, *132*, 185–198. [CrossRef]
26. Didan, K.; Munoz, A.B.; Solano, R.; Huete, A. *MODIS Vegetation Index User's Guide (MOD13 Series)*; Vegetation Index and Phenology Lab, The University of Arizona: Tucson City, AZ, USA, 2015.
27. Jiang, Z.; Huete, A.R.; Didan, K.; Miura, T. Development of a two-band enhanced vegetation index without a blue band. *Remote Sens. Environ.* **2008**, *112*, 3833–3845. [CrossRef]
28. Karnieli, A.; Kaufman, Y.J.; Remer, L.; Wald, A. AFRI—Aerosol free vegetation index. *Remote Sens. Environ.* **2001**, *77*, 10–21. [CrossRef]
29. Breiman, L. Random forests. *Mach. Learn.* **2001**, *45*, 5–32. [CrossRef]
30. Ferreira, L.G.; Sano, E.E.; Fernandez, L.E.; Araújo, F.M. Biophysical characteristics and fire occurrence of cultivated pastures in the Brazilian savanna observed by moderate resolution satellite data. *Int. J. Remote Sens.* **2013**, *34*, 154–167. [CrossRef]
31. Ferreira, L.; Fernandez, L.; Sano, E.; Field, C.; Sousa, S.; Arantes, A.; Araújo, F. Biophysical Properties of Cultivated Pastures in the Brazilian Savanna Biome: An Analysis in the Spatial-Temporal Domains Based on Ground and Satellite Data. *Remote Sens.* **2013**, *5*, 307–326. [CrossRef]
32. Aguiar, D.; Mello, M.; Nogueira, S.; Gonçalves, F.; Adami, M.; Rudorff, B. MODIS Time Series to Detect Anthropogenic Interventions and Degradation Processes in Tropical Pasture. *Remote Sens.* **2017**, *9*, 73. [CrossRef]
33. Gorelick, N.; Hancher, M.; Dixon, M.; Ilyushchenko, S.; Thau, D.; Moore, R. Google Earth Engine: Planetary-scale geospatial analysis for everyone. *Remote Sens. Environ.* **2017**, *202*, 18–27. [CrossRef]
34. IBGE. *Base Cartográfica Contínua do Brasil, ao Milionésimo—BCIM*; Instituto Brasileiro de Geografia e Estatística: Rio de Janeiro, Brazil, 2016.
35. Pekel, J.-F.; Cottam, A.; Gorelick, N.; Belward, A.S. High-resolution mapping of global surface water and its long-term changes. *Nature* **2016**, *540*, 418–422. [CrossRef] [PubMed]
36. Jiang, Y.; Sun, M.; Yang, C. A Generic Framework for Using Multi-Dimensional Earth Observation Data in GIS. *Remote Sens.* **2016**, *8*, 382. [CrossRef]
37. Lohr, S. Sampling: Design and Analysis. *J. Chem. Inf. Model.* **2000**, *596*. [CrossRef]
38. Nogueira, S.; Parente, L.; Ferreira, L. Temporal Visual Inspection: Uma ferramenta destinada à inspeção visual de pontos em séries históricas de imagens de sensoriamento remoto. In *XXVII Congresso Brasileiro de Cartografia*; Instituto Brasileiro de Geografia e Estatística: Rio de Janeiro, Brazil, 2017.
39. IBGE. *Censo Agropecuário*; Instituto Brasileiro de Geografia e Estatística: Rio de Janeiro, Brazil, 2006.
40. Noojipady, P.; Morton, C.D.; Macedo, N.M.; Victoria, C.D.; Huang, C.; Gibbs, K.H.; Bolfe, L.E. Forest carbon emissions from cropland expansion in the Brazilian Cerrado biome. *Environ. Res. Lett.* **2017**, *12*, 25004. [CrossRef]

41. Barr, K.J.; Babcock, B.A.; Carriquiry, M.A.; Nassar, A.M.; Harfuch, L. Agricultural Land Elasticities in the United States and Brazil. *Appl. Econ. Perspect. Policy* **2011**, *33*, 449–462. [CrossRef]
42. Alkimim, A.; Sparovek, G.; Clarke, K.C. Converting Brazil's pastures to cropland: An alternative way to meet sugarcane demand and to spare forestlands. *Appl. Geogr.* **2015**, *62*, 75–84. [CrossRef]
43. Smith, P.; Gregory, P.J.; van Vuuren, D.; Obersteiner, M.; Havlík, P.; Rounsevell, M.; Woods, J.; Stehfest, E.; Bellarby, J. Competition for land. *Philos. Trans. R. Soc. Lond. B Biol. Sci.* **2010**, *365*, 2941–2957. [CrossRef] [PubMed]
44. DNIT Atlas e Mapas. Available online: http://www.dnit.gov.br/mapas-multimodais/shapefiles (accessed on 17 February 2018).
45. LAPIG Matadouros e Frigoríficos do Brasil. Available online: http://maps.lapig.iesa.ufg.br/?layers=pa_br_matadouros_e_frigorificos_na_2017_lapig (accessed on 11 March 2018).
46. Bowman, M.S.; Soares-Filho, B.S.; Merry, F.D.; Nepstad, D.C.; Rodrigues, H.; Almeida, O.T. Persistence of cattle ranching in the Brazilian Amazon: A spatial analysis of the rationale for beef production. *Land Use Policy* **2012**, *29*, 558–568. [CrossRef]
47. Ferro, A.B.; Castro, E.R. De Determinantes dos preços de terras no Brasil: Uma análise de região de fronteira agrícola e áreas tradicionais. *Rev. Econ. Sociol. Rural* **2013**, *51*, 591–609. [CrossRef]
48. Fearnside, P.M. Brazil's Cuiabá-Santarém (BR-163) Highway: The Environmental Cost of Paving a Soybean Corridor through the Amazon. *Environ. Manag.* **2007**, *39*, 601–614. [CrossRef] [PubMed]
49. Pokorny, B.; de Jong, W.; Godar, J.; Pacheco, P.; Johnson, J. From large to small: Reorienting rural development policies in response to climate change, food security and poverty. *For. Policy Econ.* **2013**, *36*, 52–59. [CrossRef]
50. Mullan, K.; Sills, E.; Pattanayak, S.K.; Caviglia-Harris, J. Converting Forests to Farms: The Economic Benefits of Clearing Forests in Agricultural Settlements in the Amazon. *Environ. Resour. Econ.* **2017**, 1–29. [CrossRef]
51. UNDP. *Human Development Report 2016: Human Development for Everyone*; United Nations Development Programme: New York, NY, USA, 2016.
52. Salame, C.W.; Queiroz, J.C.B.; de Miranda Rocha, G.; Amin, M.M.; da Rocha, E.P. Use of spatial regression models in the analysis of burnings and deforestation occurrences in forest region, Amazon, Brazil. *Environ. Earth Sci.* **2016**, *75*, 274. [CrossRef]
53. De Castro Solar, R.R.; Barlow, J.; Andersen, A.N.; Schoereder, J.H.; Berenguer, E.; Ferreira, J.N.; Gardner, T.A. Biodiversity consequences of land-use change and forest disturbance in the Amazon: A multi-scale assessment using ant communities. *Biol. Conserv.* **2016**, *197*, 98–107. [CrossRef]
54. Rudorff, B.F.T.; Aguiar, D.A.; Silva, W.F.; Sugawara, L.M.; Adami, M.; Moreira, M.A. Studies on the Rapid Expansion of Sugarcane for Ethanol Production in São Paulo State (Brazil) Using Landsat Data. *Remote Sens.* **2010**, *2*, 1057–1076. [CrossRef]
55. Silva, A.A.; Miziara, F. Avanço do Setor Sucroalcooleiro e Expansão da Fronteira Agrícola em Goiás. *Pesqui. Agropecu. Trop.* **2011**, *41*, 399–407. [CrossRef]
56. Perpetua, G.M.; Thomaz Junior, A. Dinâmica Geográfica da Mobilidade do Capital na Produção de Celulose e Papel em Três Lagoas (MS). *Rev. Anpege* **2013**, *9*, 55–69. [CrossRef]
57. Pedrosa, B.C.; de Souza, T.C.L.; Turetta, A.P.D.; da Costa Coutinho, H.L. Feasibility Assessment of Sugarcane Expansion in Southwest Goiás, Brazil Based on the GIS Technology. *J. Geogr. Inf. Syst.* **2016**, *8*, 149–162. [CrossRef]
58. Richards, P.; Pellegrina, H.; Van Wey, L.; Spera, S. Soybean Development: The Impact of a Decade of Agricultural Change on Urban and Economic Growth in Mato Grosso, Brazil. *PLoS ONE* **2015**, *10*, e0122510. [CrossRef] [PubMed]
59. Latawiec, A.E.; Strassburg, B.B.N.; Valentim, J.F.; Ramos, F.; Alves-Pinto, H.N. Intensification of cattle ranching production systems: Socioeconomic and environmental synergies and risks in Brazil. *Animal* **2014**, *8*, 1255–1263. [CrossRef] [PubMed]
60. Gil, J.; Siebold, M.; Berger, T. Adoption and development of integrated crop–livestock–forestry systems in Mato Grosso, Brazil. *Agric. Ecosyst. Environ.* **2015**, *199*, 394–406. [CrossRef]

Technical Note

Towards Global-Scale Seagrass Mapping and Monitoring Using Sentinel-2 on Google Earth Engine: The Case Study of the Aegean and Ionian Seas

Dimosthenis Traganos [1,*], Bharat Aggarwal [1], Dimitris Poursanidis [2], Konstantinos Topouzelis [3], Nektarios Chrysoulakis [2] and Peter Reinartz [4]

1 German Aerospace Center (DLR), Remote Sensing Technology Institute, Rutherfordstraße 2, 12489 Berlin, Germany; Bharat.Aggarwal@dlr.de
2 Foundation for Research and Technology—Hellas (FORTH), Institute of Applied and Computational Mathematics, N. Plastira 100, Vassilika Vouton, 70013 Heraklion, Greece; dpoursanidis@iacm.forth.gr (D.P.); zedd2@iacm.forth.gr (N.C.)
3 Department of Marine Science, University of the Aegean, University Hill, 81100 Mytilene, Greece; topouzelis@marine.aegean.gr
4 German Aerospace Center (DLR), Earth Observation Center (EOC), 82234 Weßling, Germany; peter.reinartz@dlr.de
* Correspondence: dimosthenis.traganos@dlr.de; Tel.: +49-(0)30-6705-5545

Received: 29 June 2018; Accepted: 2 August 2018; Published: 5 August 2018

Abstract: Seagrasses are traversing the epoch of intense anthropogenic impacts that significantly decrease their coverage and invaluable ecosystem services, necessitating accurate and adaptable, global-scale mapping and monitoring solutions. Here, we combine the cloud computing power of Google Earth Engine with the freely available Copernicus Sentinel-2 multispectral image archive, image composition, and machine learning approaches to develop a methodological workflow for large-scale, high spatiotemporal mapping and monitoring of seagrass habitats. The present workflow can be easily tuned to space, time and data input; here, we show its potential, mapping 2510.1 km^2 of *P. oceanica* seagrasses in an area of 40,951 km^2 between 0 and 40 m of depth in the Aegean and Ionian Seas (Greek territorial waters) after applying support vector machines to a composite of 1045 Sentinel-2 tiles at 10-m resolution. The overall accuracy of *P. oceanica* seagrass habitats features an overall accuracy of 72% following validation by an independent field data set to reduce bias. We envision that the introduced flexible, time- and cost-efficient cloud-based chain will provide the crucial seasonal to interannual baseline mapping and monitoring of seagrass ecosystems in global scale, resolving gain and loss trends and assisting coastal conservation, management planning, and ultimately climate change mitigation.

Keywords: seagrass; habitat mapping; image composition; machine learning; support vector machines; Google Earth Engine; Sentinel-2; Aegean; Ionian; global scale

---

1. Introduction

Seagrasses are marine flowering plants that hold important ecological roles in coastal ecosystems since they can form extensive meadows that support high biodiversity. Their habitats are found in temperate and tropical ranges [1]; the temperate bioregions include the temperate North Atlantic, North Pacific and Southern Oceans, and the Mediterranean. The global species diversity of seagrasses is low, but species can extend over thousands of kilometers of coastline while spanning depths down to 50 m. Seagrasses form a critical marine ecosystem for carbon storage, fisheries production, sediment accumulation, and stabilization [2]. They contribute to the function of ocean ecosystems by providing an important nursery area for many species that support offshore fisheries and for adjacent habitats such as salt marshes, shellfish beds, coral reefs, and mangrove forests [3]. Seagrass ecosystems are critical for threatened species—i.e., sirenians (dugong and manatee), sea turtles, and seahorses—all are perceived to have high cultural, aesthetic, and intrinsic values. The ecosystem functions of the seagrass meadows include: maintenance of genetic variability, resilience of the coastal environment through protection from erosion, and carbon sequestration by removing carbon dioxide from the atmosphere and binding it as organic matter [4]. Their high productivity attributes them a disproportionate influence on oceanwide primary productivity, typically producing considerably more organic carbon than the seagrass ecosystem requires [5]. Carbon storage by seagrasses is essentially an effective removal of carbon dioxide from the ocean–atmosphere system which plays a significant role in the amelioration of climate change impacts [6].

The existence of seagrass datasets at regional/global scale can support resource management, strengthen decision-making, and facilitate tracking of progress towards global conservation targets set by multilateral environmental agreements, such as the Aichi Biodiversity Targets of the United Nations' (UN's) Strategic Plan for Biodiversity 2011–2020, the Ramsar Convention, and the Sustainable Development Goals (SDG) of the UN's 2030 Agenda for Sustainable Development—particularly Goal 14: "Conserve and sustainably use the oceans, seas and marine resources for sustainable development" of the UN SDG 2030 [7]. Moreover, seagrass habitats are protected at the regional/continental scales through legislation. The European Union's Habitat Directive (92/43/CEE) includes Posidonia oceanica beds among priority habitats (Habitat Type 1120: *P. oceanica* beds—Posidonion oceanicae). Seagrass meadows have a dedicated action plan within the framework of the Barcelona Convention, under the "Protocol concerning Specially Protected Areas and Biological Diversity in the Mediterranean". More recently, the Marine Strategy Framework Directive (MFSD) (2008/56/EC) has established a framework according to which each Member States shall take the necessary measures to achieve or maintain "Good Environmental Status" in the marine environment.

In parallel to the mainly anthropogenic disturbance on seagrasses, these ecosystems feature a very slow growth rate that varies between the species. During the 20th century, and more especially since the 1940s, the loss of seagrass beds has been observed in several regions due to the impacts of industry, construction, boating, overfishing, dredging, mining, algal blooms from eutrophication, and climate-change-induced rising sea levels [8]. This regression has been particularly significant near major urbanized zones and port facilities [9]. A noteworthy drawback for regional- and national-wide seagrass mapping is the absence of in situ data, sufficient for use in mapping activities using satellite remote sensing tools and methods. Open access in situ datasets exist that are suitable for image analysis, but these are site-specific, do not cover large areas or countries, and/or are outdated. A search in PANGAEA data repository with the keyword "seagrass" provides 603 datasets with the majority coming from the Pacific Ocean (484 records), followed by the Atlantic Ocean (40 records) and the Mediterranean Sea (20 records). From all these records, only ~45 datasets provide potentially suitable, spatial information for use in remote sensing image analysis. We also searched in various other sources that provide simplified seagrass distribution polygons not suitable for image classification as they are not real in situ data.

The investigation of seagrass distribution at a global scale is a complex and challenging task due to the wide range of species diversity patterns and areas where seagrasses are undocumented, but also the fact that seagrass habitats are ever-changing, as is water quality/water column clarity due to sediment processes [10]. Remote sensing has been previously used for seagrass mapping in large areas [11–14]. These studies focused on submerged habitats and have been performed during the last 10 years using mainly Landsat imagery at regional/local scales with the traditional approach of image selection and classification at local infrastructures. Nowadays, in the domain of Earth Observation-derived big data, cloud computing infrastructures can provide the platform for the analysis of multitemporal satellite data of medium to high spatial resolution. As such, Copernicus Sentinel-2 (S2) and USGS/NASA

Landsat-8 have been used in cloud computing environments such as the Google Earth Engine (GEE) platform (http://earthengine.google.org). By harnessing machine learning algorithms within the GEE environment, scientists have mapped the extent and status of global forests [15], the distribution of croplands in Africa [16], and the occurrence of the global surface water [17]. These global studies provide insights into easy-to-be-identified elements of the Earth's surface; yet, they performed the validation of the products in limited areas due to the absence of high quality and updated in situ data. Focusing on the aquatic context, the preprocessing steps for the derivation of satellite-derived bathymetry have been also implemented in GEE [18].

Our main objective through this paper is to develop and establish a cloud-based, scalable workflow for the mapping and monitoring of seagrass habitats globally. We will present all necessary preprocessing, processing and analysis tools developed within the GEE environment by exploiting the open and free access dataset of Copernicus S2 constellation within the GEE environment. A secondary aim is to highlight the importance of the synergy between the S2 image archive and the GEE platform for the global monitoring of seagrass meadows. We showcase both aims using as case studies the Aegean and Ionian Seas (Greece, eastern Mediterranean), where seagrass meadows cover large proportions of the coastal zone. Containing simple, widely used and comprehensive algorithms, the established workflow can be scaled up to map other subtidal, coastal habitats as well as coral reefs, kelp forests etc., elsewhere.

2. Materials and Methods

2.1. Study Site

The study area includes the Greek territorial waters—i.e., Ionian Sea and Greek part of the Aegean Sea—covering a total surface of 40,951 km^2 (Figure 1). We define this area on the basis of an edited 3-km buffer zone along the Greek coastline of approximately 18,000 km, following the results of [13] (see Section 2.3.2). The Aegean Sea features an intricate geomorphology that reflects past geologic history and recent geodynamic processes. Shallow shelves, deep basins, and troughs alternate throughout the buffer-defined area whose deepest point is ~2500 m (deepest area in the Aegean trench is ~5000 m). It has also a distinctive insular character with more than 1400 islands or islets, while its extensive coastline consists of several landforms, including sandy beaches, rocky shores, cliffs, coastal lagoons and deltaic systems, as well as a notable variety of coastal and marine habitat types. On the other hand, the Ionian Sea is considered an oligotrophic area, based both on low nutrient concentrations and primary production. Biological data reflect a very oligotrophic area, dominated by a microbial food web, where new production mostly derives from limited events in space and time, mainly driven by climatological factors generating mesoscale instabilities. In terms of biology, the Ionian Sea is influenced by the Adriatic water in the northern part of the subbasin, with higher phytoplankton biomass (particularly diatoms), while water of Atlantic origin makes up its southern part with the exception of the whole eastern side [19,20]. Seagrass meadows exist in protected bays and gulfs while their mean maximum depths of presence are between 25 and 35 m depending on the local conditions [21].

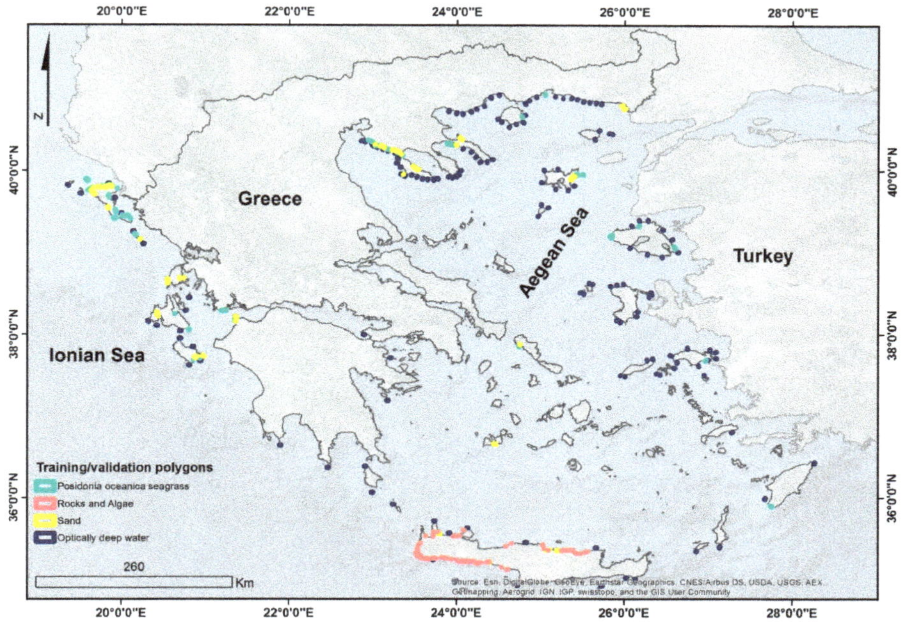

Figure 1. Geographical location of survey site and training polygons of the herein considered classes. All polygons are in Geographic Coordinate System (CGS) World Geodetic System (WGS) 84 World Geodetic System.

2.2. *Satellite Data*

We use Copernicus Sentinel-2 Level-1C (L1C) top of atmosphere (TOA) reflectance satellite data, the standard S2 archive in GEE. The available data extends from 23 June 2015 (date of launch of Sentinel-2A) to today with a 5-day temporal resolution (with the use of Sentinel-2B). For the present study, we choose a period between 1 September and 1 October 2017, which satisfies availability of both S2-A and S2-B imagery, and more importantly, a better-stratified water column in the study areas of Aegean and Ionian seas [18]. Our data input is the seven S2 bands: b1—coastal aerosol, b2—blue, b3—green, b4—red, b8—NIR, b11—SWIR1, and the QA60 band, the bitmask band which contains information on cloud mask. Bands b1 and QA60 are in 60-m spatial resolution, while b11 is in 20-m spatial resolution. GEE reprojects all to the 10-m native resolution of the b2, b3, b4, and b8. In total, 1045 S2 tiles—100 × 100 km^2 sub-images—compose our initial image dataset. All available datasets, from satellite images to field and auxiliary data are projected in the GCS WGS84 World Geodetic System.

2.3. *Field Data*

2.3.1. Training and Validation Data

We manually digitized all training data on very high spatial resolution images using the ArcGIS World Imagery base map (<60 cm pixel) (Figure 1). To ensure data quality, we selected areas in relative shallow waters where seagrass was easily interpreted. The digitization took place in areas where we have a high level of knowledge on the seabed cover and composition due to past and ongoing fieldwork activities. The nonseagrass class contained rocky, sandy, and deep-sea areas. To ensure consistency across all classes, we implement 4 × 4 homogeneous polygons (16 S2 pixels). In ArcGIS

Desktop, the digitization tool allows to design polygon data with specific dimension and size. Thus, the size of each polygon is the same and the homogeneity is related to the seabed cover as this is identified by experienced image interpreters. In total, we used 1457 homogeneous polygons (23,312 pixels) for two classes (seagrass/nonseagrass) (Figure 2; Table 1). The homogeneity of all polygons is vital because their absence may cause misclassification. The validation data consists of 322 independent field data points based on unpublished data provided by terraSolutions m.e.r. (http://www.terrasolutions.eu/) to reduce bias.

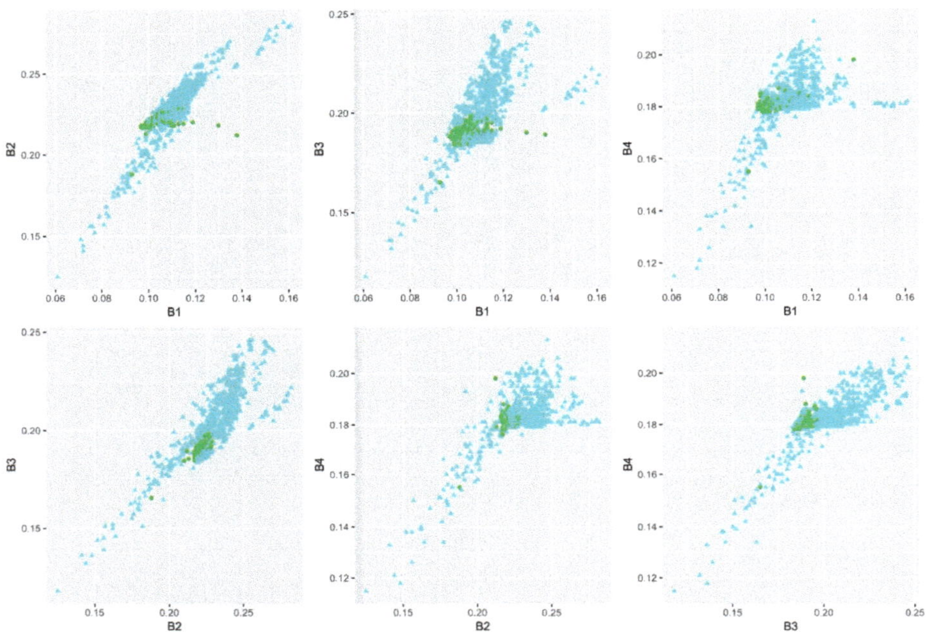

Figure 2. Scatter plots of the first four, sunglint-corrected Sentinel-2 bands depicting waveband reflectivity of the herein 1457 polygons for the whole extent of the study area. Seagrasses are in green circles and nonseagrasses are in light blue triangles.

Table 1. Number of implemented polygons and pixels per class for our ~40,951 km^2 survey site.

Class	Polygons	Pixels	%
Seagrass	329	5264	22.6
Nonseagrass	1128	18,048	77.4
Sum	1457	23,312	100

2.3.2. Auxiliary Data

We utilize two additional auxiliary datasets to aid both the time (thus computing) efficiency of our processing chain and the distillation of its results. First, to mask out both land and deeper waters, we edit a shapefile of the Greek coastline at scale of 1:90,000 [22] by the Hellenic Navy Hydrographic Service (HNHS) in a two-step way; (a) we create a buffer of 3 km to encompass the whole optically shallow extent (where there is remote sensing signal from the seabed), (b) we manually edit the buffer of 3 km to include the deeper and/or larger meadows (according to the seagrass polygons of [13]) and to delete all the vertices over land. The resulted coastal area comprises our survey site of a total area of ~40,951 km^2. In addition, we use the bathymetry of the Aegean Sea for the post-classification stage (see Section 2.4.3). The Aegean Sea bathymetry depth zones (5-m intervals until 50-m depth) are fused products of HNHS, EMODnet (the European Marine Observation and Data Network), and in situ data collected during the MARItime Spatial planning for the protection and Conservation of the biodiversity in the Aegean sea (MARISCA) project [19].

2.4. Methodology

In addition to the easy and parallel access to the satellite image archives, GEE offers quick and adaptable computational tools for remotely sensed data processing and analysis. We exploit GEE tools to build our methodological chain, which is divided into three parts: (a) preclassification, (b) classification, (c) post-classification. Figure 3 displays this chain, while Figure 4 depicts its various successive stages.

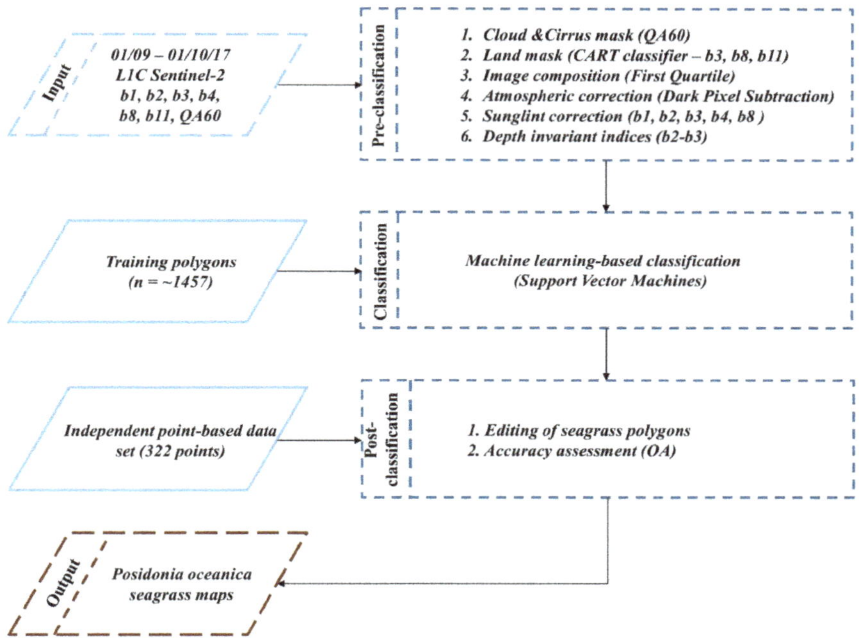

Figure 3. Methodological workflow of the present study within Google Earth Engine. OA denotes overall accuracy. In the present study we did not implement step 2 (due to the use of a coastline buffer), however, we include it as it is an important component of the methodological chain.

2.4.1. Preclassification

Our preclassification part consists of six steps:

1. Cloud mask: We use the QA60 bitmask band to mask opaque and cirrus clouds and scale S2 L1C TOA images by 10,000 (Figure 4a,b).
2. Land mask: Although here we utilize the buffered coastline shapefile of Greek waters to mask out terrestrial Greece, we include a classification and regression tree (CART) classifier [23] in the GEE code that future users could employ to mask out their terrestrial part. The classifier is applied on a b3-b8-b11 composite and the user should train it with relevant pixels over land and water.

3. Image composition: We apply image composition which yields a new pseudo-image composite whose pixels are the first quartile (Q1) of the median values of the cloud corrected and masked for land images of step 2. The purpose of this approach is to decrease noncorrected image artefacts by the previous steps.
4. Atmospheric correction: We implement a modified dark pixel subtraction method following [24] to empirically address path radiance and noise in all sbands; this method subtracts the average reflectance and two standard deviations of optically deep water (>40 m) (Figure 4b).
5. Sunglint correction: We further correct the atmospherically corrected image composites with the sunglint correction algorithm of [25]. Following a user-defined set of pixels of variable sunglint intensity, the algorithms equals the corrected for sunglint composite to the initial first quartile composite minus the product of the regression slope of b8 against b1-b4 and the difference between b8 and its minimum value (Figure 4c).
6. Depth invariant indices calculation: To compensate the influence of variable depth on seabed habitats, we derive the depth invariant indices [26,27] for each pair of bands with reasonable water penetration (b1-b2, b2-b3, b1-b3) with the statistical analysis of [28]. Prior to the machine learning-based classification, we apply a 3 $\times$ 3 low pass filter in the depth invariant as well as the sunglint-corrected input to minimize remaining noise over the optically deep water extent which would have caused misclassified seagrass pixels otherwise (Figure 4d).

2.4.2. Classification

Although we experimented with three machine learning classifiers, support vector machines (SVM) [29], random forests (RF) [30], and CART [23], we end up using only SVM due to their better yielded classification output, both qualitatively and quantitatively (Figure 4e). Based on the statistical learning theory of [29], SVM solve linear and nonlinear classification problems by fitting a separating hyperplane to the training data of the studied classes; they take their name from the support vectors, namely the points closest to the hyperplane—the only ones that ultimately define it. A small number of studies have utilized SVM to map optically shallow habitats [31–34]. Here, we run SVM with a Gaussian radial basis function kernel (RBF), a parameter g, the width of the RBF, of 1000, and a regularization parameter, C, which governs the degree of acceptable misclassification, of 100. We empirically select the adequate pair values for g and C based on experiments that we run setting their range between a minimum of 0.01 and a maximum of 1000 using a multiplier of 10. The input to the classifier is the sunglint-corrected S2 composite of b1, b2, b3, b4 and the depth invariant index b2-b3.

2.4.3. Post-Classification

The post-classification part consists of two components: the editing of seagrass polygons due to misclassified pixels as seagrass in deep water and the accuracy assessment of the these machine learning derived edited seagrass polygons. The accuracy assessment employs an independent validation data set (unpublished data from terraSolutions m.e.r.) consisting of 322 data points to reduce general bias. We report the overall accuracy of seagrass habitats in the extent of the Aegean and Ionian seas which is the ratio of the entire number of correctly classified pixels to the total number of validation pixels. The post-classification correction of seagrass pixels over deep water serves a better visualization of the classification output, but more significantly, it decreases the overprediction tendency of the classification results, namely seagrass area in the Aegean and Ionian seas; a single misclassified 10-m S2 pixel as *P. oceanica* seagrass would cause an overprediction of 100 m^2.

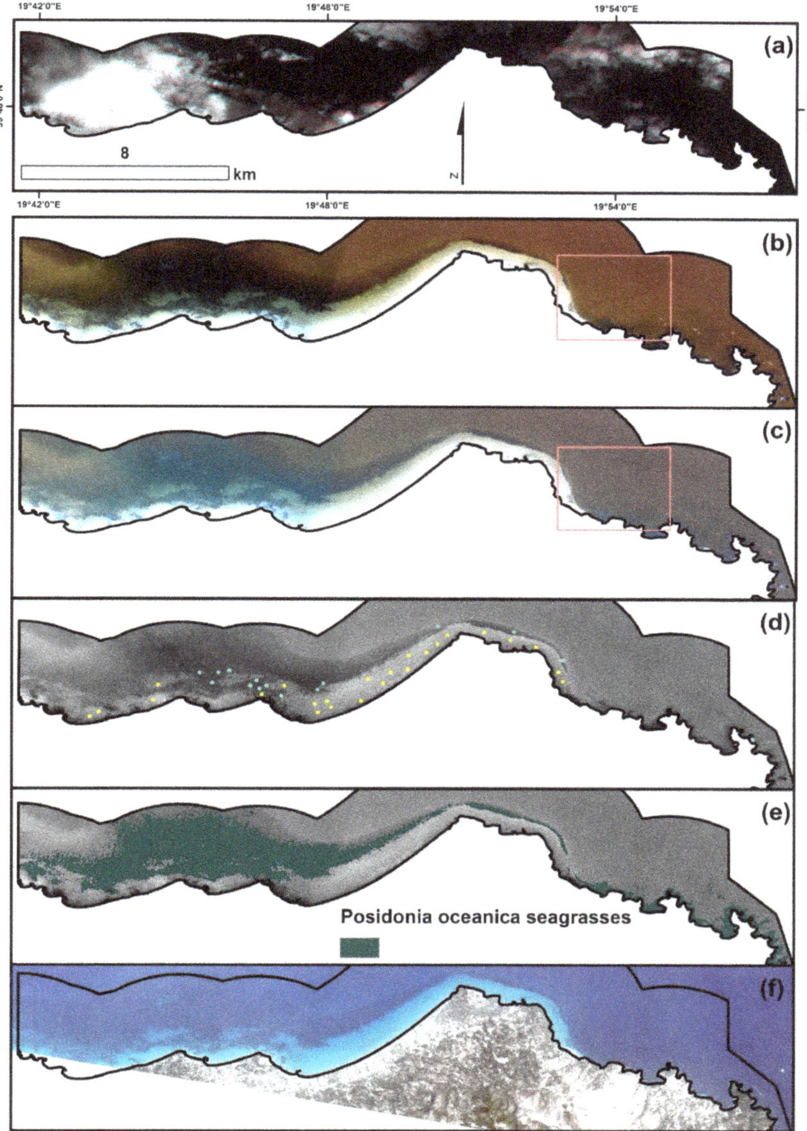

Figure 4. Successive stages of the developed workflow through the resulting false-color (b1-b2-b3) Sentinel-2 composites. (**a**) Initial S2, (**b**) cloud- and atmospherically-corrected, (**c**) sunglint-corrected, (**d**) depth invariant index b2-b3, (**e**) support vector machines-based classified product draped over the b2-b3 depth invariant index layer, (**f**) PlanetScope surface reflectance product (as imaged by Planet's Doves) in natural color for high resolution reference (3 m) (ID: 20170828_084352_100e/20170828_084352_100e_3B_AnalyticMS_SR). The pink squares indicate sunglint presence in (**b**) and its correction in (**c**). The green and yellow polygons show employed seagrass and sand pixels in the machine learning classification. All panels are in GCS WGS84 World Geodetic System.

3. Results

3.1. Preclassification

Regarding the preclassification steps, r^2 of regression between sunglint-polluted b8 and b1-b4 composites are b1: 0.3, b2: 0.3, b3: 0.39, and b4: 0.39. As regards to the only herein employed depth invariant index of b2-b3 bands, the ratio of water attenuation coefficient is 0.57.

3.2. Classification

Our methodological workflow reveals that the seagrass area of the Greek Seas is 2510.1 km^2; 1885 km^2 in the Aegean Sea and 625 km^2 in the Ionian Sea (Figure 5). As regards to individual geographical areas, the ones with the largest seagrass area (and maximum observed depth, where available, in parentheses) are Limnos (Figure 5d) with 254 km^2 (40 m), NW Peloponissos (Figure 5c) with 99 km^2, Corfu with 90 km^2, Crete with 70 km^2 (40 m), Thasos (Figure 5b) with 53 km^2 (25 m), and the Thermaikos Gulf (Figure 5a) with 49 km^2 (20 m). Across both seas, according to the employed auxiliary depth information, the depth of seagrass habitats is between 0 and 40 m.

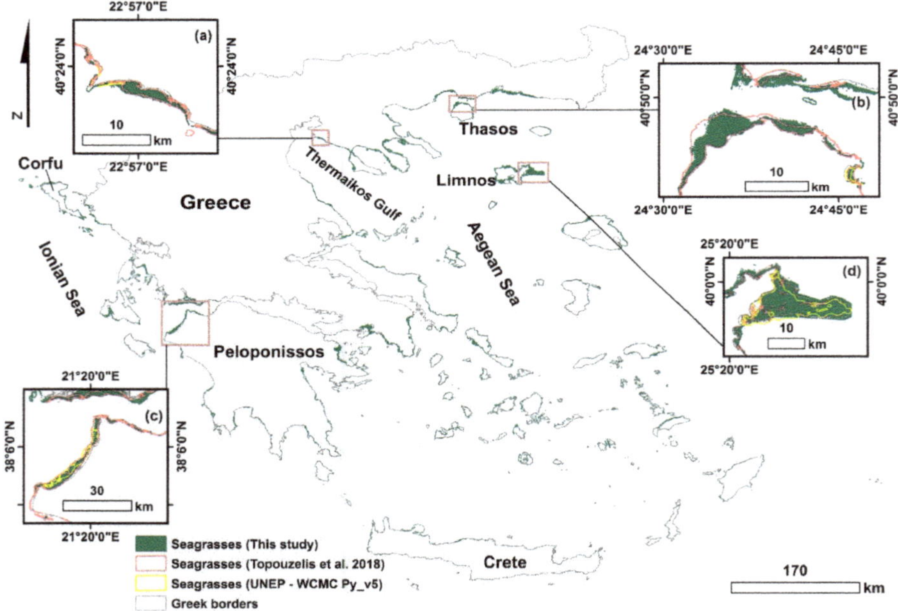

Figure 5. Distribution of seagrasses in the Greek Seas. (**a**) Thermaikos Gulf, (**b**) Thasos Island, (**c**) NE Peloponissos, (**d**) Limnos Island. Inset maps contain results from [13,35] for reference and further validation of our results. All panels are in GCS WGS84 World Geodetic System.

3.3. Post-Classification

The observed overall accuracy of *P. oceanica* seagrass in both the Aegean and Ionian seas is 72% based on the independent field data set.

4. Discussion

4.1. On Global Mapping and Monitoring of Seagrasses, and the Results of the Present Case Study

The exponential increase in cloud-based computing power, machine learning algorithms, and freely available satellite image archives has rendered the vision of global-scale mapping and monitoring of seagrasses and their neighboring habitats more feasible than ever before. However, seagrass environments vary in regards to species composition and abundance, but also water clarity (secchi depth) which will influence the capability to apply approach globally. At present, while there are tangible efforts for the global monitoring of coral habitats [36], similar efforts focusing on seagrass ecosystems have been yet not envisioned.

Here, we have developed a methodological workflow within Google Earth Engine which employs a plethora of universally used algorithms in coastal aquatic habitat remote sensing along with image composition and machine learning that could potentially be applied to map and monitor seagrasses globally. We demonstrate its power along with its issues in the Greek Seas, namely the Aegean and Ionian Sea; in a total coastline extent of 40,951 km^2, we map 2510.1 km^2 of Greek seagrasses (Posidonia oceanica species) between 0 and 40 m of depth applying Support Vector Machines in a pixel-based fashion on 1045 Sentinel-2 tiles. In comparison to existing mapping efforts and known distribution of seagrasses in our study site, our area findings are ~4.2% less than the respective coverage estimations of [13] (2619.3 km^2). This can be attributed to the different methods and data in use: pixel vs. object-based approach, different type of in situ data and difference in spatial resolution e.g., Sentinel-2 vs. Landsat-8 spatial resolution (10 vs. 30 m, respectively, resulting in a minimum mapping unit of 100 vs. 900 m^2). A near-future regional comparison of pixel- to object-based approaches in the same context could shed further light upon the nature of their discrepancies—e.g., statistical vs. environmental. On the other hand, in comparison to the United Nations Environment Programme (UNEP) World Conservation Monitoring Centre (WCMC) Version 5 seagrass distribution, the herein seagrass area are nearly four times more (639.5 km^2) [36].

4.2. The Good, the Bad, and the Best Practices of the Proposed Cloud-Based Workflow

The strength of our methodological chain lies mainly in the fact that it can be easily adjusted in space, time and data input. In comparison to the pre-processing chain of [18] and the therein use of the median value, we implement the first quartile here which yields less noisy image composites because it filters higher reflectances (=clouds and sunglint). The chain requires specific input to run:

(a) Selection of a suitable time range; the suitability relates to possible available in situ data to run the machine learning classifiers, the atmospheric, water surface and column conditions of the study area, but also the season of maximum growth of the seagrass species of interest, especially for change detection studies. Here, we have chosen one month of Sentinel-2 imagery within the period of better water column stratification of the Greek Seas.

(b) Selection of suitable points that will represent land and water for land masking (if needed), polygons over deep water (for atmospheric correction), variable sunglint intensity (for sunglint correction), and sandy seabed of variable depth (for the depth invariant index calculation).

(c) Accurate in situ data that will cover all the existing habitats within the study extent for training of the machine learning classifications and validation with an ideally independent data set to reduce potential bias; here, we design remotely sensed, homogenous 4×4 (1600 m^2) polygons for the training of the machine learning model and employ an independent point-based data set for the validation. We also decided to design deep-water polygons to minimize possible misclassifications with seagrasses.

In contrast, the weaknesses of the present workflow the following:

1. Method-wise, the herein image-based, empirical algorithms (e.g., dark pixel subtraction, sunglint correction, depth invariant indices) contain inherent assumptions in their nature and necessitate a sufficient selection of pixels to produce valuable results. Concerning the sunglint correction, specifically, an image composition spanning a large period of time can amplify the artificiality of the produced pseudo-composite, causing the sunglint correction algorithm to be unable to capture any existing interference by this phenomenon.
2. Data-wise, there is a threefold problem with Sentinel-2 applications in the remote sensing of optically shallow benthos and broadly aquatic extent. First, the tile limits of Sentinel-2 data are visible due to differences in viewing angles (odd and even detectors feature a different viewing angle) that produce striping. In turn, this artifact could severely impact classification output as it alters neighboring reflectances. A first possible solution for striping could be the application of pseudo-invariant feature normalization using a tile as a reference image and all the others as the slave ones—a theoretically, computationally expensive operation within GEE. A second solution is to split the initial study area into subareas—ultimately every tile within the visible stripes—where we could select polygons and run the classifier. The second data-related issue is the coastal aerosol band 1, which is originally in 60-m resolution in comparison to the 10-m resolution of all the other visible bands. Although on-the-fly reprojected to 10-m for visual purposes and integral towards coastal habitat mapping and SDB due to its great penetration, it causes artefacts upon application of the depth invariant indices of [26,27]. Therefore, we only utilise b2-b3 index during the classification step. This could be solved through the implementation of a downscaling approach of band 1 [37] into the existing workflow which is under exploration in terms of computation time efficiency. The third and last data-wise issue is the selection of training and validation data. We designed as homogeneous as possible polygons that represent seagrasses, sands, rocks, and deep water based on very high spatial resolution images; however, these will be as accurate as our experienced eye will dictate to us. Figure 2 shows that at all band-to-band scatterplots, the designed polygons of seagrass and non-seagrass beds are not well-differentiated and may have caused misclassifications. Generally, the collection of field data for the classification of remote sensing of aquatic habitats is expensive, time-consuming, and sparse today. More efforts should be driven towards allocating funding for accurate and high resolution in situ data and/or advocating the sharing of open datasets that would permit regional to global projects. The search for open access data on seagrass from relevant data repositories reveals a high number, however a fraction of these are potentially suitable for use in the remote sensing domain. Therefore, it is mandatory to urge a collaborative action between seagrass and remote sensing scientists, which will galvanize the development of a protocol that could be easily adapted in any seagrass bioregion for the designation of accurate and well documented with metadata, in situ data for seagrass mapping using the present workflow.

4.3. Future Endeavours

In addition to the much-needed availability of accurate in situ datasets suitable for image analysis of variable scales (from 50-cm to 30-m pixel size), we discuss three future endeavors following the use of the present cloud-based workflow:

1. Basin- (Mediterranean) to global-scale mapping and monitoring of seagrasses and related biophysical variables (specifically the climate change-related carbon sequestration): The expected lifespan of Sentinel-2 and its succeeding complementary twin mission (7.25 + 7.25 years) would unravel issues related to open and free, high spatial resolution data availability and allow intra-annual (seasonal) to interannual monitoring activities in the optically shallow grounds of seagrasses for 14.5 years by 2029, which marks the end of the announced UN decade of ocean science [38].

2. Improvement of certain stages of the present workflow: (a) Incorporation of a more sophisticated atmospheric correction algorithm like Py6S [39], (b) Implementation of optimization approaches for simultaneous derivation of benthic reflectance and bathymetry based on the semianalytical inversion model of [40,41], (c) Inclusion of best available pixel (BAP) approach within span of well-stratified column period which use pixel-based scores, according to both atmospheric-, season- and sensor-related issues, to produce a composite with the best available pixel [42], (d) Incorporation of object-based segmentation and classification methods to improve classified outputs. The main drawback of the first three improvements is they would possibly lower the time efficiency of the present version of the chain due to the higher demand in computational power based on the need to implement look up tables and/or run radiative transfer codes.
3. Integration of seagrasses and other coastal habitats to the analysis ready data (ARD) era: Recent advances in optical multispectral remote sensing (e.g., Sentinel-2, Landsat 8, Planet's Doves), cloud computing and machine learning classifiers can enable multiscale, multitemporal and sensor-agnostic approaches where all the aforementioned data will be preprocessed to a high scientific standard (Cloud Optimized GeoTIFF; [43]), further harnessing past, present and future remotely sensed big data and facilitating the near real-time measurements of physical changes of these immensely valuable habitats for Earth.

5. Conclusions

The present study introduces a complete methodological workflow for large-scale, high spatial and temporal mapping and monitoring of seagrasses and other optically shallow habitats. The workflow can be easily tuned to spatial, timely and data input; here, we showcase its large spatiotemporal and time efficiency, mapping 2510.1 km^2 of *P. oceanica* seagrasses in 40,951 km^2 of the Greek Seas utilizing a 10-m Sentinel-2 based composite of 1045 tiles in seconds. The workflow could also ingest the freely available image archive of Landsat-8 surface reflectance as input. We envisage that the herein adaptable, accurate, and time- and cost efficient cloud-based workflow will provide the vital seasonal to interannual baseline mapping and the monitoring of seagrass ecosystems on a global scale, identifying problematic areas, resolving current trends, and assisting coastal conservation, management planning, and ultimately climate change mitigation.

Author Contributions: D.T. conceived the idea; D.T., D.P. and B.A. designed the remotely sensed in situ data; D.T., D.P., and K.T. wrote the paper; B.A. wrote the code in Google Earth Engine; K.T., N.C. and P.R supervised the development of the present workflow and paper, from start to finish.

Funding: D.P. and N.C. are supported by the European H2020 Project 641762 ECOPOTENTIAL: Improving future ecosystem benefits through Earth Observations. D.T. is supported by a DLR-DAAD Research Fellowship (No. 57186656). B.A. is supported by GF-KTR 2472040.

Acknowledgments: This work contributed to and was partially supported by the European H2020 Project 641762 ECOPOTENTIAL: Improving future ecosystem benefits through Earth Observations. Dimosthenis Traganos is supported by a DLR-DAAD Research Fellowship (No. 57186656).

Conflicts of Interest: The authors declare no conflicts of interest.

References

1. Short, F.T.; Dennison, W.C.; Carruthers, T.J.B.; Waycott, M. Global seagrass distribution and diversity: A bioregional model. *J. Exp. Mar. Biol. Ecol.* **2007**, *350*, 3–20. [CrossRef]
2. Green, E.P.; Short, F.T. *World Atlas of Seagrasses*; University of California Press: Berkeley, CA, USA, 2003.
3. Duarte, C.M.; Borum, J.; Short, F.T.; Walker, D.I. Seagrass Ecosystems: Their Global Status and Prospects. In *Aquatic Ecosystems: Trends and Global Prospects*; Polunin, N., Ed.; Cambridge University Press: Cambridge, UK, 2008. [CrossRef]
4. Nordlund, M.L.; Koch, E.W.; Barbier, E.B.; Creed, J.C. Seagrass Ecosystem Services and Their Variability across Genera and Geographical Regions. *PLoS ONE* **2016**, *11*, e0163091. [CrossRef] [PubMed]

5. Githaiga, M.N.; Kairo, J.G.; Gilpin, L.; Huxham, M. Carbon storage in the seagrass meadows of Gazi Bay, Kenya. *PLoS ONE* **2017**, *12*, e0177001. [CrossRef] [PubMed]
6. Fourqurean, J.W.; Duarte, C.M.; Kennedy, H.; Marbà, N.; Holmer, M.; Mateo, M.A.; Apostolaki, E.T.; Kendrick, G.A.; Krause-Jensen, D.; McGlathery, K.J.; et al. Seagrass ecosystems as a globally significant carbon stock. *Nat. Geosci.* **2012**, *5*, 505–509. [CrossRef]
7. UN General Assembly, Transforming Our World: The 2030 Agenda for Sustainable Development. 21 October 2015. Available online: http://www.refworld.org/docid/57b6e3e44.html (accessed on 3 August 2018).
8. Waycott, M.; Carlos, M.D.; Carruthers, T.J.B.; Orth, R.J.; Dennison, W.C.; Olyarnik, S.; Calladine, A.; Fourqurean, J.W.; Heck, K.L.J.; Hughes, A.R.; et al. Accelerating loss of seagrasses across the globe threatens coastal ecosystems. *Proc. Natl. Acad. Sci. USA* **2009**, *106*, 12377–12381. [CrossRef] [PubMed]
9. Orth, R.J.; Carruthers, T.J.; Dennison, W.C.; Duarte, C.M.; Fourqurean, J.W.; Heck, K.L.; Hughes, A.R.; Kendrick, G.A.; Kenworthy, W.J.; Olyarnik, S.; et al. A global crisis for seagrass ecosystems. *Bioscience* **2006**, *56*, 987–996. [CrossRef]
10. Hossain, M.S.; Bujang, J.S.; Zakaria, M.H.; Hashim, M. The application of remote sensing to seagrass ecosystems: An overview and future research prospects. *Int. J. Remote Sens.* **2015**, *36*, 61–114. [CrossRef]
11. Hedley, J.D.; Roelfsema, C.M.; Chollett, I.; Harborne, A.R.; Heron, S.F.; Weeks, S.; Skirving, W.J.; Strong, A.E.; Eakin, C.M.; Christensen, T.R.L.; et al. Remote Sensing of Coral Reefs for Monitoring and Management: A Review. *Remote Sens.* **2016**, *8*, 118. [CrossRef]
12. Wabnitz, C.; Andréfouët, S.; Torres-Pulliza, D.; Müller-Karger, F.; Kramer, P. Regional-scale seagrass habitat mapping in the wider Caribbean region using Landsat sensors: Applications to conservation and ecology. *Remote Sens. Environ.* **2008**, *112*, 3455–3467. [CrossRef]
13. Topouzelis, K.; Makri, D.; Stoupas, N.; Papakonstantinou, A.; Katsanevakis, S. Seagrass mapping in Greek territorial waters using Landsat-8 satellite images. *Int. J. Appl. Earth Obs. Geoinform.* **2018**, *67*, 98–113. [CrossRef]
14. Roelfsema, C.M.; Kovacs, E.; Phinn, S.R.; Lyons, M.; Saunders, M.; Maxwell, P. Challenges of Remote Sensing for Quantifying Changes in Large Complex Seagrass Environments. *Estuar. Coast. Shelf Sci.* **2013**, *133*, 161–171. [CrossRef]
15. Hansen, M.C.; Potapov, P.V.; Moore, R.; Hancher, M.; Turubanova, S.A.A.; Tyukavina, A.; Thau, D.; Stehman, S.V.; Goetz, S.J.; Loveland, T.R.; et al. High-Resolution Global Maps of 21st-Century Forest Cover Change. *Science* **2013**, *342*, 850–853. [CrossRef] [PubMed]
16. Xiong, J.; Thenkabail, P.S.; Gumma, M.K.; Teluguntla, P.; Poehnelt, J.; Congalton, R.G.; Yadav, K.; Thau, D. Automated cropland mapping of continental Africa using Google earth engine cloud computing. *ISPRS J. Photogramm. Remote Sens.* **2017**, *126*, 225–244. [CrossRef]
17. Pekel, J.F.; Cottam, A.; Gorelick, N.; Belward, A.S. High-resolution mapping of global surface water and its long-term changes. *Nature* **2016**, *540*, 418–422. [CrossRef] [PubMed]
18. Traganos, D.; Poursanidis, D.; Aggarwal, B.; Chrysoulakis, N.; Reinartz, P. Estimating Satellite-Derived Bathymetry (SDB) with the Google Earth Engine and Sentinel-2. *Remote Sens.* **2018**, *10*, 859. [CrossRef]
19. Sini, M.; Katsanevakis, S.; Koukourouvli, N.; Gerovasileiou, V.; Dailianis, T.; Buhl-Mortensen, L.; Damalas, D.; Dendrinos, P.; Dimas, X.; Frantzis, A.; et al. Assembling Ecological Pieces to Reconstruct the Conservation Puzzle of the Aegean Sea. *Front. Mar. Sci.* **2017**, *4*. [CrossRef]
20. Casotti, R.; Landolfi, A.; Brunet, C.; Ortenzio, F.D.; Mangoni, O.; Ribera d'Alcala, M. Composition and dynamics of the phytoplankton of the Ionian Sea (eastern Mediterranean). *J. Geophys. Res.* **2003**, *108*. [CrossRef]
21. Gerakaris, V.; Panayotidis, P.; Tsiamis, K.; Nikolaidou, A.; Economou-Amili, A. Posidonia oceanica meadows in Greek seas: Lower depth limits and meadow densities. In Proceedings of the 5th Mediterranean Symposium on Marine Vegetation, Portorož, Slovenia, 27–28 October 2014.
22. HNHS. Greek Coastline at Scale 1:90,000. 2018. Available online: https://www.hnhs.gr/en/?option=com_opencart&Itemid=268&route=product/product&path=86&producp_id=271 (accessed on 3 August 2018).
23. Breiman, L.; Friedman, J.H.; Olshen, R.A.; Stone, C.J. Classification and Regression Trees. In *Chapman and Hall/CRC*; CRC Press: Boca Raton, FL, USA, 1984.
24. Armstrong, R.A. Remote sensing of submerged vegetation canopies for biomass estimation. *Int. J. Remote Sens.* **1993**, *14*, 621–627. [CrossRef]

25. Hedley, J.D.; Harborne, A.R.; Mumby, P.J. Technical note: Simple and robust removal of sun glint for mapping shallow-water benthos. *Int. J. Remote Sens.* **2005**, *26*, 2107–2112. [CrossRef]
26. Lyzenga, D.R. Passive remote sensing techniques for mapping water depth and bottom features. *Appl. Opt.* **1978**, *17*, 379–383. [CrossRef] [PubMed]
27. Lyzenga, D.R. Remote sensing of bottom reflectance and water attenuation parameters in shallow water using aircraft and Landsat data. *Int. J. Remote Sens.* **1981**, *2*, 71–82. [CrossRef]
28. Green, E.P.; Mumby, P.J.; Edwards, A.J.; Clark, C.D. *Remote Sensing Handbook for Tropical Coastal Management*; UNESCO Press: Paris, France, 2000; pp. 1–316.
29. Vapnik, V. *The Nature of Statistical Learning Theory*; Springer Press: New York, NY, USA, 1995.
30. Breiman, L. Random forests. *Mach. Learn.* **2001**, *45*, 5–32. [CrossRef]
31. Zhang, C. Applying data fusion techniques for benthic habitat mapping and monitoring a coral reef ecosystem. *ISPRS J. Photogramm. Remote Sens.* **2015**, *104*, 213–223. [CrossRef]
32. Traganos, D.; Reinartz, P. Mapping Mediterranean Seagrasses with Sentinel-2 Imagery. Available online: https://www.sciencedirect.com/science/article/pii/S0025326X17305726 (accessed on 3 August 2018).
33. Traganos, D.; Cerra, D.; Reinartz, P. CubeSat-Derived Detection of Seagrasses Using Planet Imagery Following Unmixing-Based Denoising: Is Small the Next Big? Available online: https://www.int-arch-photogramm-remote-sens-spatial-inf-sci.net/XLII-1-W1/283/2017/isprs-archives-XLII-1-W1-283-2017.pdf (accessed on 3 August 2018).
34. Poursanidis, D.; Topouzelis, K.; Chrysoulakis, N. Mapping coastal marine habitats and delineating the deep limits of the Neptune's seagrass meadows using VHR earth observation data. *Int. J. Remote Sens.* **2018**. [CrossRef]
35. The UN Environment World Conservation Monitoring Centre (UNEP-WCMC). Global Distribution of Seagrasses (Version 5.0). Fourth Update to the Data Layer Used in Green and Short (2003). 2017. Available online: http://data.unep-wcmc.org/datasets/7 (accessed on 3 August 2018).
36. Partnership to Develop First-Ever Global Map and Dynamic Map of Coral Reefs. 2018. Available online: https://carnegiescience.edu/node/2354 (accessed on 3 August 2018).
37. Brodu, N. Super-Resolving Multiresolution Images with Band-Independent Geometry of Multispectral Pixels. *IEEE Trans. Geosci. Remote Sens.* **2017**, *55*, 4610–4617. [CrossRef]
38. UNESCO. United Nations Decade of Ocean Science for Sustainable Development (2021–2030). 2018. Available online: https://en.unesco.org/ocean-decade (accessed on 3 August 2018).
39. Wilson, R. Py6S—A Python Interface to 6S. 2012. Available online: https://py6s.readthedocs.io/en/latest/ (accessed on 3 August 2018).
40. Lee, Z.; Carder, K.L.; Mobley, C.D.; Steward, R.G.; Patch, J.S. Hyperspectral remote sensing for shallow waters: 1. A semianalytical model. *Appl. Opt.* **1998**, *37*, 6329–6338. [CrossRef] [PubMed]
41. Lee, Z.; Carder, K.L.; Mobley, C.D.; Steward, R.G.; Patch, J.S. Hyperspectral remote sensing for shallow waters: 2. Deriving bottom depths and water properties by optimization. *Appl. Opt.* **1999**, *38*, 3831–3843. [CrossRef] [PubMed]
42. White, J.C.; Wulder, M.A.; Hobart, G.W.; Luther, J.E.; Hermosilla, T.; Griffiths, P.; Coops, N.C.; Hall, R.J.; Hostert, P.; Dyk, A.; et al. Pixel-based image compositing for large-area dense time series applications and science. *Can. J. Remote Sens.* **2014**, *40*, 192–212. [CrossRef]
43. Cloud Optimized GeoTIFF. An Imagery Format for Cloud-Native Geospatial Processing. 2018. Available online: http://www.cogeo.org/ (accessed on 3 August 2018).

Article

BULC-U: Sharpening Resolution and Improving Accuracy of Land-Use/Land-Cover Classifications in Google Earth Engine

Jacky Lee [1], Jeffrey A. Cardille [1,*] and Michael T. Coe [2]

1 Department of Natural Resource Sciences, McGill School of Environment, Montreal, QC H9X 3V9, Canada; jacky.hy.lee@gmail.com
2 The Woods Hole Research Center, Falmouth, MA 02540, USA; mtcoe@whrc.org
* Correspondence: jeffrey.cardille@mcgill.ca

Received: 30 June 2018; Accepted: 6 September 2018; Published: 12 September 2018

Abstract: Remote sensing is undergoing a fundamental paradigm shift, in which approaches interpreting one or two images are giving way to a wide array of data-rich applications. These include assessing global forest loss, tracking water resources across Earth's surface, determining disturbance frequency across decades, and many more. These advances have been greatly facilitated by Google Earth Engine, which provides both image access and a platform for advanced analysis techniques. Within the realm of land-use/land-cover (LULC) classifications, Earth Engine provides the ability to create new classifications and to access major existing data sets that have already been created, particularly at global extents. By overlaying global LULC classifications—the 300-m GlobCover 2009 LULC data set for example—with sharper images like those from Landsat, one can see the promise and limits of these global data sets and platforms to fuse them. Despite the promise in a global classification covering all of the terrestrial surface, GlobCover 2009 may be too coarse for some applications. We asked whether the LULC labeling provided by GlobCover 2009 could be combined with the spatial granularity of the Landsat platform to produce a hybrid classification having the best features of both resources with high accuracy. Here we apply an improvement of the Bayesian Updating of Land Cover (BULC) algorithm that fused unsupervised Landsat classifications to GlobCover 2009, sharpening the result from a 300-m to a 30-m classification. Working with four clear categories in Mato Grosso, Brazil, we refined the resolution of the LULC classification by an order of magnitude while improving the overall accuracy from 69.1 to 97.5%. This "BULC-U" mode, because it uses unsupervised classifications as inputs, demands less region-specific knowledge from analysts and may be significantly easier for non-specialists to use. This technique can provide new information to land managers and others interested in highly accurate classifications at finer scales.

Keywords: land cover; deforestation; Brazilian Amazon; Bayesian statistics; BULC-U; Mato Grosso; spatial resolution; Landsat; GlobCover

1. Introduction

Land use and land cover (LULC) change is a principal contributor to global greenhouse gas emissions and can have extensive indirect effects including biodiversity loss and regional hydrologic change [1–3]. Increasing global demands for agricultural commodities and other forest resources are expected to continue to put pressure on remaining forests [2,4,5]. Monitoring LULC change is critical in identifying priority conservation and restoration areas [6] and helping nations achieve their national carbon emissions targets [7,8]. LULC change in the tropics often occurs at small scales and as a result, an accurate accounting of LULC types requires data at correspondingly fine scales.

With the opening of the Landsat satellite archive a decade ago [9], a new era in remote sensing began, characterized by free data and the rapid development of time-series analysis algorithms for tracking LULC change [10–14]. There are now many potential satellite-based imagery sources spanning across more than 4 decades [9,12]. While Landsat represents the longest-running time series, additional sensors also provide free imagery [15,16]. More recently, Sentinel-2 satellites were launched in 2015 and 2017 with even finer spatial resolution and revisit times [17,18]. The increase in data at fine resolutions and in time increases the potential benefit of algorithms to incorporate evidence from large numbers of satellite images into useful maps for monitoring landscape changes.

An earlier generation of global LULC classifications was developed by both academic and governmental organizations to represent LULC at a static point in time. Some examples include: The IGBP-DISCOVER classification using MODIS imagery [19]; the Global 1-km Consensus Land-cover Product [20]; and GlobCover, using data from MERIS in two different campaigns: 2000 and 2009 [21–23]. Made and verified with great effort, these classifications are a valuable source of LULC information that have been applied for identifying patterns of land use change [24–26], agriculture inventory [27,28], and modeling species distribution [29–31]. Despite the power of these classifications, the relatively coarse spatial resolution can limit their usefulness, especially at finer scales [28,32,33]. Surprisingly few algorithms are available to sharpen the spatial resolution of moderate-resolution LULC classifications in light of finer-scale imagery.

The Bayesian Updating of Land Cover (BULC) algorithm [34] was originally devised to use Bayesian logic to create time series of land use and land cover classifications. In its original conception, BULC processes classifications to estimate the probability for each class for each pixel at each time step based on the strength of the agreement between consecutive candidate classifications. This approach allows each pixel to incorporate information from a series of land cover maps to create ongoing classifications, either to update a given classification through time or to confirm the estimated LULC class of each pixel in a study area at a given time. The effect is to blend relative candidate classifications according to their shared properties, while tracking stability and change through time as more images are analyzed. A given pixel's per-class probabilities reflect the ability to consistently label a pixel with its (assumed proper) class. Despite its utility for creating time series, initial applications of BULC were limited by the effort and foreknowledge needed to produce relatively high-quality LULC classifications across a large number of images: prospective users need to know the study area well enough to discern whether a given prospective classification is good enough for inclusion based on the identified LULC categories.

In this manuscript, we modify BULC's ability to incorporate new information by extending its potential inputs to include unsupervised classifications. We use this enhancement, which we call "Bayesian Updating of Land Cover: Unsupervised" (BULC-U) to refine the resolution of a relatively coarse global data set by an order of magnitude in a heterogeneous, finely structured landscape in Mato Grosso, Brazil. Using 13 Landsat 5 images near in time to the nominal 2009 date of the GlobCover global data set, BULC-U blends Landsat's finer-scale spatial information with the coarse labels of GlobCover 2009 to produce a higher-resolution land-cover classification with GlobCover 2009's labels and Landsat 5's spatial resolution. We conduct an accuracy assessment comparing the GlobCover 2009 and the BULC-U 2009 classification products, demonstrating that the new classification has both finer spatial resolution and improved accuracy.

2. Methods

2.1. Study Area

The study area is a 2×10^5 km^2 (166 km $\times$ 121 km) region of Mato Grosso, Brazil, located within Landsat path 224 row 69 and centered near 51.884°W, 12.601°S. (Figure 1).

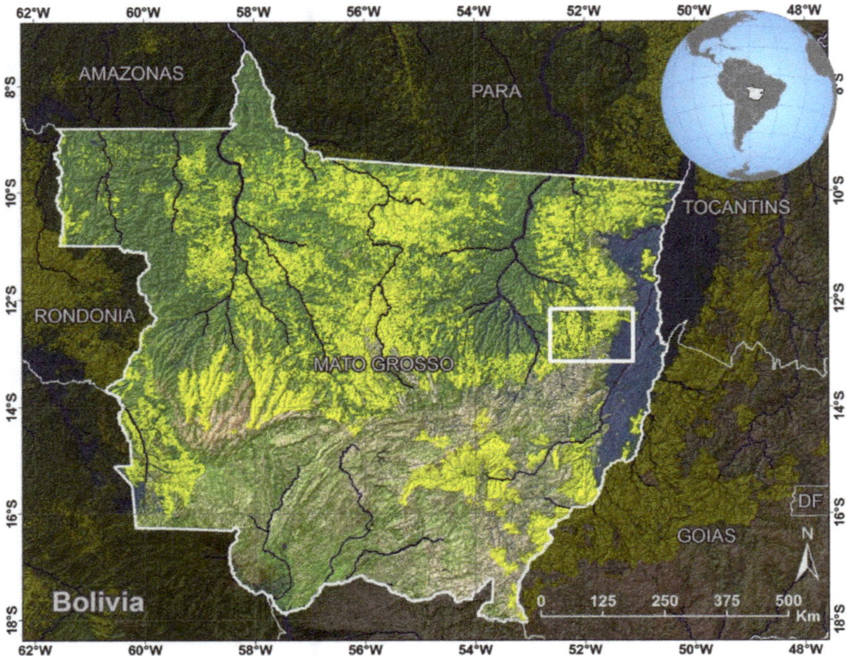

Figure 1. Location of study area in eastern Mato Grosso, Brazil.

The study area has a mix of grasslands, rainforest, hilly shrubland and extensive agricultural areas. The varied land cover types and clearly visible agriculture make this area a good location for testing the ability of BULC-U to sharpen a coarser classification. In particular, 30-m satellite imagery shows well-demarcated edges between cropland and forest that are difficult to capture with 300-m resolution imagery. In addition, the small rivers that run through the area, visible in 30-m imagery, are not discernible at coarser scales.

2.2. LULC Categories

For this study, we were interested in illustrating the study area during its conversion from a landscape that had recently been mostly dense-canopy Forest or some type of Grassland or Shrubland to one increasingly dominated by Cropland. We identified four fundamental LULC categories of interest: Cropland; Forest; Shrubland or Grassland (referred to hereafter as Shrubland); and Water. These four classes are easy to visually identify in this area in Landsat-scale satellite images, and are useful in this study for detecting active agriculture that may be missed in coarsely grained classifications.

2.3. BULC-U Algorithm

BULC-U is intended to track classes that can be reliably identified in a sequence of images that have been categorized into Events. The BULC-U algorithm can sharpen an existing LULC classification by incorporating higher-resolution information from unsupervised classifications of finer-resolution satellite images. Like the BULC algorithm on which it is based, BULC-U ingests a time-ordered series of classified images and creates a land cover classification at each time step (Figure 2).

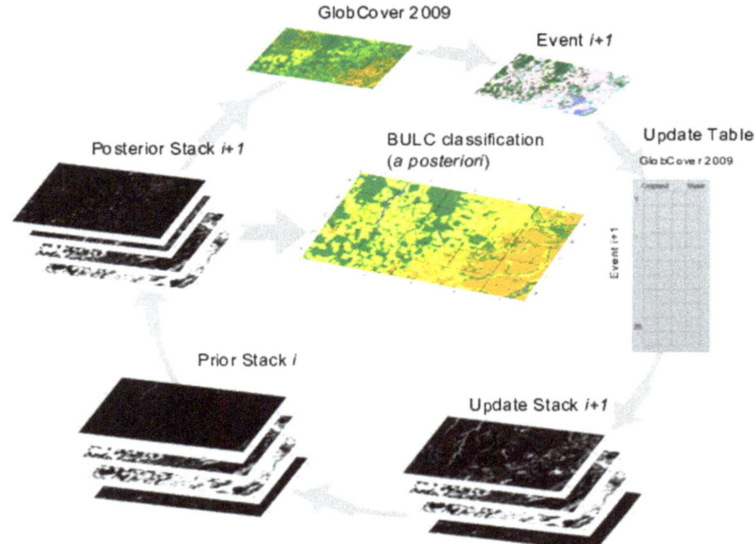

Figure 2. Schematic of BULC-U. The BULC-U processing is driven by unsupervised classifications and update tables relating them to a common reference classification (here, the GlobCover 2009 classification). As in BULC, evidence from each new Event is used to update the estimates of each pixel's LULC using Bayes' Theorem.

At each time step, BULC-U tracks an estimate of the probability of each land cover class for each pixel. Like BULC, BULC-U initializes the land cover estimate using a reference classification that represents the best existing estimate of LULC in the study area. The series of classified images, which we term 'Events', is created using unsupervised classification algorithms. BULC-U creates an 'update table' (formed like an accuracy assessment table) to compare the evidence from the Events to the reference classification following the methods outlined in Cardille and Fortin [34]. In this application, the reference classification was the 300-m GlobCover 2009 classification and Events were made from 30-m Landsat images, both prepared as described in the subsections immediately below. The BULC-U process begins with the *a priori* proposition that the LULC in the study area is exactly as seen in the reference classification, with a moderate level of confidence. The evidence from an Event is combined with the *a priori* estimate using Bayes' Theorem to create an *a posteriori* vector of probabilities for each pixel, which is used as the new *a priori* estimate for incorporating evidence from the next Event. The highest probability class of each pixel in the *a posteriori* stack can be assessed to create a BULC-U classification at any time step in every location given the information that has been seen to that point in the process. BULC and BULC-U differ in two small but important ways. The first difference is in the nature and shape of the update table—where the BULC table is square ($n \times n$ for n tracked classes), the update table in BULC-U is $m \times n$, where m is the number of classes in the unsupervised classification for an Event. The two methods also differ in the classification used as the nominal 'reference' classification in making the update table: In BULC-U, the update table is made for an Event by cross-tabulating the Event with the reference classification, not another Event as is done in BULC.

Another very substantial difference between BULC and BULC-U is its implementation in Google Earth Engine [35]. BULC was implemented as experimental code in R and could take up to several days for a run, making troubleshooting difficult and severely limiting the number of images that could be processed, the amount of data that could be retained at each time step, and the area that could be analyzed. BULC-U was implemented in Google Earth Engine's JavaScript platform, which permits easy prototyping, parameter exploration, and interactive visualization. It is also much faster: The same BULC logic of Cardille and Fortin [34], running in Earth Engine, takes a few moments for its calculations. The BULC and BULC-U methodology also take advantage of a sequential iterator tool that distinguishes it from most other work in Earth Engine. In contrast to studies that use Earth Engine's power to compute, say, the maximum greenness value for a year, BULC and BULC-U process an Event, update the state of each pixel, then process another Event, for a known finite set of Events. The iteration of BULC-U through a series of events is described below.

In this context, BULC-U operates as follows: At the beginning of a given iteration i of BULC-U, each pixel has an estimated probability vector for each LULC category, which reflects the evidence seen to that point in the process about LULC in that pixel. An update table is formed for Event i (an unsupervised Landsat classification) by cross-tabulation with the reference GlobCover classification. This table quantifies the extent to which each unsupervised class coincides with one of the LULC categories. If, say, class 7 of Event i very strongly coincides with Agriculture pixels in the GlobCover classification, the probability vectors of class 7 pixels change in the direction of Agriculture. Each pixel maintains its own history as described in Cardille and Fortin [34]. In the next iteration, a new Event $i + 1$ is considered—say, with class 11 of Event $i + 1$ coinciding with Agriculture pixels in the GlobCover classification, though not as strongly as class 7 had in iteration i. When updated, the probability vectors of those class 7 pixels would move again toward Agriculture, though not as strongly as class 11 pixels had in iteration i. Pixels that were in class 7 in Event i and class 11 in Event $i + 1$ would have moved considerably toward agriculture; pixels in class 7 in i and then a different class in $i + 1$ would have a different probability vector that reflected their own history. The preparation of the Events and the reference classification are described below.

2.4. GlobCover 2009

GlobCover is a global LULC classification with 300-m resolution and 22 potential categories, created with a nominal date of 2009 using data from the MERIS sensor [23]. Within the study area, the GlobCover 2009 LULC classification had 14 categories. Of these, seven covered more than 1% of the study area. Several of the classes were too specific for our purposes and were reclassified to one of the four categories to begin the BULC-U process. Specifically, "Rainfed Cropland (5% of the study area)" and "Mosaic Cropland" (mixed pixels strongly dominated by Cropland, 14% of the area), were reclassified as "Cropland" for the BULC-U reference layer. Second, the Forest-dominant categories "Closed to Open Broadleaved Evergreen" (33%), "Closed Broadleaved" (13%), "Flooded Broadleaved" (<1%), and "Open Broadleaved" (<0.1%) were reclassified as "Forest" for the BULC-U process. The Shrubland category was comprised of GlobCover classes "Closed to Open Shrubland" (17%), "Flooded Closed to Open Vegetation" (4%), "Closed to Open Grassland" (<0.1%), and "Sparse" (<0.1%). The "Water" LULC category was made of the "Open Water" GlobCover category. Importantly, some GlobCover categories potentially contained elements of two BULC-U target classes. These included: "Mosaic Vegetation", which comprised 13% of the study area in GlobCover; "Mosaic Forest or Shrubland" (1% of the study area), and a few pixels of "Mosaic Grassland". These were initialized as "Mosaic/Unknown" for the purposes of creating BULC-U's *a priori* classification. The effect was for BULC-U to not use spectral information of those classes to refine GlobCover. Rather, they were treated as areas whose LULC was not known clearly before the study, to be filled in with one of the four tracked LULC categories at the 30-m resolution during the BULC-U refinement process. The effect of the remapping was to condense original GlobCover categories into a set of LULC categories that could be reliably distinguished on Landsat imagery, and that were known to be accurate in the GlobCover validation report [23].

As it met our purposes of tracking the development of cropland in the area, there were indications that GlobCover was properly used only for these Level 1 categories within our study area. The GlobCover validation report [23] (which assessed points worldwide) considered the two cropland classes within our study area to be interchangeable for judging user's accuracy. Meanwhile, significant confusion was noted in the GlobCover report between the classes of evergreen broadleaf forest and closed deciduous forest, the two Level 2 forest categories that dominated forest in our study area. For shrubland, the two categories that comprised the shrubland found in our study area had user's accuracy of 50% and 20%, with the smaller of the two noted in the report for its "classification instability". Perhaps more importantly, it would have been unfair to ask the GlobCover data, a global coverage, to capture LULC to such a fine degree in an area quite small compared to its global reach. Fortunately, the desired Level 1-analogous LULC labels of our study were of sufficient quality to be used in BULC-U, as described below.

The remapped GlobCover was used in two important ways in the fusion process. First, it was used as the *a priori* classification of the area for the 86% of the study area estimated to be in one of the four tracked LULC categories. For the 14% of the pixels labeled in this way as "Mosaic/Unknown", BULC-U began with equal *a priori* probabilities and gradually refined the estimate of the LULC based on evidence from the Events as describe above. Second, the remapped GlobCover layer was used as the reference classification for the update tables. The Events with which it was compared are described below.

2.5. Landsat Imagery

Thirteen Events for BULC-U were created from clear Landsat 5 images (<10% cloud cover) spanning 2008 to 2010 (Table 1). BULC-U uses unsupervised classifications as its Events, meaning that the multidimensional color space of Landsat needed to be reduced into groupings with similar spectral characteristics. Exploratory efforts to classify images revealed considerable speckling that was not greatly improved with smoothing techniques. Since the distinctive edges of agricultural fields are often amenable to image segmentation methods [36], we segmented the images. Unsupervised classification techniques were unavailable in Google Earth Engine, and so we downloaded the images into ArcGIS for analysis. Each Landsat 5 image was segmented using the ArcGIS implementation of the Segmented Mean Shift algorithm (ESRI 2014) using bands 4, 5, and 7, which were clear and informative for segmenting the images. The segments of each unsupervised classification were then classified using the well-known ISODATA unsupervised classification tool [37,38] in ArcGIS with 20 unsupervised classes. Each of the resulting Events represented groups of pixels that mostly followed apparent LULC distinctions in the landscape, with much of the speckling removed during the segmentation (Figure 3). The degree to which a given unsupervised class was entirely within a single GlobCover LULC class varied among unsupervised classes and across the landscape. Determining the amount and meaning of this overlap was the work of the BULC-U algorithm. Events were then introduced in time order to the BULC-U algorithm's implementation in Earth Engine.

Table 1. Dates of Landsat images classified as Events in BULC-U.

Year	Day
2008	1 June 3 July 19 July 4 August
2009	18 June 4 July 20 July 5 August 22 September
2010	17 May 2 June 20 July 6 September

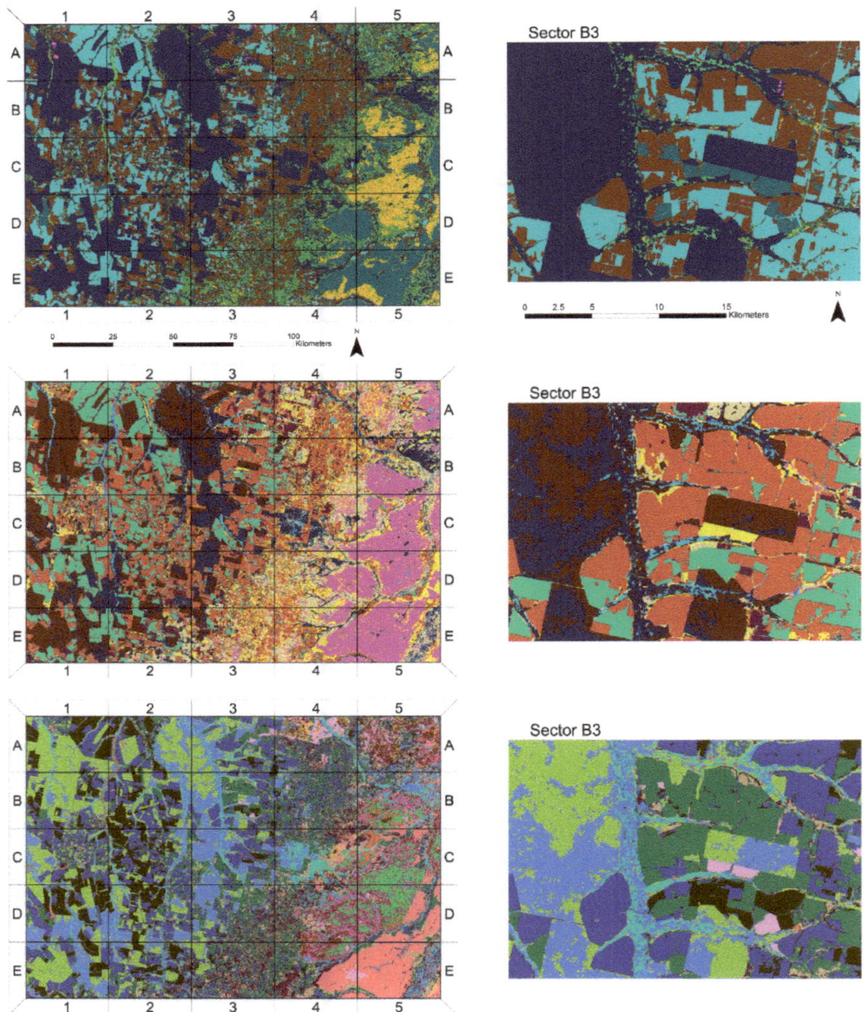

Figure 3. Three views of the study area on three different dates (18 June 2009, 5 August 2009, and 22 September 2009), as shown in the unsupervised Landsat classifications used as Events by BULC-U to refine the GlobCover 2009 classification. Left panel shows the complete study area with a reference grid superimposed; right panels show sector B3 of each classification. To create Events, each Landsat image was first segmented into relatively homogeneous regions. The band means of these regions were then clustered with unsupervised classification into 20 categories for processing as Events in BULC-U. The classifications are similar in that they are mostly successful in distinguishing LULC categories, such as Forest and Agriculture, from each other. As described in the text, BULC-U uses the degree and nature of the correspondence between GlobCover 2009 and each of these unsupervised classes to inform the probability that each of these classes is each of the tracked LULC categories. Based on this correspondence, BULC-U updates the probabilities of each class for each Landsat-sized pixel as each Event is introduced into BULC-U.

2.6. Validation

Validation of the GlobCover and BULC-U classification was done by a researcher unassociated with this project, with experience classifying time series in Mato Grosso and accuracy assessment. The simple LULC classification categories enabled straightforward comparison of the GlobCover and BULC-U classifications while avoiding complex ontological questions about the specific meaning of similar LULC types, and without being needlessly demanding of labeling accuracy in a 300-m classification. We validated both GlobCover 2009 and BULC-U 2009 with the same protocol and classes. The assessment was made by determining land cover from visual inspection of a clear Landsat 5 image that was not used in the BULC-U process. The image was from 12 September 2010 and is shown at various scales in Figure 4 and later. We identified 400 points across the study area, using 100 points each for the Cropland, Forest, Shrubland and Water. To identify reference points for the Cropland, Forest, and Shrubland classes, several thousand candidate reference points were first generated at random locations across the study area using the Create Random Points tool in ArcGIS. Points were viewed on the Landsat validation image and evaluated as being a member of one of the three terrestrial classes; any points within 30 m of an edge of two LULC classes were discarded and the next random point considered, until 100 points were found for each of the three terrestrial categories. Points that appeared in locations that had been labeled as Mosaic/Unknown when preparing the reference classification were discarded. Reference points for Water, which was a much rarer LULC class, were identified using the mask of permanent water bodies from Hansen et al. [39]. We generated a large number of candidate points randomly (again with the Create Random Points tool in ArcGIS) located within that data set's Water mask, retaining the first 100 that were identifiable as open water on the reference Landsat image and more than 30 m from an edge of two LULC categories. A confusion matrix was then created between the 400 reference points and both the GlobCover 2009 classification and the BULC-U classification to determine standard assessment values of Overall Accuracy, Producer's Accuracy, and User's Accuracy [40].

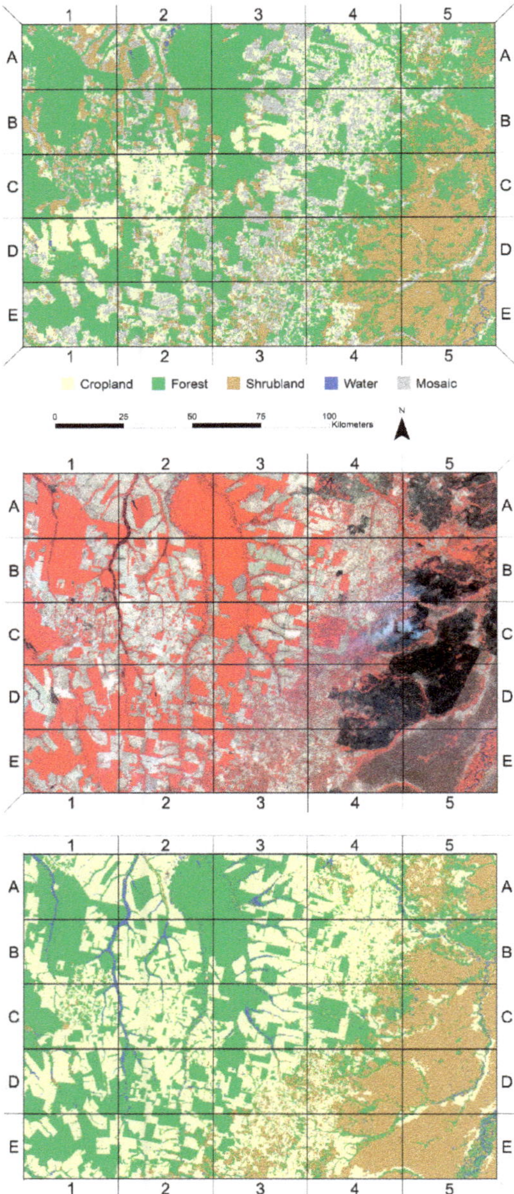

Figure 4. Comparison of GlobCover 2009 (**top**), Landsat 5 (**middle**) and BULC-U 2009 (**bottom**), showing the coarse resolution of GlobCover, the increased level of spatial detail available with Landsat, and the resulting BULC-U 2009 image created by fusing Landsat imagery and GlobCover 2009. The Landsat image is from 12 September 2010.

3. Results

Compared to the GlobCover 2009 300-m classification on which it was based, the resulting BULC-U 2009 classification represents LULC in the year 2009 (Figure 4) with a finer spatial resolution and considerably greater accuracy in all four validation categories (Table 2).

Table 2. Accuracy assessments of GlobCover and BULC-U 2009.

Land Cover Classification Accuracy				
		GlobCover	**BULC-U 2009**	**Improvement**
	Overall	69.1%	97.5%	28.4%
Producer's	Agriculture	71.0%	99.0%	28.0%
	Forest	98.0%	98.0%	0.0%
	Shrubland	67.0%	93.0%	26.0%
	Water	40.8%	100.0%	59.2%
User's	Agriculture	83.1%	95.2%	12.1%
	Forest	59.5%	98.0%	38.5%
	Shrubland	64.4%	97.9%	33.5%
	Water	100.0%	99.0%	-1.0%

The Overall Accuracy of the BULC-U 2009 classification (97.5%) is substantially higher than that of the GlobCover 2009 classification (69.1%) in the study area (Table 2). Within individual classes, the Producer's Accuracy values of BULC-U 2009 were substantially higher than that of GlobCover 2009 for three of the LULC categories: 28.0% higher for Cropland, 26.0% higher for Shrubland, 59.2% higher for Water.

Producer's Accuracy of the Forest category was equally high (98%) in both classifications. User's Accuracy values of BULC-U 2009 were also high for each class in the BULC-U 2009 classification, with all four accuracies above 95%. Meanwhile, through the incorporation of classifications based on 30-m imagery, the resulting BULC-U 2009 classification resolution appears as spatially refined as the Landsat data itself, creating a classification sharpened by a factor of 10 when compared to GlobCover.

From its initialization as the GlobCover classification, the introduction of Events changed the intermediate BULC-U classifications greatly over the first iterations and stabilized after ingesting several unsupervised classifications (Figure 5). The progression from the GlobCover *a priori* map to the BULC-U classification can be seen in Figure 6 and, in a closer view, in Figure 7, as well as the Video Abstract. GlobCover's fusion with the information from Landsat can be directly seen in those figures as the eye moves from the GlobCover panel, to Iteration 1, then Iteration 2. The GlobCover panel (Figures 6 and 7, upper left) is the best estimate of the area before any new data is considered. Figures 6 and 7's Iteration 1 panel is the best estimate by BULC-U of the area after the GlobCover classification was fused with the first Event via BULC-U. Iteration 2 shows an intermediate product that is a fused set based on GlobCover, Event 1, and Event 2. It mostly "looks like" the GlobCover classification, but is in the process of sharpening the classification in light of both the GlobCover reference and the Events—for example, reclassifying some of the pixels that will be eventually be called Water in the finer-scale BULC-U classification after all Events have been processed.

The convergence of the process suggests a spin up time of about 6 Events, with minimal differences thereafter. The convergence was evident both for the proportions (Figure 5) and, importantly, of the maps themselves between iterations (Figure 6). In observation of the LULC maps at both large (Figure 6) and much smaller (Figure 7) extents, later iterations were only slightly different from each other, with fewer than 1% of the pixels changing between iterations after BULC-U had ingested several Events.

Although the BULC-U and GlobCover 2009 classifications appear at first view to be quite similar, BULC-U revealed that the GlobCover map substantially overestimated the amount of Forest and under-reported Cropland in its assessment of the land use and land cover of the area (Table 3).

Before the BULC-U refinement process began, 45.7% of the study area was estimated as Forest by GlobCover, even without including the Forest implied by the Mosaic categories that we had recategorized to be an unknown LULC. This contrasts with what is indicated (35.1%) by the more accurate and finer-scale BULC-U 2009 classification, an estimate that is 23% lower. Closer inspection indicates that much of what was called Forest in the GlobCover classification was either mislabeled (as in sector B2), or, more often, labeled properly at the 300-m scale but contained substantial Cropland within. The amount of Cropland was more than double that estimated by GlobCover (40.0% vs. 19.2%). This very substantial difference is more than a distinction between LULC labels: even if all pixels in the GlobCover mosaic categories had been Cropland in truth, the BULC-U estimate of Cropland was even higher than what could be detected in the GlobCover reference set. In practice, the higher estimate of Cropland came both from better labeling of some GlobCover Forest pixels (Figures 8 and 9) and from splitting and labeling the 14% of the GlobCover 300-m classification that had been labeled as a Mosaic category (see especially Figure 9).

Table 3. Proportions of each category compared between GlobCover 2009 and derived BULC-U 2009. Note that GlobCover, a 300-m classification, had 14% of its area labeled as Mosaic classes that were relabeled as NoData for the BULC-U process. BULC-U's 30-m classification eliminated those mosaic categories in favor of their component LULC classes.

	Percentage Cover		
	GlobCover	**BULC-U 2009**	**Amount Loss/Gain**
Forest	45.7%	35.1%	−10.6%
Cropland	19.2%	40.0%	20.9%
Shrubland	20.8%	23.1%	2.3%
Water	0.2%	1.8%	1.6%
Mosaic	14.1%	0.0%	−14.1%

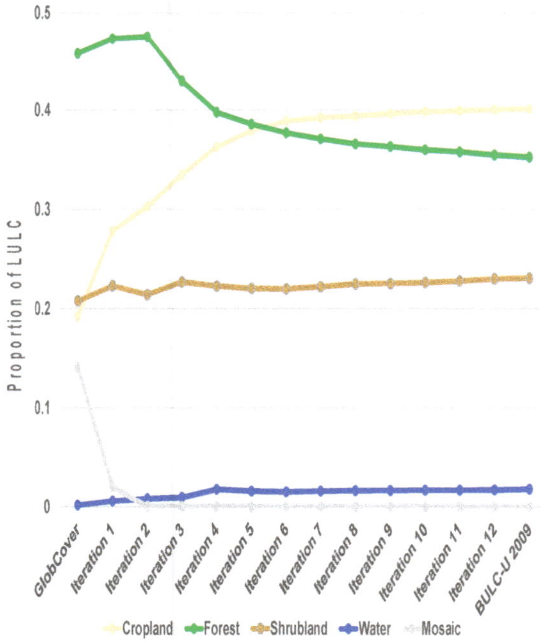

Figure 5. Convergence of the BULC-U classification across 13 iterations.

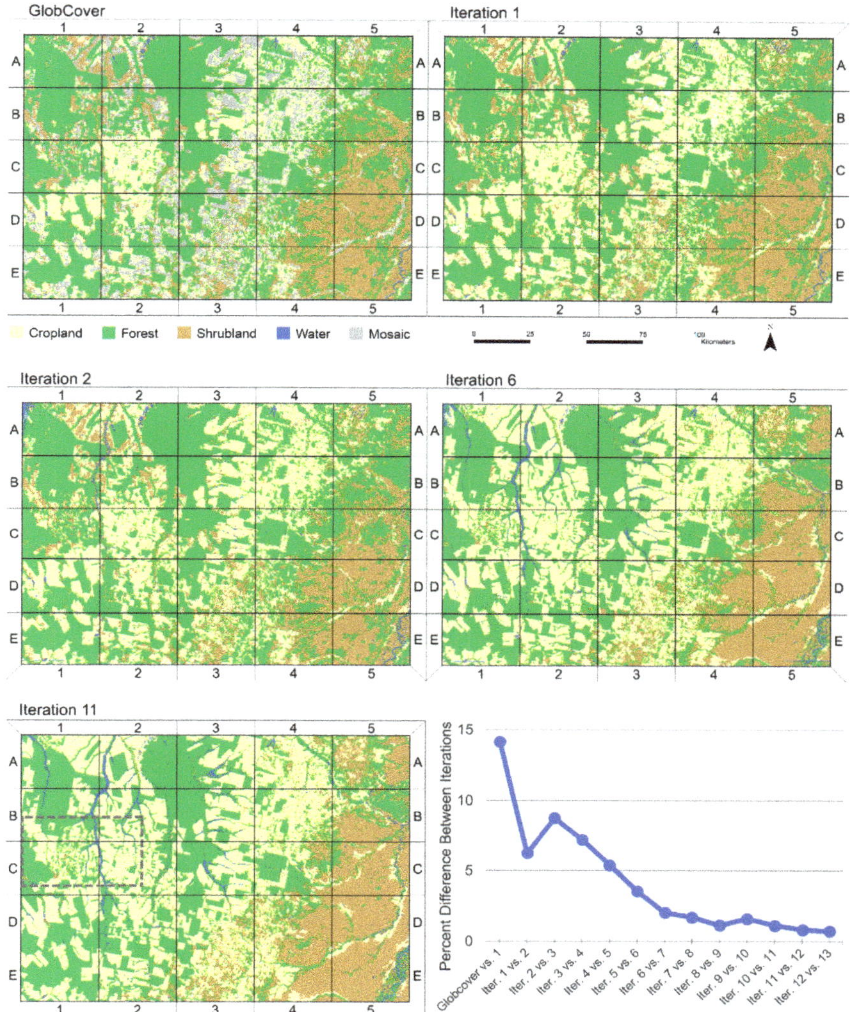

Figure 6. Progression from GlobCover 2009 LULC classification (upper left panel) toward the BULC-U 2009 classification. Shown are intermediate iterations of BULC-U as the map converges to the final classification seen in Figure 4, lower panel. Lower right panel: percentage of differing LULC labels between subsequent iterations of BULC-U, showing that iterations rapidly converge as new data is added. Fewer than 1% of pixels are different between the later iterations and the final BULC-U 2009 classification, which is taken from Iteration 13 (Figure 4, lower panel). The dashed box superimposed on Iteration 11's classifications (with the upper right corner visible on B2) is seen in Figure 7.

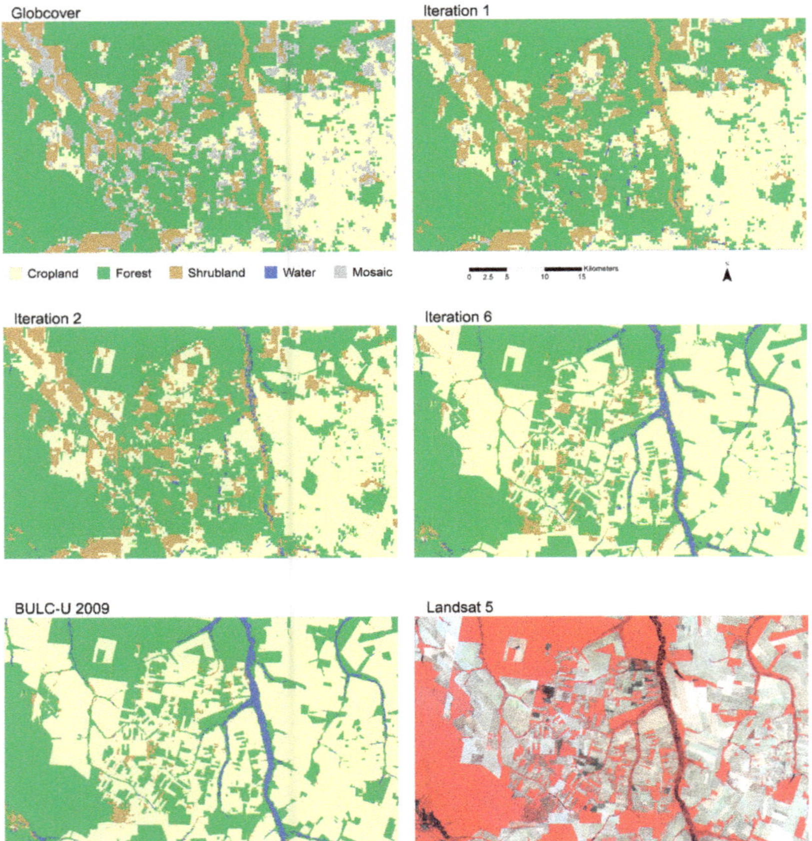

Figure 7. Progression from GlobCover 2009 LULC classification toward the BULC-U 2009 classification. The area centers on the town of Querência, whose location is shown in the dashed box on Figure 6. The BULC-U classification compares favorably with the Landsat image from 12 September 2010.

The other two categories of the BULC-U classification were also different than their counterparts in the GlobCover base image. The amount of detected open Water increased by an order of magnitude, from 0.2 to 1.8%, principally due to an increase in the number of pixels along river courses as the resolution improved with the incorporation of 30-m data. For the Shrubland category, it has roughly the same proportion in both the GlobCover and BULC-U classifications, although the location and distribution of the pixels of the class were somewhat different. This appears to have been due to the spatial configuration of Shrubland in the study area: With fewer mixed pixels than in the GlobCover 2009 Forest class, the refinement of the coarser classification tended to resolve labeling errors, rather than uncovering previously undetected pockets of Shrubland within 300-m GlobCover pixels. This can be seen most clearly in Figure 10, where the large expanse of Shrubland is mostly consistent between GlobCover and BULC-U. In most cases where Shrubland in the GlobCover category was changed by BULC-U, (e.g., in the southwest portion of Figure 8), BULC-U labeled the area Cropland. Meanwhile, some larger areas marked as Forest (e.g., in the northeast portion of Figure 10) were found by BULC-U as being more properly labeled Shrubland. These two relatively independent phenomena resulted in a Shrubland class that was more precise and accurate, with a similar amount between the two classifications (Table 3).

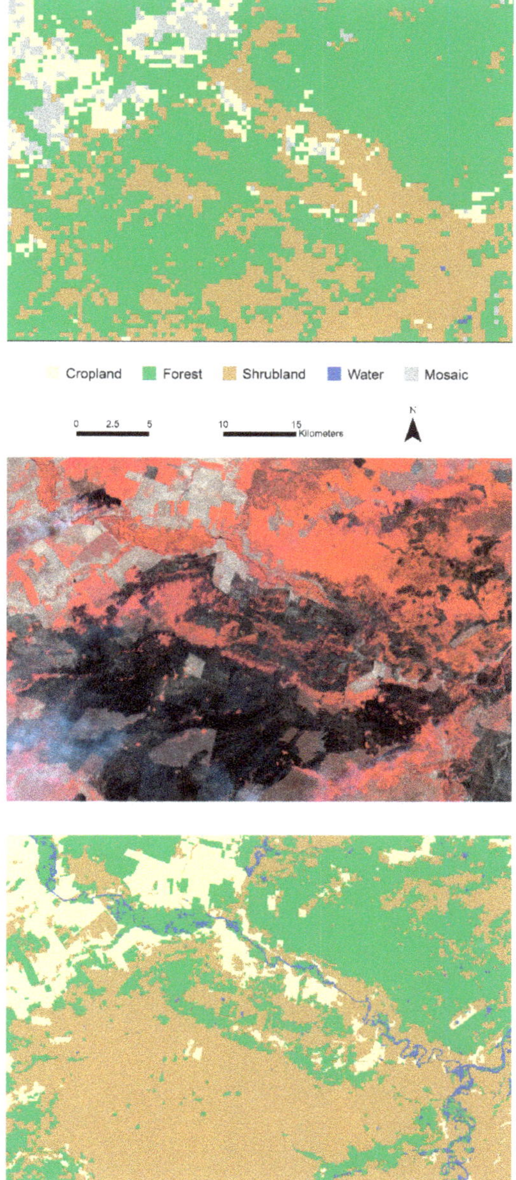

Figure 8. Closer view of Sector B5 on Figure 4, comparing GlobCover 2009 (**top**), Landsat 5 (**middle**) and BULC-U 2009 (**bottom**). The Landsat image is from 12 September 2010.

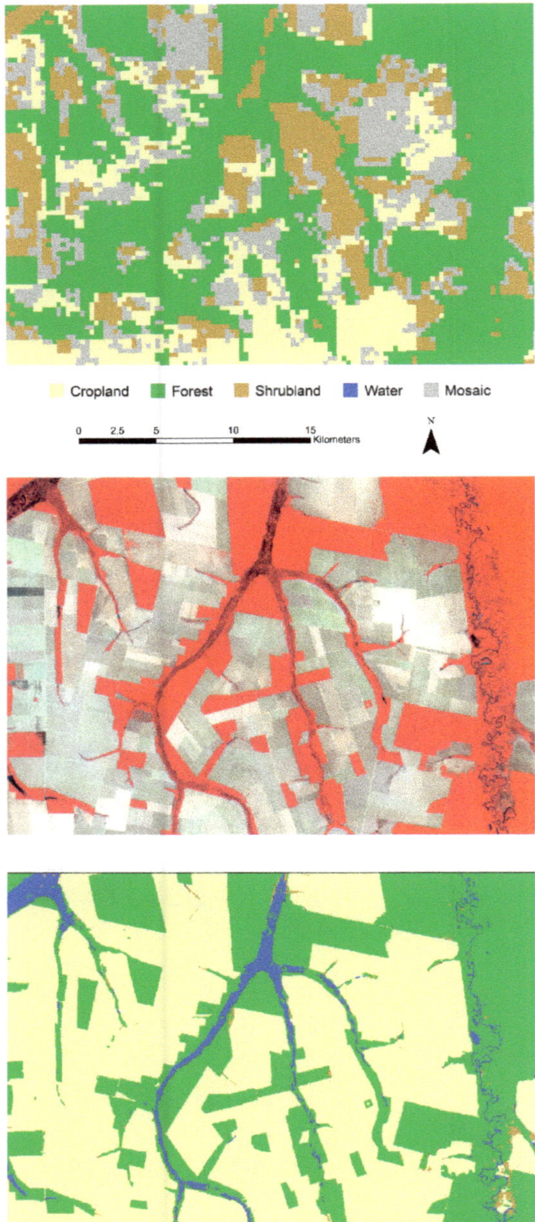

Figure 9. Closer view of Sector B2 on Figure 4, comparing GlobCover 2009 (**top**), Landsat 5 (**middle**) and BULC-U 2009 (**bottom**). The Landsat image is from 12 September 2010.

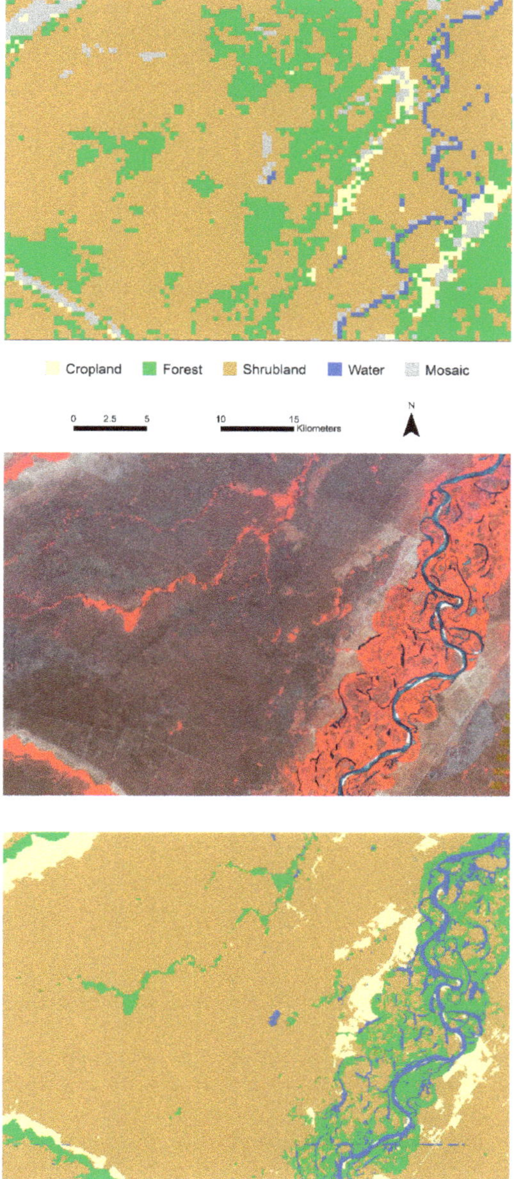

Figure 10. Closer view of Sector E5 on Figure 4, comparing GlobCover 2009 (**top**), Landsat 5 (**middle**) and BULC-U 2009 (**bottom**). The Landsat image is from 12 September 2010.

4. Discussion

The BULC-U 2009 classification represents a successful fusion of the high-quality labeling of the GlobCover project with the finer-resolution spatial information available from Landsat-class imagery. In this aspect, the BULC-U algorithm is reminiscent of the well-known pan-sharpening process, but for sharpening classifications rather than raw imagery. The specific effect of BULC-U is to tap scene-wide pixel-label information that is encoded in the relatively coarse GlobCover 2009 information and fuse it to higher-resolution spatial information taken from Landsat imagery. In this study, the spatial structure was interpreted using multiple Landsat color bands; although we have not attempted it here, it is conceivable that an even finer-scale product could be attempted by fusing GlobCover imagery with pan-sharpened Landsat or Sentinel-2 data.

4.1. Accuracy Improvements from BULC-U

The accuracy of the final BULC-U land cover product is considerably higher when compared to that of the original GlobCover 2009 classification from which it was built (Table 2). The improvement in accuracy appears to be due to two factors: First, the BULC-U algorithm was able to effectively 'repair' many inaccurate GlobCover 300-m scale pixel classification labels. In ingesting information from an Event, BULC-U treats the pixels of a given unsupervised cluster as likely being from a single LULC class. Over the course of ingesting the Events, this allowed pixels with incorrect labels in GlobCover to be gradually labeled like other pixels that shared their spectral characteristics. BULC-U also had a similar effect for areas labeled as Mosaic in GlobCover and marked as NoData by us when preparing the BULC-U run. These areas were gradually labeled like other pixels that shared their spectral characteristics at the finer 30-m scale. Second, the improved resolution of the Events used in this demonstration of BULC-U also played a role in the increased accuracy, by allowing finer-scale delineation of land cover within pixels whose LULC was heterogeneous at the 300-m scale. This was particularly relevant for 300-m pixels labeled by GlobCover as "Cropland" that contained, say, 80% Cropland and 20% Forest.

The specific ways that BULC-U improved accuracy are illustrated by a closer view of the study area in the GlobCover and BULC-U classifications, in sectors B5, B2, and E5 (Figures 8–10). Sector B5, which GlobCover had classified as predominantly Forest (with some Shrubland), BULC-U classified as predominantly Shrubland (Figure 8). BULC-U correctly identified most of Sector B2 as Cropland (Figure 9), including a substantial area that had been misclassified Shrubland or Forest in GlobCover. BULC-U also labeled much of the central part of B2 as more highly fragmented forest, and properly classified the GlobCover's Mosaic classes. GlobCover 2009 classified most of E5 as Forest, while BULC-U identified it as being predominantly Shrubland (Figure 10).

4.2. Fusing Information from Different Sensors and Projects

The BULC-U algorithm allows users to create and update land cover classifications using the same legend as a previously created classification, but with higher resolution and, at least in the case illustrated here, considerably improved accuracy. Using BULC-U it should be possible to refine the GlobCover classification elsewhere or to downscale other high-quality coarse classifications. Once a classification is refined to a sharper resolution it should be possible, as in Cardille & Fortin (2016), to roll the classification forward in time to show updated land cover at finer resolution in areas that are changing—or backward to show LULC history in earlier periods. Although we have used segmented ISODATA-classified Landsat data here to drive BULC-U, the algorithm can be driven with classifications created by any viable method. As a result, other sensors (e.g., Sentinel-2) could also be included alongside Landsat to refine and update a BULC-U classification.

4.3. GlobCover 2009 as a High-Quality Data Source

The comparison between GlobCover 2009 and BULC-U 2009 should not be misconstrued as criticism of any aspect of the GlobCover approach or result. In fact, the refinement described here would not have been possible without the GlobCover 2009 serving as a base. The global classification GlobCover 2009 is a source of high-quality global LULC data that, although imperfect at a fine scale, provided a statistical and spatial framework that could be refined by BULC-U to create an even higher-quality classification of the study area. GlobCover may well be the best product available for some locations, but even if its resolution is coarse for a given application, it can still be useful. Here, it is valuable in two roles: First, as the reference classification against which to compare unsupervised classifications; and second, as the *a priori* estimate of LULC for the study area, before the incorporation of Landsat-based Events. As seen in this study, although GlobCover 2009 was not correct on a pixel-by-pixel basis, and was substantially incorrect in its assessment of LULC proportions, it was sufficiently correct to be used as a reference to create probabilities of each class for BULC-U.

4.4. Number of Unsupervised Classes for BULC-U

The number of classes in the unsupervised Event classifications must be identified by the user of BULC-U and adapted to the distinct spectral classes in the satellite image and the number of classes being tracked. In this study, choosing the number of classes to create the Events in BULC-U was done using trial and error. During the stage of determining how many classes to use for Events, we asked whether: (a) for each of the unsupervised classes, each segment of a given class appeared to be the same LULC class in the Landsat image; and (b) the number of unsupervised classes was not excessively high to create instability in the transition matrices used by BULC. In our experiments, a too-large number of classes caused stray errors to appear unacceptably often in the output BULC-U classification, (e.g., stray clouds were displayed as Water). An Event classification with a too-small number of classes (e.g., 5 or fewer unsupervised classes) caused the resulting BULC-U classification to be less accurate, due to Event classes spanning multiple base classification classes—for example, containing both Forest and Shrubland. We advise that other users of BULC-U start with 20 unsupervised classes in a process of trial and error for creating Events.

4.5. Strengths of the BULC-U Method

The BULC-U methodology presents a considerable improvement to the overall BULC process, substantially reducing the time needed to produce an accurate time series. The original BULC algorithm required supervised classifications having the same legends for each Event, a process that demanded substantial amounts of human intervention, knowledge of the study area, and processing time. The adaptation of BULC to use unsupervised classifications allows the rapid creation of Events needed for BULC's Bayesian logic, by allowing the spectral characteristics of the imagery to directly drive the resulting time-series classification. The ease with which BULC-U generated a plausible classification can be a useful complement or starting point for more labor-intensive efforts like MapBiomas [41], which leverage a large amount of regional expert opinion and close observation to produce classifications with more detailed classes, though with substantially more effort.

One possible fruitful use of BULC-U is to update classifications to different time periods than that of the base image. Because most of a sufficiently large landscape does not change across the span of a few years or even decades, it should be possible to use the GlobCover 2009 classification to update classifications to years other than 2009. Because the vast majority of LULC pixels would not have changed their proper label in the intervening year, the same process described here could be able to be applied to earlier or later images. In the simplest case, there is nothing in this methodology that should inhibit the use of GlobCover 2009 to inform a BULC-U built using 2011 or 2012 Landsat imagery. Using this same logic, it is worth exploring BULC-U's potential to "leapfrog" to earlier dates of interest—2002 or 1986, for example. Although this is outside the scope of this manuscript, exploring the limits of that hypothesis will be the subject of future work.

5. Conclusions

Using unsupervised segmented Landsat classifications and the BULC-U algorithm, we were able to create a spatially refined map that was consistent with, but considerably improved from, the LULC labeling of the GlobCover 2009 classification. Although this technique has been shown here using Landsat 5 images and GlobCover 2009, the BULC-U algorithm is robust and general enough to use any classification legend and any satellite data. Future studies with BULC-U could include data from multiple sensors, as was done by Cardille & Fortin [34] and in review by Fortin et al. [42]. Additionally, this study used the unsupervised ISODATA classification algorithm to produce Events, and future work might compare and contrast the influence of other classification algorithms on the resulting BULC-U results.

A significant strength of the BULC-U process is that the resulting product is not entirely new, but is instead built on a foundation of an older expert-created classification. BULC-U uses data that may not have been readily available at the time of the original LULC classification to create a hybrid product with finer resolution and greater accuracy that is still compatible with the original classification. Those who have been using global classifications can continue to do so (with the same categories as before, if desired) with a finer-resolution data set that is highly consistent with the coarser source. It also opens the door to using accurate older, region-specific classifications to create new or extended time series.

As the era of open data continues, much more satellite data are available to researchers than even in the very recent past. At the same time, researchers now have decades of experience using existing classifications in a range of studies, including LULC change simulation models, carbon accounting analyses, and hydrologic studies. As more and more data emerges from archives for researchers, there will be a high priority on creating new data products that extend existing work but preserve data continuity. As a straightforward process that requires only a modest amount of expert remote sensing knowledge, BULC-U may be useful for a large set of applications.

Author Contributions: J.L. initiated this work as part of his MS thesis at McGill University, and did the initial development and coding of BULC-U. J.L. produced the first report of the method. M.T.C. and J.A.C. redeveloped the draft for publication with J.L., including figure/table design and execution. All three authors contributed to revisions.

Funding: This research was funded by a Google Earth Engine Research Award to develop BULC-U in Earth Engine.

Conflicts of Interest: The authors declare no conflicts of interest.

References

1. Brawn, J.D. Implications of agricultural development for tropical biodiversity. *Trop. Conserv. Sci.* **2017**, *10*. [CrossRef]
2. DeFries, R.S.; Rudel, T.; Uriarte, M.; Hansen, M. Deforestation driven by urban population growth and agricultural trade in the twenty-first century. *Nat. Geosci.* **2010**, *3*, 178–181. [CrossRef]
3. Costa, M.H.; Botta, A.; Cardille, J.A. Effects of large-scale changes in land cover on the discharge of the Tocantins River, Southeastern Amazonia. *J. Hydrol.* **2003**, *283*, 206–217. [CrossRef]
4. Cardille, J.A.; Bennett, E.M. Tropical teleconnections. *Nat. Geosci.* **2010**, *3*, 154–155. [CrossRef]
5. Turner, B.L., 2nd; Lambin, E.F.; Reenberg, A. The emergence of land change science for global environmental change and sustainability. *Proc. Natl. Acad. Sci. USA* **2007**, *104*, 20666–20671. [CrossRef] [PubMed]
6. Budiharta, S.; Meijaard, E.; Erskine, P.D.; Rondinini, C.; Pacifici, M.; Wilson, K.A. Restoring degraded tropical forests for carbon and biodiversity. *Environ. Res. Lett.* **2014**, *9*, 114020. [CrossRef]
7. Soares-Filho, B.; Rajao, R.; Macedo, M.; Carneiro, A.; Costa, W.; Coe, M.; Rodrigues, H.; Alencar, A. Cracking Brazil's Forest Code. *Science* **2014**, *344*, 363–364. [CrossRef] [PubMed]
8. Baccini, A.; Walker, W.; Carvalho, L.; Farina, M.; Sulla-Menashe, D.; Houghton, R.A. Tropical forests are a net carbon source based on aboveground measurements of gain and loss. *Science* **2017**, *358*, 230–233. [CrossRef] [PubMed]
9. Woodcock, C.E.; Allen, R.; Anderson, M.; Belward, A.; Bindschadler, R.; Cohen, W.; Gao, F.; Goward, S.N.; Helder, D.; Helmer, E.; et al. Free access to Landsat imagery. *Science* **2008**, *320*, 1011. [CrossRef] [PubMed]
10. Coppin, P.R.; Bauer, M.E. Digital change detection in forest ecosystems with remote sensing imagery. *Remote Sens. Rev.* **1996**, *13*, 207–234. [CrossRef]
11. Wulder, M.A.; Coops, N.C.; Roy, D.P.; White, J.C.; Hermosilla, T. Land cover 2.0. *Int. J. Remote Sens.* **2018**, *39*, 4254–4284. [CrossRef]

12. Wulder, M.A.; Masek, J.G.; Cohen, W.B.; Loveland, T.R.; Woodcock, C.E. Opening the archive: How free data has enabled the science and monitoring promise of Landsat. *Remote Sens. Environ.* **2012**, *122*, 2–10. [CrossRef]
13. Zhu, Z. Change detection using Landsat time series: A review of frequencies, preprocessing, algorithms, and applications. *ISPRS J. Photogramm. Remote Sens.* **2017**, *130*, 370–384. [CrossRef]
14. Verbesselt, J.; Zeileis, A.; Herold, M. Near real-time disturbance detection using satellite image time series. *Remote Sens. Environ.* **2012**, *123*, 98–108. [CrossRef]
15. Giri, C.; Pengra, B.; Long, J.; Loveland, T.R. Next generation of global land cover characterization, mapping, and monitoring. *Int. J. Appl. Earth Obs. Geoinf.* **2013**, *25*, 30–37. [CrossRef]
16. Yüksel, A.; Akay, A.E.; Gundogan, R. Using ASTER imagery in land use/cover classification of eastern Mediterranean landscapes according to CORINE land cover project. *Sensors* **2008**, *8*, 1237–1251. [CrossRef] [PubMed]
17. Drusch, M.; Del Bello, U.; Carlier, S.; Colin, O.; Fernandez, V.; Gascon, F.; Hoersch, B.; Isola, C.; Laberinti, P.; Martimort, P. Sentinel-2: ESA's optical high-resolution mission for GMES operational services. *Remote Sens. Environ.* **2012**, *120*, 25–36. [CrossRef]
18. Li, J.; Roy, D.P. A global analysis of Sentinel-2a, Sentinel-2b and Landsat-8 data revisit intervals and implications for terrestrial monitoring. *Remote Sens.* **2017**, *9*, 902. [CrossRef]
19. Loveland, T.R.; Reed, B.C.; Brown, J.F.; Ohlen, D.O.; Zhu, Z.; Yang, L.; Merchant, J.W. Development of a global land cover characteristics database and IGBP DISCover from 1 km AVHRR data. *Int. J. Remote Sens.* **2000**, *21*, 1303–1330. [CrossRef]
20. Tuanmu, M.N.; Jetz, W. A global 1-km consensus land-cover product for biodiversity and ecosystem modelling. *Glob. Ecol. Biogeogr.* **2014**, *23*, 1031–1045. [CrossRef]
21. Bicheron, P.; Amberg, V.; Bourg, L.; Petit, D.; Huc, M.; Miras, B.; Brockmann, C.; Hagolle, O.; Delwart, S.; Ranera, F.; et al. Geolocation Assessment of MERIS GlobCover Orthorectified Products. *IEEE Trans. Geosci. Remote Sens.* **2011**, *49*, 2972–2982. [CrossRef]
22. Bicheron, P.; Henry, C.; Bontemps, S.; Partners, G. *Globcover Products Description Manual*; MEDIAS-France: Toulouse, France, 2008; p. 25.
23. Bontemps, S.; Defourny, P.; Bogaert, E.; Arino, O.; Kalogirou, V.; Perez, J. *GLOBCOVER 2009–Products Description and Validation Report*; Université catholique de Louvain and European Space Agency: Louvain-la-Neuve, Belgium, 2011.
24. De Sy, V.; Herold, M.; Achard, F.; Beuchle, R.; Clevers, J.; Lindquist, E.; Verchot, L. Land use patterns and related carbon losses following deforestation in South America. *Environ. Res. Lett.* **2015**, *10*, 124004. [CrossRef]
25. Hu, S.; Niu, Z.; Chen, Y.; Li, L.; Zhang, H. Global wetlands: Potential distribution, wetland loss, and status. *Sci. Total Environ.* **2017**, *586*, 319–327. [CrossRef] [PubMed]
26. Sloan, S.; Jenkins, C.N.; Joppa, L.N.; Gaveau, D.L.; Laurance, W.F. Remaining natural vegetation in the global biodiversity hotspots. *Biol. Conserv.* **2014**, *177*, 12–24. [CrossRef]
27. Fritz, S.; See, L.; McCallum, I.; You, L.; Bun, A.; Moltchanova, E.; Duerauer, M.; Albrecht, F.; Schill, C.; Perger, C. Mapping global cropland and field size. *Glob. Chang. Biol.* **2015**, *21*, 1980–1992. [CrossRef] [PubMed]
28. Liu, J.; Liu, M.; Tian, H.; Zhuang, D.; Zhang, Z.; Zhang, W.; Tang, X.; Deng, X. Spatial and temporal patterns of China's cropland during 1990–2000: An analysis based on Landsat TM data. *Remote Sens. Environ.* **2005**, *98*, 442–456. [CrossRef]
29. Seppälä, S.; Henriques, S.; Draney, M.L.; Foord, S.; Gibbons, A.T.; Gomez, L.A.; Kariko, S.; Malumbres-Olarte, J.; Milne, M.; Vink, C.J. Species conservation profiles of a random sample of world spiders I: Agelenidae to Filistatidae. *Biodivers. Data J.* **2018**. [CrossRef] [PubMed]
30. Truong, T.T.; Hardy, G.E.S.J.; Andrew, M.E. Contemporary remotely sensed data products refine invasive plants risk mapping in data poor regions. *Front. Plant Sci.* **2017**, *8*, 770. [CrossRef] [PubMed]
31. Wilting, A.; Cheyne, S.M.; Mohamed, A.; Hearn, A.J.; Ross, J.; Samejima, H.; Boonratana, R.; Marshall, A.J.; Brodie, J.F.; Giordiano, A. Predicted distribution of the flat-headed cat Prionailurus planiceps (Mammalia: Carnivora: Felidae) on Borneo. *Raffles Bull. Zool.* **2016**, *33*, 173–179.
32. Schulp, C.J.; Alkemade, R. Consequences of uncertainty in global-scale land cover maps for mapping ecosystem functions: An analysis of pollination efficiency. *Remote Sens.* **2011**, *3*, 2057–2075. [CrossRef]

33. Schulp, C.J.; Alkemade, R.; Klein Goldewijk, K.; Petz, K. Mapping ecosystem functions and services in Eastern Europe using global-scale data sets. *Int. J. Biodivers. Sci. Ecosyst. Serv. Manag.* **2012**, *8*, 156–168. [CrossRef]
34. Cardille, J.A.; Fortin, J.A. Bayesian updating of land-cover estimates in a data-rich environment. *Remote Sens. Environ.* **2016**, *186*, 234–349. [CrossRef]
35. Gorelick, N.; Hancher, M.; Dixon, M.; Ilyushchenko, S.; Thau, D.; Moore, R. Google Earth Engine: Planetary-scale geospatial analysis for everyone. *Remote Sens. Environ.* **2017**, *202*, 18–27. [CrossRef]
36. Blaschke, T.; Burnett, C.; Pekkarinen, A. Image segmentation methods for object-based analysis and classification. In *Remote Sensing Image Analysis: Including the Spatial Domain*; De Jong, S.M., Van der Meer, F.D., Eds.; Springer: Dordrecht, The Netherlands, 2004; pp. 211–236.
37. Ball, G.H.; Hall, D.J. *ISODATA, A Novel Method of Data Analysis and Pattern Classification*; Stanford Research Institute: Menlo Park, CA, USA, 1965.
38. Richards, J.A.; Jia, X. *Remote Sensing Digital Image Analysis*; Springer: Berlin, Germany, 2006; p. 439.
39. Hansen, M.C.; Potapov, P.V.; Moore, R.; Hancher, M.; Turubanova, S.A.; Tyukavina, A.; Thau, D.; Stehman, S.V.; Goetz, S.J.; Loveland, T.R.; et al. High-resolution global maps of 21st-Century forest cover change. *Science* **2013**, *342*, 850–853. [CrossRef] [PubMed]
40. Congalton, R.G. A review of assessing the accuracy of classifications of remotely sensed data. *Remote Sens. Environ.* **1991**, *37*, 35–46. [CrossRef]
41. Souza, C.; Azevedo, T. *MapBiomas General Handbook*; MapBiomas: São Paulo, Brazil, 2017; pp. 1–23.
42. Fortin, J.A.; Cardille, J.A.; Perez, E. Multi-sensor detection of forest-cover change across five decades in Mato Grosso, Brazil. *Remote Sens. Environ.*. In Revision.

Article

Monitoring the Impact of Land Cover Change on Surface Urban Heat Island through Google Earth Engine: Proposal of a Global Methodology, First Applications and Problems

Roberta Ravanelli [1,*], Andrea Nascetti [1,2], Raffaella Valeria Cirigliano [1], Clarissa Di Rico [1], Giovanni Leuzzi [3], Paolo Monti [3] and Mattia Crespi [1]

1 Geodesy and Geomatics Division, DICEA—University of Rome La Sapienza, 00184 Rome, Italy; roberta.ravanelli@uniroma1.it (R.R.); andrea.nascetti@uniroma1.it (A.N.); valeria.cirigliano03@gmail.com (R.V.C.); clarissa.cdr@gmail.com (C.D.R.); mattia.crespi@uniroma1.it (M.C.)

2 Geoinformatics Division, Department of Urban Planning and Environment, KTH Royal Institute of Technology, 10044 Stockholm, Sweden

3 Department of Civil, Building and Environmental Engineering (DICEA)—University of Rome La Sapienza, 00184 Rome, Italy; giovanni.leuzzi@uniroma1.it (G.L.); paolo.monti@uniroma1.it (P.M.)

* Correspondence: roberta.ravanelli@uniroma1.it; Tel.: +39-06-4458-5087

Received: 27 June 2018; Accepted: 14 September 2018; Published: 18 September 2018

Abstract: All over the world, the rapid urbanization process is challenging the sustainable development of our cities. In 2015, the United Nation highlighted in Goal 11 of the SDGs (Sustainable Development Goals) the importance to "Make cities inclusive, safe, resilient and sustainable". In order to monitor progress regarding SDG 11, there is a need for proper indicators, representing different aspects of city conditions, obviously including the Land Cover (LC) changes and the urban climate with its most distinct feature, the Urban Heat Island (UHI). One of the aspects of UHI is the Surface Urban Heat Island (SUHI), which has been investigated through airborne and satellite remote sensing over many years. The purpose of this work is to show the present potential of Google Earth Engine (GEE) to process the huge and continuously increasing free satellite Earth Observation (EO) Big Data for long-term and wide spatio-temporal monitoring of SUHI and its connection with LC changes. A large-scale spatio-temporal procedure was implemented under GEE, also benefiting from the already established Climate Engine (CE) tool to extract the Land Surface Temperature (LST) from Landsat imagery and the simple indicator *Detrended Rate Matrix* was introduced to globally represent the net effect of LC changes on SUHI. The implemented procedure was successfully applied to six metropolitan areas in the U.S., and a general increasing of SUHI due to urban growth was clearly highlighted. As a matter of fact, GEE indeed allowed us to process more than 6000 Landsat images acquired over the period 1992–2011, performing a long-term and wide spatio-temporal study on SUHI vs. LC change monitoring. The present feasibility of the proposed procedure and the encouraging obtained results, although preliminary and requiring further investigations (calibration problems related to LST determination from Landsat imagery were evidenced), pave the way for a possible global service on SUHI monitoring, able to supply valuable indications to address an increasingly sustainable urban planning of our cities.

Keywords: SDG; surface urban heat island; Geo Big Data; Google Earth Engine; global monitoring service

1. Introduction

Cities all over the world are experiencing very fast growth and changes due to the current rapid urbanization. This is the reason why the United Nations included Goal 11: Make cities inclusive, safe, resilient and sustainable within the 17 Sustainable Development Goals (SDGs) issued in 2015. A key issue after the definition of the SDGs is the monitoring of their progress; with regard to SDG 11, there is a need for well globally-defined, easy computable and comparable indicators, representing different aspects of city conditions, obviously including the Land Cover (LC) changes and the urban climate with its most distinct feature, the Urban Heat Island (UHI) [1,2].

Specifically, the term UHI refers to the mesoscale phenomenon associated with higher atmospheric and surface temperatures occurring in urban environments compared to in the surrounding rural areas due to urbanization [3]. It is characterized by a large expanse of non-evaporating impervious materials covering a majority of urban areas with a consequent increase in sensible heat flux at the expense of latent heat flux. Therefore, the UHI and its spatio-temporal evolution are connected to the LC changes, mainly to the transformation from non-urban to urban LC.

Globally, urban footprints only account for approximately 2% of the planetary surface [4]. It was well documented that with the rapid urbanization and population growth, the increase in built-up land and the natural land replacement with artificial buildings has altered the surface energy budgets and the hydrological cycle and may affect local and regional climate by changing the physical properties of the Earth's surface [5]. Over the past decades, worldwide urbanization and climate change have been two interconnected processes describing the role of human activities in altering and modifying the climate system, which, in turn, may result in uncertain feedback for human welfare.

The UHI effects are exacerbated by the anthropogenic heat generated by traffic, industry and domestic heating, influencing the local climate through the compact mass of buildings that affect the balances of momentum, heat and moisture. The higher temperatures in UHIs increase air conditioning demands [6,7], with a retrofit effect of increasing the outdoor temperature even more, and may modify precipitation patterns. As a result, the magnitude and pattern of UHI effects have been major concerns of many urban climatology studies [8].

The UHI has been a well-known phenomenon since the XIX Century, and it has been studied for many decades; an excellent review of the state-of-the-art (measurements, models, result interpretations) is given in [9], which addresses the four different aspects of UHI, including the Surface Urban Heat Island (SUHI) besides others (canopy urban heat island, boundary urban heat island and subsurface urban heat island).

Traditionally, the UHI is investigated using ground-based in situ measurements of air temperature acquired by automatic weather stations (e.g., see [10]). Air temperature measurements allow one indeed to characterize the fine spatial and temporal variations of the UHI, and they are suitable for studying the impacts of local indicators of urbanization (e.g., sky view factor, floor area ratio) and climate factors (e.g., wind, cloud) contributing to the phenomenon [11–14]. Nonetheless, the urban meteorological networks are not always as complete as could be desired, and frequently, their stations are not uniformly distributed within the territory of the investigated cities. Consequently, large zones may remain without coverage, and it is not possible to analyze the different spatial pattern.

Another possible approach is to investigate the UHI phenomenon through numerical simulations. For instance, ongoing analyses about the UHI at the mesoscale level investigate the effectiveness of superficial properties by using the Weather Research and Forecasting (WRF) mesoscale simulation [15,16]. Anyway, additional information and/or local measurements are still required for estimating the parameters of the simulation models [17].

On the other hand, remote sensors on board satellites are able to acquire thermal infrared data from which it is possible to retrieve the urban Land Surface Temperature (LST), allowing thus the study of the SUHI [18]. The major advantage of the SUHI is that it can be calculated easily for a large number of cities across large spatial domains. This allows to investigate the role of urbanization in influencing the SUHI over large regions [19–26]. A wide literature about the determination of the urban LST is therefore available, based on remote sensing data and its connection with SUHI.

At the same time, though, it is not easy to handle and process the huge amount of Earth Observation (EO) data actually available because of the continuously increasing number of sensors, spatial and temporal resolutions and this will be even more evident in the near future with the development of the micro-satellite constellations. As a matter of fact, even today, much of the archived EO imagery is underutilized despite modern computing and analysis infrastructures, mainly due to processing limitations when faced with a huge amount of data on standard computers and of the high technical expertise needed to use traditional supercomputers or large-scale commodity cloud computing resources [27]. In the present era of Geo Big Data, new computing instruments to support remote sensing investigations are therefore necessary, in order to fully exploit and make available the information content of the acquired data.

This work is precisely included in this background: the aim is to show the present potential of Google Earth Engine (GEE) to process the huge and continuously increasing free EO Big Data (i.e., Landsat and Sentinels' imagery) for long-term spatio-temporal monitoring SUHI and its connection with the LC changes.

GEE is indeed the computing platform recently released by Google "for petabyte-scale scientific analysis and visualization of geospatial datasets". Through a dedicated High Performance Computing (HPC) infrastructure, it enables researchers to easily and quickly access more than thirty years of free public data archives, including historical imagery and scientific datasets, for developing global and large-scale remote sensing applications. In this way, many of the limitations related to the data downloading, storage and processing, which usually occur when such a large amount of Geo Big Data is analyzed, are effortlessly overcome [28,29]. These features have attracted the attention of researchers from different disciplines, which have integrated GEE into several third-party applications [28]. Among these, Climate Engine (CE) [30] can play an important role in the field of large-scale climate monitoring: it is a free web application powered by GEE that enables users to process, visualize, download and share several global and regional climate and remote sensing datasets and products in real time.

Therefore, by continuing the work started in [31], a general methodology for the large-scale monitoring of the SUHI was implemented through the joint use of GEE with CE and tested on six different U.S. cities. The work is organized as follows. In Section 2, an overview of the EO Big Data context is given, paying specific attention to the most promising analysis platforms and underlining the reasons why GEE was selected among them. In Section 3, the proposed methodology is described in detail. A large-scale spatio-temporal procedure was implemented thanks to GEE, also benefiting from the already established CE tool, to extract the LST from Landsat imagery, and a simple indicator was introduced to represent the SUHI evolution and its dependence on LC changes (Figure 1).

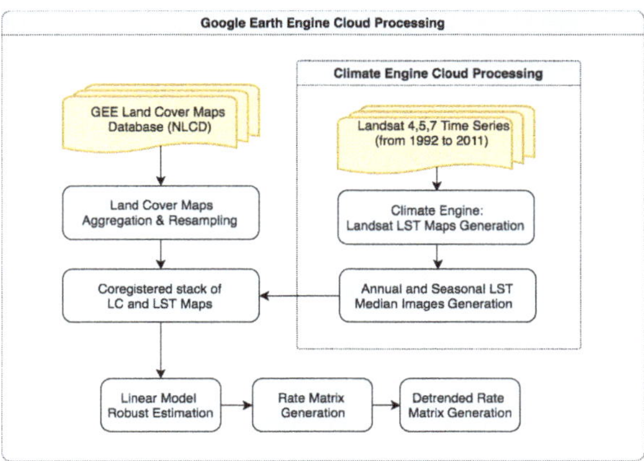

Figure 1. The workflow of the methodology.

In Section 4, the results obtained for the investigated six US cities are presented and analyzed: more details are outlined and discussed for the metropolitan area of Atlanta, then the main methodological conclusions derived with regard to Atlanta are applied to the other five metropolitan areas. Finally, in Section 5, conclusions are drawn and future prospects are outlined.

2. Geo Big Data Analysis Platforms

The latest generations of EO satellites are acquiring increasingly significant volumes of data with a full global coverage, higher spatial resolution and impressive temporal resolution. A new era of EO based on the Big Data paradigm is coming, where innovative tools able to support remote sensing investigations will be increasingly necessary in order to exploit all the potentially available geospatial information contained therein, to effectively contribute to study phenomena at the global scale and to monitor them in (near) real time [32,33]. Indeed, it is no longer technically feasible or financially affordable to transfer the data and use traditional local processing methods to address this data scaling challenge. The size of the data and complexities in preparation, handling, storage and analysis are the key aspects that should be taken into account [27].

This is the reason why, in the last few years, both satellite EO technology and innovative computing infrastructures have advanced significantly. Such solutions, developed by national space agencies, open source communities and private companies have a great potential to streamline data distribution and processing capabilities, while simultaneously lowering the technical barriers for users to exploit the EO Big Data and to develop innovative services.

In this context, ESA started the EO Exploitation Platforms initiative in 2014, a set of research and development activities that in the first phase (up to 2017) aimed to create an ecosystem of interconnected Thematic Exploitation Platforms (TEPs) on European footing, addressing the most important topics on remote sensing (i.e., coastal, forestry, hydrology, geohazards, polar, urban themes and food security). Briefly, TEPs are a collaborative, virtual work environment providing access to EO data and tools, processors and IT resources required using one coherent interface [34]. In 2017, the European Commission in collaboration with ESA and EUMETSAT launched an initiative within the Copernicus program to develop data and information access services that facilitate access to EO Big Data.

Earth Servers 1 and 2 are instead two EO Big Data cubes founded by the Horizon 2020 framework [35,36]. Their service interface is rigorously based on the Open Geospatial Consortium "Big Geo Data" standards, web coverage service and web coverage processing service [37].

Considering the high potential of EO Big Data also in commercial applications, Amazon included the full archive of Copernicus Sentinel-2 data in the Amazon Web Service platform through a collaboration with the start-up Sinergise [38,39]. They developed the Sentinel-Hub, a service-oriented satellite imagery infrastructure that takes care of all the complexity of handling of the satellite imagery archive and makes it available for end-users via easy-to-integrate web services [40].

Since 2015, the innovative GEE processing platform dramatically transformed and pushed one step ahead the EO satellite data user community. GEE consists of a multi-petabyte analysis-ready data catalog with a high-performance, intrinsically parallel computation service. It is accessed and controlled through an Internet-accessible application programming interface and an associated web-based interactive development environment that enables rapid prototyping and visualization of results [28]. Following the user demands, GEE has created a technological solution that removes the burden of data preparation, yields rapid results and maintains an active and engaged global community of contributors. Extensive research and development activity has been performed using GEE, and new applications and services can be deployed to deliver great impact to important environmental, economic and social challenges, including at the local, regional and global scales [41–43]. Such applications highlight the value of exploiting the EO data, pointing out that the main challenge is in providing the proper connection between data, applications and users. Nowadays, GEE appears as the Geo Big Data analysis platform as more functional from an operational perspective, powerful and at the same time easy to use, and this is the reason why it was selected to carry out the investigations herein presented and discussed.

3. Data and Methodology

As mentioned before, in this work, a general methodology able to investigate the temporal variations of the SUHI as a whole and its connection with LC changes was implemented through the GEE platform. The main novelty of this approach lies indeed in its easy applicability to the analysis of a huge amount of data, thanks to the Big Data processing capabilities of GEE.

Specifically, a large-scale spatio-temporal analysis was carried out in order to investigate the possible connections between the Land Surface Temperature (LST) trends and Land Use/Land Cover (LULC) changes.

The study was performed through the joint use of GEE and CE, focusing the analysis on six different U.S. Metropolitan Areas (MAs) (Figure 2), characterized by different climate conditions and the availability of a long time series of EO data: Atlanta (Georgia), Boston (Massachusetts), Chicago (Illinois), Houston (Texas), Phoenix (Arizona) and San Francisco (California).

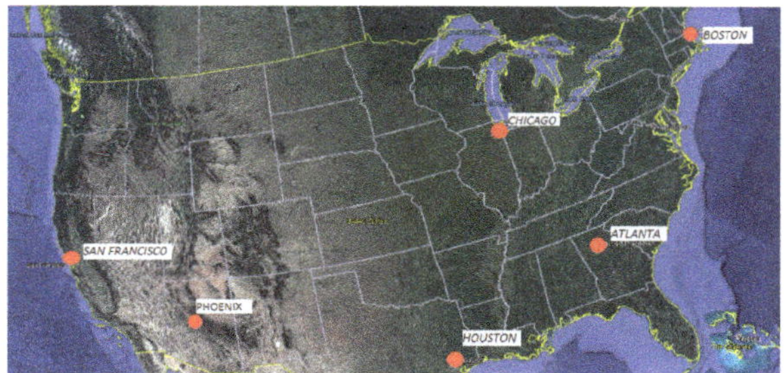

Figure 2. The six investigated U.S. Metropolitan Areas (MAs).

In the last few decades, all these cities have experienced a significant urban expansion, which has profoundly modified the main features of their territories. Residential suburbs, commercial and industrial areas have replaced forests and/or agricultural hinterlands surrounding the traditional downtown urban centers: soils that were once permeable and wet have been transformed into waterproof and dry surfaces. Such growth in the peri-urban fringe has resulted in a loss of cultivated areas and natural open spaces: the low values of albedo, vegetative cover and moisture availability in combination with the presence of high levels of anthropogenic heating can thus strengthen the UHI intensity. The selected U.S. MAs therefore represent valuable test sites to assess the effectiveness of the proposed methodology, hereafter described.

3.1. Land Surface Temperature

The CE web application was used to compute the annual median of the LST over the Landsat 4/5/7 top of atmosphere reflectance data, for every year of the two decades comprised between 1992 and 2011 in each of the Regions Of Interest (ROIs) corresponding to the six above-mentioned U.S. MAs. Each ROI was selected in order to assure enough surface variability and, at the same time, to consider surface portions remained unaltered and purely rural or natural.

Specifically, more than 6000 Landsat images were processed within CE. In general, their seasonal distribution is well balanced between the cold and the warm period, as reported in Table 1. Furthermore, the use of a robust estimator such as the annual median guarantees a good reliability of the methodology, especially at the beginning of the analysis period, where the number of available images is lower and sometimes slight inhomogeneities are present.

Table 1. Number of Landsat images available in the Climate Engine (CE) platform for each of the investigated U.S. MA, respectively in the warm (w) period (March, April, May, June, July, August) and cold (c) period (September, October, November, December, January, February). The orbits of Landsat satellites are solar synchronous, and passages are between 10:00 and 10:30 mean local time at the Equator, in order to provide maximum illumination with minimum water vapor (haze and cloud build-up) [44].

	1992		1993		1994		1995		1996		1997		1998		1999		2000		2001	
MA	w	c	w	c	w	c	w	c	w	c	w	c	w	c	w	c	w	c	w	c
Atlanta	13	19	27	17	16	18	19	15	28	17	25	20	30	24	33	35	41	39	40	36
Boston	17	25	22	20	22	23	22	17	18	13	15	17	21	18	35	37	42	40	42	48
Chicago	18	17	23	18	28	20	23	13	16	12	21	19	20	21	35	42	40	43	41	35
Houston	19	11	19	13	16	18	15	15	20	15	21	19	19	14	23	34	36	34	38	31
San Francisco	18	17	23	18	28	20	23	13	16	12	21	19	20	21	35	42	40	43	41	45
Phoenix	20	14	28	25	23	21	21	21	34	34	33	29	31	25	33	43	53	50	47	50
	2002		**2003**		**2004**		**2005**		**2006**		**2007**		**2008**		**2009**		**2010**		**2011**	
MA	w	c	w	c	w	c	w	c	w	c	w	c	w	c	w	c	w	c	w	c
Atlanta	31	34	42	37	51	42	49	31	44	38	48	31	45	38	49	42	56	50	56	41
Boston	38	34	43	40	43	40	50	40	47	43	50	37	45	34	46	40	48	40	46	41
Chicago	36	42	48	42	43	40	48	45	46	37	53	37	44	31	45	40	52	37	46	28
Houston	35	34	33	31	40	33	41	32	36	34	33	30	39	40	38	33	39	33	38	31
San Francisco	36	42	48	42	43	40	48	45	46	37	53	37	44	31	45	40	52	37	46	28
Phoenix	41	42	56	57	59	54	63	49	57	52	57	42	67	57	67	61	64	60	68	57

Hence, for every MA, a set of 20 thermal images (one per year) was retrieved from CE: the value stored in each pixel is the LST annual median computed over all the Landsat images available in CE for the considered year.

As an example, Figure 3 gives an overall look at the investigated long-term phenomenon, as detected by considering the largest available set of Landsat imagery. In particular, it shows two of the LST annual median maps obtained for the Atlanta MA, respectively for the years 1994 (left panel) and 2010 (right); these two years were selected (as close as possible to the start and the end of the investigated period 1992–2011) since for there is a good balance between the number of images acquired during the cold and the warm seasons (see Table 1), and in this way, the seasonal effects are filtered out. Just a simple visual comparison of these two panels highlights a global increase of the LST.

Thus, with the aim of seeking the simplest suitable model for the temporal variation of the LST, an analysis of the surface temperature was performed at the pixel scale over the 20 years, on the basis of the 20 mentioned computed LST annual median maps. The analysis showed that a linear model can be considered as a reasonable approximation of the phenomenon. Thus, for every pixel of each ROI, the parameters of a simple linear model:

$$T = T_0 + r\,(t - t_0) \quad (1)$$

describing the LST trend as a function of time were robustly estimated; here, T is the LST, T_0 is the estimated LST referring to t_0 (the analysis starting year, here 1992), t is the current time and r is the rate of the temperature variation per year. In particular, the original LST images retrieved from CE were undersampled with a factor of 10, in order to speed up the estimation of the parameters of the linear model (Equation (1)), computed through the robust regression method [45,46]. Therefore, since the LST images have a Ground Sampling Distance (GSD) of 30 m, as the raw Landsat images, the linear model parameters were estimated at a 300-m spatial resolution. Figure 4 reports the maps of the T_0 (left panel) and r (right panel) parameters obtained for the Atlanta MA.

Hence, it is worth trying to investigate the possible relation of such observed LST trends, and their relative growth rates, with the corresponding LC transformations.

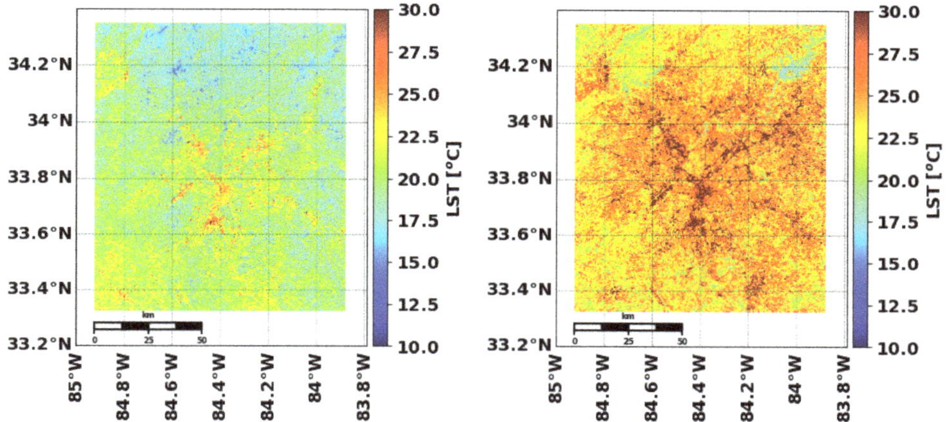

Figure 3. Maps of the LST annual median for 1994 (**left**) and 2010 (**right**), computed through CE over all the Landsat images available for the considered year within the ROI of Atlanta MA. The pixel size is 30 m, as the raw Landsat images from which they derive.

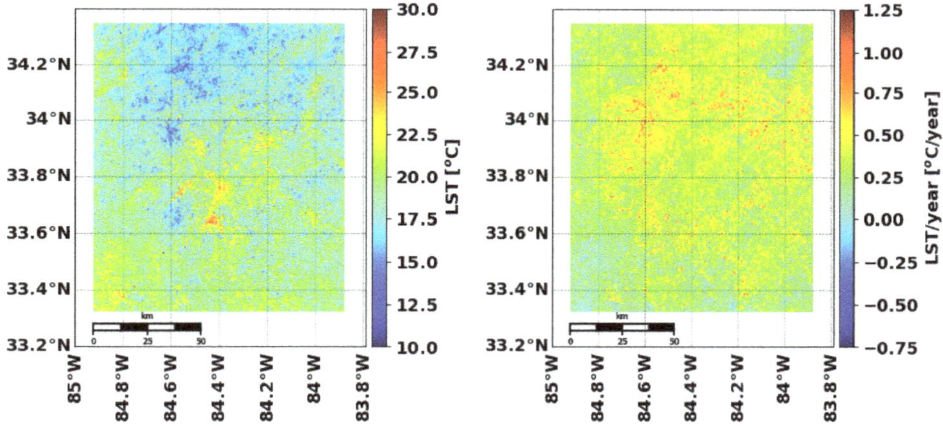

Figure 4. Constant (left panel) and rate (right panel) parameters of the LST linear model of Equation (1) for the Atlanta MA (pixel size of 300 m).

3.2. National Land Cover Database

The USGS National Land Cover Database (NLCD) [47] was used to assess the LC changes that occurred in the six investigated MAs during the twenty years under consideration. The NLCD is a 30-m Landsat-based land cover database spanning four periods (1992, 2001, 2006 and 2011). It was directly retrieved from GEE on the ROIs corresponding to the six MAs for the years 1992 and 2011.

Specifically, the database provides an LC classification for functional classes whose identification codes suffered some modifications during the considered interval 1992–2011: indeed, the NLCD classification scheme used for the NLCD1992 product differs slightly from that adopted in the more recent products (i.e., NLCD2001, NLCD2006, NLCD2011). In order to overcome this problem and to compare LC homogeneous data, the various LC classes were grouped into main three classes related to urbanized, cultivated and forest/shrubland areas as follows:

- the *Urbanized* class includes the classes *Low Intensity Residential*, *High Intensity Residential* and *Commercial/Industrial/Transportation* of the NLCD1992 and the classes *Developed Open Space*, *Developed Low Intensity*, *Developed Medium Intensity* and *Developed High Intensity* of the NLCD2011;
- the *Cultivated* class includes the classes *Pasture/Hay*, *Row Crops*, *Small Grains* and *Fallow* of the NLCD1992 and the classes *Pasture/Hay* and *Cultivated Crops* of the NLCD2011;
- the *Forest/Shrubland* classes include the classes *Deciduous Forest*, *Evergreen Forest* and *Mixed Forest* for both the considered years and the classes *Shrubland* of the NLCD1992 and *Dwarf Scrub*, *Shrub/Scrub* of the NLCD2011.

Moreover, after grouping into the chosen three main classes, an undersampling with the same factor of 10 as before was performed on the LC database; in this way, the spatial resolution of the derived LC database for the investigated MAs was equal (300 m GSD) to the one of the derived parameter maps, allowing a one-to-one pixel correspondence. This is a simplification, but it is convenient to highlight the main influences of the long-term relevant LC changes (from rural-cultivated and forest/shrubland to urban areas and vice versa) on the variations experienced by the LST over the considered period.

3.3. Correlation between Land Surface Temperature Data and Land Use Data

A spatial analysis was performed to find a possible correlation between the observed LST trends and the corresponding changes in the LC data. In other words, the specific aim was to verify whether the ROI pixels that experienced the same LC changes during the twenty years showed similar LST growth rates or not. Operatively, on the basis of the adopted linear temperature growth model (Equation (1)), a (3×3) matrix (*Rate Matrix*) was computed for each MA investigated for the rate (r) parameter. In particular, each cell (i, j) of this matrix (Figure 5):

- aggregates all the ROI pixels characterized by the LC class i in the initial year (1992) and the LC class j in the final year (2011);
- stores in three different layers the hereafter described statistical parameters (mean, median, numerosity, Permanence/Variation Index), computed considering the values of r of all the ROI pixels that underwent the LC change from class i to class j.

A schematic representation of the obtained *Rate Matrix* is shown in Figure 5.

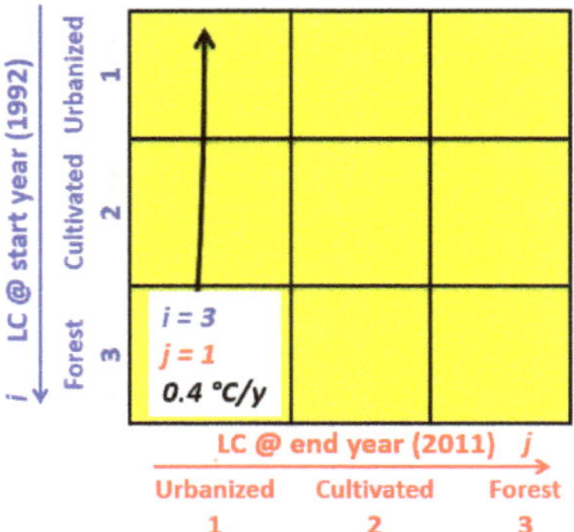

Figure 5. Representation of the LC changes within the *Rate Matrix*: the row index (i) represents the initial LC class and the column index (j) the final LC class. As an example, the highlighted cell represents the transition from the *Forest* class ($i = 3$) to the *Urbanized* class ($j = 1$), with an estimated LST increasing rate of 0.4 °C/year.

3.3.1. Numerosity

This layer quantifies the number of ROI pixels effectively involved in the change from the LC class i to the LC class j. Therefore, the values on the main diagonal indicate the number of stable pixels, which did not change their original LC and which constitutes, as expected, the largest portion. Then, there are the pixels that from purely rural classes (*Cultivated* or *Forest/Shrubland*) turned into urban ones (lower triangle) or vice versa, (upper triangle); finally, there are also pixels changing within rural classes (from *Cultivated* to *Forest/Shrubland* or vice versa). Recalling that the pixels have a 300-m GSD, it is possible to infer the surface of the ROI that underwent a specific LC change from their numerosity.

3.3.2. Mean, Median

In these layers, the mean and the median of the parameters of the linear model describing the LST trend for the considered LC change from class i to class j are stored. The values obtained are very similar for both statistical parameters, pointing out the substantial absence of outliers. Therefore, for the sake of brevity, only the mean value was considered in the analysis.

3.3.3. Permanence/Variation Index

The overall evaluation of the LC changes, which occurred in the investigated ROIs, has been represented through the herein introduced temporal Permanence/Variation Index, defined as follows:

$$I_v = \frac{n_{ij}}{\sum_j n_{ij}} \quad (2)$$

where n_{ij} is the number of the ROI pixels involved in the change from the LC class i to the LC class j. Therefore, I_v represents an index of permanence for the elements belonging to the main diagonal (here, usually the highest values are observed), whereas it represents an index of variation for the elements in the lower and upper triangles, defining the frequency of occurrence of the changes experienced by each individual LC.

4. Study Regions of Interest

A preliminary deeper analysis was developed for the Atlanta MA, following the above described procedure and performing also a comparison between the LST and the air temperature measured on the ground at the time of each image acquisition. Then, considering the results obtained on Atlanta MA, the analysis was replicated for all the other five MAs.

4.1. Metropolitan Area of Atlanta

It is widely acknowledged that for the past 30 years, the city of Atlanta has undergone heavy alterations in its LC: urban growth, sprawl, loss of forest and cropland have modified drastically its territorial features [48]. Such a tendency is confirmed by a visual analysis of the aggregated LC database computed as described in Section 3.2 and shown in Figure 6, where the *Urbanized*, *Forest/Shrubland* and *Cultivated* LC classes are depicted. Obviously, also the I_v reflects such urban growth: over only 20 years, about 40% of *Cultivated* and *Forest/Shrubland* pixels became *Urbanized*. Specifically, the developed methodology was applied to an ROI covering an area of about 11,000 km^2, corresponding to the city of Atlanta and its surroundings.

The elements of the *Rate Matrices* (Figure 7a) show the presence of the already mentioned global LST increasing trend, which is independent of the specific LC variation. Nonetheless, it can be observed how the highest increasing rate corresponds precisely to the LC change from the class *Forest* to *Urbanized* (0.41 °C/year), immediately followed by the *Cultivated* to *Urbanized* transformation (0.39 °C/year). Indeed, the substitution of rural and natural areas with dry, impervious surfaces (roofs, asphalted surfaces, etc.) results in less moisture available to keep the soil cool.

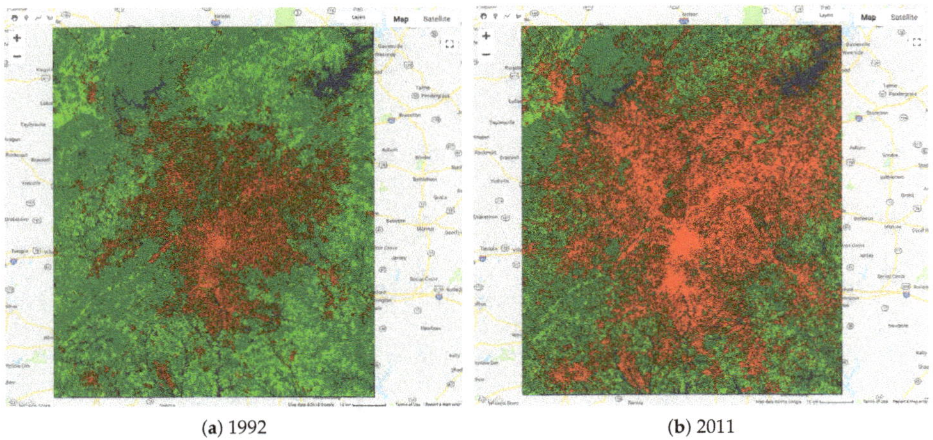

(**a**) 1992 (**b**) 2011

Figure 6. Atlanta MA: LC situation in 1992 (**a**) and in 2011 (**b**). In red, the *Urbanized* class, in light green, the *Cultivated* class, in dark green, the *Forest* and *Shrubland* classes.

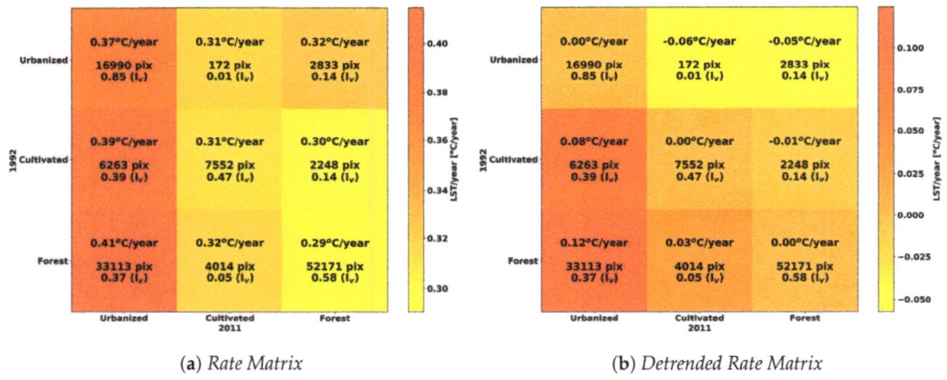

(**a**) *Rate Matrix* (**b**) *Detrended Rate Matrix*

Figure 7. Atlanta MA: *Rate* and *Detrended Rate Matrices*.

On the other hand, even considering the LC stable pixels, it is possible to highlight quite high increasing rates, in the range of 0.29–0.37 °C/year, which appear unrealistic even in the presence of the well-known global warming phenomenon. Therefore, considering a small portion (about 13.7 km^2, Figure 8) of the urban area located around a permanent weather station for air temperature measurement [49], an extensive comparison was performed between the LST and the air temperature measured at the time (with less than one hour approximation) of each image acquisition (around 10:00, local time) over the period 1992–2011. It is known [9] that in the morning, the difference between LST and air temperature in urban areas can be positive for up to 10 °C; anyway, the long-term trends of both temperatures should be equal. Actually, the median temperatures difference for 1992 is around 6 °C, whereas the increasing rates were respectively estimated as equal to 0.22 °C/year for LST (probably due to the fact that small green areas are included in this urban area) and to 0.08 °C/year for air temperature, which appears definitely more realistic (Figure 9). A calibration issue in the LST retrieval procedure from Landsat imagery was therefore supposed to explain such a difference. In addition, the LSTs intra-comparison retrieved from Landsat 5 and Landsat 7 in the available common interval

(1999–2011) over the same small portion of the urban area evidenced an anomalous increasing rate difference, up to 0.38 °C/year.

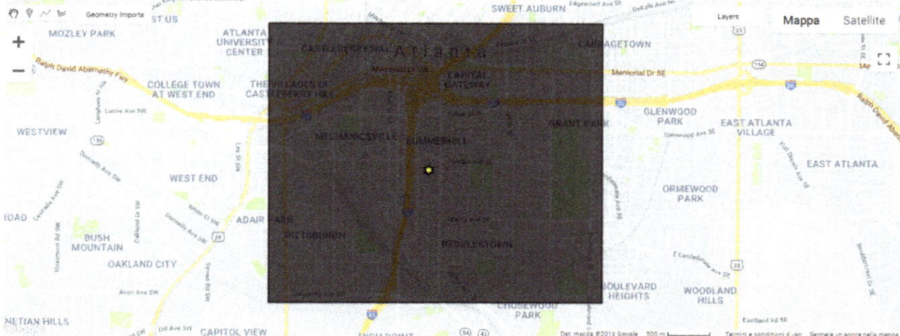

Figure 8. The ROI around the Georgia State University weatherSTEM station, Peoplestown (GA), considered for the LST vs. air temperature comparison (the star denotes the location of the permanent weather station).

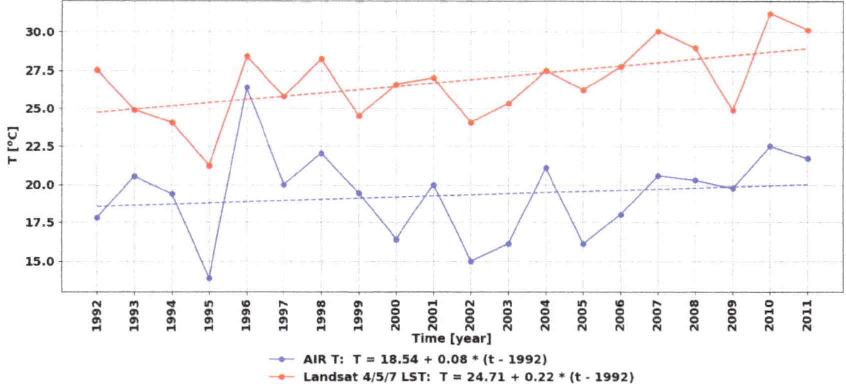

Figure 9. Comparison of the LST retrieved from the Landsat data with the air temperature measured at the Georgia State University weather station.

Consequently, the results related to the LST increasing rates were judged as unreliable; this is the reason why the *Rate Matrix* was detrended for the general LST increase, in order to highlight the net effects of the LC changes and the specific role of the urbanization. In detail, the LST trend was separately removed from each row of the *Rate Matrices*, by subtracting the values pertaining to the main diagonal (i.e., the cell representing the permanence in the same LC class during the investigated period) from those related to the other two cells (which represent the changes among the LC classes). In this way, the so-called *Detrended Rate Matrix* was computed, in which only the LST increase effectively due to the LC variation is considered, and obviously, the values on the main diagonal are zero. It is worth underlining once more that the lower triangle cells (2,1), (3,1) and (3,2) (according to the description given for the *Rate Matrix* and represented in Figure 5) of the *Detrended Rate Matrix* represent respectively the net effects of temperature increase due to the LC changes from *Cultivated* to *Urbanized*, from *Forest/Shrubland* to *Urbanized* and from *Forest/Shrubland* to *Cultivated*. On the contrary, the *Detrended Rate Matrix* upper triangle cells (1,2), (1,3) and (2,3) represent the net effects of the inverse LC changes.

Hence, the values obtained for the *Detrended Rate Matrix* are shown in Figure 7b, where all the changes represented in the lower triangle of the matrix are positive (0.03 °C/year for *Forest/Shrubland* to *Cultivated* LC change) or strongly positive (0.08 °C/year for *Cultivated* to *Urbanized* LC change and 0.12 °C/year for *Forest/Shrubland* to *Urbanized* LC change). Moreover, as regards the reverse LC changes from *Urbanized* to the *Cultivated* or *Forest/Shrubland* classes, it is possible to notice a decreasing LST trend, as expected (Figure 7b); anyway, given the small number of pixels involved in those two transformations, the results are less significant.

It is worth underlining that the meaning of the *Rate Matrix* is doubtful for the discussed possible calibration issues of the procedure implemented in CE to retrieve the LST, whereas the *Detrended Rate Matrix* is clearly able to highlight the connection between LC changes and the increase of urban LST and SUHI. Therefore, only the *Detrended Rate Matrix* has been considered significant to represent the results of the long-term monitoring of the impact of LC changes on the SUHI.

Finally, since the number of available images is slightly unbalanced between the warmer and the colder periods (Table 1), the analysis was repeated considering them separately. However, the results (Figures 10 and 11) suggest that the seasonal effect does not have a significant influence on the results in the case of the Atlanta MA.

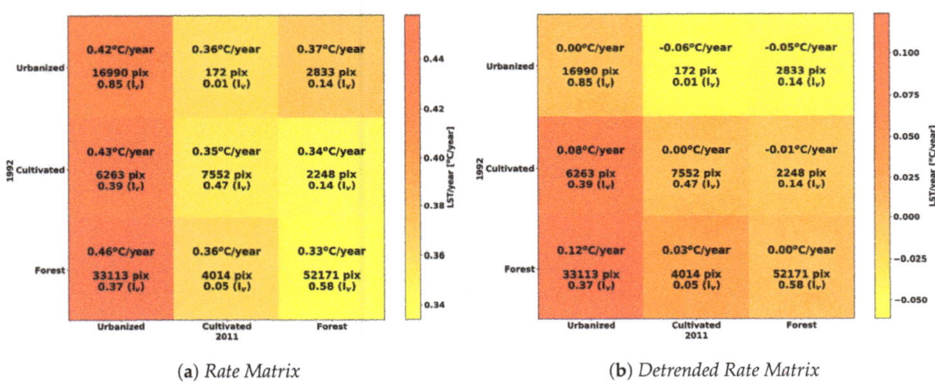

(**a**) *Rate Matrix* (**b**) *Detrended Rate Matrix*

Figure 10. Atlanta MA warm period: *Rate* and *Detrended Rate Matrices*.

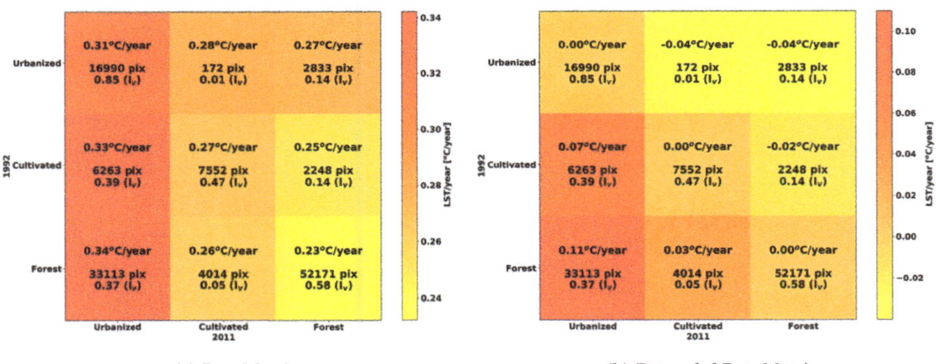

(**a**) *Rate Matrix* (**b**) *Detrended Rate Matrix*

Figure 11. Atlanta MA cold period: *Rate* and *Detrended Rate Matrices*.

4.2. Metropolitan Area of Boston

In the case of Boston MA, the investigated ROI includes both the cities of Boston and Providence (Figure 12). Boston MA is characterized by a much colder climate than Atlanta, and it was selected in order to understand the behavior of the SUHI in a region surrounded by the sea and to analyze if and how such proximity affects the phenomenon.

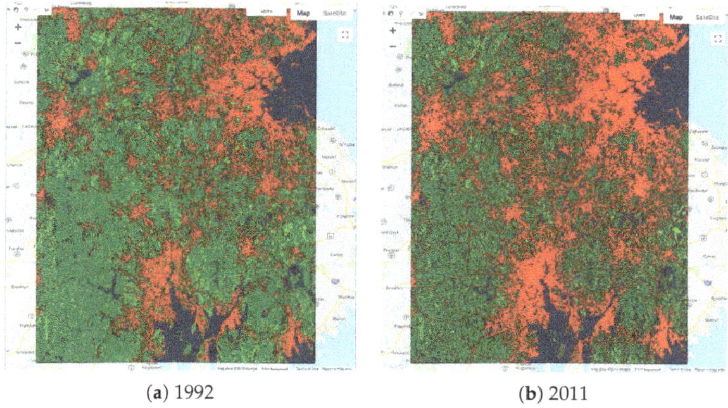

(a) 1992 (b) 2011

Figure 12. Boston MA: LC situation in 1992 (a) and in 2011 (b). In red, the *Urbanized* class, in light green, the *Cultivated* class, in dark green, the *Forest* and *Shrubland* classes.

The *Detrended Rate Matrix* (Figure 13) reveals that the LST again increases for the LC transitions from *Cultivated* or *Forest* to *Urbanized*. On the contrary, when the LC changes from *Urbanized* to *Forest*, the LST decreases. Nevertheless, a comparison with Atlanta MA highlights that in the Boston MA, the LST increase due to urbanization is significantly lower, despite the values of the Permanence/Variation Index I_v being quite similar (except for the LC change from *Forest* to *Urbanized*, which is lower for Boston). This is likely due to the effect of the Atlantic Ocean.

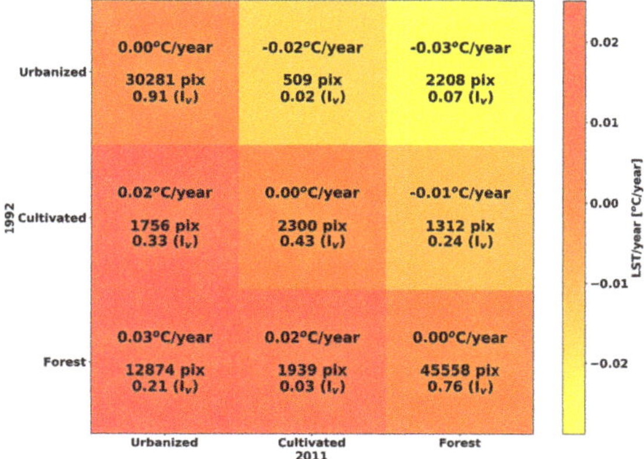

Figure 13. Boston MA: *Detrended Rate Matrix*.

4.3. Metropolitan Area of Chicago

In the case of Chicago MA, the developed methodology was applied to an ROI covering an area of about 11,000 km^2, corresponding to the city of Chicago and its surroundings (Figure 14).

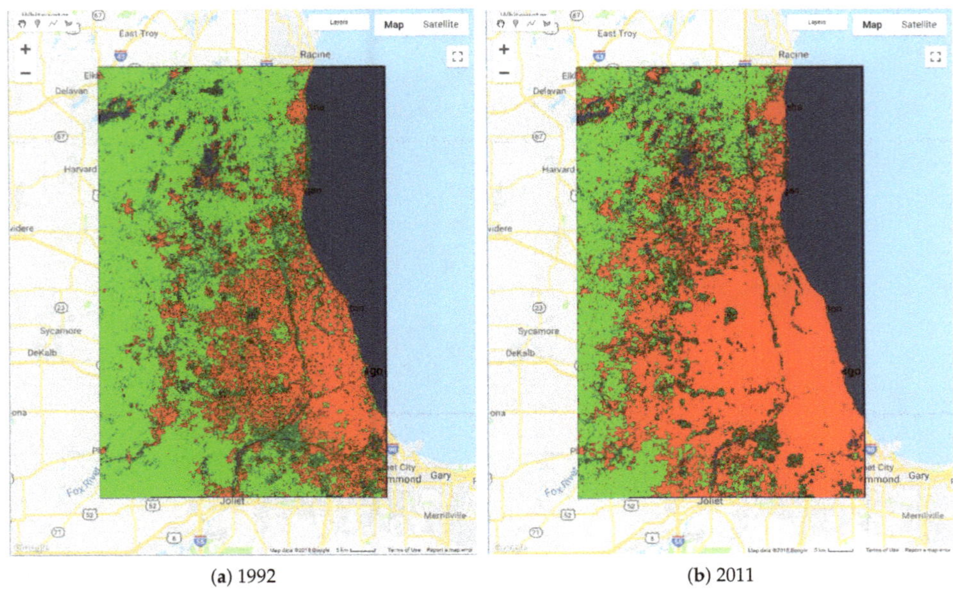

(**a**) 1992 (**b**) 2011

Figure 14. Chicago MA: LC situation in 1992 (**a**) and in 2011 (**b**). In red, the *Urbanized* class, in light green, the *Cultivated* class, in dark green, the *Forest* and *Shrubland* classes.

Considering the *Detrended Rate Matrix* (Figure 15), the highest increasing rate (0.04 °C/year) occurs in the case of the LC change from *Cultivated* to *Urbanized*, characterized by a high Permanence/Variation Index ($I_v = 0.34$). The highest I_v occurs for the *Forest*/*Urbanized* change ($I_v = 0.51$), which, however, involves a lower number of pixels; in this case, the LST increasing rate is lower, amounting to 0.03 °C/year (Figure 15). Finally, the changes from urban LC to rural/natural LCs are not relevant since they involve a very limited number of pixels.

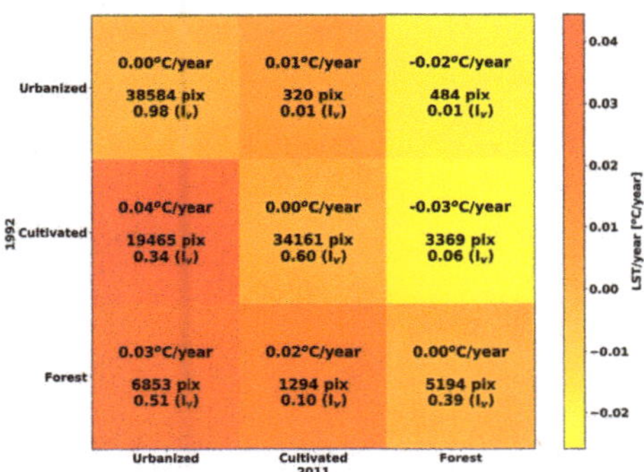

Figure 15. Chicago MA: *Detrended Rate Matrix*.

4.4. Metropolitan Area of Houston

In the last few decades, Houston has been subjected to a significant urban growth [50] that has altered considerably its territorial features, as shown in the aggregated LC maps (Figure 16).

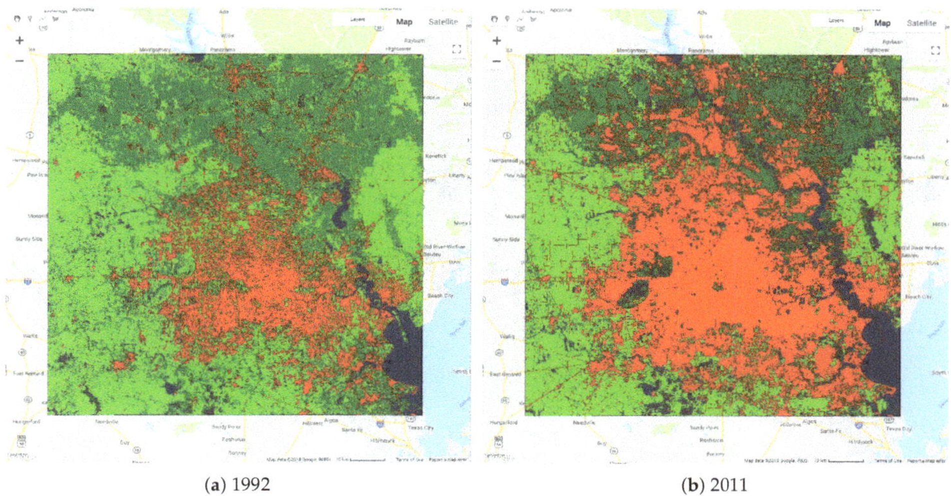

Figure 16. Houston MA: LC situation in 1992 (**a**) and in 2011 (**b**). In red, the *Urbanized* class, in light green, the *Cultivated* class, in dark green, the *Forest* class.

For Houston MA, the results are similar to those obtained for Atlanta MA, but with lower increasing rates (Figure 17): the LST increases in the cells aggregating the ROI pixels in which the rural/natural-urban transition occurs and decreases in the opposite transformation, even if in this particular case, the number of pixels involved is low. Indeed, in this case, the LST increasing rate is equal to 0.02 °C/year for the *Cultivated*/*Urbanized* changes, and it is much lower than the value obtained in the *Forest*/*Urbanized* changes, which amounts to 0.10 °C/year. As regards the Permanence/Variation Index I_v, it differs by about 10% in the two cases, even if the final altered surface is slightly wider for the *Cultivated*/*Urbanized* transition; indeed, in 20 years, about 40% of the forest has been urbanized against 30% of the cultivated areas, which is evident also in Figure 16.

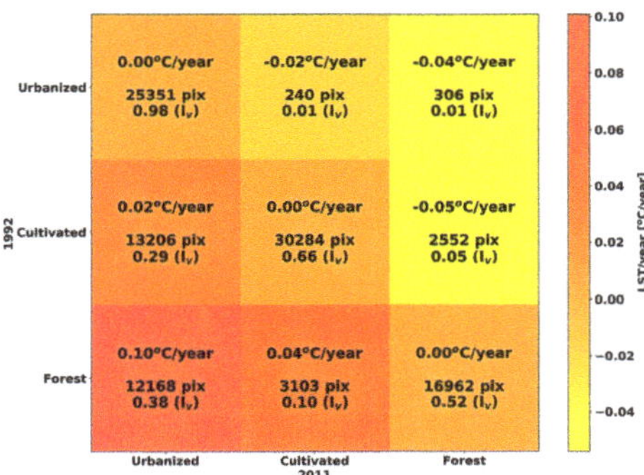

Figure 17. Houston MA: *Detrended Rate Matrix.*

4.5. Metropolitan Area of Phoenix

The developed methodology was applied to the MA of Phoenix [51–57]. Owing to its particular location (it is located in the northeastern reaches of the Sonoran Desert), the Phoenix MA climatic conditions are significantly different from those observed for the other investigated areas. Indeed, the *Forest/Shrubland* LC class, here mainly related to shrublands, covers wide portions of the ROI (Figure 18). Nonetheless, the developed methodology is once again able to detect effectively the LST trends (Figure 19) connected to LC changes (Figure 19). Indeed, the *Cultivated/Urbanized* change results are characterized by the highest increasing LST rate (0.11 °C/year).

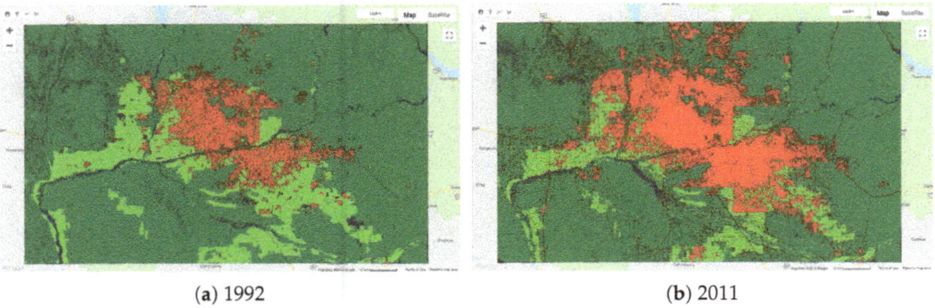

(**a**) 1992 (**b**) 2011

Figure 18. Phoenix MA: LC situation in 1992 (**a**) and in 2011 (**b**). In red, the *Urbanized* class, in light green, the *Cultivated* class, in dark green, the *Forest* and *Shrubland* classes.

On the contrary, a negative LST increasing rate (−0.06 °C/year) can be observed in the LC change from *Shrubland* (desert) to *Cultivated*, while a positive increasing rate (0.09 °C/year) can be noticed in the reverse change. The first situation corresponds to the ROI pixels that become cultivated and thus irrigated, whereas the latter is related to the abandonment of the rural areas, which thus were no longer irrigated.

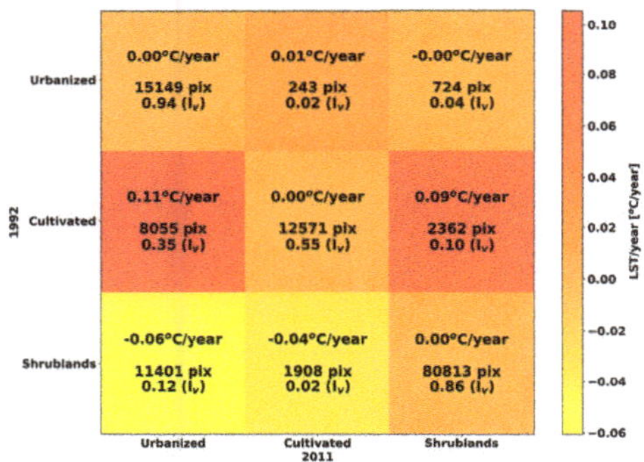

Figure 19. Phoenix MA: *Detrended Rate Matrix.*

4.6. Metropolitan Area of San Francisco

In the case of San Francisco MA, the investigated ROI includes both the San Francisco Bay area and the city of San Jose (Figure 20). The results show that the number of the ROI pixels involved in the LC change from the *Forest* or *Cultivated* classes to the *Urbanized* class is lower than that observed in the previously investigated MAs (Figure 20). In other words, the urban growth was not so strong during the period considered.

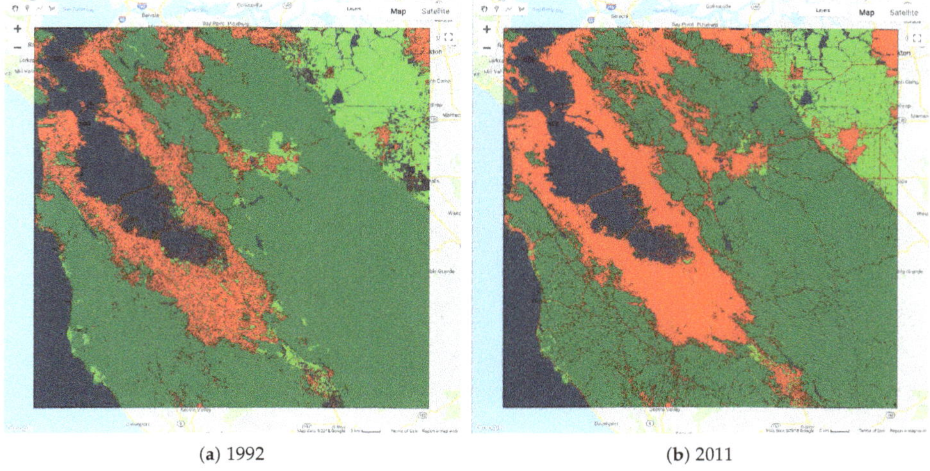

(a) 1992 (b) 2011

Figure 20. San Francisco MA: LC situation in 1992 (**a**) and in 2011 (**b**). In red, the *Urbanized* class, in light green, the *Cultivated* class, in dark green, the *Forest* and *Shrubland* classes.

The elements of the *Detrended Rate Matrix* (Figure 21) show the familiar increasing of the LST observed in the transition from rural/natural into urban LU. The highest and significant increasing rate occurs for the *Cultivated* to *Urbanized* LC change (0.12 °C/year), anyway involving a small portion of the initially cultivated area ($I_v = 0.18$); I_v for the *Forest* to *Urbanized* LC transition is even lower, amounting to 0.06, and the increasing rate is also lower (0.05 °C/year). Furthermore, there are not significant reverse LC changes.

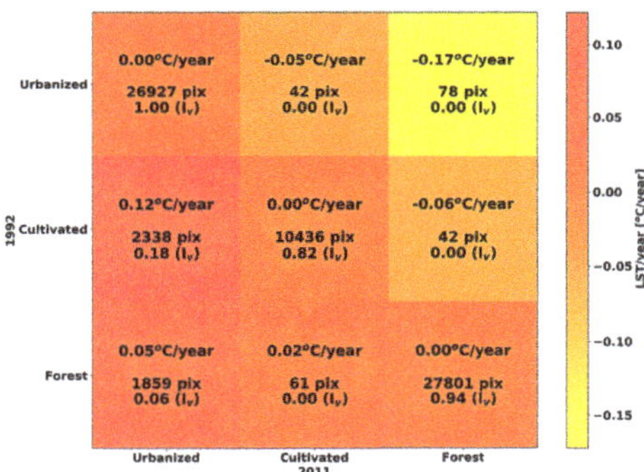

Figure 21. San Francisco MA: *Detrended Rate Matrix.*

5. Conclusions and Prospects

The aim of this work was to show that it is presently possible to monitor SUHI and its connection with LC changes on a long-term and wide spatio-temporal basis by leveraging the large-scale analysis capabilities of GEE. A general procedure was defined and implemented, thanks to the joint use of GEE and CE, and the indicator *Detrended Rate Matrix* was introduced to globally represent the net effect of LC changes on SUHI.

The procedure was applied to six diverse U.S. MAs, characterized by different climatic conditions and significant urban expansions over the 1992–2011 period.

Three main simplifications were introduced in this first application of the proposed procedure to test its feasibility:

- the spatial analysis was performed at a spatial resolution 10-times lower than the original available information, both regarding LST and LC (a GSD equal to 300 m instead of the original 30-m GSD);
- only three main LC classes were considered, suitably grouping those related to *Urbanized*, *Cultivated* and *Forest/Shrubland* areas, which were differently defined in the 1992 and 2011 versions of the USGS NLCD;
- a linear model for the LST temporal evolution was adopted.

It is however important to underline that these simplifications are not intrinsic to the procedure and can be released in the future.

A starting analysis to tune the procedure was performed on the Atlanta MA. A general LST increase was detected, also for LC stable pixels, which kept their original LC over the whole period 1992–2011; anyway, the quite high estimated increasing rates, in the range 0.29–0.37 °C/year, appeared unrealistic. Therefore, considering a small portion of the urban area located around a permanent station for air temperature measurement, an extensive comparison was performed between the LST and the air temperature measured at the time of each image acquisition over the period 1992–2011, and a significant inconsistency (0.22 °C/year for LST vs. 0.08 °C/year for air temperature) was highlighted. A calibration issue in the LST retrieval procedure from Landsat imagery was supposed to explain such a difference; this was the main reason to introduce the *Detrended Rate Matrix*, in order to eliminate the effect of the biased LST increases and to highlight the net effect of the LC changes on the SUHI.

The application of the *Detrended Rate Matrix* evidenced a strong correlation between the highest increasing LST rates and the transformations from rural to urbanized LC found, common to all the studied ROIs. Hence, the results clearly show how urbanization heavily influences the magnitude of the SUHI effects with significant increases in the LST, up to 0.12 °C/year, and demonstrate the effectiveness of the proposed methodology, whose real strength lies in its simplicity, ease of use and rapidity, thanks to the GEE computing capabilities. Indeed, it was possible to analyze a huge amount of data (see Table 1) in a short time (few minutes), not even imaginable until recently. Furthermore, the analysis can be easily extended to other urban areas, provided that a sufficient number of Landsat images is available.

The present feasibility of the proposed procedure and the encouraging obtained results, although preliminary and requiring further investigations, pave the way for a possible global service on SUHI monitoring, able to provide valuable indications to address an increasingly sustainable urban planning of our cities.

In this respect, some possible prospects for the future can be outlined, in order to enhance the proposed procedure:

- to understand the better and feasible way(s) to release the above-mentioned simplifications;
- to deeply investigate the possible calibration issue of the LST retrieval procedure, widening the comparison with ground air temperature measurements in different areas;
- to extend the analysis using also the Sentinel-2 and Sentinel-3 imagery and the CORINE land cover database, already available in the GEE archive.

In this way, one step ahead towards a deeper interpretation of the SUHI phenomena related to specific areas will be possible, but this was not within the objectives of the present investigation.

Author Contributions: Conceptualization, R.R., A.N., M.C., P.M. Methodology, R.R., A.N., M.C. Software, R.R., A.N. Validation, R.R., A.N., C.D.R., R.V.C., P.M. Formal analysis, R.R., A.N. Investigation, R.R., A.N., C.D.R., R.V.C., P.M., G.L. Resources, R.R., A.N. Data curation, R.R., A.N. Writing, original draft preparation, R.R., A.N., C.D.R., R.V.C., P.M. Writing, review and editing, R.R., A.N., P.M., M.C. Visualization, R.R., A.N. Supervision, R.R., A.N., P.M., M.C. Project administration, R.R. Funding acquisition, M.C.

Funding: This work was partially supported by URBAN-GEO BIG DATA, a Project of National Interest (PRIN) funded by the Italian Ministry of Education, University and Research (MIUR) id. 20159CNLW8.

Conflicts of Interest: The authors declare no conflict of interest.

References

1. Oke, T.R. *Boundary Layer Climates*; Routledge: London, UK, 1987.
2. Souch, C.; Grimmond, S. Applied climatology: Urban climate. *Prog. Phys. Geogr.* **2006**, *30*, 270–279. [CrossRef]
3. Voogt, J.A.; Oke, T.R. Thermal remote sensing of urban climates. *Remote Sens. Environ.* **2003**, *86*, 370–384. [CrossRef]
4. Grimm, N.B.; Grove, J.G.; Pickett, S.T.; Redman, C.L. Integrated approaches to long-term studies of urban ecological systems: Urban ecological systems present multiple challenges to ecologists—Pervasive human impact and extreme heterogeneity of cities, and the need to integrate social and ecological approaches, concepts, and theory. *AIBS Bull.* **2000**, *50*, 571–584.
5. Pachauri, R.K.; Allen, M.R.; Barros, V.R.; Broome, J.; Cramer, W.; Christ, R.; Church, J.A.; Clarke, L.; Dahe, Q.; Dasgupta, P.; et al. *Climate Change 2014: Synthesis Report. Contribution of Working Groups I, II and III to the Fifth Assessment Report of the Intergovernmental Panel on Climate Change*; IPCC: Geneva, Switzerland, 2014.
6. Krpo, A.; Salamanca, F.; Martilli, A.; Clappier, A. On the impact of anthropogenic heat fluxes on the urban boundary layer: a two-dimensional numerical study. *Bound. Layer Meteorol.* **2010**, *136*, 105–127. [CrossRef]
7. Salvati, A.; Roura, H.C.; Cecere, C. Assessing the urban heat island and its energy impact on residential buildings in Mediterranean climate: Barcelona case study. *Energy Build* **2017**, *146*, 38–54. [CrossRef]
8. Stewart, I.D.; Oke, T.R.; Krayenhoff, E.S. Evaluation of the 'local climate zone'scheme using temperature observations and model simulations. *Int. J. Climatol.* **2014**, *34*, 1062–1080. [CrossRef]
9. Oke, T.R.; Mills, G.; Christen, A.; Voogt, J.A. *Urban Climates*; Cambridge University Press: Cambridge, UK, 2017.
10. Barlow, J.F. Progress in observing and modelling the urban boundary layer. *Urban Clim.* **2014**, *10*, 216–240. [CrossRef]
11. Morris, C.; Simmonds, I.; Plummer, N. Quantification of the influences of wind and cloud on the nocturnal urban heat island of a large city. *J. Appl. Meteorol.* **2001**, *40*, 169–182. [CrossRef]
12. Oke, T.R. City size and the urban heat island. *Atmos. Environ.* **1973**, *7*, 769–779. [CrossRef]
13. Schatz, J.; Kucharik, C.J. Seasonality of the urban heat island effect in Madison, Wisconsin. *J. Appl. Meteorol. Climatol.* **2014**, *53*, 2371–2386. [CrossRef]
14. Zhao, C.; Fu, G.; Liu, X.; Fu, F. Urban planning indicators, morphology and climate indicators: A case study for a north-south transect of Beijing, China. *Build. Environ.* **2011**, *46*, 1174–1183. [CrossRef]
15. Morini, E.; Touchaei, A.G.; Rossi, F.; Cotana, F.; Akbari, H. Evaluation of albedo enhancement to mitigate impacts of urban heat island in Rome (Italy) using WRF meteorological model. *Urban Clim.* **2018**, *24*, 551–566. [CrossRef]
16. Morini, E.; Touchaei, A.G.; Castellani, B.; Rossi, F.; Cotana, F. The Impact of Albedo Increase to Mitigate the Urban Heat Island in Terni (Italy) Using the WRF Model. *Sustainability* **2016**, *8*, 999. [CrossRef]
17. Cantelli, A.; Monti, P.; Leuzzi, G. Numerical study of the urban geometrical representation impact in a surface energy budget model. *Environmental Fluid Mechanics* **2015**, *15*, 251–273. [CrossRef]
18. Parlow, E.; Vogt, R.; Feigenwinter, C. The urban heat island of Basel–seen from different perspectives. *J. Geogr. Soc. Berl.* **2014**, *145*, 96–110.
19. Li, X.; Zhou, Y.; Asrar, G.R.; Imhoff, M.; Li, X. The surface urban heat island response to urban expansion: A panel analysis for the conterminous United States. *Sci. Total Environ.* **2017**, *605*, 426–435. [CrossRef] [PubMed]
20. Clinton, N.; Gong, P. MODIS detected surface urban heat islands and sinks: Global locations and controls. *Remote Sens. Environ.* **2013**, *134*, 294–304. [CrossRef]
21. Cui, Y.; Xu, X.; Dong, J.; Qin, Y. Influence of urbanization factors on surface urban heat island intensity: A comparison of countries at different developmental phases. *Sustainability* **2016**, *8*, 706. [CrossRef]
22. Heinl, M.; Hammerle, A.; Tappeiner, U.; Leitinger, G. Determinants of urban–rural land surface temperature differences–A landscape scale perspective. *Landsc. Urban Plan.* **2015**, *134*, 33–42. [CrossRef]

23. Imhoff, M.L.; Zhang, P.; Wolfe, R.E.; Bounoua, L. Remote sensing of the urban heat island effect across biomes in the continental USA. *Remote Sens. Environ.* **2010**, *114*, 504–513. [CrossRef]
24. Peng, S.; Piao, S.; Ciais, P.; Friedlingstein, P.; Ottle, C.; Breon, F.M.; Nan, H.; Zhou, L.; Myneni, R.B. Surface urban heat island across 419 global big cities. *Environ. Sci. Technol.* **2011**, *46*, 696–703. [CrossRef] [PubMed]
25. Zhang, P.; Imhoff, M.L.; Bounoua, L.; Wolfe, R.E. Exploring the influence of impervious surface density and shape on urban heat islands in the northeast United States using MODIS and Landsat. *Can. J. Remote Sens.* **2012**, *38*, 441–451.
26. Zhou, D.; Zhao, S.; Liu, S.; Zhang, L.; Zhu, C. Surface urban heat island in China's 32 major cities: Spatial patterns and drivers. *Remote Sens. Environ.* **2014**, *152*, 51–61. [CrossRef]
27. Cossu, R.; Petitdidier, M.; Linford, J.; Badoux, V.; Fusco, L.; Gotab, B.; Hluchy, L.; Lecca, G.; Murgia, F.; Plevier, C.; et al. A roadmap for a dedicated Earth Science Grid platform. *Earth Sci. Inf.* **2010**, *3*, 135–148. [CrossRef]
28. Gorelick, N.; Hancher, M.; Dixon, M.; Ilyushchenko, S.; Thau, D.; Moore, R. Google Earth Engine: Planetary-scale geospatial analysis for everyone. *Remote Sens. Environ.* **2017**, *202*, 18–27. [CrossRef]
29. Nascetti, A.; Di Rita, M.; Ravanelli, R.; Amicuzi, M.; Esposito, S.; Crespi, M. Free global DSM assessment on large scale areas exploiting the potentialities of the innovative Google Earth Engine platform. *Int. Arch. Photogramm. Remote Sens. Spat. Inf. Sci.* **2017**, *XLII-1/W1*, 627–633. [CrossRef]
30. Huntington, J.L.; Hegewisch, K.C.; Daudert, B.; Morton, C.G.; Abatzoglou, J.T.; McEvoy, D.J.; Erickson, T. Climate Engine: Cloud Computing and Visualization of Climate and Remote Sensing Data for Advanced Natural Resource Monitoring and Process Understanding. *Bull. Am. Meteorol. Soc.* **2017**, *98*, 2397–2410. [CrossRef]
31. Ravanelli, R.; Nascetti, A.; Cirigliano, R.V.; Di Rico, C.; Monti, P.; Crespi, M. Monitoring Urban Heat Island through Google Earth Engine: potentialities and difficulties in different cities of the United States. *Int. Arch. Photogramm. Remote Sens. Spat. Inf. Sci.* **2018**, *XLII-3*, 1467–1472. [CrossRef]
32. Hey, T.; Tansley, S.; Tolle, K.M. *The Fourth Paradigm: Data-Intensive Scientific Discovery*; Microsoft Research: Redmond, WA, USA, 2009; Volume 1.
33. Li, D. Brain Cognition and Spatial Cognition. In *Keynote Speech at ISPRS Technical Commission III Symposium, Beijing (China)*; Springer: Berlin, Germany, 2018.
34. ESA Thematic Exploitation Platform. 2017. Available online: https://tep.eo.esa.int/home (accessed on 15 December 2017).
35. EarthServer. 2017. Available online: http://www.earthserver.eu/ (accessed on 15 December 2017).
36. Merticariu, V.; Baumann, P. The EarthServer Federation: State, Role, and Contribution to GEOSS. *EGU Gen. Assembl. Conf. Abstr.* **2016**, *18*, 17298.
37. Open Geospatial Consortium (OGC). Standards and Supporting Documents, 2017. Available online: http://www.opengeospatial.org/standards (accessed on 15 December 2017).
38. Amazon Web Service. Sentinel on Amazon Web Service, 2017. Available online: http://sentinel-pds.s3-website.eu-central-1.amazonaws.com/ (accessed on 20 December 2017).
39. Sinergise, 2017. Available online: http://www.sinergise.com/ (accessed on 20 December 2017).
40. Sinergise. Sentinel-Hub, 2017. Available online: http://www.sentinel-hub.com/ (accessed on 20 December 2017).
41. Patel, N.N.; Angiuli, E.; Gamba, P.; Gaughan, A.; Lisini, G.; Stevens, F.R.; Tatem, A.J.; Trianni, G. Multitemporal settlement and population mapping from Landsat using Google Earth Engine. *Int. J. Appl. Earth Obs. Geoinform.* **2015**, *35*, 199–208. [CrossRef]
42. Pekel, J.F.; Cottam, A.; Gorelick, N.; Belward, A.S. High-resolution mapping of global surface water and its long-term changes. *Nature* **2016**, *540*, 418. [CrossRef] [PubMed]
43. Lobell, D.B.; Thau, D.; Seifert, C.; Engle, E.; Little, B. A scalable satellite-based crop yield mapper. *Remote Sens. Environ.* **2015**, *164*, 324–333. [CrossRef]
44. USGS. What Are the Acquisition Schedules for the Landsat Satellites? 2017. Available online: https://landsat.usgs.gov/what-acquisition-schedule-landsat (accessed on 25 June 2018).
45. Huber, P.J. Robust regression: asymptotics, conjectures and Monte Carlo. *Ann. Stat.* **1973**, *1*, 799–821. [CrossRef]
46. Huber, P.J. Robust statistics. In *International Encyclopedia of Statistical Science*; Springer: Berlin, Germany, 2011; pp. 1248–1251.

47. Homer, C.; Dewitz, J.; Yang, L.; Jin, S.; Danielson, P.; Xian, G.; Coulston, J.; Herold, N.; Wickham, J.; Megown, K. Completion of the 2011 National Land Cover Database for the conterminous United States–representing a decade of land cover change information. *Photogramm. Eng. Remote Sens.* **2015**, *81*, 345–354.
48. Lo, C.; Quattrochi, D.A. Land-Use and Land-Cover change, Urban Heat Island Phenomenon, and Health Implications: A Remote Sensing Approach. *Photogramm. Eng. Remote Sens.* **2003**, *69*, 1053–1063. [CrossRef]
49. Weather Underground. Georgia State University WEATHERSTEM Station, Peoplestown (GA), 1992–2011. Available online: https://www.wunderground.com/weather/us/ga/peoplestown/KGAPEOPL2 (accessed on 10 August 2018).
50. Streutker, D.R. Satellite-measured growth of the urban heat island of Houston, Texas. *Remote Sens. Environ.* **2003**, *85*, 282–289. [CrossRef]
51. Lee, T.W.; Lee, J.; Wang, Z.H. Scaling of the urban heat island intensity using time-dependent energy balance. *Urban Clim.* **2012**, *2*, 16–24. [CrossRef]
52. Fernando, H.; Lee, S.; Anderson, J.; Princevac, M.; Pardyjak, E.; Grossman-Clarke, S. Urban fluid mechanics: air circulation and contaminant dispersion in cities. *Environ. Fluid Mech.* **2001**, *1*, 107–164. [CrossRef]
53. Doran, J.; Berkowitz, C.M.; Coulter, R.L.; Shaw, W.J.; Spicer, C.W. The 2001 Phoenix Sunrise experiment: vertical mixing and chemistry during the morning transition in Phoenix. *Atmos. Environ.* **2003**, *37*, 2365–2377. [CrossRef]
54. Lee, S.M.; Fernando, H.J.; Princevac, M.; Zajic, D.; Sinesi, M.; McCulley, J.L.; Anderson, J. Transport and diffusion of ozone in the nocturnal and morning planetary boundary layer of the Phoenix valley. *Environ. Fluid Mech.* **2003**, *3*, 331–362. [CrossRef]
55. Brazel, A.; Gober, P.; Lee, S.; Grossman-Clarke, S.; Zehnder, J.; Hedquist, B.; Comparri, E. Dynamics and determinants of urban heat island change (1990–2004) with Phoenix, Arizona, USA. *Clim. Res.* **2007**, *33*, 171–182. [CrossRef]
56. Di Sabatino, S.; Hedquist, B.; Carter, W.; Leo, L.; Fernando, H. Phoenix urban heat island experiment: Effects of built elements. In Proceedings of the 8th Symposium on the Urban Environment, Phoenix, Arizona, 2009. Available online: www.ams.confex.com/ams/pdfpapers/147757.pdf (accessed on 01 June 2017).
57. Wang, C.; Myint, S.W.; Wang, Z.; Song, J. Spatio-temporal modeling of the urban heat island in the Phoenix metropolitan area: Land use change implications. *Remote Sens.* **2016**, *8*, 185. [CrossRef]

Article

Regional Crop Gross Primary Productivity and Yield Estimation Using Fused Landsat-MODIS Data

Mingzhu He [1,*], John S. Kimball [1,2], Marco P. Maneta [3], Bruce D. Maxwell [4], Alvaro Moreno [1], Santiago Beguería [5] and Xiaocui Wu [6]

1. Numerical Terradynamic Simulation Group, College of Forestry & Conservation, University of Montana, Missoula, MT 59812, USA; johnk@ntsg.umt.edu (J.S.K.); alvaro.moreno@ntsg.umt.edu (A.M.)
2. Department of Ecosystem and Conservation Sciences, College of Forestry & Conservation, University of Montana, Missoula, MT 59812, USA
3. Department of Geosciences, University of Montana, Missoula, MT 59812, USA; Marco.Maneta@mso.umt.edu
4. Department of Land Resources and Environmental Science, Montana State University, Bozeman, MT 59717, USA; bmax@montana.edu
5. Estación Experimental de Aula Dei, Consejo Superior de Investigaciones Científicas (EEAD-CSIC), 50059 Zaragoza, Spain; santiago.begueria@csic.es
6. Department of Microbiology and Plant Biology, Center for Spatial Analysis, University of Oklahoma, Norman, OK 73019, USA; xiaocui.wu@ou.edu

* Correspondence: mingzhu.he@ntsg.umt.edu

Received: 16 January 2018; Accepted: 22 February 2018; Published: 28 February 2018

Abstract: Accurate crop yield assessments using satellite remote sensing-based methods are of interest for regional monitoring and the design of policies that promote agricultural resiliency and food security. However, the application of current vegetation productivity algorithms derived from global satellite observations is generally too coarse to capture cropland heterogeneity. The fusion of data from different sensors can provide enhanced information and overcome many of the limitations of individual sensors. In thitables study, we estimate annual crop yields for seven important crop types across Montana in the continental USA from 2008–2015, including alfalfa, barley, maize, peas, durum wheat, spring wheat and winter wheat. We used a satellite data-driven light use efficiency (LUE) model to estimate gross primary productivity (GPP) over croplands at 30-m spatial resolution and eight-day time steps using a fused NDVI dataset constructed by blending Landsat (5 or 7) and Terra MODIS reflectance data. The fused 30-m NDVI record showed good consistency with the original Landsat and MODIS data, but provides better spatiotemporal delineations of cropland vegetation growth. Crop yields were estimated at 30-m resolution as the product of estimated GPP accumulated over the growing season and a crop-specific harvest index (HI_{GPP}). The resulting GPP estimates capture characteristic cropland productivity patterns and seasonal variations, while the estimated annual crop production results correspond favorably with reported county-level crop production data ($r = 0.96$, relative RMSE = 37.0%, $p < 0.05$) from the U.S. Department of Agriculture (USDA). The performance of estimated crop yields at a finer (field) scale was generally lower, but still meaningful ($r = 0.42$, relative RMSE = 50.8%, $p < 0.05$). Our methods and results are suitable for operational applications of crop yield monitoring at regional scales, suggesting the potential of using global satellite observations to improve agricultural management, policy decisions and regional/global food security.

Keywords: crop yield; gross primary productivity (GPP); data fusion; Landsat; MODIS

1. Introduction

Accurate quantification of crop yield at regional to global scales is important in supporting policy- and decision-making in agriculture [1–4]. Numerous approaches have been developed to estimate crop yield for various cropping systems [1,3–7]. Agricultural surveys provide a reliable way to estimate crop yield, but are less effective over larger regions due to excessive time and budget constraints [8]. In recent decades, satellite remote sensing has been employed for agricultural applications, including crop yield monitoring [9–11]. Current operational satellite records, including Landsat and MODIS (MODerate resolution Imaging Spectroradiometer), are sensitive to photosynthetic vegetation cover and provide frequent observations with global coverage and consistent sampling, as well as relatively long-term overlapping records.

Traditional crop yield estimation methods have used empirical relationships between vegetation biomass and remote sensing spectral vegetation indices to estimate yields [3,11,12]. For example, crop yield derived from MODIS NDVI data from 2000 to 2006 in the Canadian Prairies for barley, canola, field peas and spring wheat accounted for 48 to 90%, 32 to 82%, 53% to 89% and 47 to 80% of the variability in reported crop yield from Statistics Canada, respectively [13]. However, these empirical models are fundamentally simple and specific to the limited areas and conditions from which they were developed and cannot easily be extended to other areas.

Another approach involves estimating crop yield as the product of vegetation gross primary productivity (GPP) and an empirical harvest index (HI) specific to different crop types. GPP, representing the total carbon uptake by plant photosynthesis, can be estimated at spatial and temporal scales suitable for cropland applications using a light use efficiency (LUE) model driven by remote sensing inputs [14–16]. Two global operational GPP products are currently produced using the satellite data-driven LUE model logic, including the NASA MODIS MOD17 and SMAP (Soil Moisture Active Passive) Level 4 Carbon (L4C) products [15,17]. The MOD17 product provides continuous GPP estimates with eight-day temporal fidelity and 500-m spatial resolution (Version 6) spanning all global vegetated ecosystems and extending from 2000 to the present. The L4C product uses a similar LUE model framework driven by combined satellite information from MODIS and SMAP sensors to estimate GPP and underlying environmental constraints to vegetation growth, including soil moisture-related water supply controls; the L4C product is derived globally from 2015 to the present and provides daily temporal fidelity and 1 to 9-km resolution. However, while operational GPP products derived from global satellite observations provide consistent and frequent temporal sampling, the spatial scale of these products may be too coarse for many agricultural applications; the global LUE algorithm parameterizations for croplands used in the MOD17 and L4C products also only distinguish general crop functional types (e.g., cereal vs. broadleaf), which can degrade GPP accuracy for agricultural ecosystems [17–21]. Alternatively, GPP products derived using LUE model parameterizations that distinguish a greater number of crop types, and with finer spatial resolution (e.g., 30-m) and suitable temporal fidelity (e.g., eight-day), may overcome many of the above limitations while enhancing the utility of these data for agricultural applications. The Landsat TM and ETM+ sensors on the Landsat 5 and 7 platforms provide 30-m resolution imagery that is well suited for capturing surface spectral reflectance heterogeneity at the level of individual agricultural fields [22,23]. However, Landsat has limited temporal coverage due to a long revisit cycle (16-day), data loss from atmosphere aerosol and cloud contamination and failure of the Landsat 7 sensor Scan Line Corrector (SLC) in May 2003 [23,24]; these factors contribute to degraded Landsat utility for cropland monitoring [23,25]. MODIS provides similar spectral information as Landsat, but with more frequent eight-day composite global observations of surface conditions, albeit at a coarser (250 to 1000 m) spatial resolution that is less suitable for heterogeneous agricultural landscapes [22]. Data fusion methods have been used to reduce the constraints of single sensor remote sensing by blending similar spectral information from Landsat and MODIS to generate harmonized multi-sensor observations, providing both relatively fine-scale spatial resolution and frequent temporal sampling [22,25–30]. The spatial and temporal adaptive reflectance fusion model (STARFM), developed by [25], has been widely used for blending surface reflectance data from Landsat and MODIS. The STARFM approach was modified and improved by [23] for more complex situations. However, the STARFM model is computationally expensive, which can impose a constraint on regional applications. Alternatively, a rule-based piecewise regression model based on MODIS and Landsat data was developed by [31] to derive 30-m NDVI (Normalized Difference Vegetation Index) maps over northeastern Colorado. A simple linear relationship between NDVI and *fPAR* (fraction of photosynthetically-active radiation) was also used to downscale 250-m MODIS *fPAR* data to 30-m resolution using Landsat TM NDVI data [32]. Similar vegetation information may be used as primary inputs to a satellite-based LUE model to derive GPP with enhanced spatial and temporal resolution suitable for agricultural applications.

Although GPP represents the total amount of carbon accumulation in vegetation biomass from photosynthesis, most agricultural applications are concerned with estimating crop grain yield, which generally represents a much smaller portion of GPP. Crop yield can be estimated as the product of GPP and an empirical harvest index (HI) that defines the conversion ratio between crop GPP and grain yield, with HI varying for different crop types and environmental conditions [33–37]. MODIS GPP was converted to wheat yield for the 2001 and 2002 growing seasons in Montana and North Dakota by [38]. The resulting county-level estimated wheat yields from this study showed relatively low correlations with observed wheat yield in Montana (R^2 value of 0.46 and 0.33) and North Dakota (R^2 of 0.06 and 0.16), respectively; however, the model results explained 67% and 33% of the total variance in observed wheat yield over both states [38]. Inaccurate HI settings may propagate to significant errors in the final crop yield calculations; however, the integration of regional survey data can improve HI accuracy and reduce uncertainties in crop yield predictions. The U.S. Department of Agriculture (USDA) National Agricultural Statistics Service (NASS) provides annual crop yield and harvested area information at the county level for specific crop types within the continental USA, which can be used as a benchmark for crop model calibration (e.g., HI) and validation.

Google Earth Engine (GEE) is a cloud-based platform designed for efficient planetary-scale geospatial analysis. GEE has a large data catalog co-located with massive CPU that allows for interactive data exploration, providing an easy and quick way to process and analyze data [39]. GEE has been used for mapping crop yields with satellite data spanning multiple states and years (2000 to 2013) in the Midwestern United States [40]. GEE has also been used for other hydrological and ecological applications, including mapping global forest change [41], detecting global surface water change [42] and generating a dynamic Landsat-based NDVI product for the Conterminous United States [43].

In this paper, we show that a global satellite data-driven productivity model (i.e., the MODIS MOD17 algorithm) can be adapted for regional crop assessment. Our approach is significant because it enables crop monitoring over large regions through the modification and application of existing operational satellite remote sensing data and models. We demonstrate the model approach by characterizing the spatial extent and annual variability in cropland productivity (GPP) and yield over the state of Montana in the continental USA. Montana's agriculture is mostly rainfed, and production is very sensitive to climate variability. Montana is also a major producer of wheat, barley and pulse crops and is representative of intermountain western agriculture.

We used a satellite data-driven LUE modeling approach based on the MODIS MOD17 algorithm to derive relatively high spatial (30-m) resolution GPP and crop yield predictions across the region and spanning the recent (2008 to 2015) satellite record. A data fusion approach was used for blending 30-m Landsat 5 and Landsat 7 NDVI with 250-m eight-day MODIS NDVI observations to produce harmonized 30-m eight-day fused NDVI data record for Montana extending from 2008 to 2015. The 30-m NDVI record was used as a primary LUE model input to estimate *fPAR* and GPP at a similar 30-m resolution and eight-day time step for the major Montana crop types. Both the LUE modeling framework and satellite NDVI data fusion were implemented within the GEE framework for this investigation. We derived 30-m annual crop yield maps encompassing Montana cropland areas using the resulting 30-m GPP record and crop-specific HI coefficients calibrated and validated using NASS county-level crop production data and field-level yield data. These results were used to assess regional patterns and anomalies in cropland productivity and yield over the multi-year satellite record. A detailed summary of the methods and results from this study is described below.

2. Materials and Methods

2.1. Study Area

Montana (MT, 44°~49°N, 104°W~116°W) encompasses 380,832 km^2 of the northwest continental United States and is the largest landlocked state in the country. Agriculture is the largest economic sector in MT [44]. The Cropland Data Layer (CDL) from National Agricultural Statistics Service (NASS) provides cropland classification maps over the continental U.S. domain from 2008 to the present; the CDL is a geo-referenced and crop-specific land cover data layer derived using relatively fine-scale satellite imagery (e.g., Landsat 4/5/7/8, Terra MODIS NDVI, etc.) with a decision tree classifier. Croplands encompassed 29.60 $\pm$ 1.98 percent of the entire MT land area over the eight-year (2008 to 2015) study period based on the 30-m CDL product. The MT croplands are dominated by seven major crop types, including maize, barley, durum wheat, spring wheat, winter wheat, alfalfa and peas, which were selected in this study for subsequent GPP and crop yield predictions. The MT croplands are mainly distributed in the northcentral and northeastern portions of the state and are more dispersed in central and southern areas for the 2008 to 2015 study period (Figure 1a). Alfalfa is dispersed across MT, but is concentrated in the northwest and central areas of the state. Barley is also grown in most counties, but is more prevalent in the north central portion of MT, together with spring wheat and winter wheat. Some spring wheat is also located in the northeast, where durum wheat is also planted. Peas are planted near barley, spring wheat, winter wheat and durum wheat. Maize is mainly distributed in the southeastern, northeastern and south central portions of the state over the 2008 to 2015 record. Crop types and planted areas show strong annual variations in MT from 2008 to 2015 (Figure 1b). For example, the total cropland area of the seven major MT crops ranged from an annual minimum of 4.13 $\times$ 10^6 hectares in 2009 to a maximum of 6.34 $\times$ 10^6 hectares in 2014; whereas individual crop types ranged from 0.7% (maize) to 30.7% (spring wheat) of the total planted area in MT over the study period.

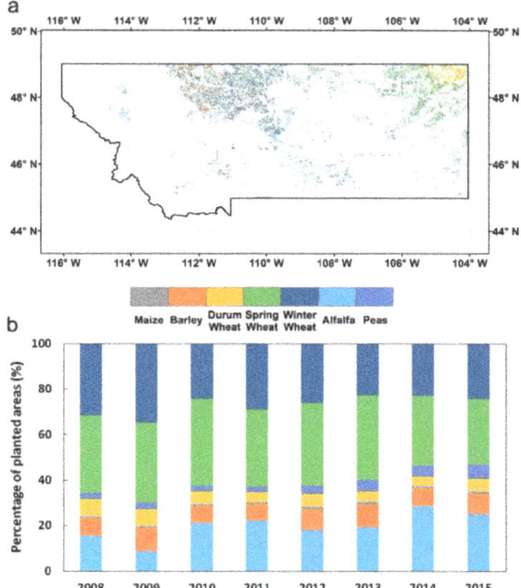

Figure 1. (**a**) Cropland distributions for the seven major crop types in Montana in 2015 based on the 30-m Cropland Data Layer from National Agricultural Statistics Service(NASS), including alfalfa, barley, maize, durum wheat, spring wheat, winter wheat and peas; areas in white denote other crop types and land cover areas; (**b**) planted areas for the different crop types per year from 2008 to 2015 are represented as a proportion (%) of the total planted area for all seven MT crop types.

MT is located in the northern temperate latitudes and has a semi-arid continental climate with an extended winter frozen season, where the potential growing season extends from the end of snow cover depletion in spring to the onset of persistent frozen temperatures in autumn. Based on the reported planting and harvest dates for field crops [45], the planting dates for the major MT crop types (except winter wheat) range from early April to mid-June, whereas harvest dates extend from late July to early October. Winter wheat is usually planted in early September and harvested during the following July to August period. Thus, for this study, we assume an annual period for seasonal crop development, harvest and yield from April to September over the 2008 to 2015 record.

2.2. Satellite Based Crop GPP and Yield Modeling

2.2.1. Developing 30-m MODIS-Landsat Fused NDVI Maps for MT from 2008 to 2015

All data processing and modeling work in this study was conducted in the Google Earth Engine (GEE) framework, which has an extensive library of useful datasets and functions, while providing an efficient and quick way to process and analyze large geospatial datasets. The geospatial data used in this study are summarized in Table 1. The 30-m resolution surface spectral reflectance data records generated from the Landsat Ecosystem Disturbance Adaptive Processing System (LEDAPS, [46]) from Landsat 5 (2008 to 2011) and Landsat 7 (2012 to 2015) were extracted and processed over the MT domain. Through LEDAPS, lower order spectral imagery from Landsat 5 (or 7) were calibrated, converted to top-of-atmosphere reflectance and then atmospherically corrected using the MODIS/6S methodology to produce the surface reflectance product [46]. Cloud screening was applied to the reflectance data to remove cloud-contaminated pixels according to cloud mask information provided by the surface reflectance data products [47]. The surface reflectance data for Landsat Bands 3 (red: 0.63 to 0.69 μm) and 4 (near-infrared: 0.76 to 0.90 μm) were then selected to calculate the 30-m resolution NDVI for each image sequence. Simultaneously, the MODIS 250-m 8-day global surface reflectance product (MOD09Q1), after masking out cloud-contaminated pixels, was used to derive the MODIS NDVI record over the MT domain and 2008 to 2015 study period. Both Landsat and MODIS NDVI data were reprojected to a consistent geographic projection using the WGS84 datum. Since the MODIS (250-m 8-day) and Landsat (30-m) records spanned the same MT domain and period, we assumed that the NDVI retrievals from the different sensor records are consistent and comparable. Therefore, a pixel-wise linear regression model was applied to blend the Landsat and MODIS NDVI records to construct a fused NDVI record with consistent 30-m spatial resolution and 8-day temporal fidelity over the MT domain and for each year of record from 2008 to 2015 (https://code.earthengine.google.com/1fb49cd39a59bf1d6d653bc04499c690). The linear regression model describing the relationship between Landsat and MODIS NDVI was developed and applied on a pixel basis. The fused 30-m NDVI record was derived using the resulting slopes and intercepts from the linear regression model and the MODIS NDVI data. The underlying assumption of Landsat and MODIS spectral consistency used to derive the fused NDVI record was determined to have a suitable signal-to-noise ratio for our agricultural application, but may be insufficient for distinguishing subtler environmental trends.

Table 1. Descriptions of data used in this study.

Data	Time Period	Spatial Resolution	Temporal Resolution
Landsat 5 surface reflectance	2008 to 2011	30-m	16 day
Landsat 7 surface reflectance	2012 to 2015	30-m	16 day
MOD09Q1	2008 to 2015	250-m	8 day
MOD17A2H GPP	2008 to 2015	500-m	8 day
Flux tower based GPP	2000 to 2006	1-km	daily
Cropland Date Layer	2008 to 2015	30-m	annual
Gridded Surface Meteorological Dataset	2008 to 2015	4-km	daily
USDA NASS crop yield/production data	2008 to 2015	County	annual
Crop yield field measurements	2008 to 2015	10-m	annual

2.2.2. Cropland GPP Estimation

GPP for the seven major crop types in MT during 2008 to 2015 was calculated using an LUE framework similar to the MODIS MOD17 model [14]:

$$GPP = LUEmax \times f(VPD) \times f(T) \times fPAR \times SWrad \times 0.45 \quad (1)$$

$$f(T) = \begin{cases} 0 & T \le T_{min} \\ \frac{T - T_{min}}{T_{max} - T_{min}} & T_{min} < T < T_{max} \\ 1 & T \ge T_{max} \end{cases} \quad (2)$$

$$f(VPD) = \begin{cases} 0 & VPD \ge VPD_{max} \\ \frac{VPD_{max} - VPD}{VPD_{max} - VPD_{min}} & VPD_{min} < VPD < VPD_{max} \\ 1 & VPD \le VPD_{min} \end{cases} \quad (3)$$

where *LUEmax* is a prescribed maximum light use efficiency (g C MJ^{-1}), which is specific for different biome types (Table 2); photosynthetically-active radiation (*PAR*) is estimated as 45% of shortwave radiation ($SWrad \times 0.45$; [48]); *VPD* is the mean daily vapor pressure deficit (Pa); T is the minimum daily surface air temperature (°C); VPD_{max}, VPD_{min}, T_{max} and T_{min} represent respective maximum and minimum *VPD* and temperature (*T*) levels for plant photosynthesis, which vary for different plant types (Table 2); *f(VPD)* and *f(T)* are dimensionless (0 to 1) scalar linear ramp functions ranging between optimal (1) and fully constrained (0) levels under unfavorable atmospheric moisture deficit (*VPD*) and minimum daily temperature conditions; the product of $f(VPD)$ and *f(T)* describes the daily LUE and GPP reduction from potential (*LUEmax*) conditions due to environmental stress. The LUE model environmental response functions are prescribed for different biome types within a global Biome Properties Look-Up Table (BPLUT) that distinguishes two major cropland functional types for cereal and broadleaf growth forms (Table 2). For estimating GPP, the seven major MT crop types were grouped into each of these two general plant functional types as either cereal (maize, barley, durum wheat, spring wheat, winter wheat) or broadleaf (peas, alfalfa) BPLUT categories.

Table 2. LUE model Biome Properties Look-Up Table (BPLUT) used for the MT cropland GPP calculations.

Crop Type	*LUEmax* (g C MJ^{-1})	VPD_{max} (Pa)	VPD_{min} (Pa)	T_{max} (°C)	T_{min} (°C)
Cereal crop	2.55	6940	1	45.85	−23.15
Broadleaf crop	2.5	7000	1500	27.85	−2.15

The daily minimum air temperature (T) and shortwave radiation (SWrad) inputs to the LUE model were obtained from a Gridded Surface Meteorological dataset (Gridmet, [49]), which provides relatively high spatial resolution (~4 km) daily surface meteorological data from 1979 to 2017 over the continental United States. The daily *VPD* inputs were derived using relative humidity and air temperature data from Gridmet. The daily surface meteorology for each 30-m pixel used for GPP estimation was subsampled from the overlying 4-km resolution Gridmet cell. The BPLUT parameters (VPD_{max}, VPD_{min}, T_{min} and T_{max}) for the cereal and broadleaf crop types were obtained from the Soil Moisture Active Passive (SMAP) L4C model GPP product [16]. The L4C GPP product uses a similar LUE model framework and MODIS vegetation observations as the MOD17 GPP product, but the L4C BPLUT parameters are calibrated using historical tower eddy covariance CO_2 flux observations from a global tower carbon-flux network (FLUXNET) that includes multiple cropland tower sites.

The fraction of photosynthetically-active radiation (*fPAR*) absorbed by crops was calculated from the 30-m 8-day fused NDVI as [1,50]:

$$fPAR = \frac{NDVI - NDVI_{min}}{NDVI_{max} - NDVI_{min}} * (fPAR_{max} - fPAR_{\min}) \quad (4)$$

where $fPAR_{max}$ and $fPAR_{min}$, representing the maximum and minimum $fPAR$ values of vegetation in the domain during the growing season, were assigned respective values of 0.95 and 0.01, following [1] and [50]. $NDVI_{max}$ and $NDVI_{min}$ represent the NDVI values corresponding to 98% and 2% of the NDVI frequency distributions for all seven MT crop types each year (April to September) over the entire study period.

Tower eddy covariance CO_2 flux measurement-based daily GPP observations were used to evaluate the satellite-based GPP estimates from this study. However, since no available flux tower sites were located in MT cropland areas, an alternative tower site representing a natural grassland from Fort Peck, MT (48.3077°N, 105.1019°W; US-FPe), was selected for the comparison. The US-FPe tower measurement-based daily GPP observations were compared against overlying satellite-based GPP estimates representing a cereal crop (Table 2) for this location. The US-FPe tower and model data records represented different time periods (2000 to 2006 vs. 2008 to 2015), so we compared GPP 8-day climatology records derived from the model (GPP_M) and flux tower observations (GPP_F). We also compared GPP_M at the US-FPe tower location with the MODIS GPP Version 6 (MOD17H2) product, which has 500-m resolution and 8-day temporal fidelity. The GPP_M and MOD17H2 8-day climatology records were compared for randomly selected locations representing different MT crop types over the 2008 to 2015 period. Both the GPP_M and MOD17H2 products were extracted as the mean values within 1×1 km^2 windows centered over selected locations representing uniform crop type conditions to minimize potential geolocation errors.

2.2.3. Crop Production and Yield Estimation

Crop yield was calculated as the product of the estimated cumulative 8-day crop GPP for the growing season between April and September for each year of the 2008 to 2015 study period and an empirical harvest index (HI) specific to each of the seven MT crop types. Crop production was obtained by integrating the estimated yield over the area planted for each crop type at both county and MT state levels. Modeled crop production (P_M) was compared with respective MT county-level annual production values for each crop type reported by the USDA. The range of HI values for each crop type was assembled from the literature and summarized in Table 3; this range includes HI values used to convert aboveground biomass, GPP or net primary production to crop yield. However, in this study, we only consider HI as the conversion ratio from GPP to crop yield (HI_{GPP}). The Monte Carlo Markov chain (MCMC) method was used to calibrate HI_{GPP} (Table 3) by minimizing the root mean square error between modeled crop production (P_M) and two-thirds of the NASS reported county scale crop production (P_N) records for each crop type in MT.

Table 3. Estimated annual harvest index (HI) values for the seven major MT crop types.

Crop Type	HI from Literature *	Calibrated HI_{GPP}
Alfalfa	0.07 to 0.18	0.55
Barley	0.30 to 0.62	0.42
Maize	0.25 to 0.58	0.44
Durum Wheat	0.31 to 0.43	0.22
Peas	0.33 to 0.59	0.28
Spring Wheat	0.31 to 0.53	0.24
Winter Wheat	0.33 to 0.53	0.35

* HI from the literature includes HI values derived for converting biomass, gross and net primary production to crop yield.

We also compared the 30-m model crop yield estimates ($Yield_M$) against independent field-scale crop yield observations ($Yield_F$) obtained from 10 m by 10 m plots at four different MT farms. The $Yield_F$ observations were obtained from GPS linked yield monitors on combine harvesters and represent four different crop types, including barley, peas, spring wheat and winter wheat. Here, the collocated spatial mean $Yield_M$ results and $Yield_F$ observations for individual crop types were compared over coarser 90 m by 90 m windows to reduce potential geolocation errors for each year of record from 2008 to 2015.

2.3. Statistical Metrics

The correlation coefficient (r) was used to characterize the correspondence between the model results (Mod) and validation (Val) datasets at a 0.05 p-value significance threshold. Bias, calculated as the difference between Mod and Val (bias = Mod-Val), and the root mean square error (RMSE) were used to evaluate model performance in relation to the GPP and yield observations. The model bias and RMSE metrics were also expressed as a relative percentage of the validation datasets.

3. Results

3.1. NDVI Fusion

An example comparison showing the baseline Landsat and MODIS NDVI records relative to the fused 30-m NDVI record developed from this study is shown for the Valley County sub-region of MT (Figure 2). The Landsat 5 (or 7) NDVI record has extensive gaps in spatial and temporal coverage relative to the MODIS record due to less frequent Landsat temporal sampling (16-day revisit cycle) and data loss from atmosphere aerosol and cloud contamination effects, as well as the SLC-off issues in Landsat 7 (Figure 2a). Nevertheless, the Landsat data provide approximately eight-fold improved spatial resolution NDVI observations relative to MODIS, which enhances the delineation of agricultural fields and heterogeneous crop types. However, the MODIS (MOD09Q1) NDVI record (Figure 2b) provides complete spatial coverage and continuous eight-day sampling. The resulting fused NDVI record (Figure 2c) benefits from the combined qualities of both Landsat and MODIS, while minimizing the limitations of the individual sensor records. For Valley County and other MT areas, the fused NDVI 30-m record shows similar spatial patterns and magnitudes as the component Landsat and MODIS NDVI records. The fused NDVI record also provides complete spatial coverage with continuous eight-day sampling at 30-m resolution over the entire MT study domain and 2008 to 2015 record.

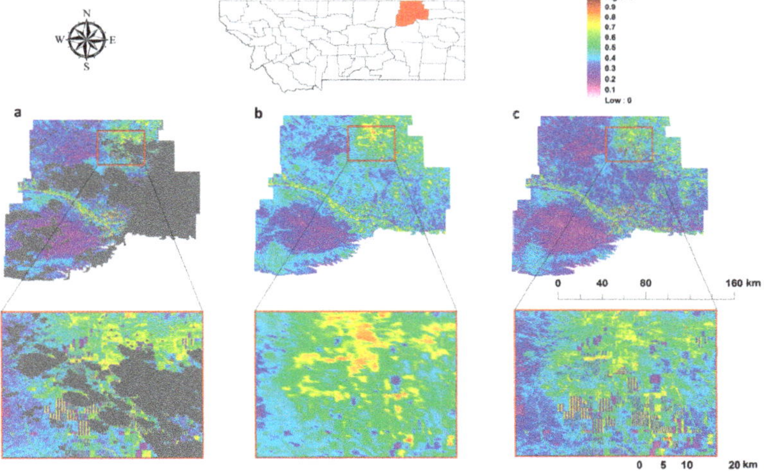

Figure 2. Examples of (**a**) 30-m NDVI derived from Landsat 5 surface reflectance data after removing cloud-contaminated pixels; (**b**) 250-m NDVI from the MODIS MOD09Q1 eight-day temporal composite product; and (**c**) 30-m NDVI created by fusing Landsat and MODIS data from (**a**,**b**). The NDVI metric ranges from zero to one between respective low to high levels of vegetation greenness. The grey area represents no NDVI data. All three images are obtained for the selected period during 1 to 8 July 2008 over Valley County, Montana.

For the entire MT domain, the Landsat 5 and MODIS NDVI records were favorably correlated ($r = 0.78 \pm 0.22$, $p < 0.05$) for the 2008 to 2011 portion of the study period overlapping with Landsat 5. Similar favorable correlations were also found between Landsat 7 and MODIS NDVI records for the 2012 to 2015 portion of the study period ($r = 0.77 \pm 0.21$, $p < 0.05$). These results indicate suitable conditions for deriving a fused NDVI record using the linear regression model relationship between the Landsat and MODIS NDVI records.

One pixel for each crop type, including alfalfa, barley, maize, peas, durum wheat, spring wheat and winter wheat, was randomly selected within the MT domain and over the 2008 to 2015 study period to illustrate the NDVI seasonal variations from Landsat 5 (LS5), Landsat 7 (LS7), MODIS and the fused NDVI records derived from MODIS and Landsat 5 (LS5_fused) and Landsat 7 (LS7_fused) inputs (Figure 3). The NDVI time series from the different sensor products in each plot in Figure 3 have consistent geographic coordinates and overlapping time periods, but have varying footprints due to the different pixel sizes represented from each data record. All of the NDVI records show similar seasonality for each of the seven crop types, with seasonal maximum NDVI values generally occurring in the summer (June to August) and early spring (February to April) and seasonal minimum NDVI values occurring in late autumn (September to October) and winter (December to January) periods. The LS5 and LS7 records have relatively sparse growing season temporal coverage from 2008 to 2015 due to cloud contamination and long revisit cycle (16-day), while MODIS NDVI and the two fused NDVI records provide continuous eight-day time series spanning the entire study period. The LS5 and MODIS NDVI records were well correlated for each MT crop type during the overlapping period from 2008 to 2011 ($0.91 \leq r \leq 0.98$, $p < 0.05$). The LS7 and MODIS NDVI records for the seven MT crop types also showed strong correspondence for the overlapping period from 2012 to 2015 ($0.93 \leq r \leq 0.99$, $p < 0.05$). Strong correlations were found between the LS5 and LS5_fused ($r \geq 0.91$, $p < 0.05$) and LS7 and LS7_fused ($r \geq 0.95$, $p < 0.05$) records. The MODIS NDVI record was also strongly correlated with the LS5_fused ($r \geq 0.99$, $p < 0.05$) and LS7_fused ($r \geq 0.97$, $p < 0.05$) results. These results indicated that the LS5_fused and LS7_fused records preserve similar variability as the native LS5, LS7 and MODIS NDVI records. There was no significant difference between LS5_fused and LS7_fused NDVI records from the different crop types for the 2008 to 2011 period of overlapping LS5 and LS7 observations. These results indicated the reasonable accuracy and performance for our agricultural application, allowing production of a continuous (2008 to 2015) 30-m, eight-day NDVI record for MT derived using the linear regression model developed between overlapping 250-m MODIS and 30-m Landsat 5 (2008 to 2011) and Landsat 7 (2012 to 2015) NDVI records.

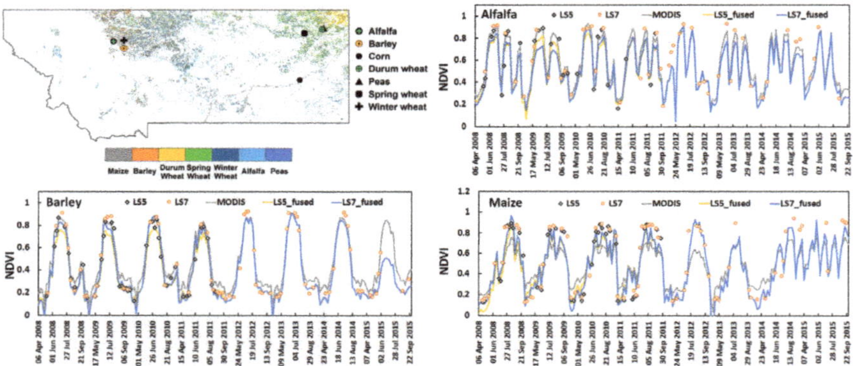

Figure 3. *Cont.*

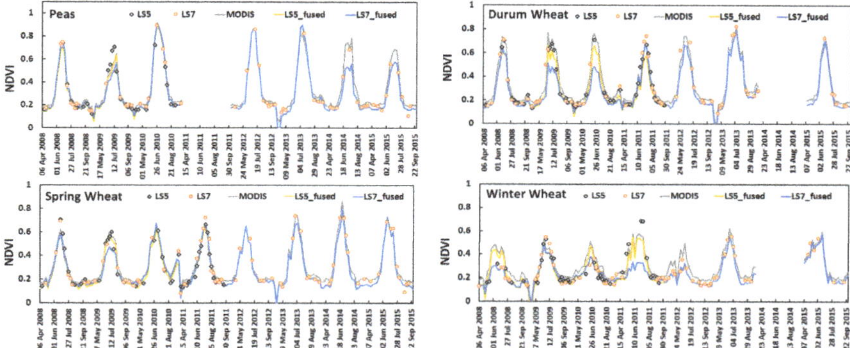

Figure 3. Seasonal NDVI variations for randomly-selected pixels representing the major MT crop types as derived from the different sensor records for the 2008 to 2015 growing season (April to September), including: Landsat 5 (LS5) and Landsat 7 (LS7) after removing cloud contamination; the MODIS MOD09Q1 product (MODIS); fusion of MODIS MOD09Q1 with Landsat 5 (LS5_fused) and Landsat 7 (LS7_fused). The MT crop types represented include alfalfa, barley, maize, peas, durum wheat, spring wheat and winter wheat, while the locations of the selected pixels representing each crop type are indicated in the upper left, with the background image representing the seven major MT crop types depicted from the 2015 Cropland Data Layer (CDL).

3.2. Regional GPP Estimation over Montana

The fused NDVI record was used with Gridmet daily meteorological data and the yearly NASS CDL as primary inputs to the LUE model (Equations (1) to (4)) to derive 30-m eight-day GPP (GPP_M) across MT from April to September for the seven major crop types and eight-year (2008 to 2015) study period. The mean GPP_M (2008 to 2015) results across the MT domain showed distinct spatial patterns corresponding to the different crop types. The GPP_M results showed higher productivity (>4 g C m^{-2} day^{-1}) in the northwest and central areas, where alfalfa and barley were located. Higher GPP_M values were also found in the southeast portion of MT associated with maize production. The central and northeast areas of MT showed slightly lower GPP_M (<2.8 g C m^{-2} day^{-1}) associated with durum wheat, spring wheat and winter wheat production.

The GPP_M mean eight-day climatology derived from April to September over the 2008 to 2015 study period is compared with the MODIS GPP (MOD17A2H) mean eight-day climatology for the same period from randomly-selected pixels representing the different MT crop types in Figure 4. Here, both GPP_M and MODIS GPP were extracted as the spatial mean values within 1×1 km^2 windows representing each major crop type. The GPP_M results showed consistent seasonal variations, but different magnitudes among the different MT crop types relative to the MOD17A2H global GPP product (Figure 4). The GPP_M for alfalfa showed stronger annual variations than the other crop types during the 2008 to 2015 study period, indicated by a larger temporal standard deviation. The GPP_M results also showed less temporal variability than MODIS GPP for barley, peas, spring wheat and winter wheat, relative to the other crop types. At US-FPe, the GPP_M mean eight-day climatology derived from the 2008 to 2015 record was compared with the GPP_F during the growing season ($r = 0.77$), but with a relatively large bias (1.73 g C m^{-2} day^{-1}) and RMSE (2.16 g C m^{-2} day^{-1}, Figure 4); these results are consistent with the GPP_M depiction of a cereal cropland for this site relative to the GPP_F observations representing a less productive natural grassland [51]. The GPP_M results also showed stronger correspondence with the MODIS GPP record ($r = 0.97$, $p < 0.05$) at the US-FPe site than with the GPP_F, including smaller positive GPP_M bias.

The eight-day 30-m GPP_M results were compared with the eight-day MODIS (MOD17A2H) 500-m GPP record over the entire 2008 to 2015 study period for each selected MT crop type location, and summarized in Table 4. The eight-day GPP_M results showed generally favorable correspondence ($0.38 \leq r \leq 0.92$) with the MODIS GPP record for all seven major MT crop types.). For alfalfa, the GPP_M results were relatively consistent with the MODIS GPP record from April to mid-June, but with a more productive bias between mid-July and early September (e.g., Figure 4). The seasonal variations and magnitudes of GPP_M for barley were relatively close to the MODIS GPP record (r = 0.87; bias = −0.14 g C m^{-2} day^{-1}). Significant differences in magnitude between GPP_M and MODIS GPP occurred during the peak growing season (mid-June to August) for maize, peas and the three wheat species (e.g., Figure 4). However, these crop types all showed strong correlations between GPP_M and MODIS GPP during the 2008 to 2015 growing seasons ($r \geq 0.89$, Table 4). Moreover, the GPP_M results showed the lowest correlations (r = 0.38, $p < 0.05$) and largest positive bias (2.69 g C m^{-2} day^{-1}) with MODIS GPP for maize compared to the other six MT crop types. These results are consistent with a generally low LUE and productivity bias in the MODIS GPP record for croplands [17,51,52].

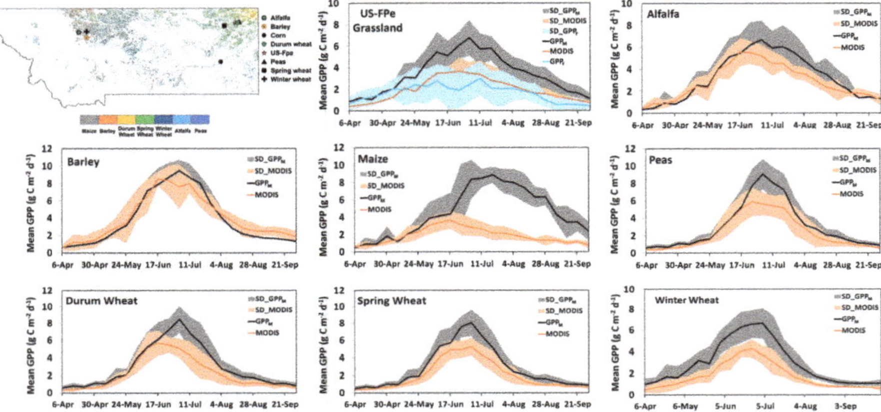

Figure 4. The mean (2008 to 2015) eight-day climatologies of modeled GPP (GPP_M) and the MOD17A2H v006 GPP (MODIS) for the selected major crop type locations and Fort Peck (US-FPe) natural grassland tower site in Montana. The eight-day climatology (2000 to 2006) of GPP estimated from the US-FPe flux tower measurements (GPP_F) is also shown. The grey shading denotes the temporal standard deviation of GPP_M (SD_GPP_M), and the light orange shading indicates the standard deviation of MODIS GPP (SD_MODIS) for the different crop types, while the light blue shading represents the standard deviation of GPP estimated from flux tower measurements (SD_GPP_F) at US-FPe. The locations of the randomly-selected pixels representing the different crop types are indicated in the upper left map, which also shows the CDL cropland distribution map in 2015 as the background.

Table 4. Summary of statistical metrics describing relations between the modeled GPP (GPP_M) and Version 6 MODIS GPP product (MOD17A2H) for the seven crop types in Montana from April to September during 2008 to 2015 (all of the correlations are significant with $p < 0.05$).

Crop Type	r	Bias (g C m^{-2} day^{-1})	RMSE (g C m^{-2} day^{-1})
Alfalfa	0.81	0.45	1.48
Barley	0.87	−0.14	1.51
Maize	0.38	2.69	3.92
Durum wheat	0.89	0.80	1.47
Peas	0.91	0.74	1.44
Spring wheat	0.92	0.82	1.39
Winter wheat	0.90	1.20	1.66

3.3. Crop Yield Monitoring in Montana

Using the calibrated HI_{GPP} values for each crop type (Table 3), crop yield ($Yield_M$) was estimated for each MT crop type over the 2008 to 2015 study period. Crop production was then calculated as crop yield multiplied by the total planted area each year for each crop type across MT. The comparisons of the estimated crop production (P_M) with reported annual county-level NASS crop production (P_N) values for the 2008 to 2015 record and MT domain are summarized in Table 5 and Figure 5; these results indicated strong correspondence between P_M and P_N at the MT county level ($r \geq 0.85$, $p < 0.05$). The P_M results also showed large spatial and annual variability in MT crop yields, which were generally consistent with the NASS crop production survey record. The P_M results showed the strongest correlation with P_N for durum wheat ($r \approx 1.00$, $p < 0.05$), with moderate model bias (−10.2%) and low relative RMSE (21.9%). The P_M relationship with P_N for alfalfa had the lowest correlation (r = 0.85), the second largest relative bias (−11.0%) and the largest relative RMSE (42.3%) difference due to larger annual fluctuations in model estimated production than the county survey record. Alfalfa for hay production is mixed with grasses and occurs on both irrigated and non-irrigated lands across nearly all MT counties, which may explain the relatively larger estimation errors for this crop. The P_M and P_N relationship for winter wheat showed the strongest correspondence (r = 0.99, $p < 0.05$), with relatively low bias (−1.5%) and RMSE (21.0%). Crops that are mostly rainfed in MT (barley, peas, durum wheat, winter wheat) showed overall lower average estimation uncertainties (Table 5; Figure 5). The model crop production results showed strong correlation (r = 0.96) and low relative bias (−7.8%) with the reported NASS county-level data when the model results were combined for all major MT crops during the 2008 to 2015 study period. However, the model results showed a lower RMSE (37.0%) performance, mainly due to the relatively large difference between P_M and P_N for alfalfa.

The spatial pattern of estimated 30-m resolution $Yield_M$ results is illustrated in Figure 6 for winter wheat in 2015. The 30-m results reveal the heterogeneous distribution of field dimensions and crops within the region, as well as the large variation of $Yield_M$ levels for winter wheat driven largely by spatial and temporal variations in NDVI. Winter wheat accounted for approximately 52.2% of the total cropland area in Liberty County and 24.1% of the total cropland area across MT in 2015. However, this crop type is distributed with other crops (shown in grey) in a complex spatial mosaic that is effectively delineated by the 30-m results.

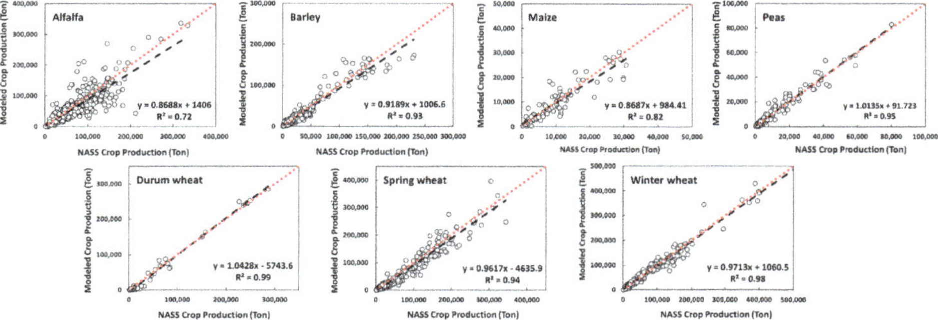

Figure 5. The comparisons between modeled annual crop production and USDA NASS reported county-level crop production data during the study period of 2008 to 2015 for the major crop types across Montana, including alfalfa, barley, maize, peas, durum wheat, spring wheat and winter wheat. The red dotted line is the 1:1 line, while the black dashed line is the linear regression relationship.

Table 5. Comparisons between mean county-level crop production from the USDA NASS survey (P_N) and mean modeled crop production (P_M) for each county in Montana from 2008 to 2015 (all of the correlations are significant with $p < 0.05$).

Crop Type	Mean of P_N (10^3 Ton)	Mean of P_M (10^3 Ton)	r	Bias (10^3 Ton)	Relative Bias	RMSE (10^3 Ton)	Relative RMSE
Alfalfa	67.22 ± 48.97	59.81 ± 50.25	0.85	−7.41	−11.0%	28.44	42.3%
Barley	33.19 ± 44.81	31.50 ± 42.77	0.96	−1.69	−5.1%	12.22	36.8%
Maize	10.27 ± 8.19	9.91 ± 7.84	0.91	−0.36	−3.5%	3.46	33.6%
Peas	10.70 ± 13.87	10.94 ± 14.39	0.98	0.24	2.2%	3.10	29.0%
Durum Wheat	39.55 ± 68.91	35.50 ± 72.21	1.00	−4.05	−10.2%	8.67	21.9%
Spring Wheat	59.41 ± 70.94	52.49 ± 70.45	0.97	−6.91	−11.6%	19.05	32.1%
Winter Wheat	70.84 ± 110.11	69.87 ± 107.93	0.99	−0.97	−1.4%	14.87	21.0%
All Crops	51.08 ± 69.22	47.10 ± 68.20	0.96	−3.98	−7.8%	18.89	37.0%

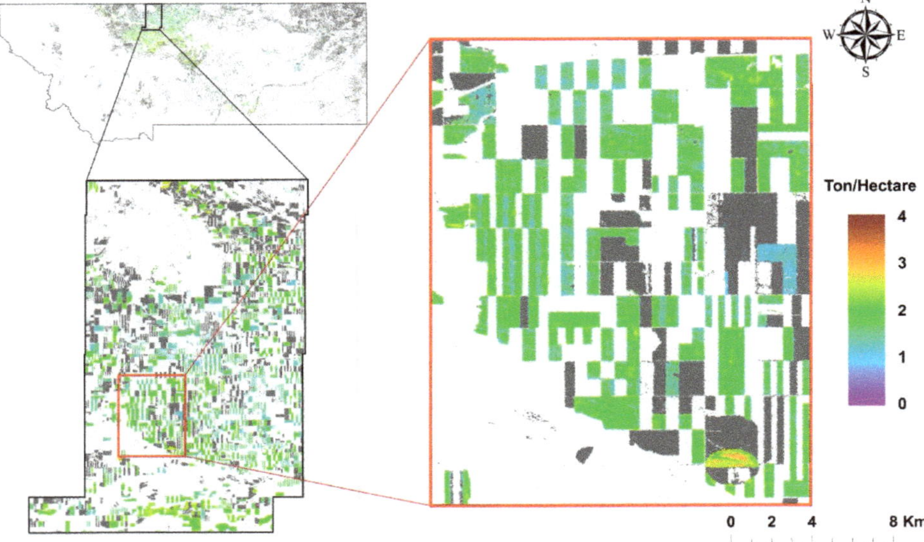

Figure 6. Example spatial distribution of estimated crop yield for winter wheat across MT and Liberty County in 2015; grey areas denote other CDL-defined crop types represented in this study, while white areas denote other land cover types excluded from the analysis.

Despite the enhanced delineation of field characteristics, the 30-m resolution $Yield_M$ results showed relatively low correspondence with the field-scale crop yield data ($Yield_F$) obtained at four farms representing barley, peas, spring wheat and winter wheat. These results showed moderate correspondence between $Yield_M$ and $Yield_F$ values for winter wheat ($r = 0.57$) and barley ($r = 0.68$), but much lower correlations for spring wheat and peas ($r < 0.10$), including large negative model biases (−12.0–−71.4%) and degraded RMSE (32.9–75.3%) relative to the available observations. The small number of field-scale yield observations makes it difficult to draw conclusions about the performance of the algorithm to retrieve field-scale yields. For instance, the $Yield_F$ observations for peas only represented one year of data from a single farm, generating the largest relative model bias (−71.4%) and RMSE (75.3%) among the four crop types represented; whereas, there were only three site-years of data for spring wheat from three different farms, inducing large relative model bias (−40.1%) and RMSE (50.5%). When all of the field-scale crop yield data were combined, the $Yield_M$ performance showed improved correspondence ($r = 0.44$, $p < 0.05$), relative bias (−36.1%) and RMSE (48.8%).

Although inconclusive, these results indicate meaningful, but lower model accuracy in delineating field level variations in crop yields relative to the county-level results.

The spatial distribution of the model uncertainties for alfalfa, barley, spring wheat and winter wheat at the county level are presented in Figure 7 because these crops were grown in most counties during 2008 to 2015 in Montana and permit the evaluation of the spatial variability of estimation errors. Estimation errors for alfalfa production (Figure 7a) tended to be larger in the more mountainous western portion of MT and in the eastern end of the state near the border with North Dakota. In these two regions, irrigated agriculture has increased significantly in the last few decades, and the mix of dryland and irrigated farming has the potential to increase the variance in production in these counties. Additionally, western Montana's complex terrain produces highly variable climatic conditions, which also translates into high production variability. Variability in farming practices (dryland versus irrigated) also applies to spring and winter wheat (Figure 7b,d). For these crops, the most accurate model predictions are concentrated in the north-central part of the state, termed the 'Golden Triangle' and known for highly productive dryland grain farming; model accuracy is generally lower moving in any cardinal direction from this region. The spatial patterns of model uncertainty for barley production (Figure 7c) is moderate, with generally better model performance in the north-central portions of MT around Pondera, Teton, Choteau and Glacier counties. However, the model results exhibit somewhat larger uncertainties in the eastern part of the state, where counties are less specialized in barley production and farming practices are more variable.

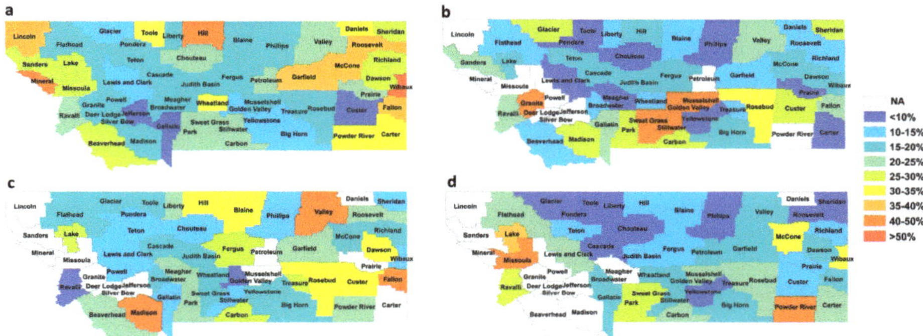

Figure 7. Standard error of model estimated annual crop production compared with USDA NASS reported crop production records per county for (**a**) alfalfa, (**b**) spring wheat, (**c**) barley and (**d**) winter wheat across MT. The NA term and associated white areas denote MT counties with no reported NASS crop production data for the crop types examined in this study.

The eight-year study period was too short for effectively quantifying and diagnosing regional trends in planted area and crop yield/production. However, the total planted area for all seven MT crop types showed a significant increasing trend from 2008 (4.36×10^6 hectares) to 2015 (5.95×10^6 hectares). The statewide averaged crop production for alfalfa was 3.01×10^6 ton during the study period, with two annual production peaks in 2010 (3.83×10^6 tons) and 2013 (3.21×10^6 tons) (Figure 8). The amount of land allocated to alfalfa increased during the study period at a rate of 1.52×10^5 hectares per year. Alfalfa is relatively productive (3.77 ± 1.06 tons/hectare) and has a larger planted area (3.63 to 18.05×10^5 hectares) in MT during 2008 to 2015 than the other six crop types. Barley showed a strong increasing trend in crop production from 2011 to 2015, together with a significant increase in planted area (Figure 8). Maize showed an increasing trend in planted area from 2008 to 2014, but did not show a congruent increase in crop production; maize production also peaked in 2013 (2.00×10^5 ton) relative to the other years examined (Figure 8). Though maize represented the lowest planted area of the major MT crops from 2008 to 2015, peas showed significant and consistent increasing trends in crop production and planted area during the study period (Figure 8). However, the crop yield for peas was 1.24 ± 0.28 tons/hectare, covering approximately 3.7% of the total cropland area in MT. There were no obvious trends in planted area or production for MT wheat crops (Figure 8). Spring wheat and winter wheat were planted over large areas in MT (>1.10 × 10^6 hectare), while estimated annual production and yield for these two cereal crops averaged 1.27 ± 0.27 and 1.90 ± 0.39 ton/hectare during the study period, respectively.

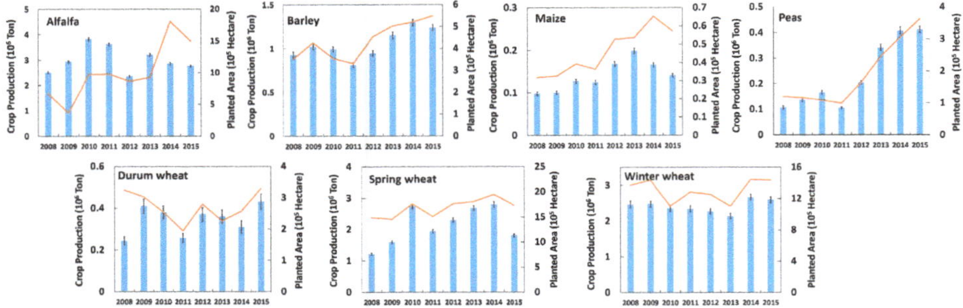

Figure 8. The variations of estimated annual crop yields from this study (bars) and USDA NASS reported planted areas (lines) for each major crop type from 2008 to 2015 in Montana; vertical error bars denote spatial variability (standard deviation) in estimated yields.

4. Discussion

Data fusion approaches have been applied to synthesize images from multiple satellite remote sensing sources at different spatial and temporal scales [23–28]. The simple linear regression model used in this study blended overlapping 250-m eight-day MODIS NDVI with finer spatial resolution (30-m), but less frequent Landsat NDVI data to generate a continuous NDVI record with 30-m resolution and eight-day temporal fidelity extending across MT from 2008 to 2015. A primary assumption of the data fusion approach is that the spatial pattern of the relationships between MODIS and Landsat NDVI are consistent over the sensor records. However, variations of MODIS and Landsat NDVI relationships may vary due to different acquisition dates, cloud contamination, aerosols, viewing angles and temporal compositing [24], or from sub-grid-scale spatial heterogeneity that is not resolved by the coarser MODIS footprint observations [26]; these factors can introduce uncertainties into the NDVI fusion results, which can propagate into model-derived *fPAR* error. However, the *fPAR* uncertainty is only one of several potential sources of LUE model GPP error. Other potential error sources include daily surface meteorological inputs and model BPLUT parameterizations defining maximum productivity rates and environmental response characteristics for different crop types.

The LUE model BPLUT parameterization used in this study distinguishes two major cropland categories (cereal and broadleaf), while the model parameters were calibrated using a global network of tower eddy covariance measurement sites and are consistent with the SMAP L4C operational GPP product [16]. The SMAP L4C GPP product shows favorable accuracy and performance in global croplands [16], indicating that the BPLUT parameters are also suitable for MT croplands. Nevertheless, variability in plant characteristics can occur within these general cropland categories, which can lead to significant model GPP error [52]. However, both the MODIS MOD17 and SMAP L4C operational GPP products utilize a static land cover (LC) classification and 500-m resolution MODIS (MCD12Q1) *fPAR* record as primary inputs to derive GPP at respective 500-m and 1-km spatial scales. In this study, a finer (30-m) resolution NASS Cropland Data Layer (CDL) and fused NDVI record was used to derive GPP. The GPP results from this study are thus expected to provide enhanced regional accuracy for agricultural applications due to the finer spatial delineation of cropland heterogeneity and model representation of dynamic cropland areas from the NASS CDL record. For example, approximately 1.16 × 10^5 hectares changed from cereal to broadleaf crop types, and 3.50 × 10^5 hectares changed from broadleaf to cereal crops, while cropland area varied by 3.90 × 10^5 hectares in MT from 2014 to 2015 based on the NASS CDL record; these changes are represented by our model results and influence the seasonal phenology, pattern and magnitude of estimated productivity and crop yields across the region. The resulting 30-m GPP simulations over MT croplands produced results that were generally similar, but more productive than the MODIS MOD17 (MOD17A2H, Collection 6) operational GPP product (Figure 4). However, the MODIS GPP product has been reported to have low productivity in croplands [18,20,53], indicating that the model results from this study benefit from a finer spatial resolution and a refined model cropland calibration. Though the Gridmet database shows favorable accuracy in relation to in situ weather station network measurements [49], the daily meteorological inputs may contribute uncertainties that propagate to cropland productivity model errors. Moreover, finer scale heterogeneity in surface meteorology not resolved by the 4-km Gridmet spatial resolution may contribute to model GPP errors derived at 30-m resolution. Management factors, including fertilization and irrigation practices, are not directly considered in our model and may also introduce uncertainties into the resulting GPP_M calculations.

Despite the potential model uncertainties, the modeled crop production (P_M) results were favorably correlated with the county-scale NASS annual crop production (P_N) records for all seven major crop types in MT during the 2008 to 2015 study period (r = 0.96, relative bias = −7.8%, relative RMSE = 37.0%). The county level model performance is robust, and the spatial fidelity achieved with the fusion of sensors allows delineation of production variability within a county, which is an important indicator of potential imbalances within a region. However, the estimated crop yields ($Yield_M$) showed lower correspondence with field-scale yield ($Yield_F$) records (r = 0.44). Potential reasons for the lower model performance in estimated field-scale yields are discussed below, but when interpreting the performance metrics, it is important to bear in mind that the state-wise harvest index (HI_{GPP}) used to derive $Yield_M$ from the 30-m GPP calculations was calibrated using county-level NASS annual crop yield data ($Yield_N$), which represents a major source of model $Yield_M$ uncertainty at finer spatial scales. A second consideration is that the $Yield_F$ observations used for model validation were obtained from a very limited set of 10 m by 10 m plots at four farm locations, which may not adequately reflect the 30-m resolution $Yield_M$ estimates derived across MT using coarser scale satellite NDVI observations and ancillary data. Uncertainty in the CDL files used to delineate annual crop types and planted areas in this study may also contribute to low $Yield_M$ and $Yield_F$ correspondence. Indeed, the reported CDL accuracy from 2008 to 2015 across MT croplands varies from 69.6% (2010) to 85.0% (2014), while the CDL accuracy levels may be inflated because they do not account for edge effects [54]. CDL accuracy may also be degraded in regions with sparse or complex agriculture [55], which is characteristic of MT croplands.

Alternatively, reported crop yields for spring wheat in Lethbridge, Canada, near the MT border, ranged from 1.5 to 1.8 tons/hectare [56]; reported spring wheat yields also ranged from 0.5 to less than 3 tons/hectare in arid and semi-arid areas of the Canadian Prairies [22]. These reported yields are similar to our model $Yield_m$ results for MT spring wheat (1.27 ± 0.27 tons/hectare). The mean modeled crop yield during 2008 to 2015 for barley in this study is 2.48 ± 0.61 tons/hectare, which is also similar to reported barley yields on the Canadian Prairies (0.9 to 4.5 tons/hectare; [22]). For alfalfa, the resulting crop yield estimates range from 1.58 to 8.06 tons/hectare, while another study reported alfalfa yields from 2.97 to 14.72 tons/hectare [57], similar to our model results. Moreover, alfalfa yields reported by the Food and Agricultural Organization range from 5 to 17 tons/hectare under rainfed conditions with 500 to 800 mm annual rainfall [58]; these values are similar to our model yields for MT alfalfa and indicate that the calibrated HI_{GPP} parameter used in this investigation is reasonable, though it is higher than the HI values for this crop type reported from prior limited studies. Thus, our model estimates of crop yield and production may be suitable for regional cropland productivity and yield assessments that can inform agricultural decisions. These findings are consistent with a similar regional analysis of smallholder agricultural productivity across Africa indicating that high-resolution satellite imagery can be used to make crop yield predictions that are roughly as accurate as survey-based measurements [59].

A limitation of our current calibration strategy that contributes to modest estimation performance at the field scale is that sub-grid scale heterogeneity in environmental conditions, including soil moisture, nutrient availability and resource competition from weeds, is not accounted for by the regional model, and probably most importantly, calibration does not differentiate between irrigated and rainfed crops because spatially explicit information about this practice is currently not available for MT. More specific model calibrations that account for irrigated and non-irrigated crops are expected to increase the precision of production estimates at sub-county scales. Moreover, the growing season in this study is assumed to extend from April to September for all seven MT crop types examined, which may not be accurate for each crop type. For example, winter wheat is planted in September or October and harvested the following July or August [45]. All of these factors likely contribute to the lower $Yield_M$ and $Yield_F$ correspondence.

For the seven major crop types examined over the 2008 to 2015 study period, although the total planted area in MT is increasing (2.79×10^5 hectares year^{-1}), the planted area of each crop type shows large annual variability (Figure 8), which contributes to variations in annual crop production in addition to other influential factors including climate variability. Spring wheat, winter wheat and alfalfa are the three largest components of the total cropland area in MT (34.4%, 26.8% and 19.7%, respectively) from 2008 to 2015. The land dedicated to alfalfa increased by approximately 1.52×10^5 hectares year^{-1}, and was also the largest contributor to annual variability in MT total planted area according to the NASS CDL record. The planted areas and estimated annual crop production for spring wheat and winter wheat did not show consistent trends during the study period (Figure 8). Simultaneously, barley, maize and peas showed increasing trends in planted area, but at relatively low levels (2.86, 0.48 and 3.73×10^4 hectares year^{-1} respectively), although the combined extent of all three crops occupied only 13.5% of the total planted cropland area and made only a small contribution to changes in total planted area and crop production for the state.

Because Montana's agriculture is mostly rainfed and has relatively low diversification, planted areas and crop production are very sensitive to climate variations and to changes in agricultural pricing. The MT crop production for all selected crops showed peaks in 2010 and 2013, which had the largest growing season precipitation volumes during the eight-year study period (351.6 mm in 2010 and 336.4 mm in 2013 and approximately 24.6% and 19.2% above mean precipitation from 2008 to 2015), as well as large ratios of precipitation to potential evapotranspiration (0.45 and 0.40 for the years 2010 and 2013, respectively), indicating wet climate anomalies for these two years. Further research is needed to understand the sensitivity of annual crop yield and planted land variations to external factors such as climate fluctuations or changes in crop prices or cost of agricultural inputs.

In recent decades, agriculture has continuously outpaced all other industries as a proportion of total gross domestic product (GDP) in MT, while agricultural production accounted for over 30% of the state's basic industry employment, labor income and gross sales [44], suggesting the importance for obtaining consistent crop yield assessments across the entire state. This study provides a potential approach for continuous crop yield monitoring at 30-m resolution, which may better inform agricultural management and policy decisions. However, further studies are needed to improve crop yield estimation accuracy and performance and clarify the role of socioeconomic factors, management practices and climate variability on crop yield patterns and trends. Promising areas for model improvement include the use of additional satellite observations to construct more accurate estimates of key vegetation parameters and drivers [60] and the use of high temporal frequency estimates of crop yield at key decision points during the growing season to determine the projected value of alternative decisions. More detailed cropland inventory data capable of resolving field and farm level heterogeneity may improve the delineation of HI_{GPP} and yield variations for different crop conditions and consequently improve the utility of yield estimates. In situ monitoring of GPP from flux tower sites representing major MT crop types would provide an effective means to improve model calibration and performance. The results of this study and other ongoing efforts are being used to improve crop monitoring capabilities using global satellite observations to inform resource management and policy decisions and enhance national and global food security.

5. Conclusions

We applied a satellite data-driven LUE model framework to produce 30-m resolution daily GPP and annual crop yield estimates for the seven major crop types in Montana from 2008 to 2015. A fused 30-m eight-day NDVI record was developed using an empirical regression model combining similar spectral information from overlapping MODIS and Landsat imagery. The fused NDVI record was used with 4-km resolution gridded daily surface meteorology as primary inputs to the LUE model to derive *fPAR* and GPP at 30-m resolution and an eight-day time step over MT cropland areas. The fused NDVI record overcomes many of the limitations of the contributing Landsat and MODIS sensor records, including Landsat data gaps and relatively coarse (250-m) MODIS spatial resolution (Figure 2). The fused record also shows generally consistent NDVI magnitudes and seasonal variations as the original Landsat and MODIS data (Figure 3), while providing complete spatial coverage, eight-day temporal fidelity and 30-m spatial resolution. The resulting 30-m GPP simulations over MT croplands are produced benefiting from the finer resolution (30-m, eight-day) NDVI and the refined model cropland calibration. Estimated annual crop production and yields are derived from the 30-m eight-day GPP simulations over MT agricultural areas defined from the annual USDA NASS cropland data layer (CDL) and calibrated HI_{GPP} values for the seven major MT crop types. The model results corresponded favorably to NASS county-level crop production data (Table 5), indicating that the modeling approach captures regional patterns and annual variations in annual crop yield across MT and that obtaining regional-scale crop production assessments using existing global production models is possible.

Acknowledgments: This work was conducted at the University of Montana with funding provided by the USDA NIFA (National Institute of Food and Agriculture) contract 658 2016-67026-25067, USDA 365063, NASA EPSCoR (Established Program to Stimulate Competitive Research) contract 80NSSC18M0025 and NASA (NNX14AI50G, NNX14A169G, NNX08AG87A). The field-level crop yield observations were funded by the Montana Research and Economic Development Initiative.

Author Contributions: M.H., J.S.K. and M.P.M. developed this study. M.H. wrote the manuscript, and all authors discussed and reviewed the manuscript. B.D.M. provided field scale crop yield data.

Conflicts of Interest: The authors declare no conflict of interest.

References

1. Lobell, D.B.; Asner, G.P.; Ortiz-Monasterio, J.I.; Benning, T.L. Remote sensing of regional crop production in the Yaqui Valley, Mexico: Estimates and uncertainties. *Agric. Ecosyst. Environ.* **2003**, *94*, 205–220. [CrossRef]
2. Lobell, D.B.; Cassman, K.G.; Field, C.B. Crop yield gaps: Their importance, magnitudes and causes. *Annu. Rev. Environ. Resour.* **2009**, *34*, 179–204. [CrossRef]
3. Moriondo, M.; Maselli, F.; Bindi, M. A simple model for regional wheat yield based on NDVI data. *Eur. J. Agron.* **2007**, *26*, 266–274. [CrossRef]
4. Yuan, W.; Cai, W.; Nguy-Robertson, A.L.; Fang, H.; Suyker, A.E.; Chen, Y.; Dong, W.; Liu, S.; Zhang, H. Uncertainty in simulating gross primary production of cropland ecosystem from satellite-based models. *Agric. For. Meteorol.* **2015**, *207*, 48–57. [CrossRef]
5. Ines, A.V.M.; Das, N.N.; Hansen, J.W.; Njoku, E.G. Assimilation of remotely sensed soil moisture and vegetation with a crop simulation model for maize yield prediction. *Remote Sens. Environ.* **2013**, *138*, 149–164. [CrossRef]
6. Singh, R.; Semwal, D.P.; Rai, A.; Chhikara, R.S. Small area estimation of crop yield using remote sensing satellite data. *Int. J. Remote Sens.* **2002**, *23*, 49–56. [CrossRef]
7. Sakamoto, T.; Gitelson, A.A.; Arkebauer, T.J. MODIS-based cron grain yield estimation model incorporating crop phenology information. *Remote Sens. Environ.* **2013**, *131*, 215–231. [CrossRef]

8. Fang, H.; Liang, S.; Hoogenboom, G.; Teasdale, J.; Cavigelli, M. Corn-yield estimation through assimilation of remotely sensed data into the CSM-CERES-Maize model. *Int. J. Remote Sens.* **2008**, *29*, 3011–3032. [CrossRef]
9. Ren, J.; Chen, Z.; Zhou, Q.; Tang, H. Regional yield estimation for winter wheat with MODIS-NDVI data in Shandong, China. *Int. J. Appl. Earth Obs. Geoinf.* **2008**, *10*, 403–413. [CrossRef]
10. Lobell, D.B. The use of satellite data for crop yield gap analysis. *Field Crops Res.* **2013**, *143*, 56–64. [CrossRef]
11. Prasad, A.K.; Chai, L.; Singh, R.P.; Kafatos, M. Crop yield estimation model for Iowa using remote sensing and surface parameters. *Int. J. Appl. Obs. Geoinf.* **2006**, *8*, 26–33. [CrossRef]
12. Bolton, D.K.; Friedl, M.A. Forecasting crop yield using remotely sensed vegetation indices and crop phenology metrics. *Agric. For. Meteorol.* **2013**, *173*, 74–84. [CrossRef]
13. Mkhabela, M.S.; Bullock, P.; Raj, S.; Wang, S.; Yang, Y. Crop yield forecasting on the Canadian Prairies using MODIS NDVI data. *Agric. For. Meteorol.* **2011**, *151*, 385–393. [CrossRef]
14. Xiao, X.M.; Hollinger, D.; Aber, J.; Goltz, M.; Davidson, E.A.; Zhang, Q.Y.; Moore, B., III. Satellite-based modeling of gross primary production in an evergreen needleleaf forest. *Remote Sens. Environ.* **2004**, *89*, 519–534. [CrossRef]
15. Running, S.W.; Nemani, R.R.; Heinsch, F.A.; Zhao, M.S.; Reeves, M.; Hashimoto, H. A continuous satellite-drived measure of global terrestrial primary production. *BioScience* **2004**, *54*, 547–560. [CrossRef]
16. Hilker, T.; Coops, N.C.; Wulder, M.A.; Black, T.A.; Guy, R.D. The use of remote sensing in light use efficiency based models of gross primary production: A review of current status and future requirements. *Remote Sens. Environ.* **2008**, *404*, 411–423. [CrossRef] [PubMed]
17. Jones, L.A.; Kimball, J.S.; Reichle, R.H.; Madani, N.; Glassy, J.M.; Ardizzone, J.V.; Colliander, A.; Cleverly, J.; Desai, A.R.; Eamus, D.; et al. The SMAP Level 4 carbon product for monitoring ecosystem Land-Atmosphere CO_2 exchange. *IEEE Trans. Geosci. Remote Sens.* **2017**, *99*, 1–16. [CrossRef]
18. Turner, D.P.; Ritts, W.D.; Cohen, W.B.; Gower, S.T.; Running, S.W.; Zhao, M.; Costa, M.H.; Kirschbaum, A.A.; Ham, J.M.; Saleska, S.R.; et al. Evaluation of MODIS NPP and GPP products across multiple biomes. *Remote Sens. Environ.* **2006**, *102*, 282–292. [CrossRef]
19. Zhang, Y.; Yu, Q.; Jiang, J.; Tang, Y. Calibration of Terra/MODIS gross primary production over an irrigated cropland on the North China Plain and an alpine meadow on the Tibetan Plateau. *Glob. Chang. Biol.* **2008**, *14*, 757–767. [CrossRef]
20. Zhao, M.; Heinsch, F.A.; Nemani, R.R.; Running, S.W. Improvements of the MODIS terrestrial gross and net primary prodcution global data set. *Remote Sens. Environ.* **2005**, *95*, 164–176. [CrossRef]
21. Zhang, F.; Chen, J.M.; Chen, J.; Gough, C.M.; Martin, T.A.; Dragoni, D. Evaluating spatial and temporal patterns of MODIS GPP over the conterminous U.S. against flux measurements and a process model. *Remote Sens. Environ.* **2012**, *124*, 717–729. [CrossRef]
22. Gao, F.; Hilker, T.; Zhu, X.; Anderson, M.C.; Masek, J.G.; Wang, P.; Yang, Y. Fusing Landsat and MODIS data for vegetation monitoring. *IEEE Geosci. Remote Sens. Mag.* **2015**, *3*, 47–60. [CrossRef]
23. Hilker, T.; Wulder, M.A.; Coops, N.C.; Seitz, N.; White, J.C.; Gao, F.; Masek, J.G.; Stenhouse, G. Generation of dense time series synthetic Landsat data through data blending with MODIS using a spatial and temporal adaptive reflectance fusion model. *Remote Sens. Environ.* **2009**, *113*, 1988–1999. [CrossRef]
24. Ju, J.C.; Roy, D.P. The availability of cloud-free Landsat ETM+ data over the conterminous United States and globally. *Remote Sens. Environ.* **2007**, *112*, 1196–1211. [CrossRef]
25. Gao, F.; Masek, J.; Schwaller, M.; Hall, F. On the blending of the Landsat and MODIS surface reflectance: Predicting daily Landsat surface reflectance. *IEEE Trans. Geosci. Remote Sens.* **2006**, *44*, 2207–2218.
26. Gao, F.; Anderson, M.C.; Zhang, X.; Yang, Z.; Alfieri, J.G.; Kustas, W.P.; Mueller, R.; Johnson, D.M.; Prueger, J.H. Toward mapping crop progress at field scales through fusion of Landsat and MODIS imagery. *Remote Sens. Environ.* **2017**, *188*, 9–25. [CrossRef]
27. Zhu, X.; Helmer, E.H.; Gao, F.; Liu, D.; Chen, J.; Lefsky, M.A. A flexible spatiotemporal method for fusing satellite images with different resolutions. *Remote Sens. Environ.* **2016**, *172*, 165–177. [CrossRef]
28. Schmidt, M.; Udelhoven, T.; Gill, T.; Roder, A. Long term data fusion for a dense time series analysis with MODIS and Landsat imagery in an Australian savanna. *J. Appl. Remote Sens.* **2012**, *6*, 063512.
29. Zhang, W.; Li, A.; Jin, H.; Bian, J.; Zhang, Z.; Lei, G.; Qin, Z.; Huang, C. An enhanced spatial and temporal data fusion model for fusing Landsat and MODIS surface reflectance to generate high temporal Landsat-like data. *Remote Sens.* **2013**, *5*, 5346–5368. [CrossRef]

30. Doraiswamy, P.C.; Hatfield, J.L.; Jackson, T.J.; Akhmedov, B.; Prueger, J.; Stern, A. Crop condition and yield simulations using Landsat and MODIS. *Remote Sens. Environ.* **2004**, *92*, 548–559. [CrossRef]
31. Gu, Y.; Wylie, B.K. Downscaling 250-m MODIS growing season NDVI based on multiple-date Landsat images and data mining approaches. *Remote Sens.* **2015**, *7*, 3489–3506. [CrossRef]
32. Hwang, T.; Song, C.; Bolstad, P.V.; Band, L.E. Downscaling real-time vegetation dynamics by fusing multi-temporal MODIS and Landsat NDVI in topographically complex terrain. *Remote Sens. Environ.* **2011**, *115*, 2499–2512. [CrossRef]
33. Prince, S.D.; Haskett, J.; Steininger, M.; Strand, H.; Wright, R. Net Primary production of U.S. Midwest croplands from agricultural harvest yield data. *Ecol. Appl.* **2001**, *11*, 1194–1205. [CrossRef]
34. Lorenz, A.J.; Gustafson, T.J.; Coors, J.G.; de Leon, N. Breeding maize for a bioeconomy: A literature survey examining harvest index and stover yield and their relationship to a grain yield. *Crop Sci.* **2010**, *50*, 1–12. [CrossRef]
35. Kemanian, A.R.; Stöckle, C.O.; Huggins, D.R.; Viega, L.M. A simple method to estimate harvest index in grain crops. *Field Crops Res.* **2007**, *103*, 208–216. [CrossRef]
36. Peltonen-Sainio, P.; Muurinen, S.; Rajala, A.; Jauhiainen, L. Variation in harvest index of modern spring barley, oat and wheat cultivars adapted to northern growing conditions. *J. Agric. Sci.* **2008**, *146*, 35–47. [CrossRef]
37. Iannucci, A.; Di Fonzo, N.; Martiniello, P. Alfalfa (*Medicago sativa* L.) seed yield and quality under different forage management systems and irrigation treatments in a Mediterranean environment. *Field Crops Res.* **2002**, *78*, 65–74. [CrossRef]
38. Reeves, M.C.; Zhao, M.; Running, S.W. Usefulness and limits of MODIS GPP for estimating wheat yield. *Int. J. Remote Sens.* **2005**, *26*, 1403–1421. [CrossRef]
39. Gorelick, N.; Hancher, M.; Dixon, M.; Ilyushchenko, S.; Thau, D.; Moore, R. Google earth engine: Planetary-scale geospatial analysis for everyone. *Remote Sens. Environ.* **2017**, *202*, 18–27. [CrossRef]
40. Lobell, D.B.; Thau, D.; Seifert, C.; Engle, E.; Little, B. A scalable satellite-based crop yield mapper. *Remote Sens. Environ.* **2015**, *164*, 324–333. [CrossRef]
41. Hansen, M.C.; Potapov, P.V.; Moore, R.; Hancher, M.; Turubanova, S.A.; Tyukavina, A.; Thau, D.; Stehman, S.V.; Goetz, S.J.; Loveland, T.R.; et al. High-resolution global maps of 21st-century forest cover change. *Science* **2013**, *342*, 850–853. [CrossRef] [PubMed]
42. Pekel, J.-F.; Cottam, A.; Gorelick, N.; Belward, A.S. High-resolution mapping of global surface water and its long-term changes. *Nature* **2016**, *540*, 418–422. [CrossRef] [PubMed]
43. Robinson, N.P.; Allread, B.W.; Jones, M.O.; Moreno, A.; Kimball, J.S.; Naugle, D.E.; Erickson, T.A.; Richardson, A.D. A dynamic Landsat derived Normalized Difference Vegetation Index (NDVI) product for the Conterminous United States. *Remote Sens.* **2017**, *9*, 863. [CrossRef]
44. Labus, M.P.; Nielsen, G.A.; Lawrence, R.L.; Engel, R.; Long, D. Wheat yield estimates using multi-temporal NDVI satellite imagery. *Int. J. Remote Sens.* **2002**, *23*, 4169–4180. [CrossRef]
45. NASS Field Crops. Usual Planting and Harvesting Dates. 2010. Available online: http://usda.mannlib.cornell.edu/usda/current/planting/planting-10--29--2010.pdf (accessed on 24 February 2018).
46. Masek, J.G.; Vermote, E.F.; Saleous, N.E.; Wolfe, R.; Hall, F.G.; Huemmrich, K.F.; Gao, F.; Kutler, J.; Lim, T. A Landsat surface reflectance dataset for North America, 1990–2000. *IEEE Geosci. Remote Sens. Lett.* **2006**, *3*, 68–72. [CrossRef]
47. U.S. Geological Survey. *Landsat 4–7 Surface Reflectance Product Guide*; U.S. Geological Survey: Reston, WV, USA, 2017.
48. Heinsch, F.A.; Zhao, M.; Running, S.W.; Kimball, J.S.; Nemani, R.R.; Davis, K.J.; Bolstad, P.V.; Cook, B.D.; Desai, A.R.; Ricciuto, D.M.; et al. Evaluation of Remote Sensing based terrestrial productivity from MODIS using regional tower eddy flux network observations. *IEEE Trans. Geosci. Remote Sens.* **2006**, *44*, 1908–1925. [CrossRef]
49. Abatzoglou, J.T. Development of gridded surface meteorological data for ecological applications and modelling. *Int. J. Climatol.* **2013**, *33*, 121–131. [CrossRef]
50. Peng, D.; Zhang, B.; Liu, L.; Fang, H.; Chen, D.; Hu, Y.; Liu, L. Characteristics and drivers of global NDVI-based FPAR from 1982 to 2006. *Glob. Biogeochem. Cycles* **2012**, *26*. [CrossRef]

51. Zhang, L.; Wylie, B.; Loveland, T.; Fosnight, E.; Tieszen, L.L.; Ji, L.; Gilmanov, T. Evaluation and comparison of gross primary production estimates for the Northern Great Plains grassland. *Remote Sens. Environ.* **2007**, *106*, 173–189. [CrossRef]
52. Madani, N.; Kimball, J.S.; Affleck, D.L.R.; Kattge, J.; Graham, J.; van Bodegom, P.M.; Reich, P.B.; Running, S.W. Improving ecosystem productivity modeling through spatially explicit estimation of optimal light use efficiency. *J. Geophys. Res. Biogeosci.* **2014**, *119*, 1755–1769. [CrossRef]
53. Xin, Q.; Broich, M.; Suyker, A.E.; Yu, L.; Gong, P. Multi-scale evaluation of light use efficiency in MODIS gross primary productivity for croplands in the Midwestern United States. *Agric. For. Meteorol.* **2015**, *201*, 111–119. [CrossRef]
54. Lark, T.J.; Mueller, R.M.; Johnson, D.M.; Gibbs, H.K. Measuring land-use and land-cover change using the U.S. department of agriculture's cropland data layer: Cautions and recommendations. *Int. J. Appl. Earth Obs. Geoinf.* **2017**, *62*, 224–235. [CrossRef]
55. Larsen, A.E.; Hendrickson, B.T.; Dedeic, N.; MacDonald, A.J. Taken as a given: Evaluating the accuracy of remotely sensed crop data in the USA. *Agric. Syst.* **2015**, *141*, 121–125. [CrossRef]
56. Smith, W.N.; Grant, B.B.; Desjardins, R.L.; Kroebel, R.; Li, C.; Qian, B.; Worth, D.E.; McConkey, B.G.; Drury, C.F. Assessing the effects of climate change on crop production and GHG emissions in Canada. *Agric. Ecosyst. Environ.* **2013**, *179*, 139–150. [CrossRef]
57. Li, Y.; Hunag, M. Pasture yield and soil water depletion of continuous growing alfalfa in the Loess Plateau of China. *Agric. Ecosyst. Environ.* **2008**, *124*, 24–32. [CrossRef]
58. Steduto, P.; Hsiao, T.C.; Fereres, E.; Raes, D. *Crop Yield Response to Water*; Food and Agriculture Organization of the United Nations: Rome, Italy, 2012.
59. Burke, M.; Lobell, D.B. Satellite-based assessment of yield variation and its determinants in smallholder African systems. *Proc. Natl. Acad. Sci. USA* **2017**, *114*, 2189–2194. [CrossRef] [PubMed]
60. Guan, K.; Wu, J.; Kimball, J.S.; Anderson, M.C.; Frolking, S.; Li, B.; Hain, C.R.; Lobell, D. The shared and unique values of optical, fluorescence, thermal and microwave satellite data for estimating large-scale crop yields. *Remote Sens. Environ.* **2017**, *199*, 333–349. [CrossRef]

Article

The First Wetland Inventory Map of Newfoundland at a Spatial Resolution of 10 m Using Sentinel-1 and Sentinel-2 Data on the Google Earth Engine Cloud Computing Platform

Masoud Mahdianpari [1,2,*], Bahram Salehi [3], Fariba Mohammadimanesh [1,2], Saeid Homayouni [4] and Eric Gill [2]

1 C-CORE, 1 Morrissey Rd, St. John's, NL A1B 3X5, Canada
2 Department of Electrical and Computer Engineering, Memorial University of Newfoundland, St. John's, NL A1C 5S7, Canada; f.mohammadimanesh@mun.ca (F.M.); ewgill@mun.ca (E.G.)
3 Environmental Resources Engineering, College of Environmental Science and Forestry, State University of New York, NY 13210, USA; bsalehi@esf.edu
4 Department of Geography, Environment, and Geomatics, University of Ottawa, Ottawa, ON K1N 6N5, Canada; Saeid.Homayouni@uottawa.ca
* Correspondence: m.mahdianpari@mun.ca; Tel.: +1-709-986-0110

Received: 22 October 2018; Accepted: 20 December 2018; Published: 28 December 2018

Abstract: Wetlands are one of the most important ecosystems that provide a desirable habitat for a great variety of flora and fauna. Wetland mapping and modeling using Earth Observation (EO) data are essential for natural resource management at both regional and national levels. However, accurate wetland mapping is challenging, especially on a large scale, given their heterogeneous and fragmented landscape, as well as the spectral similarity of differing wetland classes. Currently, precise, consistent, and comprehensive wetland inventories on a national- or provincial-scale are lacking globally, with most studies focused on the generation of local-scale maps from limited remote sensing data. Leveraging the Google Earth Engine (GEE) computational power and the availability of high spatial resolution remote sensing data collected by Copernicus Sentinels, this study introduces the first detailed, provincial-scale wetland inventory map of one of the richest Canadian provinces in terms of wetland extent. In particular, multi-year summer Synthetic Aperture Radar (SAR) Sentinel-1 and optical Sentinel-2 data composites were used to identify the spatial distribution of five wetland and three non-wetland classes on the Island of Newfoundland, covering an approximate area of 106,000 km^2. The classification results were evaluated using both pixel-based and object-based random forest (RF) classifications implemented on the GEE platform. The results revealed the superiority of the object-based approach relative to the pixel-based classification for wetland mapping. Although the classification using multi-year optical data was more accurate compared to that of SAR, the inclusion of both types of data significantly improved the classification accuracies of wetland classes. In particular, an overall accuracy of 88.37% and a Kappa coefficient of 0.85 were achieved with the multi-year summer SAR/optical composite using an object-based RF classification, wherein all wetland and non-wetland classes were correctly identified with accuracies beyond 70% and 90%, respectively. The results suggest a paradigm-shift from standard static products and approaches toward generating more dynamic, on-demand, large-scale wetland coverage maps through advanced cloud computing resources that simplify access to and processing of the "Geo Big Data." In addition, the resulting ever-demanding inventory map of Newfoundland is of great interest to and can be used by many stakeholders, including federal and provincial governments, municipalities, NGOs, and environmental consultants to name a few.

Keywords: wetland; Google Earth Engine; Sentinel-1; Sentinel-2; random forest; cloud computing; geo-big data

1. Introduction

Wetlands cover between 3% and 8% of the Earth's land surface [1]. They are one of the most important contributors to global greenhouse gas reduction and climate change mitigation, and they greatly affect biodiversity and hydrological connectivity [2]. Wetland ecosystem services include flood- and storm-damage protection, water-quality improvement and renovation, aquatic and plant-biomass productivity, shoreline stabilization, plant collection, and contamination retention [3]. However, wetlands are being drastically converted to non-wetland habitats due to both anthropogenic activities, such as intensive agricultural and industrial development, urbanization, reservoir construction, and water diversion, as well as natural processes, such as rising sea levels, thawing of permafrost, changing in precipitation patterns, and drought [1].

Despite the vast expanse and benefits of wetlands, there is a lack of comprehensive wetland inventories in most countries due to the expense of conducting nation-wide mapping and the highly dynamic, remote nature of wetland ecosystems [4]. These issues result in fragmented, partial, or outdated wetland inventories in most countries worldwide, and some have no inventory available at all [5]. Although North America and some parts of Western Europe have some of the most comprehensive wetland inventories, these are also incomplete and have considerable limitations related to the resolution and type of data, as well as to developed methods [6]. These differences make these existing inventories incomparable [1] and highlight the significance of long-term comprehensive wetland monitoring systems to identify conservation priorities and sustainable management strategies for these valuable ecosystems.

Over the past two decades, wetland mapping has gained recognition thanks to the availability of remote sensing tools and data [6,7]. However, accurate wetland mapping using remote sensing data, especially on a large-scale, has long proven challenging. For example, input data should be unaffected/less affected by clouds, haze, and other disturbances to obtain an acceptable classification result [4]. Such input data can be generated by compositing a large volume of satellite images collected during a specific time period. This is of particular concern for distinguishing backscattering/spectrally similar classes (e.g., wetland), wherein discrimination is challenging using a single image. Historically, the cost of acquiring multi-temporal remote sensing data precluded such large-scale land cover (e.g., wetland) mapping [8]. Although Landsat sensors have been collecting Earth Observation (EO) data at frequent intervals since the mid-1980s [9], open-access to its entire archive has occurred since 2008 [8]. This is of great benefit for land cover mapping on a large-scale. However, much of this archived data has been underutilized to date. This is because collecting, storing, processing, and manipulating multi-temporal remote sensing data that cover a large geographic area over three decades are infeasible using conventional image processing software on workstation PC-based systems [10]. This is known as the "Geo Big Data" problem and it demands new technologies and resources capable of handling such a large volume of satellite imagery from the data science perspective [11].

Most recently, the growing availability of large-volume open-access remote sensing data and the development of advanced machine learning tools have been integrated with recent implementations of powerful cloud computing resources. This offers new opportunities for broader sets of applications at new spatial and temporal scales in the geospatial sciences and addresses the limitation of existing methods and products [12]. Specifically, the advent of powerful cloud computing resources, such as NASA Earth Exchange, Amazon's Web Services, Microsoft's Azure, and Google cloud platform has addressed these Geo Big Data problems. For example, Google Earth Engine (GEE) is an open-access, cloud-based platform for parallel processing of petabyte-scale data [13]. It hosts a vast pool of satellite imagery and geospatial datasets, and allows web-based algorithm development and results visualization in a reasonable processing time [14–16]. In addition to its computing and storage capacity, a number of well-known machine learning algorithms have been implemented, allowing batch processing using JavaScript on a dedicated application programming interface (API) [17].

Notably, the development of advanced machine learning tools further contributes to handling large multi-temporal remote sensing data [18]. This is because traditional classifiers, such as maximum likelihood, insufficiently manipulate complicated, high-dimensional remote sensing data. Furthermore, they assume that input data are normally distributed, which may not be the case [19]. However, advanced machine learning tools, such as Decision Tree (DT), Support Vector Machine (SVM), and Random Forest (RF), are independent of input data distribution and can handle large volumes of remote sensing data. Previous studies have demonstrated that both RF [20] and SVM [21] outperformed DT for classifying remote sensing data. RF and SVM have also relatively equal strength in terms of classification accuracies [22]. However, RF is much easier to execute relative to SVM, given that the latter approach requires the adjustment of a large number of parameters [23]. RF is also insensitive to noise and overtraining [24] and has shown high classification accuracies in various wetland studies [19,25].

Over the past three years, several studies have investigated the potential of cloud-computing resources using advanced machine learning tools for processing/classifying the Geo Big Data in a variety of applications. These include global surface water mapping [26], global forest-cover change mapping [27], and cropland mapping [28], as well as studies focusing on land- and vegetation-cover changes on a smaller scale [29,30]. They demonstrated the feasibility of characterizing the elements of the Earth surface at a national and global scale through advanced cloud computing platforms.

Newfoundland and Labrador (NL), a home for a great variety of flora and fauna, is one of the richest provinces in terms of wetlands and biodiversity in Canada. Most recently, the significant value of these ecosystems has been recognized by the Wetland Mapping and Monitoring System (WMMS) project, launched in 2015. Accordingly, a few local wetland maps, each covering approximately 700 km^2 of the province, were produced. For example, Mahdianpari et al. (2017) introduced a hierarchical object-based classification scheme for discriminating wetland classes in the most easterly part of NL, the Avalon Peninsula, using Synthetic Aperture Radar (SAR) observations obtained from ALOS-2, RADARSAT-2, and TerraSAR-X imagery [19]. Later, Mahdianpari et al. (2018) proposed the modified coherency matrix obtained from quad-pol RADARSAT-2 imagery to improve wetland classification accuracy. They evaluated the efficiency of the proposed method in three pilot sites across NL, each of which covers 700 km^2 [31]. Most recently, Mohammadimanesh et al. (2018) investigated the potential of interferometric coherence for wetland classification, as well as the synergy of coherence with SAR polarimetry and intensity features for wetland mapping in a relatively small area in NL (the Avalon Peninsula) [32]. These local-scale wetland maps exhibit the spatial distribution patterns and the characteristics of wetland species (e.g., dominant wetland type). However, such small-scale maps have been produced by incorporating different data sources, standards, and methods, making them of limited use for rigorous wetland monitoring at the provincial, national, and global scales.

Importantly, precise, comprehensive, provincial-level wetland inventories that map small to large wetland classes can significantly aid conservation strategies, support sustainable management, and facilitate progress toward national/global scale wetland inventories [33]. Fortunately, new opportunities for large-scale wetland mapping are obtained from the Copernicus programs by the European Space Agency (ESA) [34]. In particular, concurrent availability of 12-days SAR Sentinel-1 and 10-days optical Sentinel-2 (multi-spectral instrument, MSI) sensors provides an unprecedented opportunity to collect high spatial resolution data for global wetland mapping. The main purpose of these Sentinel Missions is to provide full, free, and open access data to facilitate the global monitoring of the environment and to offer new opportunities to the scientific community [35]. This highlights the substantial role of Sentinel observations for large-scale land surface mapping. Accordingly, the synergistic use of Sentinel-1 and Sentinel-2 EO data offers new avenues to be explored in different applications, especially for mapping phenomena with highly dynamic natures (e.g., wetland).

Notably, the inclusion of SAR data for land and wetland mapping is of great significance for monitoring areas with nearly permanent cloud-cover. This is because SAR signals are independent of solar radiation and the day/night condition, making them superior for monitoring geographic regions with dominant cloudy and rainy weather, such as Newfoundland, Canada. Nevertheless, multi-source satellite data are advantageous in terms of classification accuracy relative to the accuracy achieved by a single source of data [36]. This is because optical sensors are sensitive to the reflective and spectral characteristics of ground targets [37,38], whereas SAR sensors are sensitive to their structural, textural, and dielectric characteristics [39,40]. Thus, a synergistic use of two types of data offers complementary information, which may be lacking when utilizing one source of data [41,42]. Several studies have also highlighted the great potential of fusing optical and SAR data for wetland classification [25,36,41].

This study aims to develop a multi-temporal classification approach based on open-access remote sensing data and tools to map wetland classes as well as the other land cover types with high accuracy, here piloting this approach for wetland mapping in Canada. Specifically, the main objectives of this study were to: (1) Leverage open access SAR and optical images obtained from Sentinel-1 and Sentinel-2 sensors for the classification of wetland complexes; (2) assess the capability of the Google Earth Engine cloud computing platform to generate custom land cover maps, which are sufficient in discriminating wetland classes as standard land cover products; (3) compare the efficiency of both pixel-based and object-based random forest classification; and (4) produce the first provincial-scale, fine resolution (i.e., 10 m) wetland inventory map in Canada. The results of this study demonstrate a paradigm-shift from standard static products and approaches toward generating more dynamic, on-demand, large-scale wetland coverage maps through advanced cloud computing resources that simplify access to and processing of a large volume of satellite imagery. Given the similarity of wetland classes across the country, the developed methodology can be scaled-up to map wetlands at the national-scale.

2. Materials and Methods

2.1. Study Area

The study area is the Island of Newfoundland, covering an approximate area of 106,000 km^2, located within the Atlantic sub-region of Canada (Figure 1). According to the Ecological Stratification Workings Group of Canada, "each part of the province is characterized by distinctive regional ecological factors, including climatic, physiography, vegetation, soil, water, fauna, and land use" [43].

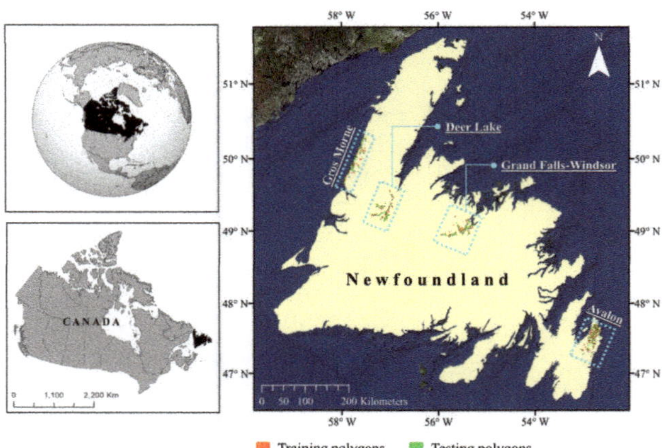

Figure 1. The geographic location of the study area with distribution of the training and testing polygons across four pilot sites on the Island of Newfoundland.

In general, the Island of Newfoundland has a cool summer and a humid continental climate, which is greatly affected by the Atlantic Ocean [43]. Black spruce forests that dominate the central area, and balsam fir forests that dominate the western, northern, and eastern areas, are common on the island [44]. Based on geography, the Island of Newfoundland can be divided into three zones, namely the southern, middle, and northern boreal regions, and each is characterized by various ecoregions [45]. For example, the southern boreal zone contains the Avalon forest, Southwestern Newfoundland, Maritime Barrens, and South Avalon-Burin Oceanic Barrens ecoregions. St. John's, the capital city, is located at the extreme eastern portion of the island, in the Maritime Barren ecoregion, and is the foggiest, windiest, and cloudiest Canadian city.

All wetland classes characterized by the Canadian Wetland Classification System (CWCS), namely bog, fen, marsh, swamp, and shallow-water [1], are found throughout the island. However, bog and fen are the most dominant classes relative to the occurrence of swamp, marsh, and shallow-water. This is attributed to the island climate, which facilitates peatland formation (i.e., extensive agglomeration of partially-decomposed organic peat under the surface). Other land cover classes are upland, deep-water, and urban/bare land. The urban and bare land classes, both having either an impervious surface or exposed soil [46], include bare land, roads, and building facilities and, thus, are merged into one single class in the final classification map.

Four pilot sites, which are representative of regional variation in terms of both landscape and vegetation, were selected across the island for in-situ data collection (see Figure 1). The first pilot site is the Avalon area, located in the south-east of the island in the Maritime Barren ecoregion, which experiences an oceanic climate of foggy, cool summers, and relatively mild winters [47]. The second and third pilot sites are Grand Falls-Windsor, located in the north-central area of the island, and Deer Lake, located in the northern portion of the island. Both fall within the Central Newfoundland ecoregion and experience a continental climate of cool summers and cold winters [47]. The final pilot site is Gros Morne, located on the extreme west coast of the island, in the Northern Peninsula ecoregion, and this site experiences a maritime-type climate with cool summers and mild winters [47].

2.2. Reference Data

In-situ data were collected via an extensive field survey of the sites mentioned above in the summers and falls of 2015, 2016 and 2017. Using visual interpretation of high resolution Google Earth imagery, as well as the CWCS definition of wetlands, potential and accessible wetland sites were flagged across the island. Accessibility via public roads, the public or private ownership of lands, and prior knowledge of the area were also taken into account for site visitation. In-situ data were collected to cover a wide range of wetland and non-wetland classes with a broad spatial distribution across NL. One or more Global Positioning System (GPS) points, depending on the size of each wetland, along with the location's name and date were recorded. Several digital photographs and ancillary notes (e.g., dominant vegetation and hydrology) were also recorded to aid in preparing the training samples. During the first year of data collection (i.e., 2015), no limitation was set on the size of the wetland, and this resulted in the production of several small-size classified polygons. To move forward with a larger size, wetlands of size >1 ha (where possible) were selected during the years 2016 and 2017. Notably, a total of 1200 wetland and non-wetland sites were visited during in-situ data collection at the Avalon, Grand Falls-Windsor, Deer Lake, and Gros Morne pilot sites over three years. Such in-situ data collection over a wide range of wetland classes across NL captured the variability of wetlands and aided in developing robust wetland training samples. Figure 1 depicts the distribution of the training and testing polygons across the Island.

Recorded GPS points were then imported into ArcMap 10.3.1 and polygons illustrating classified delineated wetlands were generated using a visual analysis of 50 cm resolution orthophotographs and 5 m resolution RapidEye imagery. Next, polygons were sorted based on their size and alternately assigned to either training or testing groups. Thus, the training and testing polygons were obtained from independent samples to ensure robust accuracy assessment. This alternative assignment also ensured that both the training (~50%) and testing (~50%) polygons had equal numbers of small and large polygons, allowing similar pixel counts and taking into account the large variation of intra-wetland size. Table 1 presents the number of training and testing polygons for each class.

Table 1. Number of training and testing polygons in this study.

Class	Training Polygons	Testing Polygons
bog	92	91
fen	93	92
marsh	75	75
swamp	78	79
shallow-water	55	56
deep-water	17	16
upland	92	92
urban/bare land	99	98
total	601	599

2.3. Satellite Data, Pre-Processing, and Feature Extraction

2.3.1. SAR Imagery

A total of 247 and 525 C-band Level-1 Ground Range Detected (GRD) Sentinel-1 SAR images in ascending and descending orbits, respectively, were used in this study. This imagery was acquired during the interval between June and August of 2016, 2017 and 2018 using the Interferometric Wide (IW) swath mode with a pixel spacing of 10 m and a swath of 250 km with average incidence angles varying between 30° and 45°. As a general rule, Sentinel-1 collects dual- (HH/HV) or single- (HH) polarized data over Polar Regions (i.e., sea ice zones) and dual- (VV/VH) or single- (VV) polarized data over all other zones [48]. However, in this study, we took advantage of being close to the Polar regions and thus, both HH/HV and VV/VH data were available in our study region. Accordingly, of 247 SAR ascending observations (VV/VH), 12, 120 and 115 images were collected in 2016, 2017 and 2018, respectively. Additionally, of 525 descending observations (HH/HV), 111, 260, and 154 images were acquired in 2016, 2017 and 2018, respectively. Figure 2 illustrates the number of SAR observations over the summer of the aforementioned years.

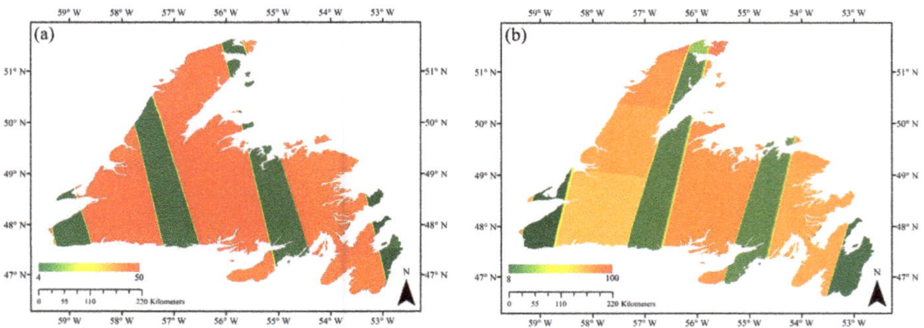

Figure 2. The total number of (**a**) ascending Synthetic Aperture Radar (SAR) observations (VV/VH) and (**b**) descending SAR observations (HH/HV) during summers of 2016, 2017 and 2018. The color bar represents the number of collected images.

Sentinel-1 GRD data were accessed through GEE. We applied the following pre-processing steps, including updating orbit metadata, GRD border noise removal, thermal noise removal, radiometric calibration (i.e., backscatter intensity), and terrain correction (i.e., orthorectification) [49]. These steps resulted in generating the geo-coded backscatter intensity images. Notably, this is similar to the pre-processing steps implemented in the ESA's SNAP Sentinel-1 toolbox. The unitless backscatter intensity images were then converted into normalized backscattering coefficient (σ^0) values in dB (i.e., the standard unit for SAR backscattering representation). Further pre-processing steps, including incidence angle correction [50] and speckle reduction (i.e., 7 $\times$ 7 adaptive sigma Lee filter in this study) [51,52], were also carried out on the GEE platform.

Following the procedure described above, σ^0_{VV}, σ^0_{VH}, σ^0_{HH}, and σ^0_{HV} (i.e., backscatter coefficient images) were extracted. Notably, σ^0_{VV} observations are sensitive to soil moisture and are able to distinguish flooded from non-flooded vegetation [53], as well as various types of herbaceous wetland classes (low, sparsely vegetated areas) [54]. This is particularly true for vegetation in the early stages of growing when plants have begun to grow in terms of height, but have not yet developed their canopy [53]. σ^0_{VH} observations can also be useful for monitoring wetland herbaceous vegetation. This is because cross-polarized observations are produced by volume scattering within the vegetation canopy and have a higher sensitivity to vegetation structures [55]. σ^0_{HH} is an ideal SAR observation for wetland mapping due to its sensitivity to double-bounce scattering over flooded vegetation [41,56]. Furthermore, σ^0_{HH} is less sensitive to the surface roughness compared to σ^0_{VV}, making the former advantageous for discriminating water and non-water classes. In addition to SAR backscatter coefficient images, a number of other polarimetric features were also extracted and used in this study. Table 2 represents polarimetric features extracted from the dual-pol VV/VH and HH/HV Sentinel-1 images employed in this study. Figure 3a illustrates the span feature, extracted from HH/HV data, for the Island of Newfoundland.

Table 2. A description of extracted features from SAR and optical imagery.

Data	Feature Description	Formula
Sentinel-1	vertically transmitted, vertically received SAR backscattering coefficient	σ^0_{VV}
	vertically transmitted, horizontally received SAR backscattering coefficient	σ^0_{VH}
	horizontally transmitted, horizontally received SAR backscattering coefficient	σ^0_{HH}
	horizontally transmitted, vertically received SAR backscattering coefficient	σ^0_{HV}
	Span or total scattering power	$\lvert S_{VV}\rvert^2 + \lvert S_{VH}\rvert^2$, $\lvert S_{HH}\rvert^2 + \lvert S_{HV}\rvert^2$
	difference between co- and cross-polarized observations	$\lvert S_{VV}\rvert^2 - \lvert S_{VH}\rvert^2$, $\lvert S_{HH}\rvert^2 - \lvert S_{HV}\rvert^2$
	ratio	$\frac{\lvert S_{VV}\rvert^2}{\lvert S_{VH}\rvert^2}$, $\frac{\lvert S_{HH}\rvert^2}{\lvert S_{HV}\rvert^2}$
Sentinel-2	spectral bands 2 (blue), 3 (green), 4 (red) and 8 (NIR)	B_2, B_3, B_4, B_8
	the normalized difference vegetation index (NDVI)	$\frac{B_8-B_4}{B_8+B_4}$
	the normalized difference water index (NDWI)	$\frac{B_3-B_8}{B_3+B_8}$
	modified soil-adjusted vegetation index 2 (MSAVI2)	$\frac{2B_8+1-\sqrt{(2B_8+1)^2-8(B_8-B_4)}}{2}$

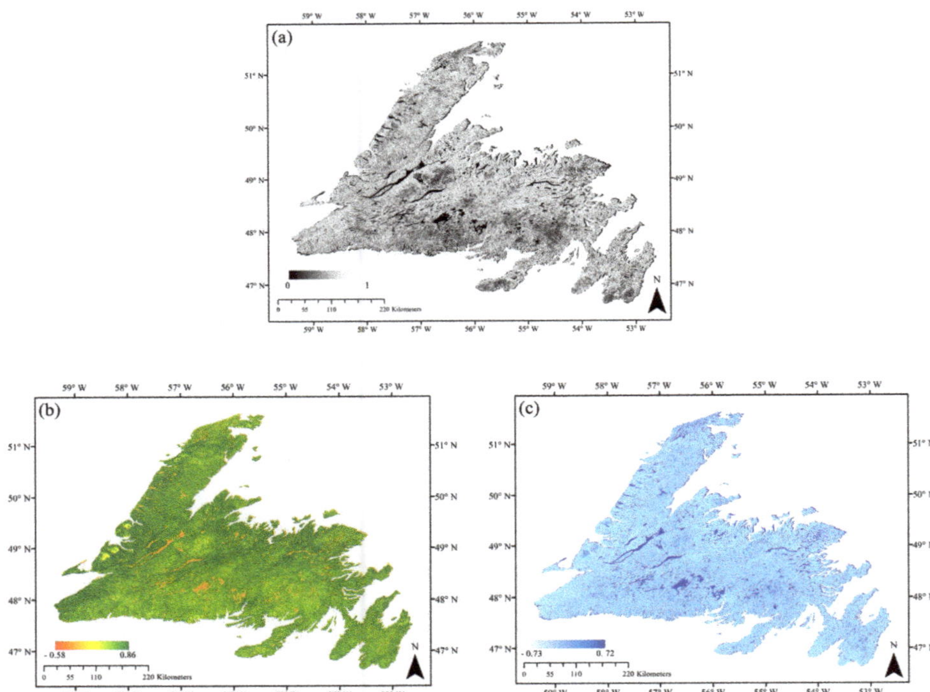

Figure 3. Three examples of extracted features for land cover classification in this study. The multi-year summer composite of (**a**) span feature extracted from HH/HV Sentinel-1 data, (**b**) normalized difference vegetation index (NDVI), and (**c**) normalized difference water index (NDWI) features extracted from Sentinel-2 data.

2.3.2. Optical Imagery

Creating a 10 m cloud-free Sentinel-2 composition for the Island of Newfoundland over a short period of time (e.g., one month) is a challenging task due to chronic cloud cover. Accordingly, the Sentinel-2 composite was created for three-months between June and August, during the leaf-on season for 2016, 2017 and 2018. This time period was selected since it provided the most cloud-free data and allowed for maximum wall-to-wall data coverage. Furthermore, explicit wetland phenological information could be preserved by compositing data acquired during this time period. Accordingly, monthly composite and multi-year summer composite were used to obtain cloud-free or near-cloud-free wall-to-wall coverage.

Both Sentinel-2A and Sentinel-2B Level-1C data were used in this study. There were a total of 343, 563 and 1345 images in the summer of 2016, 2017 and 2018, respectively. The spatial distribution of all Sentinel-2 observations during the summers of 2016, 2017 and 2018 are illustrated in Figure 4a. Notably, a number of these observations were affected by cloud coverage. Figure 4b depicts the percentage of cloud cover distribution during these time periods. To mitigate the limitation that arises due to cloud cover, we applied a selection criteria to cloud percentage (<20%) when producing our cloud-free composite. Next, the QA60 bitmask band (a quality flag band) provided in the metadata was used to detect and mask out clouds and cirrus. Sentinel-2 has 13 spectral bands at various spatial resolutions, including four bands at 10 m, six at 20 m, and three bands at 60 m spatial resolution. For this study, only blue (0.490 µm), green (0.560 µm), red (0.665 µm), and near-infrared (NIR, 0.842 µm) bands were used. This is because the optical indices selected in this study are based on the above mentioned optical bands (see Table 2) and, furthermore, all these bands are at a spatial resolution of 10 m.

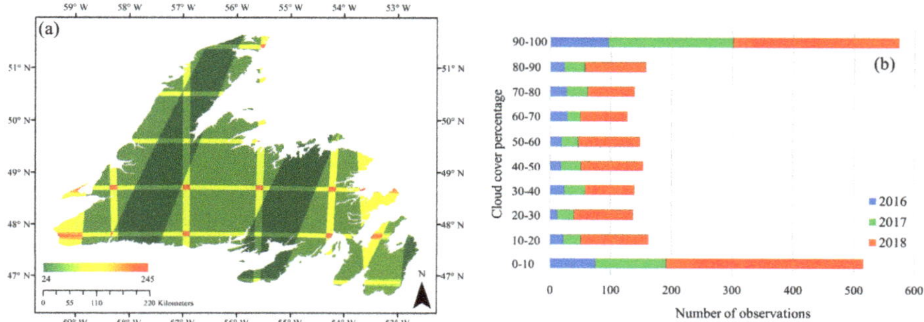

Figure 4. (**a**) Spatial distribution of Sentinel-2 observations (total observations) during summers of 2016, 2017 and 2018 and (**b**) the number of observations affected by varying degrees of cloud cover (%) in the study area for each summer.

In addition to optical bands (2, 3, 4 and 8), NDVI, NDWI and MSAVI2 indices were also extracted (see Table 2). NDVI is one of the most well-known and commonly used vegetation indices for the characterization of vegetation phenology (seasonal and inter-annual changes). Using the ratioing operation (see Table 2), NDVI decreases several multiplicative noises, such as sun illumination differences, cloud shadows, as well as some atmospheric attenuation and topographic variations, within various bands of multispectral satellite images [57]. NDVI is sensitive to photosynthetically active biomasses and can discriminate vegetation/non-vegetation, as well as wetland/non-wetland classes. NDWI is also useful, since it is sensitive to open water and can discriminate water from land. Notably, NDWI can be extracted using different bands of multispectral data [58], such as green and shortwave infrared (SWIR) [59], red and SWIR [60], as well as green and NIR [61]. Although some studies reported the superiority of SWIR for extracting the water index due to its lower sensitivity to the sub-pixel non-water component [58], we used the original NDWI index proposed by [61] in this study. This is because it should provide accurate results at our target resolution and, moreover, it uses green and NIR bands of Sentinel-2 data, both of which are at a 10 m spatial resolution. Finally, MSAVI2 was used because it addresses the limitations of NDVI in areas with a high degree of exposed soil surface. Figure 3b,c demonstrates the multi-year summer composite of NDVI and NDWI features extracted from Sentinel-2 optical imagery.

2.4. Multi-Year Monthly and Summer Composite

Although several studies have used the Landsat archive to generate nearly-cloud-free Landsat composites of a large area (e.g., [62–64]), to the best of our knowledge, such an investigation has not yet been thoroughly examined for Sentinel-2 data. This is unfortunate since the latter data offer both improved temporal and spatial resolution relative to Landsat imagery, making them advantageous for producing high resolution land cover maps on a large-scale. For example, Roy et al. (2010) produced monthly, seasonally, and yearly composites using maximum NDVI and brightness temperature obtained from Landsat data for the conterminous United States [64]. Recent studies also used different compositing approaches, such as seasonally [62] and yearly [63] composites obtained from Landsat data in their analysis.

In this study, two different types of image composites were generated: Multi-year monthly and summer composites. Due to the prevailing cloudy and rainy weather conditions in the study area, it was impossible to collect sufficient cloud-free optical data to generate a full-coverage monthly composite of Sentinel-2 data for classification purposes. However, we produced the monthly composite (optical) for spectral signature analysis to identify the month during which the most semantic information of wetland classes could be obtained. A multi-year summer composite was produced to capture explicit phenological information appropriate for wetland mapping. As suggested by recent research [65], the multi-year spring composite is advantageous for wetland mapping in the Canada's boreal regions. This is because such time-series data capture within-year surface variation. However, in this study, the multi-year summer composite was used given that the leaf-on season begins in late spring/early summer on the Island of Newfoundland.

Leveraging the GEE composite function, 10 m wall-to-wall, cloud-free composites of Sentinel-2 imagery, comprising original optical bands (2, 3, 4 and 8), NDVI, NDWI, and MSAVI2 indices, across the Island of Newfoundland were produced. SAR features, including σ^0_{VV}, σ^0_{VH}, σ^0_{HH}, σ^0_{HV}, span, ratio, and difference between co- and cross-polarized SAR features (see Table 2), were also stacked using GEE's array-based computational approach. Specifically, each monthly and summer season group of images were stacked into a single median composite on a per-pixel, per band basis.

2.5. Separability Between Wetland Classes

In this study, the separability between wetland classes was determined both qualitatively, using box-and-whiskers plots, and quantitatively, using Jeffries–Matusita (JM) distance. The JM distance indicates the average distance between the density function of two classes [66]. It uses both the first order (mean) and second order (variance) statistical variables from the samples and has been illustrated to be an efficient separability measure for remote sensing data [67,68]. Given normal distribution assumptions, the JM distance between two classes is represented as

$$JM = 2\left(1 - e^{-B}\right) \tag{1}$$

where B is the Bhattacharyya (BH) distance given by

$$B = \frac{1}{8}\left(\mu_i - \mu_j\right)^T \left(\frac{\Sigma_i + \Sigma_j}{2}\right)^{-1} \left(\mu_i - \mu_j\right) + \frac{1}{2}\ln\left(\frac{\left|\left(\Sigma_i + \Sigma_j\right)/2\right|}{\sqrt{|\Sigma_i||\Sigma_j|}}\right) \tag{2}$$

where μ_i and Σ_i are the mean and covariance matrix of class i and μ_j and Σ_j are the mean and covariance matrix of class j. The JM distance varies between 0 and 2, with values that approach 2 demonstrating a greater average distance between two classes. In this study, the separability analysis was limited to extracted features from optical data. This is because a detailed backscattering analysis of wetland classes using multi-frequency SAR data, including X-, C-, and L-band, has been presented in our previous study [19].

2.6. Classification Scheme

2.6.1. Random Forest

In this study, the random forest (RF) algorithm was used for both pixel-based and object-based wetland classifications. RF is a non-parametric classifier, comprised of a group of tree classifiers, and is able to handle high dimensional remote sensing data [69]. It is also more robust compared to the DT algorithm and easier to execute relative to SVM [23]. RF uses bootstrap aggregating (bagging) to produce an ensemble of decision trees by using a random sample from the given training data, and determines the best splitting of the nodes by minimizing the correlation between trees. Assigning a label to each pixel is based on the majority vote of trees. RF can be tuned by adjusting two input parameters [70], namely the number of trees (*Ntree*), which is generated by randomly selecting samples from the training data, and the number of variables (*Mtry*), which is used for tree node splitting [71]. In this study, these parameters were selected based on (a) direction from previous studies (e.g., [56,69,72]) and (b) a trial-and-error approach. Specifically, *Mtry* was assessed for the following values (when *Ntree* was adjusted to 500): (a) One third of the total number of input features; (b) the square root of the total number of input features; (c) half of the total number of input features; (d) two thirds of the total number of input features; and (e) the total number of input features. This resulted in marginal or no influence on the classification accuracies. Accordingly, the square root of the total number of variables was selected for *Mtry*, as suggested by [71]. Next, by adjusting the optimal value for *Mtry*, the parameter *Ntree* was assessed for the following values: (a) 100; (b) 200; (c) 300; (d) 400; (e) 500; and (f) 600. A value of 400 was then found to be appropriate in this study, as error rates for all classification models were constant beyond this point. The 601 training polygons in different categories were used to train the RF classifier on the GEE platforms (see Table 1).

2.6.2. Simple Non-Iterative Clustering (SNIC) Superpixel Segmentation

Conventional pixel-based classification algorithms rely on the exclusive use of the spectral/backscattering value of each pixel in their classification scheme. This results in "salt and pepper" noise in the final classification map, especially when high-resolution images are employed [73]. An object-based algorithm, however, can mitigate the problem that arises during such image processing by taking into account the contextual information within a given imaging neighborhood [74]. Image segmentation divides an image into regions or objects based on the specific parameters (e.g., geometric features and scaled topological relation). In this study, simple non-iterative clustering (SNIC) algorithm was selected for superpixel segmentation (i.e., object-based) analysis [75]. The algorithm starts by initializing centroid pixels on a regular grid in the image. Next, the dependency of each pixel relative to the centroid is determined using its distance in the five-dimensional space of color and spatial coordinates. In particular, the distance integrates normalized spatial and color distances to produce effective, compact and approximately uniform superpixels. Notably, there is a trade-off between compactness and boundary continuity, wherein larger compactness values result in more compact superpixels and, thus, poor boundary continuity. SNIC uses a priority queue, 4- or 8-connected candidate pixels to the currently growing superpixel cluster, to select the next pixels to join the cluster. The candidate pixel is selected based on the smallest distance from the centroid. The algorithm takes advantage of both priority queue and online averaging to evolve the centroid once each new pixel is added to the given cluster. Accordingly, SNIC is superior relative to similar clustering algorithms (e.g., Simple Linear Iterative Clustering) in terms of both memory and processing time. This is attributed to the introduction of connectivity (4- or 8-connected pixels) that results in computing fewer distances during centroid evolution [75].

2.6.3. Evaluation Indices

Four evaluation indices, including overall accuracy (OA), Kappa coefficient, producer accuracy, and user accuracy were measured using the 599 testing polygons held back for validation purposes (see Table 1). Overall accuracy determines the overall efficiency of the algorithm and can be measured by dividing the total number of correctly-labeled samples by the total number of the testing samples. The Kappa coefficient indicates the degree of agreement between the ground truth data and the predicted values. Producer's accuracy represents the probability that a reference sample is correctly identified in the classification map. User's accuracy indicates the probability that a classified pixel in the land cover classification map accurately represents that category on the ground [76].

Additionally, the McNemar test [77] was employed to determine the statistically significant differences between various classification scenarios in this study. Particularly, the main goals were to determine: (1) Whether a statistically significant difference exists between pixel-based and object-based classifications based on either SAR or optical data; and (2) whether a statistically significant difference exists between object-based classifications using only one type of data (SAR or optical data) and an integration of two types of data (SAR and optical data). The McNemar test is non-parametric and is based on the classification confusion matrix. The test is based on a chi-square (χ^2) distribution with one degree of freedom [78,79] and assumes the number of correctly and incorrectly identified pixels are equal for both classification scenarios [77],

$$\chi^2 = \frac{(f_{12} - f_{21})^2}{f_{12} + f_{21}} \tag{3}$$

where f_{12} and f_{21} represent the number of pixels that were correctly identified by one classifier as compared to the number of pixels that the other method incorrectly identified, respectively.

2.7. Processing Platform

The GEE cloud computing platform was used for both the pixel-based and superpixel RF classification in this study. Both Sentinel-1 and Sentinel-2 data hosted within the GEE platform were used to construct composite images. The zonal boundaries and the reference polygons were imported into GEE using Google fusion tables. A JavaScript API in the GEE code editor was used for pre-processing, feature extraction, and classification in this study. Accordingly, we generated 10 m spatial resolution wetland maps of Newfoundland for our multi-year seasonal composites of optical, SAR, and integration of both types of data using pixel-based and object-based approaches.

3. Results

3.1. Spectral Analysis of Wetland Classes Using Optical Data

To examine the discrimination capabilities of different spectral bands and vegetation indices, spectral analysis was performed for all wetland classes. Figures 5–7 illustrate the statistical distribution of reflectance, NDVI, NDWI, and MSAVI2 values for the multi-year monthly composites of June, July, and August, respectively, using box-and-whisker plots.

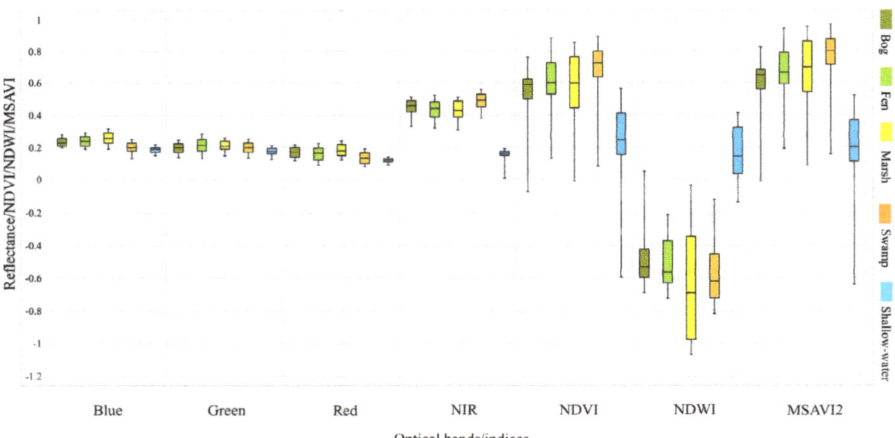

Figure 5. Box-and-whisker plot of the multi-year June composite illustrating the distribution of reflectance, NDVI, NDWI, and MSAVI2 for wetland classes obtained using pixel values extracted from training datasets. Note that black, horizontal bars within boxes illustrate median values, boxes demonstrate the lower and upper quartiles, and whiskers extend to minimum and maximum values.

As shown, all visible bands poorly distinguish spectrally similar wetland classes, especially the bog, fen, and marsh classes. The shallow-water class, however, can be separated from other classes using the red band in August (see Figure 7). Among the original bands, NIR represents clear advantages when discriminating the shallow-water from other classes (see Figures 5–7), but is not more advantageous for classifying herbaceous wetland classes. Overall, vegetation indices are superior when separating wetland classes compared to the original bands.

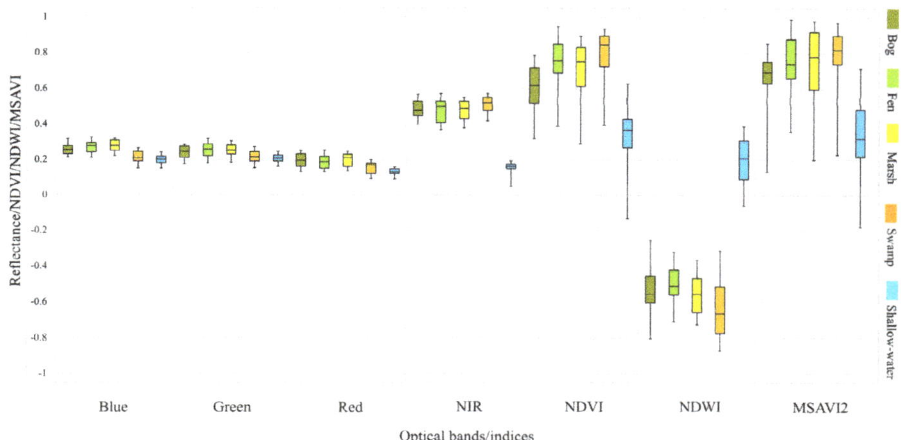

Figure 6. Box-and-whisker plot of the multi-year July composite illustrating the distribution of reflectance, NDVI, NDWI, and MSAVI2 for wetland classes obtained using pixel values extracted from training datasets.

As illustrated in Figures 5–7, the shallow-water class is easily distinguishable from other classes using all vegetation indices. The swamp and bog classes are also separable using the NDVI index from all three months. Although both NDVI and MSAVI2 are unable to discriminate herbaceous wetland classes using the June composite, the classes of bog and fen are distinguishable using the NDVI index obtained from the July and August composites.

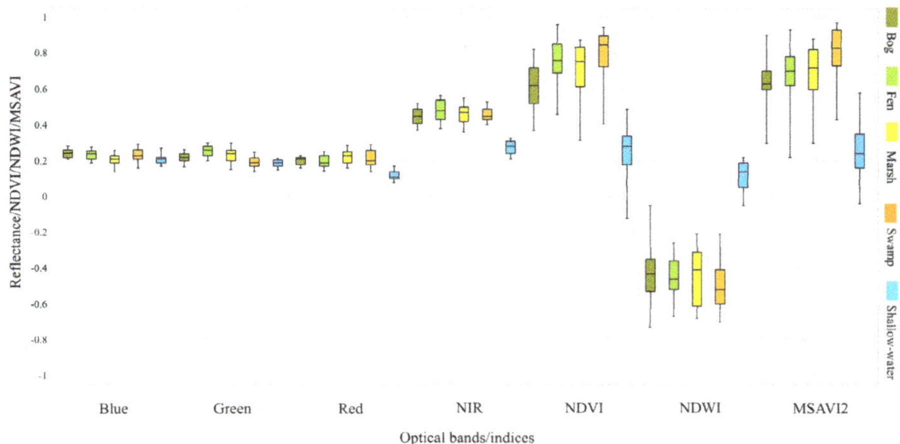

Figure 7. Box-and-whisker plot of the multi-year August composite illustrating the distribution of reflectance, NDVI, NDWI, and MSAVI2 for wetland classes obtained using pixel values extracted from training datasets.

The mean JM distances obtained from the multi-year summer composite for wetland classes are represented in Table 3.

Table 3. Jeffries–Matusita (JM) distances between pairs of wetland classes from the multi-year summer composite for extracted optical features in this study.

Optical Features	d_1	d_2	d_3	d_4	d_5	d_6	d_7	d_8	d_9	d_{10}
blue	0.002	0.204	0.470	1.153	0.232	0.299	1.218	0.520	1.498	0.380
green	0.002	0.331	0.391	0.971	0.372	0.418	1.410	0.412	1.183	0.470
red	0.108	0.567	0.570	1.495	0.546	0.640	1.103	0.634	1.391	0.517
NIR	0.205	0.573	0.515	1.395	0.364	0.612	1.052	0.649	1.175	1.776
NDVI	0.703	0.590	0.820	1.644	0.586	0.438	1.809	0.495	1.783	1.938
NDWI	0.268	0.449	0.511	1.979	0.643	0.519	1.792	0.760	1.814	1.993
MSAVI2	0.358	0.509	0.595	1.763	0.367	0.313	1.745	0.427	1.560	1.931
all	1.098	1.497	1.561	1.999	1.429	1.441	1.999	1.614	1.805	1.999

Note: d_1: Bog/Fen, d_2: Bog/Marsh, d_3: Bog/ Swamp, d_4: Bog/Shallow-water, d_5: Fen/Marsh, d_6: Fen/Swamp, d_7: Fen/Shallow-water, d_8: Marsh/Swamp, d_9: Marsh/Shallow-water, and d_{10}: Swamp/Shallow-water.

According to the JM distance, shallow-water is the most separable class from other wetland classes. In general, all wetland classes, excluding shallow-water, are hardly distinguishable from each other using single optical feature and, in particular, bog and fen are the least separable classes. However, the synergistic use of all features considerably increases the separability between wetland classes, with JM values exceeding 1.4 in most cases; however, bog and fen remain hardly discernible in this case.

3.2. Classification

The overall accuracies (OA) and Kappa coefficients of different classification scenarios are presented in Table 4. Overall, the classification results using optical imagery were more advantageous relative to SAR imagery. As illustrated, the optical imagery resulted in approximately 4% improvements in both the pixel-based and object-based approaches. Furthermore, object-based classifications were found to be superior to pixel-based classifications using optical (~6.5% improvement) and SAR (~6% improvements) imagery in comparative cases. It is worth noting that the accuracy assessment in this study was carried out using the testing polygons well distributed across the whole study region.

Table 4. Overall accuracies and Kappa coefficients obtained from different classification scenarios in this study.

Classification	Data Composite	Scenario	Overall Accuracy (%)	Kappa Coefficient
pixel-based	SAR	S1	73.12	0.68
	Optic	S2	77.16	0.72
object-based	SAR	S3	79.14	0.74
	Optic	S4	83.79	0.80
	SAR + optic	S5	88.37	0.85

The McNemar test revealed that the difference between the accuracies of pixel-based and object-based classifications was statistically significant when either SAR ($p = 0.023$) or optical ($p = 0.012$) data were compared (see Table 5). There was also a statistically very significant difference between object-based classifications using SAR vs. SAR/optical data ($p = 0.0001$) and optical vs. SAR/optical data ($p = 0.008$).

Table 5. The results of McNemar test for different classification scenarios in this study.

Scenarios	χ^2	p-Value
S1 vs. S3	5.21	0.023
S2 vs. S4	6.27	0.012
S3 vs. S5	9.27	0.0001
S4 vs. S5	7.06	0.008

Figure 8 demonstrates the classification maps using SAR and optical multi-year summer composites for Newfoundland obtained from pixel- and object-based RF classifications. They illustrate the distribution of land cover classes, including both wetland and non-wetland classes, identifiable at a 10 m spatial resolution. In general, the classified maps indicate fine separation of all land cover units, including bog and fen, shallow- and deep-water, and swamp and upland, as well as other land cover types.

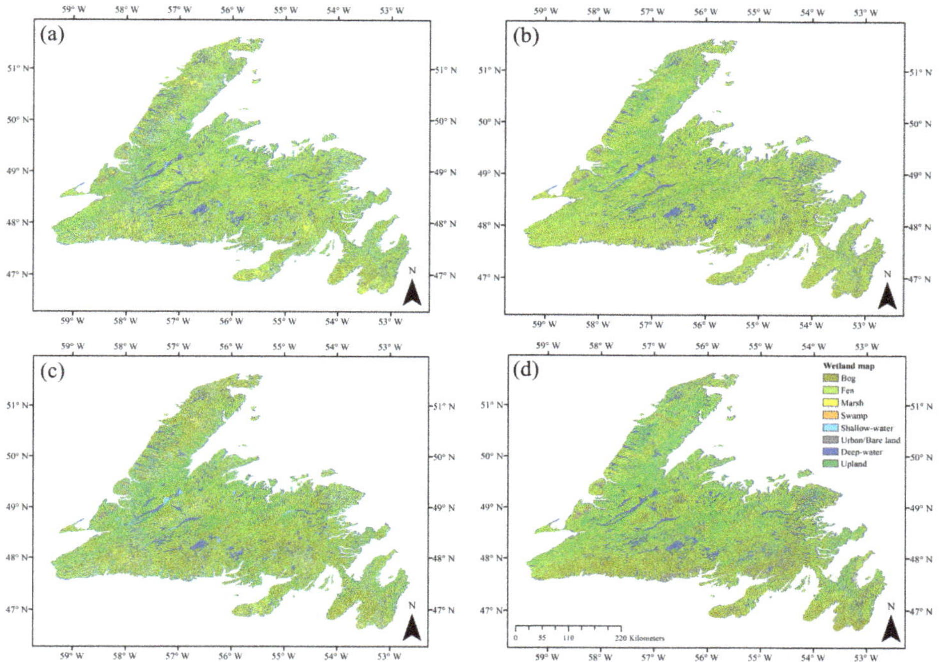

Figure 8. The land cover maps of Newfoundland obtained from different classification scenarios, including (**a**) S1, (**b**) S2, (**c**) S3 and (**d**) S4 in this study.

Figure 9 depicts the confusion matrices obtained from different methods, wherein the diagonal elements are the producer's accuracies. The user's accuracies of land cover classes using different classification scenarios are also demonstrated in Figure 10. Overall, the classification of wetlands have lower accuracies compared to those of the non-wetland classes. In particular, the classification of swamp has the lowest producer's and user's accuracies among wetland (and all) classes in this study. In contrast, the classification accuracies of bog and shallow-water are higher (both user's and producer's accuracies) than the other wetland classes.

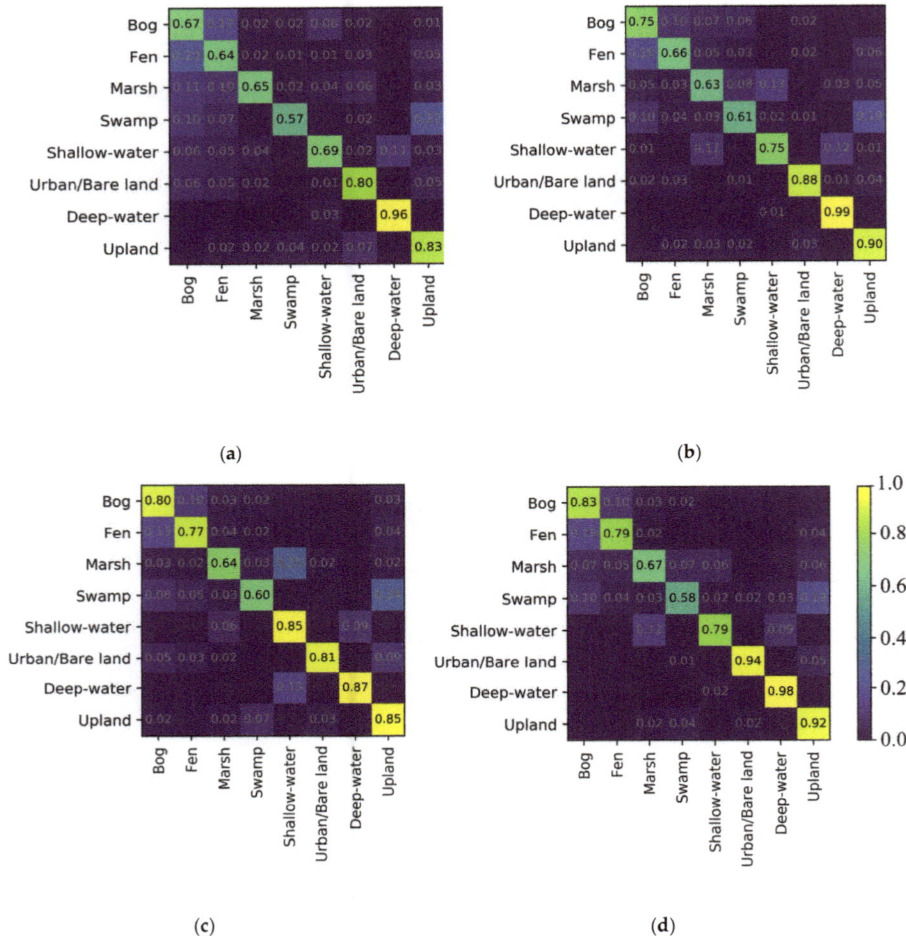

Figure 9. The confusion matrices obtained from different classification scenarios, including (**a**) S1, (**b**) S2, (**c**) S3 and (**d**) S4 in this study.

Notably, all methods successfully classified the non-wetland classes with producer's accuracies beyond 80%. Among the first four scenarios, the object-based classification using optical imagery (i.e., S4) was the most successful approach for classifying the non-wetland classes, with producer's and user's accuracies exceeding 90% and 80%, respectively. The wetland classes were also identified with high accuracies in most cases (e.g., bog, fen, and shallow-water) in S4.

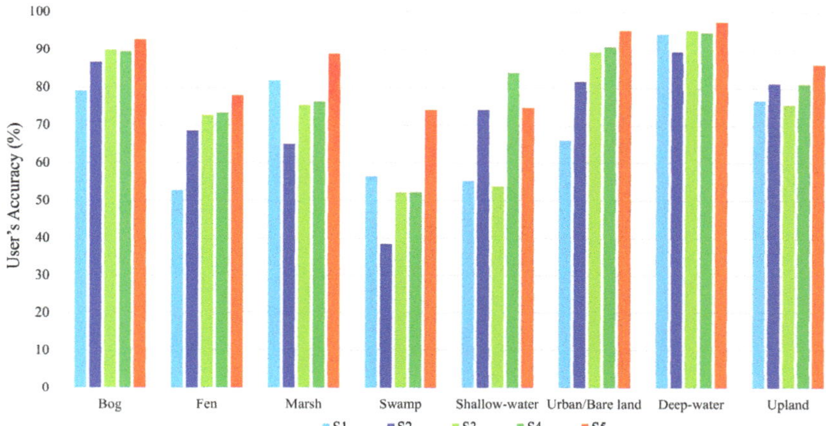

Figure 10. The user's accuracies for various land cover classes in different classification scenarios in this study.

The object-based approach, due to its higher accuracies, was selected for the final classification scheme in this study, wherein the multi-year summer SAR and optical composites were integrated (see Figure 11).

The final land cover map is noiseless and accurately represents the distribution of all land cover classes on a large-scale. As shown, the classes of bog and upland are the most prevalent wetland and non-wetland classes, respectively, in the study area. These observations agree well both with field notes recorded by biologists during the in-situ data collection and with visual analysis of aerial and satellite imagery. Figure 11 also illustrates several insets from the final land cover map in this study. The visual interpretation of the final classified map by ecological experts demonstrated that most land cover classes were correctly distinguished across the study area. For example, ecological experts noted that bogs appear as a reddish color in optical imagery (true color composite). As shown in Figure 11, most bog wetlands are accurately identified in all zoomed areas. Furthermore, small water bodies (e.g., small ponds) and the perimeter of deep water bodies are correctly mapped belonging to the shallow-water class. The upland and urban/bare land classes were also correctly distinguished.

The confusion matrix for the final classification map is illustrated in Figure 12. Despite the presence of confusion among wetland classes, the results obtained from the multi-year SAR/optical composite were extremely positive, taking into account the complexity of distinguishing similar wetland classes. As shown in Figure 12, all non-wetland classes and shallow-water were correctly identified with producer's accuracies beyond 90%. The most similar wetland classes, namely bog and fen, were classified with producer's accuracies exceeding 80%. The other two wetland classes were also correctly identified with a producer's accuracy of 78% for marsh and 70% for swamp.

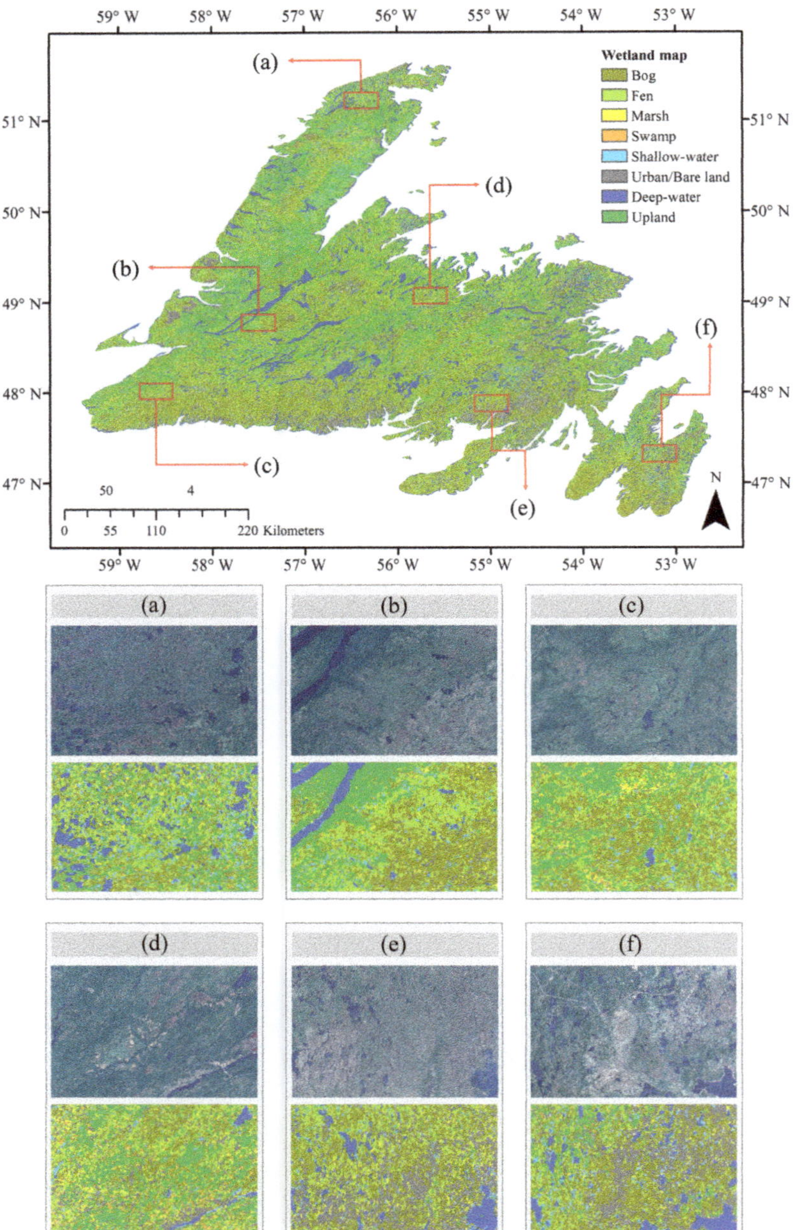

Figure 11. The final land cover map for the Island of Newfoundland obtained from the object-based Random Forest (RF) classification using the multi-year summer SAR/optical composite. An overall accuracy of 88.37% and a Kappa coefficient of 0.85 were achieved. A total of six insets and their corresponding optical images (i.e., Sentinel-2) were also illustrated to appreciate some of the classification details. Please also see Supplementary Materials for details of the final classification map.

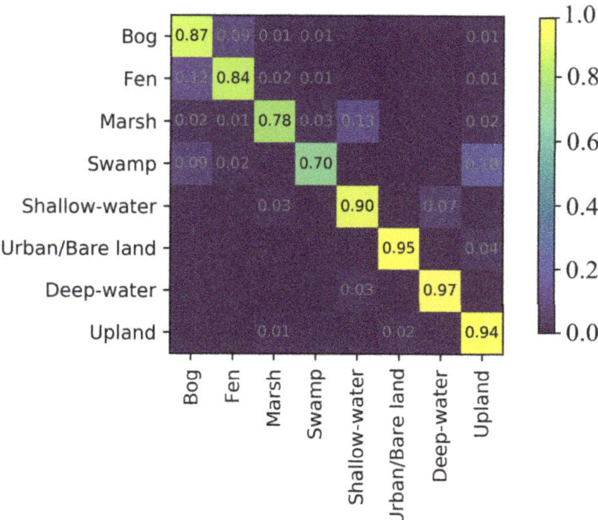

Figure 12. The confusion matrix for the final classification map obtained from the object-based RF classification using the multi-year summer SAR/optical composite (OA: 88.37%, K: 0.85).

4. Discussion

In general, the results of the spectral analysis demonstrated the superiority of the NIR band compared to the visible bands (i.e., blue, green, and red) for distinguishing various wetland classes. This was particularly true for shallow-water, which was easily separable using NIR. This is logical, given that water and vegetation exhibit strong absorption and reflection, respectively, in this region of the electromagnetic spectrum. NDVI was found to be the most useful vegetation index. This finding is potentially explained by the high sensitivity of NDVI to photosynthetically active biomasses [57]. Furthermore, the results of the spectral analysis of wetland classes indicated that class separability using the NDVI index is maximized in July, which corresponds to the peak growing season in Newfoundland. According to the box-and-whisker plots and the JM distances, the spectral similarities of wetland classes are slightly concerning, as they revealed the difficulties in distinguishing similar wetland classes using a single optical feature, which is in agreement with a previous study [80]. However, the inclusion of all optical features significantly increased the separability between wetland classes.

As shown in Figure 9, confusion errors occurred among all classes, especially those of wetlands using the pixel-based classification approach. Notably, the highest confusion was found between the swamp and upland classes in some cases. The upland class is characterized by dry forested land, and swamps are specified as woody (forested) wetland. This results in similarities in both the visual appearance and spectral/backscattering signatures for these classes. With regard to SAR signatures, for example, the dominant scattering mechanism for both classes is volume scattering, especially when the water table is low in swamp [81], which contributes to the misclassification between the two. This is of particular concern when shorter wavelengths (e.g., C-band) are employed, given their shallower penetration depth relative to that of longer wavelengths (e.g., L-band).

Confusion was also common among the herbaceous wetland classes, namely bog, fen, and marsh. This is attributable to the heterogeneity of the landscape in the study area. As field notes suggest, the herbaceous wetland classes were found adjacent to each other without clear cut borders, making them hardly distinguishable. This is particularly severe for bog and fen, since both have very similar ecological and visual characteristics. For example, both are characterized by peatlands, dominated by ecologically similar vegetation types of *Sphagnum* in bogs and *Graminoid* in fens.

Another consideration when interpreting the classification accuracies for different wetland classes is the availability of the training samples/polygons for the supervised classification. As shown in Table 1, for example, bogs have a larger number of training polygons compared to the swamp class. This is because NL has a moist and cool climate [43], which contributes to extensive peatland formation. Accordingly, bog and fen were potentially the most visited wetland classes during in-situ data collection. This resulted in the collection of a larger number of training samples/polygons for these classes. On the other hand, the swamp class is usually found in physically smaller areas relative to those of other classes; for example, in transition zones between wetland and other land cover classes. As such, they may have been dispersed and mixed with other land cover classes, making them difficult to distinguish by the classifier.

Comparison of the classification accuracies using optical and SAR images (i.e., S1 vs. S2 and S3 vs. S4) indicated, according to all evaluation indices in this study, the superiority of the former relative to the latter for wetland mapping in most cases. This suggests that the phenological variations in vegetative productivity captured by optical indices (e.g., NDVI), as well as the contrast between water and non-water classes captured by the NDWI index are more efficient for wetland mapping in our study area than the extracted features from dual-polarimetric SAR data. This finding is consistent with the results of a recent study [12] that employed optical, SAR, and topographic data for predicting the probability of wetland occurrence in Alberta, Canada, using the GEE platform. However, it should be acknowledged that the lower success of SAR compared to optical data is, at least, partially related to the fact that the Sentinel-1 sensor does not collect full-polarimetric data at the present time. This hinders the application of advanced polarimetric decomposition methods that demand full-polarimetric data. Several studies highlighted the great potential of polarimetric decomposition methods for identifying similar wetland classes by characterizing their various scattering mechanisms using such advanced approaches [19,56].

Despite the superiority of optical data relative to SAR, the highest classification accuracy was obtained by integrating multi-year summer composites of SAR and optical imagery using the object-based approach (see Table 4(S5)). In particular, this classification scenario demonstrates an improvement of about 9% and 4.5% in overall accuracy compared to the object-based classification using the multi-year summer SAR and optical composites, respectively. This is because optical and SAR data are based on range and angular measurements and collect information about the chemical and physical characteristics of wetland vegetation, respectively [82]; thus, the inclusion of both types of observations enhances the discrimination of backscattering/spectrally similar wetland classes [41,42]. Accordingly, it was concluded that the multi-year summer SAR/optical composite is very useful for improving overall classification accuracy by capturing chemical, biophysical, structural, and phenological variations of herbaceous and woody wetland classes. This was later reaffirmed via the confusion matrix (see Figure 12) of the final classification map, wherein confusion decreased compared to classifications based on either SAR or optical data (see Figure 9). Furthermore, the McNemar test indicated that there was a very statistically significant difference ($p < 0.05$) for object-based classifications using SAR vs. optical/SAR (S3 vs. S5) and optical vs. optical/SAR (S4 vs. S5) models (see Table 5).

Notably, the multi-year summer SAR/optical composite improved the producer's accuracies of marsh and swamp classes. Specifically, the inclusion of SAR and optical data improved the producer's accuracies of marsh in the final classification map by about 14% and 11% compared to the object-based classification using SAR and optical imagery on their own, respectively. This, too, occurred to a lesser degree for swamp, wherein the producer's accuracies improved in the final classified map by about 12% and 10% compared to those of object-based classified maps using optical and SAR imagery, respectively. The accuracies for other wetland classes, namely bog and fen, were also improved by about 4% and 5%, respectively, in this case relative to the object-based classification using the multi-year optical composite.

Despite significant improvements in the producer's accuracies for some wetland classes (e.g., marsh and swamp) using the SAR/optical data composite, marginal to no improvements were obtained in this case for the non-wetland classes compared to classification based only on optical data. In particular, the use of SAR data does not offer substantial gains beyond the use of optical imagery for distinguishing typical land cover classes, such as urban and deep-water, nor does it present any clear disadvantages. Nevertheless, combining both types of observations addresses the limitation that arises due to the inclement weather in geographic regions with near-permanent cloud cover, such as Newfoundland. Therefore, the results reveal the importance of incorporating multi-temporal optical/SAR data for classification of backscattering/spectrally similar land cover classes, such as wetland complexes. Accordingly, given the complementary advantages of SAR and optical imagery, the inclusion of both types of data still offers a potential avenue for further research in land cover mapping on a large scale.

The results demonstrate the superiority of object-based classification compared to the pixel-based approach in this study. This is particularly true when SAR imagery was employed, as the producer's accuracies for all wetland classes were lower than 70% (see Figure 9a). Despite applying speckle reduction, speckle noise can remain, and this affects the classification accuracy during such processing. In contrast to the pixel-based approach, object-based classification benefits from both backscattering/spectral information, as well as contextual information within a given neighborhood. This further enhances semantic land cover information and is very useful for the classification of SAR imagery [31].

As noted in a previous study [83], the image mosaicking technique over a long time-period may increase classification errors in areas of high inter-annual change, causing a signal of seasonality to be overlooked. Although this image mosaicking technique is essential for addressing the limitation of frequent cloud cover for land cover mapping using optical remote sensing data across a broad spatial scale, this was mitigated in this study to a feasible extent. In particular, to diminish the effects of multi-seasonal observations, the mosaicked image in this study was produced from the multi-year summer composite rather than the multi-year, multi-seasonal composite. The effectiveness of using such multi-year seasonal (e.g., either spring or summer) composites has been previously highlighted, given the potential of such data to capture surface condition variations beneficial for wetland mapping [65]. The overall high accuracy of this technique obtained in this study further corroborates the value of such an approach for mapping wetlands at the provincial-level.

Although the classification accuracies obtained from our previous studies were slightly better in some cases (e.g., [19,31]), our previous studies involve more time and resources when compared with the current study. For example, our previous study [19] incorporated multi-frequency (X-, C-, and L-bands), multi-polarization (full-polarimetric RADARSAT-2) SAR data to produce local-scale wetland inventories. However, the production of such inventories demanded significant levels of labor, in terms of data preparation, feature extraction, statistical analysis, and classification. Consequently, updating wetland inventories using such methods on a regular basis for a large scale is tedious and expensive. In contrast, the present study relies on open access, regularly updated remotely sensed imagery collected by the Sentinel Missions at a 10 m spatial resolution, which is of great value for provincial- and national-scale wetland inventory maps that can be efficiently and regularly updated.

As mentioned earlier, GEE is an ideal platform that hosts Sentinel-1 and Sentinel-2 data and offers advanced processing functionally. This removes the process of downloading a large number of satellite images, which are already in "analysis ready" formats [34] and, as such, offers significant built-in time saving aspects [84]. Despite these benefits, limitations with GEE are related to both the lack of atmospherically-corrected Sentinel-2 data within its archive and the parallel method of the atmospheric correction at the time of this research. This may result in uncertainty due to the bidirectional reflectance effects caused by variations in sun, sensor, and surface geometries during satellite acquisitions [12]. Such an atmospheric correction algorithm has been carried out in local applications, such as the estimation of forest aboveground biomass [85], using the Sentinel-2 processing toolbox. Notably, Level-2A Sentinel-2 bottom-of-atmosphere (BOA) data that are atmospherically-corrected are of great value for extracting the most reliable temporal and spatial information, but such data are not yet available within GEE. Recent research, however, reported the potential of including BOA Sentinel-2 data in the near future into the GEE archive [12]. Although the high accuracies of wetland classifications in this study indicated that the effects of top-of-atmosphere (TOA) reflectance could be negligible, a comparison between TOA and BOA Sentinel-2 data for wetland mapping is suggested for future research.

In the near future, the addition of more machine learning tools and EO data to the GEE API and data catalog, respectively, will further simplify information extraction and data processing. For example, the availability of deep learning approaches through the potential inclusion of TensorFlow in the GEE platform will offer unprecedented opportunities for several remote sensing tasks [13]. Currently, however, employing state-of-the-art classification algorithms across broad spatial scales requires downloading data for additional local processing tasks and uploading data back to GEE due to the lack of functionality for such processing at present. Downloading such a large amount of remote sensing data is time consuming, given bandwidth limitations, and further, its processing demands a powerful local processing machine. Nevertheless, full exploitation of deep learning methods for mapping wetlands at hierarchical levels requires abundant, high-quality representative training samples.

The approaches presented in this study may be extended to generate a reliable, hierarchical, national-scale Canadian wetland inventory map and are an essential step toward global-scale wetland mapping. However, more challenges are expected when the study area is extended to the national-scale (i.e., Canada) with more cloud cover, more fragmented landscapes, and various dominant wetland classes across the country [86]. Notably, the biggest challenge in producing automated, national-scale wetland inventories is collecting a sufficient amount of high quality training and testing samples to support dependable coding, rapid product delivery, and accurate wetland mapping on large-scale. Although using GEE for discriminating wetland and non-wetland samples could be useful, it is currently inefficient for identifying hierarchical wetland ground-truth data. There are also challenges related to inconsistency in terms of wetland definitions at the global-scale that can vary by country (e.g., Canadian Wetland Classification System, New Zealand, and East Africa) [1]. However, given recent advances in cloud computing and big data, these barriers are eroding and new opportunities for more comprehensive and dynamic views of the global extent of wetlands are arising. For example, the integration of Landsat and Sentinel data using the GEE platform will address the limitations of cloud cover and lead to production of more accurate, finer category wetland classification maps, which are of great benefit for hydrological and ecological monitoring of these valuable ecosystems [87]. The results of this study suggest the feasibility of generating provincial-level wetland inventories by leveraging the opportunities offered by cloud-computing resources, such as GEE. The current study will contribute to the production of regular, consistent, provincial-scale wetland inventory maps that can support biodiversity and sustainable management of Newfoundland and Labrador's wetland resources.

5. Conclusions

Cloud-based computing resources and open-access EO data have caused a remarkable paradigm-shift in the field of landcover mapping by replacing the production of standard static maps with those that are more dynamic and application-specific thanks to recent advances in geospatial science. Leveraging the computational power of the Google Earth Engine and the availability of high spatial resolution remote sensing data collected by Copernicus Sentinels, the first detailed (category-based), provincial-level wetland inventory map was produced in this study. In particular, multi-year summer Sentinel-1 and Sentinel-2 data were used to map a complex series of small and large, heterogeneous wetlands on the Island of Newfoundland, Canada, covering an approximate area of 106,000 km^2.

Multiple classification scenarios, including those that were pixel- versus object-based, were considered and the discrimination capacities of optical and SAR data composites were compared. The results revealed the superiority of object-based classification relative to the pixel-based approach. Although classification accuracy using the multi-year summer optical composite was found to be more accurate than the multi-year summer SAR composite, the inclusion of both types of data (i.e., SAR and optical) significantly improved the accuracies of wetland classification. An overall classification accuracy of 88.37% was achieved using an object-based RF classification with the multi-year (2016–2018) summer optical/SAR composite, wherein wetland and non-wetland classes were distinguished with accuracies beyond 70% and 90%, respectively.

This study further contributes to the development of Canadian wetland inventories, characterizes the spatial distribution of wetland classes over a previously unmapped area with high spatial resolution, and importantly, augments previous local-scale wetland map products. Given the relatively similar ecological characteristics of wetlands across Canada, future work could extend this study by examining the value of the presented approach for mapping areas containing wetlands with similar ecological characteristics and potentially those with a greater diversity of wetland classes in other Canadian provinces and elsewhere. Further extension of this study could also focus on exploring the efficiency of a more diverse range of multi-temporal datasets (e.g., the 30 years Landsat dataset) to detect and understand wetland dynamics and trends over time in the province of Newfoundland and Labrador.

Supplementary Materials: The following are available online at http://www.mdpi.com/2072-4292/11/1/43/s1, The 10 m wetland extent product mapped complex series of small and large wetland classes accurately and precisely.

Author Contributions: M.M. and F.M. designed and performed the experiments, analyzed the data, and wrote the paper. B.S., S.H., and E.G. contributed editorial input and scientific insights to further improve the paper. All authors reviewed and commented on the manuscript.

Funding: This project was undertaken with the financial support of the Research & Development Corporation of Government of Newfoundland and Labrador (now InnovateNL) under Grant to M. Mahdianpari (RDC 5404-2108-101) and the Natural Sciences and Engineering Research Council of Canada under Grant to B. Salehi (NSERC RGPIN2015-05027).

Acknowledgments: Field data were collected by various organizations, including Ducks Unlimited Canada, Government of Newfoundland and Labrador Department of Environment and Conservation, and Nature Conservancy Canada. The authors thank these organizations for the generous financial support and providing such valuable datasets. The authors would like to thank the Google Earth Engine team for providing cloud-computing resources and European Space Agency (ESA) for providing open-access data. Additionally, the authors would like to thank anonymous reviewers for their helpful comments and suggestions.

Conflicts of Interest: The authors declare no conflict of interest.

References

1. Tiner, R.W.; Lang, M.W.; Klemas, V.V. *Remote Sensing of Wetlands: Applications and Advances*; CRC Press: Boca Raton, FL, USA, 2015.
2. Mitsch, W.J.; Bernal, B.; Nahlik, A.M.; Mander, Ü.; Zhang, L.; Anderson, C.J.; Jørgensen, S.E.; Brix, H. Wetlands, carbon, and climate change. *Landsc. Ecol.* **2013**, *28*, 583–597. [CrossRef]
3. Mitsch, W.J.; Gosselink, J.G. The value of wetlands: Importance of scale and landscape setting. *Ecol. Econ.* **2000**, *35*, 25–33. [CrossRef]
4. Gallant, A.L. The Challenges of Remote Monitoring of Wetlands. *Remote Sens.* **2015**, *7*, 10938–10950. [CrossRef]
5. Maxa, M.; Bolstad, P. Mapping northern wetlands with high resolution satellite images and LiDAR. *Wetlands* **2009**, *29*, 248. [CrossRef]
6. Tiner, R.W. Wetlands: An overview. In *Remote Sensing of Wetlands*; CRC Press: Boca Raton, FL, USA, 2015; pp. 20–35.
7. Mohammadimanesh, F.; Salehi, B.; Mahdianpari, M.; Homayouni, S. Unsupervised Wishart Classfication of Wetlands in Newfoundland, Canada Using Polsar Data Based on Fisher Linear Discriminant Analysis. *Int. Arch. Photogramm. Remote Sens. Spat. Inf. Sci.* **2016**, *41*, 305. [CrossRef]

8. Wulder, M.A.; Masek, J.G.; Cohen, W.B.; Loveland, T.R.; Woodcock, C.E. Opening the archive: How free data has enabled the science and monitoring promise of Landsat. *Remote Sens. Environ.* **2012**, *122*, 2–10. [CrossRef]
9. Xie, Y.; Sha, Z.; Yu, M. Remote sensing imagery in vegetation mapping: A review. *J. Plant Ecol.* **2008**, *1*, 9–23. [CrossRef]
10. Teluguntla, P.; Thenkabail, P.; Oliphant, A.; Xiong, J.; Gumma, M.K.; Congalton, R.G.; Yadav, K.; Huete, A. A 30-m landsat-derived cropland extent product of Australia and China using random forest machine learning algorithm on Google Earth Engine cloud computing platform. *ISPRS J. Photogramm. Remote Sens.* **2018**, *144*, 325–340. [CrossRef]
11. Shelestov, A.; Lavreniuk, M.; Kussul, N.; Novikov, A.; Skakun, S. Exploring Google earth engine platform for Big Data Processing: Classification of multi-temporal satellite imagery for crop mapping. *Front. Earth Sci.* **2017**, *5*, 17. [CrossRef]
12. Hird, J.N.; DeLancey, E.R.; McDermid, G.J.; Kariyeva, J. Google Earth Engine, open-access satellite data, and machine learning in support of large-area probabilistic wetland mapping. *Remote Sens.* **2017**, *9*, 1315. [CrossRef]
13. Gorelick, N.; Hancher, M.; Dixon, M.; Ilyushchenko, S.; Thau, D.; Moore, R. Google Earth Engine: Planetary-scale geospatial analysis for everyone. *Remote Sens. Environ.* **2017**, *202*, 18–27. [CrossRef]
14. Sazib, N.; Mladenova, I.; Bolten, J. Leveraging the Google Earth Engine for Drought Assessment Using Global Soil Moisture Data. *Remote Sens.* **2018**, *10*, 1265. [CrossRef]
15. Aguilar, R.; Zurita-Milla, R.; Izquierdo-Verdiguier, E.; de By, R.A. A Cloud-Based Multi-Temporal Ensemble Classifier to Map Smallholder Farming Systems. *Remote Sens.* **2018**, *10*, 729. [CrossRef]
16. de Lobo Lobo, F.; Souza-Filho, P.W.M.; de Moraes Novo, E.M.L.; Carlos, F.M.; Barbosa, C.C.F. Mapping Mining Areas in the Brazilian Amazon Using MSI/Sentinel-2 Imagery (2017). *Remote Sens.* **2018**, *10*, 1178. [CrossRef]
17. Kumar, L.; Mutanga, O. Google Earth Engine Applications since Inception: Usage, Trends, and Potential. *Remote Sens.* **2018**, *10*, 1509. [CrossRef]
18. Waske, B.; Fauvel, M.; Benediktsson, J.A.; Chanussot, J. Machine learning techniques in remote sensing data analysis. In *Kernel Methods for Remote Sensing Data Analysis*; Wiley Online Library: Hoboken, NJ, USA, 2009; pp. 3–24.
19. Mahdianpari, M.; Salehi, B.; Mohammadimanesh, F.; Motagh, M. Random forest wetland classification using ALOS-2 L-band, RADARSAT-2 C-band, and TerraSAR-X imagery. *ISPRS J. Photogramm. Remote Sens.* **2017**, *130*, 13–31. [CrossRef]
20. Thanh Noi, P.; Kappas, M. Comparison of random forest, k-nearest neighbor, and support vector machine classifiers for land cover classification using Sentinel-2 imagery. *Sensors* **2018**, *18*, 18. [CrossRef]
21. Huang, C.; Davis, L.S.; Townshend, J.R.G. An assessment of support vector machines for land cover classification. *Int. J. Remote Sens.* **2002**, *23*, 725–749. [CrossRef]
22. Pal, M. Random forest classifier for remote sensing classification. *Int. J. Remote Sens.* **2005**, *26*, 217–222. [CrossRef]
23. Rodriguez-Galiano, V.F.; Ghimire, B.; Rogan, J.; Chica-Olmo, M.; Rigol-Sanchez, J.P. An assessment of the effectiveness of a random forest classifier for land-cover classification. *ISPRS J. Photogramm. Remote Sens.* **2012**, *67*, 93–104. [CrossRef]
24. Gislason, P.O.; Benediktsson, J.A.; Sveinsson, J.R. Random forests for land cover classification. *Pattern Recognit. Lett.* **2006**, *27*, 294–300. [CrossRef]
25. Whyte, A.; Ferentinos, K.P.; Petropoulos, G.P. A new synergistic approach for monitoring wetlands using Sentinels-1 and 2 data with object-based machine learning algorithms. *Environ. Model. Softw.* **2018**, *104*, 40–54. [CrossRef]
26. Pekel, J.-F.; Cottam, A.; Gorelick, N.; Belward, A.S. High-resolution mapping of global surface water and its long-term changes. *Nature* **2016**, *540*, 418. [CrossRef] [PubMed]
27. Hansen, M.C.; Potapov, P.V.; Moore, R.; Hancher, M.; Turubanova, S.A.A.; Tyukavina, A.; Thau, D.; Stehman, S.V.; Goetz, S.J.; Loveland, T.R. High-resolution global maps of 21st-century forest cover change. *Science* **2013**, *342*, 850–853. [CrossRef] [PubMed]

28. Xiong, J.; Thenkabail, P.S.; Gumma, M.K.; Teluguntla, P.; Poehnelt, J.; Congalton, R.G.; Yadav, K.; Thau, D. Automated cropland mapping of continental Africa using Google Earth Engine cloud computing. *ISPRS J. Photogramm. Remote Sens.* **2017**, *126*, 225–244. [CrossRef]
29. Tsai, Y.; Stow, D.; Chen, H.; Lewison, R.; An, L.; Shi, L. Mapping Vegetation and Land Use Types in Fanjingshan National Nature Reserve Using Google Earth Engine. *Remote Sens.* **2018**, *10*, 927. [CrossRef]
30. Huang, H.; Chen, Y.; Clinton, N.; Wang, J.; Wang, X.; Liu, C.; Gong, P.; Yang, J.; Bai, Y.; Zheng, Y. Mapping major land cover dynamics in Beijing using all Landsat images in Google Earth Engine. *Remote Sens. Environ.* **2017**, *202*, 166–176. [CrossRef]
31. Mahdianpari, M.; Salehi, B.; Mohammadimanesh, F.; Brisco, B.; Mahdavi, S.; Amani, M.; Granger, J.E. Fisher Linear Discriminant Analysis of coherency matrix for wetland classification using PolSAR imagery. *Remote Sens. Environ.* **2018**, *206*, 300–317. [CrossRef]
32. Mohammadimanesh, F.; Salehi, B.; Mahdianpari, M.; Motagh, M.; Brisco, B. An efficient feature optimization for wetland mapping by synergistic use of SAR intensity, interferometry, and polarimetry data. *Int. J. Appl. Earth Obs. Geoinf.* **2018**, *73*, 450–462. [CrossRef]
33. Ozesmi, S.L.; Bauer, M.E. Satellite remote sensing of wetlands. *Wetlands Ecol. Manag.* **2002**, *10*, 381–402. [CrossRef]
34. d'Andrimont, R.; Lemoine, G.; van der Velde, M. Targeted Grassland Monitoring at Parcel Level Using Sentinels, Street-Level Images and Field Observations. *Remote Sens.* **2018**, *10*, 1300. [CrossRef]
35. Aschbacher, J.; Milagro-Pérez, M.P. The European Earth monitoring (GMES) programme: Status and perspectives. *Remote Sens. Environ.* **2012**, *120*, 3–8. [CrossRef]
36. Bwangoy, J.-R.B.; Hansen, M.C.; Roy, D.P.; De Grandi, G.; Justice, C.O. Wetland mapping in the Congo Basin using optical and radar remotely sensed data and derived topographical indices. *Remote Sens. Environ.* **2010**, *114*, 73–86. [CrossRef]
37. Mahdianpari, M.; Salehi, B.; Rezaee, M.; Mohammadimanesh, F.; Zhang, Y. Very deep convolutional neural networks for complex land cover mapping using multispectral remote sensing imagery. *Remote Sens.* **2018**, *10*, 1119. [CrossRef]
38. Rezaee, M.; Mahdianpari, M.; Zhang, Y.; Salehi, B. Deep convolutional neural network for complex wetland classification using optical remote sensing imagery. *IEEE J. Sel. Top. Appl. Earth Obs. Remote Sens.* **2018**, *11*, 3030–3039. [CrossRef]
39. Amarsaikhan, D.; Saandar, M.; Ganzorig, M.; Blotevogel, H.H.; Egshiglen, E.; Gantuyal, R.; Nergui, B.; Enkhjargal, D. Comparison of multisource image fusion methods and land cover classification. *Int. J. Remote Sens.* **2012**, *33*, 2532–2550. [CrossRef]
40. Mahdianpari, M.; Salehi, B.; Mohammadimanesh, F.; Brisco, B. An assessment of simulated compact polarimetric SAR data for wetland classification using random Forest algorithm. *Can. J. Remote Sens.* **2017**, *43*, 468–484. [CrossRef]
41. van Beijma, S.; Comber, A.; Lamb, A. Random forest classification of salt marsh vegetation habitats using quad-polarimetric airborne SAR, elevation and optical RS data. *Remote Sens. Environ.* **2014**, *149*, 118–129. [CrossRef]
42. Zhang, J. Multi-source remote sensing data fusion: Status and trends. *Int. J. Image Data Fusion* **2010**, *1*, 5–24. [CrossRef]
43. Ecological Stratification Working Group. *A National Ecological Framework for Canada*; Agriculture and Agri-Food Canada, Research Branch, Centre for Land and Biological Resources Research, and Environment Canada, State of the Environment Directorate, Ecozone Analysis Branch: Ottawa/Hull, QC, Canada, 1996.
44. South, R. *Biogeography and Ecology of the Island of Newfoundland*; Springer Science & Business Media: Berlin/Heidelberg, Germany, 1983; Volume 48, ISBN 9061931010.
45. Meades, S.J. *Ecoregions of Newfoundland and Labrador*; St. John's, Newfoundland and Labrador: Parks and Natural Areas Division, Department of Environment and Conservation, Government of Newfoundland and Labrador: Corner Brook, NL, Canada, 1990.
46. Zhang, X.; Wu, B.; Ponce-Campos, G.; Zhang, M.; Chang, S.; Tian, F. Mapping up-to-Date Paddy Rice Extent at 10 M Resolution in China through the Integration of Optical and Synthetic Aperture Radar Images. *Remote Sens.* **2018**, *10*, 1200. [CrossRef]

47. Marshall, I.B.; Schut, P.; Ballard, M. *A National Ecological Framework for Canada: Attribute Data*; Environmental Quality Branch, Ecosystems Science Directorate, Environment Canada and Research Branch, Agriculture and Agri-Food Canada: Ottawa, QC, Canada, 1999.
48. Sentinel-1-Observation Scenario—Planned Acquisitions—ESA. Available online: https://sentinel.esa.int/web/sentinel/missions/sentinel-1/observation-scenario (accessed on 13 November 2018).
49. Sentinel-1 Algorithms. Google Earth Engine API. Google Developers. Available online: https://developers.google.com/earth-engine/sentinel1 (accessed on 13 November 2018).
50. Gauthier, Y.; Bernier, M.; Fortin, J.-P. Aspect and incidence angle sensitivity in ERS-1 SAR data. *Int. J. Remote Sens.* **1998**, *19*, 2001–2006. [CrossRef]
51. Lee, J.-S.; Wen, J.-H.; Ainsworth, T.L.; Chen, K.-S.; Chen, A.J. Improved sigma filter for speckle filtering of SAR imagery. *IEEE Trans. Geosci. Remote Sens.* **2009**, *47*, 202–213.
52. Mahdianpari, M.; Salehi, B.; Mohammadimanesh, F. The effect of PolSAR image de-speckling on wetland classification: Introducing a new adaptive method. *Can. J. Remote Sens.* **2017**, *43*, 485–503. [CrossRef]
53. Mohammadimanesh, F.; Salehi, B.; Mahdianpari, M.; Brisco, B.; Motagh, M. Multi-temporal, multi-frequency, and multi-polarization coherence and SAR backscatter analysis of wetlands. *ISPRS J. Photogramm. Remote Sens.* **2018**, *142*, 78–93. [CrossRef]
54. Baghdadi, N.; Bernier, M.; Gauthier, R.; Neeson, I. Evaluation of C-band SAR data for wetlands mapping. *Int. J. Remote Sens.* **2001**, *22*, 71–88. [CrossRef]
55. Steele-Dunne, S.C.; McNairn, H.; Monsivais-Huertero, A.; Judge, J.; Liu, P.-W.; Papathanassiou, K. Radar remote sensing of agricultural canopies: A review. *IEEE J. Sel. Top. Appl. Earth Obs. Remote Sens.* **2017**, *10*, 2249–2273. [CrossRef]
56. de Almeida Furtado, L.F.; Silva, T.S.F.; de Moraes Novo, E.M.L. Dual-season and full-polarimetric C band SAR assessment for vegetation mapping in the Amazon várzea wetlands. *Remote Sens. Environ.* **2016**, *174*, 212–222. [CrossRef]
57. Jensen, J.R. *Remote Sensing of the Environment: An Earth Resource Perspective 2/e*; Pearson Education: Delhi, India, 2009.
58. Ji, L.; Zhang, L.; Wylie, B. Analysis of dynamic thresholds for the normalized difference water index. *Photogramm. Eng. Remote Sens.* **2009**, *75*, 1307–1317. [CrossRef]
59. Xu, H. Modification of normalised difference water index (NDWI) to enhance open water features in remotely sensed imagery. *Int. J. Remote Sens.* **2006**, *27*, 3025–3033. [CrossRef]
60. Rogers, A.S.; Kearney, M.S. Reducing signature variability in unmixing coastal marsh Thematic Mapper scenes using spectral indices. *Int. J. Remote Sens.* **2004**, *25*, 2317–2335. [CrossRef]
61. McFeeters, S.K. The use of the Normalized Difference Water Index (NDWI) in the delineation of open water features. *Int. J. Remote Sens.* **1996**, *17*, 1425–1432. [CrossRef]
62. Flood, N. Seasonal composite Landsat TM/ETM+ images using the medoid (a multi-dimensional median). *Remote Sens.* **2013**, *5*, 6481–6500. [CrossRef]
63. Griffiths, P.; van der Linden, S.; Kuemmerle, T.; Hostert, P. A pixel-based Landsat compositing algorithm for large area land cover mapping. *IEEE J. Sel. Top. Appl. Earth Obs. Remote Sens.* **2013**, *6*, 2088–2101. [CrossRef]
64. Roy, D.P.; Ju, J.; Kline, K.; Scaramuzza, P.L.; Kovalskyy, V.; Hansen, M.; Loveland, T.R.; Vermote, E.; Zhang, C. Web-enabled Landsat Data (WELD): Landsat ETM+ composited mosaics of the conterminous United States. *Remote Sens. Environ.* **2010**, *114*, 35–49. [CrossRef]
65. Wulder, M.; Li, Z.; Campbell, E.; White, J.; Hobart, G.; Hermosilla, T.; Coops, N. A National Assessment of Wetland Status and Trends for Canada's Forested Ecosystems Using 33 Years of Earth Observation Satellite Data. *Remote Sens.* **2018**, *10*, 1623. [CrossRef]
66. Swain, P.H.; Davis, S.M. Remote sensing: The quantitative approach. *IEEE Trans. Pattern Anal. Mach. Intell.* **1981**, 713–714. [CrossRef]
67. Padma, S.; Sanjeevi, S. Jeffries Matusita based mixed-measure for improved spectral matching in hyperspectral image analysis. *Int. J. Appl. Earth Obs. Geoinf.* **2014**, *32*, 138–151. [CrossRef]
68. Schmidt, K.S.; Skidmore, A.K. Spectral discrimination of vegetation types in a coastal wetland. *Remote Sens. Environ.* **2003**, *85*, 92–108. [CrossRef]
69. Belgiu, M.; Drăguţ, L. Random forest in remote sensing: A review of applications and future directions. *ISPRS J. Photogramm. Remote Sens.* **2016**, *114*, 24–31. [CrossRef]

70. Mahdianpari, M.; Salehi, B.; Mohammadimanesh, F.; Larsen, G.; Peddle, D.R. Mapping land-based oil spills using high spatial resolution unmanned aerial vehicle imagery and electromagnetic induction survey data. *J. Appl. Remote Sens.* **2018**, *12*, 036015. [CrossRef]
71. Breiman, L. Random forests. *Mach. Learn.* **2001**, *45*, 5–32. [CrossRef]
72. Mohammadimanesh, F.; Salehi, B.; Mahdianpari, M.; English, J.; Chamberland, J.; Alasset, P.-J. Monitoring surface changes in discontinuous permafrost terrain using small baseline SAR interferometry, object-based classification, and geological features: A case study from Mayo, Yukon Territory, Canada. *GIScience Remote Sens.* **2018**, 1–26. [CrossRef]
73. Blaschke, T. Object based image analysis for remote sensing. *ISPRS J. Photogramm. Remote Sens.* **2010**, *65*, 2–16. [CrossRef]
74. Benz, U.C.; Hofmann, P.; Willhauck, G.; Lingenfelder, I.; Heynen, M. Multi-resolution, object-oriented fuzzy analysis of remote sensing data for GIS-ready information. *ISPRS J. Photogramm. Remote Sens.* **2004**, *58*, 239–258. [CrossRef]
75. Achanta, R.; Süsstrunk, S. Superpixels and polygons using simple non-iterative clustering. In Proceedings of the 2017 IEEE Conference on Computer Vision and Pattern Recognition (CVPR), Honolulu, HI, USA, 21–26 July 2017; pp. 4895–4904.
76. Congalton, R.G. A review of assessing the accuracy of classifications of remotely sensed data. *Remote Sens. Environ.* **1991**, *37*, 35–46. [CrossRef]
77. McNemar, Q. Note on the sampling error of the difference between correlated proportions or percentages. *Psychometrika* **1947**, *12*, 153–157. [CrossRef] [PubMed]
78. de Leeuw, J.; Jia, H.; Yang, L.; Liu, X.; Schmidt, K.; Skidmore, A.K. Comparing accuracy assessments to infer superiority of image classification methods. *Int. J. Remote Sens.* **2006**, *27*, 223–232. [CrossRef]
79. Dingle Robertson, L.; King, D.J. Comparison of pixel-and object-based classification in land cover change mapping. *Int. J. Remote Sens.* **2011**, *32*, 1505–1529. [CrossRef]
80. Adam, E.; Mutanga, O.; Rugege, D. Multispectral and hyperspectral remote sensing for identification and mapping of wetland vegetation: A review. *Wetlands Ecol. Manag.* **2010**, *18*, 281–296. [CrossRef]
81. Mohammadimanesh, F.; Salehi, B.; Mahdianpari, M.; Brisco, B.; Motagh, M. Wetland Water Level Monitoring Using Interferometric Synthetic Aperture Radar (InSAR): A Review. *Can. J. Remote Sens.* **2018**, 1–16. [CrossRef]
82. Chen, B.; Xiao, X.; Li, X.; Pan, L.; Doughty, R.; Ma, J.; Dong, J.; Qin, Y.; Zhao, B.; Wu, Z. A mangrove forest map of China in 2015: Analysis of time series Landsat 7/8 and Sentinel-1A imagery in Google Earth Engine cloud computing platform. *ISPRS J. Photogramm. Remote Sens.* **2017**, *131*, 104–120. [CrossRef]
83. Kelley, L.; Pitcher, L.; Bacon, C. Using Google Earth Engine to Map Complex Shade-Grown Coffee Landscapes in Northern Nicaragua. *Remote Sens.* **2018**, *10*, 952. [CrossRef]
84. Jacobson, A.; Dhanota, J.; Godfrey, J.; Jacobson, H.; Rossman, Z.; Stanish, A.; Walker, H.; Riggio, J. A novel approach to mapping land conversion using Google Earth with an application to East Africa. *Environ. Model. Softw.* **2015**, *72*, 1–9. [CrossRef]
85. Vafaei, S.; Soosani, J.; Adeli, K.; Fadaei, H.; Naghavi, H.; Pham, T.D.; Tien Bui, D. Improving accuracy estimation of forest aboveground biomass based on incorporation of ALOS-2 PALSAR-2 and sentinel-2A imagery and machine learning: A case study of the Hyrcanian forest area (Iran). *Remote Sens.* **2018**, *10*, 172. [CrossRef]
86. Dong, J.; Xiao, X.; Menarguez, M.A.; Zhang, G.; Qin, Y.; Thau, D.; Biradar, C.; Moore, B., III. Mapping paddy rice planting area in northeastern Asia with Landsat 8 images, phenology-based algorithm and Google Earth Engine. *Remote Sens. Environ.* **2016**, *185*, 142–154. [CrossRef] [PubMed]
87. Wulder, M.A.; White, J.C.; Masek, J.G.; Dwyer, J.; Roy, D.P. Continuity of Landsat observations: Short term considerations. *Remote Sens. Environ.* **2011**, *115*, 747–751. [CrossRef]

Article

A Cloud-Based Multi-Temporal Ensemble Classifier to Map Smallholder Farming Systems

Rosa Aguilar *, Raul Zurita-Milla, Emma Izquierdo-Verdiguier and Rolf A. de By

Faculty of Geo-Information Science and Earth Observation (ITC). University of Twente, 7514 AE Enschede, The Netherlands; r.zurita-milla@utwente.nl (R.Z.-M.); e.izquierdoverdiguier@utwente.nl (E.I.-V.); r.a.deby@utwente.nl (R.A.d.B)

* Correspondence: r.m.aguilardearchila@utwente.nl; Tel.: +31-53-487-4444

Received: 30 March 2018; Accepted: 7 May 2018; Published: 9 May 2018

Abstract: Smallholder farmers cultivate more than 80% of the cropland area available in Africa. The intrinsic characteristics of such farms include complex crop-planting patterns, and small fields that are vaguely delineated. These characteristics pose challenges to mapping crops and fields from space. In this study, we evaluate the use of a cloud-based multi-temporal ensemble classifier to map smallholder farming systems in a case study for southern Mali. The ensemble combines a selection of spatial and spectral features derived from multi-spectral Worldview-2 images, field data, and five machine learning classifiers to produce a map of the most prevalent crops in our study area. Different ensemble sizes were evaluated using two combination rules, namely majority voting and weighted majority voting. Both strategies outperform any of the tested single classifiers. The ensemble based on the weighted majority voting strategy obtained the higher overall accuracy (75.9%). This means an accuracy improvement of 4.65% in comparison with the average overall accuracy of the best individual classifier tested in this study. The maximum ensemble accuracy is reached with 75 classifiers in the ensemble. This indicates that the addition of more classifiers does not help to continuously improve classification results. Our results demonstrate the potential of ensemble classifiers to map crops grown by West African smallholders. The use of ensembles demands high computational capability, but the increasing availability of cloud computing solutions allows their efficient implementation and even opens the door to the data processing needs of local organizations.

Keywords: Google Earth Engine; crop classification; multi-classifier; cloud computing; time series; high spatial resolution

---

1. Introduction

Smallholder farmers cultivate more than 80% of the cropland area available in Africa [1] where the agricultural sector provides about 60% of the total employment [2]. However, the inherent characteristics of smallholder farms such as their small size (frequently less than 1 ha and with vaguely delineated boundaries), the ir location in areas with extreme environmental variability in space and time, and the use of mixed cropping systems, have prevented a sustainable improvement on smallholder agriculture in terms of volume and quality [3]. Yet, an increase of African agricultural productivity is imperative because the continent will experience substantial population growth in the coming decades [4]. Realizing that croplands are scarce, the productivity increase should have the lowest reasonable environmental impact and should be as sustainable as possible [5]. A robust agricultural monitoring system is then a prerequisite to promote informed decisions not only at executive or policy levels but also at the level of daily field management. Such a system could, for example, help to reduce price fluctuations by deciding on import and export needs for each crop [6], to establish agricultural insurance mechanisms, or to estimate the demand for agricultural inputs [6,7].

Crop maps are a basic but essential layer of any agricultural monitoring system and are critical to achieve food security [8,9]. Most African countries, however, lack reliable crop maps. Remote sensing image classification is a convenient approach for producing these maps due to advantages in terms of cost, revisit time, and spatial coverage [10]. Indeed, remotely sensed image classification has been successfully applied to produce crop maps in homogeneous areas [11–14].

Smallholder farms, which shape the predominate crop production systems in Africa, present significant mapping challenges compared to homogeneous agricultural areas (i.e., with intensive or commercial farms) [8]. Difficulties are not only in requiring very high spatial resolution data, but also in the spectral identification of farm fields and crops because smallholder fields are irregularly shaped and their seasonal variation in surface reflectance is strongly influenced by irregular and variable farm practices in environmentally diverse areas. Because of these peculiarities, the production of reliable crop maps from remotely sensed images is not an easy task [15].

In general, a low level of accuracy in image classification is tackled by using more informative features, or by developing new algorithms or approaches to combine existing ones [16]. Indeed, several studies have shown that classification accuracy improves when combining spectral (e.g., vegetation indices), spatial (e.g., textures), and temporal (e.g., multiple images during the cropping season) features [17]. Compared to single band, spectral indices are less affected by atmospheric conditions, illumination differences, and soil background, and thus bring forward an enhanced vegetation signal that is normally easier to classify [18]. Spatial features benefit crop discrimination [19], especially in heterogeneous areas where high local variance is more relevant when very high spatial resolution images are applied [20,21]. Regarding temporal features, multi-temporal spectral indices have been exploited in crop identification because they provide information about the seasonal variation in surface reflectance caused by crop phenology [13,22–24].

The second approach to increase classification accuracy (i.e., by developing new algorithms) has been extensively used by the remote sensing community, which has rapidly adopted and adapted novel machine learning image classification approaches [25–27]. The combination of existing classifiers (ensemble of classifiers) has, however, received comparatively little attention, although it is known that ensemble classifiers increase classification accuracy because no single classifier outperforms the others [28]. A common approach to implement a classifier ensemble, also known as a multi-classifier, consists of training several "base classifiers", which are subsequently applied to unseen data to create a set of classification outputs that are next combined using various rules to obtain a final classification output [28,29]. At the expense of increased computational complexity, ensemble classifiers can handle complex feature spaces and reduce misclassifications caused by using non-optimal, overfitted, or undertrained classifiers and, hence, the y improve classification accuracy. Given the increasing availability of computing resources, various studies have shown that ensemble classifiers outperform individual classifiers [30–32]. Yet, the use of ensemble classifiers remains scarce in the context of remote sensing [33] and is limited to image subsets, mono-temporal studies, or to the combination of only a few classifiers [34–36].

Ensemble classifiers produce more accurate classification results because they can capture and model complex decision boundaries [37]. The use of ensembles for agricultural purposes as reported in various studies has shown that they outperformed individual classifiers [34,35,38]. Any classifier that provides a higher accuracy than one obtained by chance is suitable for integration in an ensemble [39], and may contribute to shape the final decision boundaries [29]. In other words, the strength of ensembles comes from the fact that the base classifiers misclassify different instances. For this purpose, several techniques can be applied. For example, by selecting classifiers that rely on different algorithms, by applying different training sets, by training on different feature subsets, or by using different parameters [40,41].

In this study, we evaluate the use of a cloud-based ensemble classifier to map African smallholder farming systems. Thanks to the use of cloud computing, various base classifiers and combination rules were efficiently tested. Moreover, it allowed training of the ensemble with a wide array of spectral, spatial, and temporal features extracted from the available set of very high spatial resolution images.

2. Materials and Methods

This section provides a description of the available images and the approach used to develop our ensemble classifiers.

2.1. Study Area and Data

The study area covers a square of 10 × 10 km located near Koutiala, southern Mali, West Africa. This site is also an ICRISAT-led site contributing to the Joint Experiment for Crop Assessment and Monitoring (JECAM) [42]. For this area, a time series of seven multi-spectral Worldview-2 images was acquired for the cropping season of 2014. Acquisition dates of the images range from May to November covering both the beginning and the end of the crop growing season [42]. The exact acquisition dates are: 22 May, 30 May, 26 June, 29 July, 18 October, 1 November, and 14 November. The images have a pixel size of about 2 m and contain eight spectral bands in the visible, red-edge and near-infrared part of the electromagnetic spectrum. Figure 1 illustrates the study area and a zoomed in view of the area with agricultural fields. All the images were preprocessed using the STARS project workflow which uses the 6S radiative transfer model for atmospheric correction [43]. The images were atmospherically and radiometrically corrected, co-registered, and trees and clouds were masked. Crop labels for five main crops namely maize, millet, peanut, sorghum, and cotton, were collected in the field. A total of 45 fields were labeled in situ in the study area indicated in Figure 1b. This ground truth data was used to train base classifiers and to assess the accuracy of both base classifiers and ensembles.

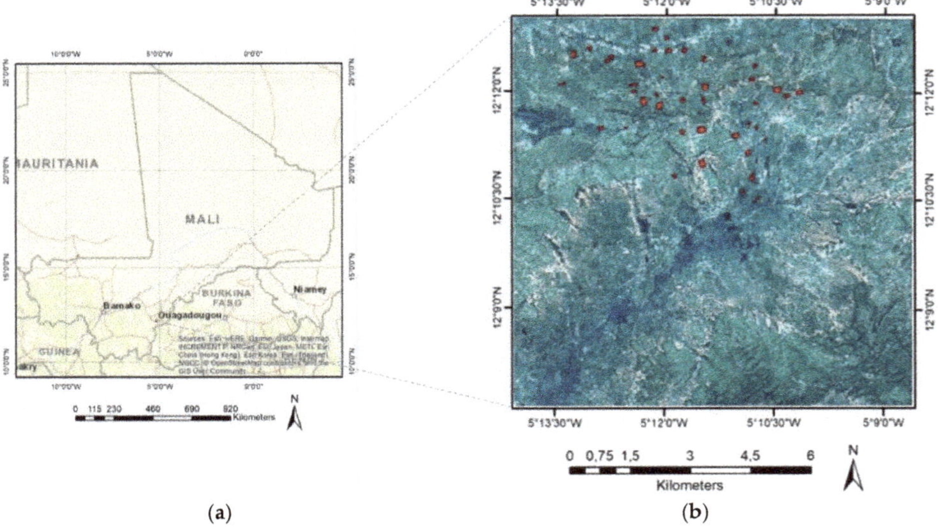

Figure 1. Study area. (**a**) Location of the study area in Mali; (**b**) The study's field plots overlapping a Worldview-2 image of the study area on 18 October 2014 using natural color composite.

2.2. Methods

Figure 2 presents a high-level view of the developed workflow. First, in the data preparation step (described more fully in Section 2.2.1), we extract a suite of spatial and spectral features from the available images and select the most relevant ones for image classification. Then, multiple classifiers are trained, tested, and applied to the images (Section 2.2.2). Finally, we test various approaches to create ensembles from the available classifiers and assess their classification accuracy using an independent test set (Section 2.2.3).

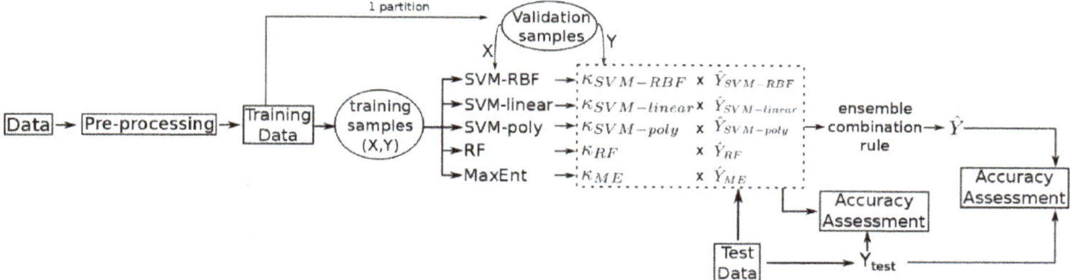

Figure 2. Overview of the ensemble classifier system. X represents the features extracted during pre-processing, Y and Y_{test} represent ground truth of training and test data, $\hat{Y}_{class}$ is the prediction of a classifier and $\hat{Y}$ the ensemble prediction. K_{class} is the kappa obtained by a classifier.

2.2.1. Data Preparation

A comprehensive set of spectral and spatial features is generated from the (multi-spectral) time series of Worldview-2 images. The spectral features include the vegetation indices listed in Table 1.

Table 1. Vegetation indices, formulas, and reference. WorldView-2 band names abbreviations are: R = Red, RE = Red edge, G = Green, B = Blue and NIR = Near IR-1.

Vegetation Index (VI)	Formula
Normalized Difference Vegetation Index (NDVI) [44]	$(NIR - R)/(NIR + R)$
Green Leaf Index (GLI) [45]	$(2 \times G - R - B)/(2 \times G + R + B)$
Enhanced Vegetation Index (EVI) [46]	$EVI = 2.5 \times (NIR - R)/(NIR + 6 \times R - 7.5 \times B + 1)$
Soil Adjusted Vegetation Index (SAVI) [47]	$(1 + L) \times (NIR - R)/(NIR + R + L)$, where $L = 0.5$
Modified Soil Adjusted Vegetation Index (MSAVI) [48]	$0.5 \times \left(2 \times NIR + 1 - \sqrt{(2 \times NIR + 1)^2 - 8 \times (NIR - R)}\right)$
Transformed Chlorophyll Absorption in Reflectance Index (TCARI) [49]	$3 \times ((RE - R) - 0.2 \times (RE - G) \times (RE/R))$
Visible Atmospherically Resistance Index (VARI) [50]	$(G - R)/(G + R - B)$

Spatial features are based on the Gray Level Co-occurrence Matrix (GLCM). Fifteen features proposed by [51] and three features from [52] are derived. This selection fits with their function availability in the Google Earth Engine (GEE) [53]. Formulas of these features are shown in Tables A1 and A2.

Textural features are calculated as the average of their values in four directions (0, 45, 90, 135), applying a window of 3 × 3 pixels to the original spectral bands of each image. This configuration corresponds to the default setup in GEE and is deemed appropriate for our study since our goal is to create an efficient ensemble and not to optimize the configuration to extract spatial features.

The extraction of spectral and spatial features, computed for each pixel, results in 140 features for a single image and in 980 features for the complete time series (Table 2). Although GEE is a scalable and cloud-based platform, a timely execution of the classifiers is not possible without reducing the number of features used. Moreover, we know and empirically see (results not shown) that many features contain similar information and are highly correlated. Thus, a guided regularized random forest (GRRF) [54] is applied to identify the most relevant features. This feature selection step helps to make our classification problem both more tractable in GEE and more interpretable. GRRF requires the optimization of two regularization parameters. The most relevant features are obtained using the criteria of gain regularized higher than zero. This optimization is done for ten subsets of training data generated by randomly splitting 2129 training samples. Each subset is fed to the GRRF to select the most relevant spectral and spatial features after optimizing the two regularization parameters. The selected features are then used to train an RF classifier using all the training samples. The best set of spatial and spectral features is determined by ranking the resulting RF classifiers according to their OA for 1258 test samples.

Table 2. Type and number of features extracted from a single multi-spectral WorldView-2 image, and from the time series of seven images. Gray Level Co-occurrence Matrix (GLCM).

Feature	Features Per Image	Total Per Image Series
Spectral bands	7	49
Vegetation indices	7	49
GLCM-based features applied to image bands	126	882
Total	140	980

2.2.2. Base Classifiers

Several classifiers are used to create our ensembles, after performing an exploratory analysis with the available classifiers in GEE. Five classifiers are selected to create our ensembles based on their algorithmic approach and overall accuracy (OA): Random Forest (RF; [55]), Maximum Entropy Model (MaxEnt; [56]), Support Vector Machine (SVM; [57]) with linear, polynomial and Gaussian kernels. A combination with other types of classifier, e.g., a deep learning algorithm could easily be allowed when such becomes available in GEE (with inclusion of TensorFlow). This is expected to happen given the active research being performed in this field. The following paragraphs briefly describe our chosen classifiers and explain how they are used in this study.

RF is a well-known machine learning algorithm [58–61] created by combining a set of decision trees. A typical characteristic of RF is that each tree is created with a random selection of training instances and features. Once the trees are created, classification results are obtained by majority voting. RF has reached around 85% OA in crop type classification using a multi-spectral time series of RapidEye images [62], and higher than 80% for a time series of Landsat7 images in homogeneous regions [13]. RF has two user-defined parameters: the number of trees and the number of features available to build each decision tree. In our study, an RF with 300 trees is created and we set the number of features per split to the square root of the total number of features. These are standard settings [63].

MaxEnt computes an approximated probability distribution consistent with the constraints (facts) observed in the data (predictor values) and as uniform as possible [64]. This provides maximum entropy while avoiding assumptions on the unknown, hence the name of the classifier. MaxEnt was proposed to estimate geographic species distribution and potential habitat [56], to classify vegetation from remote sensing images [65], and groundwater potential mapping [66]. In our study, MaxEnt was applied with default parameter values in GEE as follows: weight for L1 regularization set to 0, weight for L2 regularization set to 0.00001, epsilon set to 0.00001, minimum number of iterations set to 0, and maximum number of iterations set to 100.

SVM is another well-known machine learning algorithm that has been widely applied for crop classification [11,67]. SVM has demonstrated its robustness to outliers and is an excellent classifier when the number of input features is high [12]. The original binary version of SVM aims to find the optimal plane that separates the available data into two classes by maximizing the distance (margins) between the so-called support vectors (i.e., the closest training samples to the optimal hyperplane). Multiple binary SVMs can be combined to tackle a multi-class problem. When the training data cannot be separated by a plane, it is mapped to a multidimensional feature space in which the samples are separated by a hyperplane. This leads to a non-linear classification algorithm that, thanks to the so-called kernel trick, only needs the definition of the dot products among the training data [68]. Linear, radial, and polynomial kernels are commonly used to define these dot products. The linear SVM only requires fixing the so-called C parameter, which represents the cost of misclassifying samples, whereas the radial and polynomial kernels require the optimization of an additional parameter, respectively called gamma and the polynomial degree. In this work, all SVM parameters were obtained by 5-fold cross validation [69], Linear kernel (SVML) was tuned in a range of initial values C = [1, 10, 50, 100, 200, 300, 400, 500, 600, 700, 800, 1000]. Gaussian kernel (SVMR) used an initial values range of C = [1, 10, 100, 200, 300] and gamma = [0.001, 0.1, 0.5, 1, 5, 10]. Also, parameters for SVM polynomial (SVMP) were tuned using C = [10, 100, 300], gamma = [0.1, 1, 10], degree = [2, 3, 4] and coef0 = [1, 10, 100].

All classifiers are trained and applied separately using a modified leave-one-out method in which the training set is stratified and randomly partitioned into k (10) equally sized subsamples. Each base classifier is trained with k − 1 subsamples, leaving one subsample out [40]. Using ten different seeds to generate the subsamples, the se methods allow us to generate 100 subsets of training data that, in turn, allow 20 versions of each base classifier to be generated and a total of 100 classification models when combining the five classifiers as presented in Figure 3. This training method prevents overfitting of the base classifiers because 10% of the data is discarded each time. Overfitting prevention is desirable because the ensemble is not trainable. Metrics reported are OA and kappa coefficient. Producer accuracy (PA) per class is also computed and is used to contrast performance of individual classifiers versus ensemble classifiers.

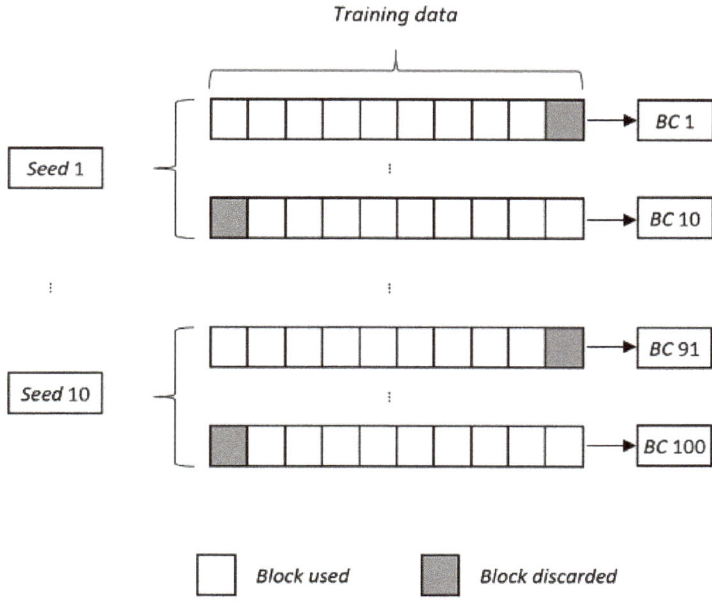

Figure 3. Leave-one-out strategy using ten seeds for generating 100 training datasets to train base classifiers (BC).

2.2.3. Ensemble Classifiers

Two combination rules, namely majority and weighted majority voting, are tested in this study to create ensemble classifiers. In the case of majority voting, the output of the ensemble is the most assigned class by classifiers, whereas in the weighted majority voting rule, a weight is assigned to each classifier to favor those classifiers with better performance in the voting decision. Both rules are easily implemented and produce results comparable to more complicated combination schemes [30,36,70]. Moreover, the se rules do not require additional training data because they are not trainable [40] which means that the required parameters for the ensemble are available as the classifiers are generated and their accuracy assessed.

Majority voting works as follows. Let x denote one of the decision problem instances, let L be the number of base classifiers used, and let C be the number of possible classes. The decision (output) of classifier i on x is represented as a binary vector $d_{x,i}$ of the form $(0, \ldots, 0, 1, 0, \ldots, 0)$, where $d_{x,i,j} = 1$ if and only if the classifier labels that instance x with class C_j. Further, we denote vector summation by $\sum$ and define the function idx@max as the index at which a maximum value is found in a vector. This function resolves ties as follow: if multiple maximal values are found, the index of the first occurrence is picked and returned. The majority voting rule of an ensemble classifier on decision problem x defines the class number D_x as:

$$D_x = \text{idx@max} \sum_{i=1}^{L} d_{x,i}, \quad (1)$$

following [29].

Weighted majority voting is an extension of the above and uses weights w_i per base classifier i.

$$D_x = \text{idx@max} \sum_{i=1}^{L} w_i d_{x,i}, \quad (2)$$

In this, we choose $w_i = \log\left(\frac{k}{1-k}\right)$, where k is the kappa coefficient of base classifier i over an independent sample set [29].

As mentioned in Section 2.2.2, our training procedure yields 20 instances of each base classifier. This allows creating two 100-classifier ensembles as well as a larger number of ensembles formed by 5, 10, 15, ... , 95 classifiers. The latter ensembles serve to evaluate the impact of the size of the ensemble. To avoid biased results, we combine the base classifiers while keeping the proportion of each type of classifier. For example, the 10-classifier ensemble is created by combining two randomly chosen base classifiers of each type. This experiment means that we evaluate the classification accuracy of 191 ensembles. Classification accuracy is assessed by means of their OA, the ir kappa coefficient and the producer's accuracy of each class. Besides, results of the most effective configuration of ensembles and the individual classifier with higher accuracy are compared to get insights into their performance. Examples of their output are analyzed by visual inspection.

3. Experiment Results and Discussion

3.1. Data Preparation

A feature selection method is applied before the classification to reduce the dimensionality of data without losing classification efficiency. In our study, we selected the GRRF method because it selects the features in a transparent and understandable way. The application of the GRRF to the expanded time series (i.e., the original bands plus spectral and spatial features), leads to the selection of 45 features as shown in Table 3; bands, spectral, and spatial features were selected. In general, spatial features were predominantly selected in almost all the images, whereas vegetation indices were selected in only five images. Vegetation indices have more influence when taken from images acquired when the crop has grown than when the field is bare.

A more detailed analysis of Table 3 shows that the selected multi-spectral bands and vegetation indices respectively represent 24.44% and 26.66% of the most relevant features. Textural features represent 48.88% of the most relevant features, which emphasizes the relevance of considering spatial context when analyzing very high spatial resolution images. As an example, Figures 4 and 5 show the temporal evolution of a vegetation index and of one of the GLCM-based spatial features. In Figure 4, changes in TCARI are presented. Figure 4a shows a low vegetation signal since the crop is at an initial stage. In Figure 4b,c, a higher vegetation signal is shown, which relates to a more advanced growth stage. TCARI was selected for three different dates underlining the importance of changes in vegetation index for crop discrimination. Similarly, Figure 5 displays a textural feature (sum average of band 8) for a specific parcel, which points at variation in spatial patterns as the growing season goes by.

Table 3. Guided regularized random forest (GRRF) selected features sorted by image date [b2: band 2, b3: band 3, b4: band 4, b5: band 5, b6: band 6, b7: band 7, b8: band 8, idm: inverse different moment, savg: sum average, dvar: difference variance, corr: correlation, diss: dissimilarity].

Image Date						
22 May 2014	**30 May 2014**	**26 June 2014**	**29 July 2014**	**18 October 2014**	**1 November 2014**	**14 November 2014**
b3	b3_savg	b4_diss	b3	SAVI	b3_diss	b2
b7	b5_savg	b5_dvar	b5_savg	VARI	b4_dvar	b2_savg
b8	b6_corr	b8_ent	b6		b4_idm	b3_dvar
b8_idm	b7_idm	GLI	b6_corr		b4_savg	b8
	b7_savg	MSAVI	b6_savg		b6	EVI
	b8_savg	TCARI	b8_diss		b6_savg	TCARI
	VARI				b7_corr	
					b7_savg	
					b8_diss	
					b8_savg	
					EVI	
					GLI	
					TCARI	
					VARI	

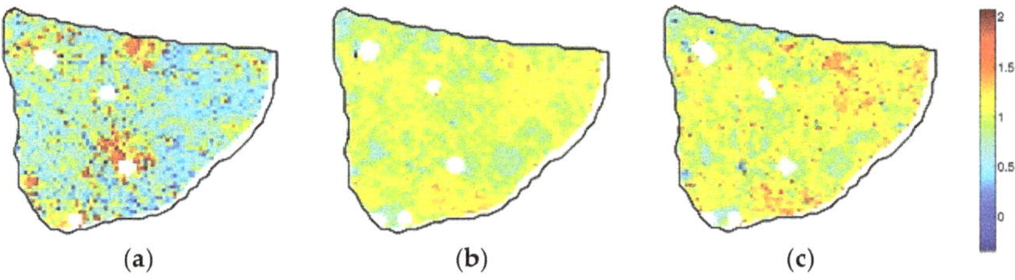

Figure 4. Vegetation Index (TCARI) for a sample parcel. Dates are: (**a**) 26 June 2014; (**b**) 1 November 2014; and (**c**) 14 November 2014.

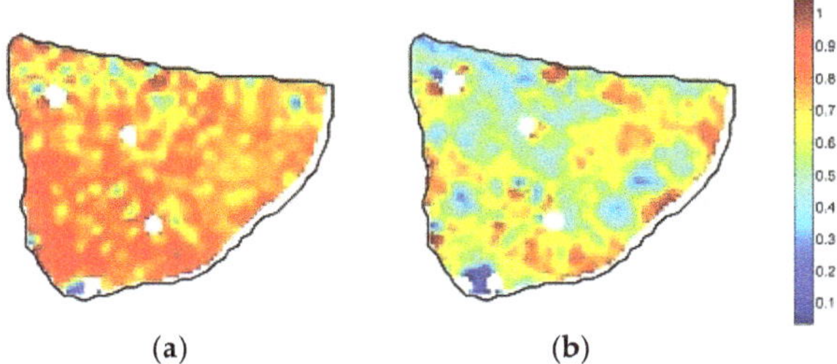

Figure 5. Sum average of band 8 (b8_avg) for a sample parcel. Dates are: (**a**) 30 May 2014; (**b**) 1 November 2014.

3.2. Base Classifiers and Ensembles

The accuracy of the 20 base classifiers created for each classification method is assessed using ground truth data. Table 4 lists the number of pixels per crop class used for the training and testing phase.

Table 4. Number of pixels per crop class for training base classifiers and assessing accuracy (testing).

Class	Crop Name	# Pixels	
		Training	Testing
1	Maize	395	234
2	Millet	531	309
3	Peanut	276	168
4	Sorghum	472	291
5	Cotton	455	256
	Total	**2129**	**1258**

Figure 6 illustrates the mean performance of all base classifiers as a boxplot. The mean OA of each classifier method ranges between 59% and 72%. SVMR obtained higher accuracy than SVMP and SVML [26,71]. Lower accuracy of SVML means that linear decision boundaries are not suitable for classifying patterns in this data [72]. RF had slightly better performance than SVMR. This result is consistent with [58]. MaxEnt presented the lowest performance confirming the need for more research before it can be operationally used in multi-class classification contexts [73].

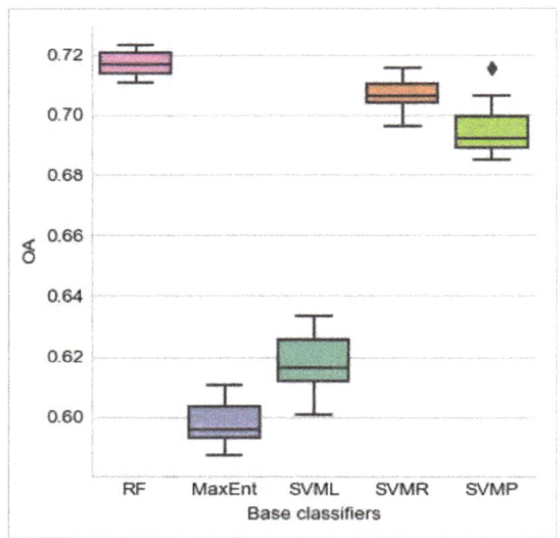

Figure 6. Boxplot of overall accuracy (OA) of base classifiers.

A comparison between the performance of base classifiers and ensembles was carried out. Thus, Table 5 summarizes minimum, mean, and maximum overall accuracy and kappa coefficient for both base classifiers and ensembles. We observe that ensemble classifiers in all cases outperform base classifiers with a rate of improvement ranging from 5.15% to 29.50%. On average, majority voting provides an accuracy that is 2.45% higher than that of the best base classifier (RF). Improvements are higher, at 4.65%, when a weighted voting rule is applied. This is because more effective base classifiers have more influence (weight) in the rule created to combine their outputs. Table 5 also reports associated statistics for kappa, but these values should be considered carefully [74].

Table 5. Summary statistics for the overall accuracy and kappa coefficient of base classifiers and ensembles. Maximum Entropy Model (MaxEnt). Random Forest (RF). Support Vector Machine (SVM) with linear kernel (SVML). SVM with polynomial kernel (SVMP). SVM with Gaussian kernel (SVMR). Majority voting (Voting). Weighted majority voting (WVoting). In bold, the maximum OA (mean) and the maximum kappa (mean) for base classifiers and ensembles.

		OA				Kappa			
	Classifier	**Mean**	**Std**	**Min**	**Max**	**Mean**	**Std**	**Min**	**Max**
Base Classifier	MaxEnt	0.5975	0.0078	0.5874	0.6105	0.4913	0.0098	0.4785	0.5070
	RF	**0.7172**	0.0041	0.7107	0.7234	**0.6412**	0.0050	0.6333	0.6480
	SVML	0.6176	0.0095	0.6010	0.6335	0.5165	0.0119	0.4958	0.5361
	SVMP	0.6951	0.0092	0.6852	0.7154	0.6151	0.0114	0.6029	0.6401
	SVMR	0.7069	0.0048	0.6963	0.7154	0.6294	0.0058	0.6172	0.6398
Ensemble	Voting	0.7348	0.0060	0.7059	0.7464	0.6642	0.0075	0.6279	0.6788
	WVoting	**0.7506**	0.0060	0.7234	0.7607	**0.6841**	0.0076	0.6497	0.6969

The number of classifiers to build an ensemble was analyzed. In Figure 7, the mean and standard deviation of the OA is presented for each ensemble size. The weighted voting scheme outperforms the simple majority voting. The accuracy of the ensembles increases as the number of classifiers grows. However, maximum accuracy is reached when the number of classifiers is 75 for weighted voting and 45 for majority voting. This means that the majority voting approach tends to saturate with fewer classifiers than the weighted majority voting approach. The standard deviation shows a decreasing trend because as the size of the ensemble increases, results become more stable. These results are congruent with the theoretical basis of ensemble learning [29,39].

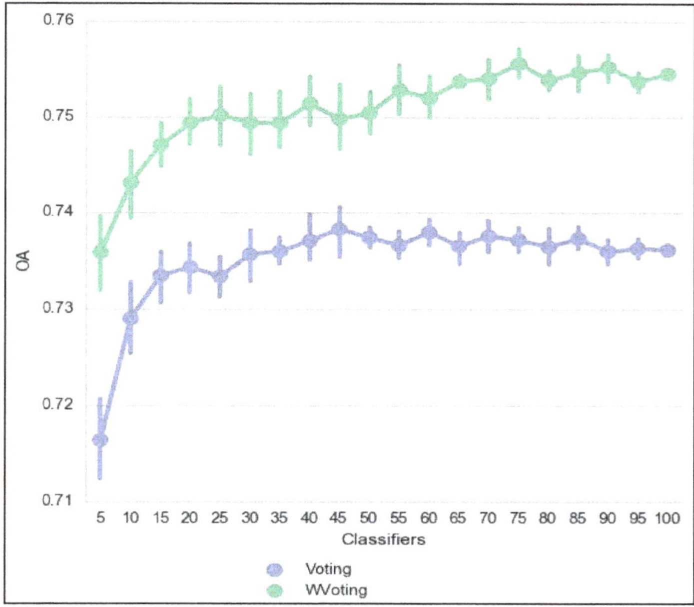

Figure 7. Mean and standard deviation of the overall accuracy using majority voting and weighted majority voting.

We contrast results of an ensemble sized 75 (hereafter called ensemble-75) with results obtained by an instance of RF because it had the best performance among base classifiers. Also, we compared the performance of ensemble-75 with the ensemble composed of 100 classifiers (hereafter called ensemble-100). OA for ensemble-75 is 0.7591, our chosen RF has an OA of 0.7170 and ensemble-100 has 0.7543. In Table 6, we present the confusion matrix obtained for the selected RF.

Table 6. Confusion matrix applying a base RF classifier, PA: Producer accuracy per class.

	Maize	Millet	Peanut	Sorghum	Cotton	PA
Maize	140	30	5	44	15	0.5983
Millet	14	239	16	33	7	0.7735
Peanut	10	17	109	24	8	0.6488
Sorghum	18	23	8	224	18	0.7698
Cotton	13	26	6	21	190	0.7422

Table 7 shows the confusion matrix of the selected ensemble and in Table 8 results of applying 100 classifiers are presented.

Table 7. Confusion matrix applying an ensemble of 75 classifiers. PA: Producer accuracy per class.

	Maize	Millet	Peanut	Sorghum	Cotton	PA
Maize	167	23	3	32	9	0.7137
Millet	19	243	13	28	6	0.7864
Peanut	5	19	120	21	3	0.7143
Sorghum	21	16	6	230	18	0.7904
Cotton	17	25	4	15	195	0.7617

Table 8. Confusion matrix applying an ensemble of 100 classifiers. PA: Producer accuracy per class.

	Maize	Millet	Peanut	Sorghum	Cotton	PA
Maize	163	25	3	34	9	0.6966
Millet	18	245	13	27	6	0.7929
Peanut	5	21	117	22	3	0.6964
Sorghum	22	13	7	229	20	0.7869
Cotton	18	24	4	15	195	0.7617

Regarding the comparison between the performance of ensemble-75 and ensemble-100, we notice that ensemble-100 has a slightly lower OA and ensemble-75 produces better results in four of five crops. The improvement of ensemble-100 in Millet is only 0.82%, whereas there is no difference in Cotton. Sorghum, Maize, and Peanut display a lower performance with 0.43%, 2.39%, and 2.5% respectively. This means that the maximum accuracy is obtained when 75 classifiers are combined, and that addition of more classifiers does not improve the performance of ensembles.

Figure 8 presents example fields to illustrate the classification results produced by ensemble-75, ensemble-100, and the selected RF. We extracted only the fields where ground truth data was available. We observe that in both ensembles, millet is less confused with peanut and cotton than in the RF classification. Cotton is less confused with sorghum as well. Besides, confusion between maize and sorghum is lower in the ensembles than in RF. This is also true for millet. Misclassifications could obey to differences in management activities in those fields (i.e., weeding) because multiple visits by various team confirmed that a single crop was grown. Moreover, by visual analysis, it can be observed that a map produced by an ensemble seems less heterogeneous than the map produced by a base classifier (RF). Differences between maps produced by ensemble-75 and ensemble-100 are visually hardly noticeable.

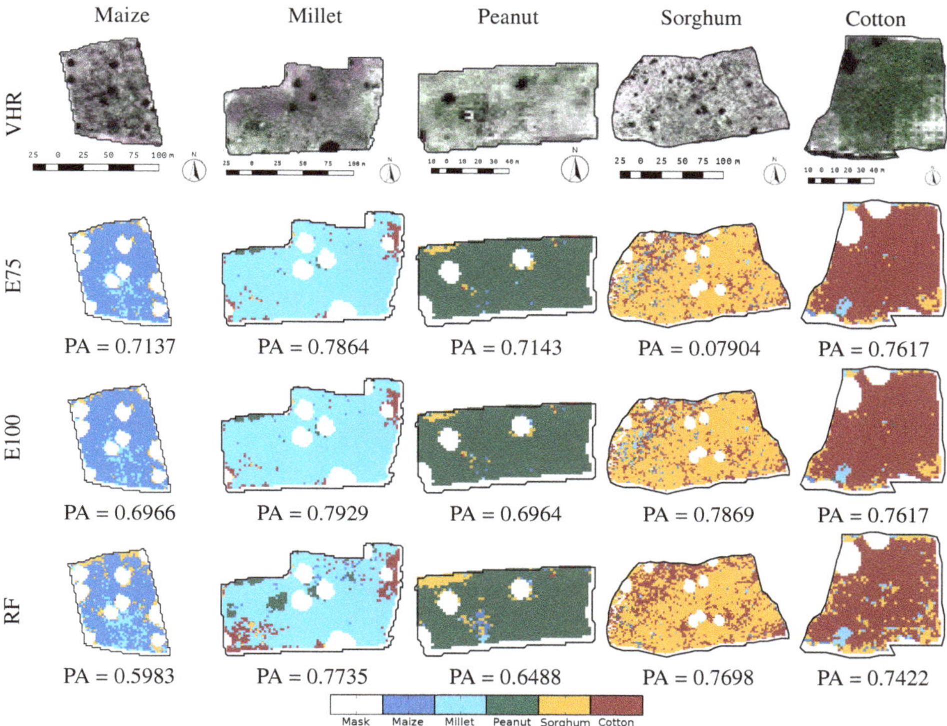

Figure 8. Comparison between field classifications produced by a 75-classifiers ensemble (E75), the 100-classifiers ensemble (E100), and a random forest classifier (RF). PA: Accuracy per class is listed below each crop. Mask corresponds to trees inside fields or clouds. VHR: overlapping area in a World-View2 image on 7 July 2014 using natural color composite.

4. Conclusions and Future Work

Reliable crop maps are fundamental to address current and future resource requirements. They support better agricultural management and consequently lead to enhanced food security. In a smallholder farming context, the production of reliable crop maps remains highly relevant because reported methods and techniques applied successfully to medium and lower spatial resolution images do not necessarily achieve the same success in heterogeneous environments. In this study, we introduced and tested a novel, and cloud-based ensemble method to map crops using a wide array of spectral and spatial features extracted from time series of very high spatial resolution images. The experiments carried out demonstrated the potential of ensemble classifiers to map crops grown by West African smallholders. The proposed ensemble obtained a higher overall accuracy (75.9%) than any individual classifier. This represents an improvement of 4.65% in comparison with the average overall accuracy values (71.7%) of the best base classifier tested in this study (random forest). The improvements over other tested classifiers like linear support vector machines and maximum entropy are larger, at 21.5% and 25.6% respectively. As theoretically expected, the weighted majority voting approach outperformed majority voting. A maximum performance was reached when the number of classifiers was 75. This indicates that at a certain point the addition of more classifiers does not lead to improvement of the classification results.

From a technical point of view, it is important to note that the generation of spectral and spatial features as well as the optimal use of ensemble learning, demand high computational capabilities. Today's approaches to image processing (big data and cloud-based) allow this concern to be overcome and hold promise for practitioners (whether academic or industrial) in developing nations, as the historic setting has often confronted them with technical barriers that were hard to overcome. Data availability, computer hardware, software, or internet bandwidth have often been in the way of a more prominent uptake of remote sensing based solutions. These barriers are slowly eroding, and opportunities are arising as a consequence. In our case, GEE was helpful in providing computational capability for data preparation and allowed the systematic creation and training of up to 100 classifiers and their combinations. Further work to extend this study includes the classification of other smallholder areas in sub-Saharan African, and the addition of new images such as Sentinel-1 and -2 as time series.

Author Contributions: R.A., R.Z.-M., E.I.-V. and R.A.d.B. conceptualized the study and designed the experiments. R.A. performed the experiments, most of the analysis and prepared the first draft of the manuscript. R.Z.-M. reviewed, expanded and edited the manuscript. E.I.-V. performed the feature selection, prepared some illustrations and reviewed the manuscript. R.A.d.B. reviewed and edited the final draft of the manuscript.

Funding: This research was partially funded by Bill and Melinda Gates Foundation via the STARS Grant Agreement (1094229-2014).

Acknowledgments: We are grateful to the four reviewers for their constructive criticism on earlier drafts, which helped to improve the paper. We wish to express our gratitude to all the STARS partners and, in particular, to the ICRISAT-led team for organizing and collecting the required field data in Mali and to the STARS ITC team for pre-processing the WorldView-2 images. We express our gratitude also towards the GEE developer team for their support and timely answers to our questions.

Conflicts of Interest: The authors declare no conflict of interest.

Appendix A. Textural Features Formulas

Table A1 lists textural features from [51] with their corresponding formulas; in these, we have used the following notational conventions:

$p(i,j)$ is the (*i,j*)th entry in a normalized gray tone matrix,

$p_x(i) = \sum_{j=1}^{N_g} P(i,j)$ is the *i*th entry in the marginal-probability matrix computed by summing the rows of $p(i,j)$, for fixed *i*,

$p_x(j) = \sum_{j=1}^{N_g} P(i,j)$, is the *j*th entry in the marginal-probability matrix computed by summing the columns of $p(i,j)$, for fixed *j*,

N_g, is the number of distinct gray levels in the quantized image,

$p_{x+y}(k) = \sum_{i=1}^{N_g} \sum_{j=1}^{N_g} p(i,j)_{i+j=k}$, and $p_{x-y}(k) = \sum_{i=1}^{N_g} \sum_{j=1}^{N_g} p(i,j)_{|i-j|=k}$

Table A2 specifies names of the textural features proposed by [52], and their formulas, in which the following notation is used:

$s(i,j,\delta,T)$ is the (*i,j*)th entry in a normalized gray level matrix, equivalent to *p(i,j)*,

T represents the region and shape used to estimate the second order probabilities, and

$\delta = (\Delta x, \Delta y)$ is the displacement vector.

Table A1. Textural feature formulas from Gray Level Co-occurrence Matrix, as described in [51].

Name/Formula	Name/Formula
Angular Second Moment $f_1 = \sum_{i=1}^{N_g}\sum_{j=1}^{N_g}\{p(i,j)\}^2$	*Contrast* $f_2 = \sum_{n=0}^{N_{g-1}} n^2\left\{\sum_{i=1}^{N_g}\sum_{j=1}^{N_g} p(i,j)_{\lvert i-j\rvert = n}\right\}$
Correlation $f_3 = \sum_{i=1}^{N_g}\sum_{j=1}^{N_g}\frac{(i,j)p(i,j)-\mu_x\mu_y}{\sigma_x\sigma_y}$	*Variance* $f_4 = \sum_{i=1}^{N_g}\sum_{j=1}^{N_g}(i-\mu)^2 p(i,j)$
Inverse Difference Moment $f_5 = \sum_{i=1}^{N_g}\sum_{j=1}^{N_g}\frac{1}{1+(i-j)^2}p(i,j)$	*Sum Average* $f_6 = \sum_{i=2}^{2N_g} i p_{x+y}(i)$
Sum Variance $f_7 = \sum_{i=2}^{2N_g}(i-f_8)^2 P_{x+y}(i)$	*Sum Entropy* $f_8 = -\sum_{i=2}^{2N_g} p_{x+y}(i)log\{p_{x+y}(i)\}$
Entropy $f_8 = -\sum_{i=2}^{2N_g} p_{x+y}(i)log\{p_{x+y}(i)\}$	*Difference Variance* f_{10} = variance of p_{x-y}
Difference Entropy $f_{11} = \sum_{i=0}^{N_{g-1}} p_{x-y}(i)log\{p_{x-y}(i)\}$	*Information Measures of Correlation 1* $f_{12} = \frac{HXY-HXY1}{\max\{HX,HY\}}$ **where,** $HXY = -\sum_{i=1}^{N_g}\sum_{j=1}^{N_g} p(i,j)\log(p(i,j))$ $HXY1 = -\sum_{i=1}^{N_g}\sum_{j=1}^{N_g} p(i,j)\log\{p_x(i)p_y(j)\}$ HX and HY are entropies of p_x and p_y
Information Measures of Correlation 2 $f_{13} = \left(1-e^{[-2.0(HXY2-HXY)]}\right)^{1/2}$, where $HXY2 = -\sum_{i=1}^{N_g}\sum_{j=1}^{N_g} p_x(i)p_y(j)\log\{p_x(i)p_y(j)\}$	*Maximal Correlation Coefficient* f_{14} = (sec ond largest eigen value of Q)$^{\frac{1}{2}}$ where $Q_{(i,j)} = \sum_{k=0}^{N_{g-1}}\frac{p(i,k)p(j,k)}{p_x(i)p_y(k)}$
Dissimilarity $f_{15} = \sum_{i=1}^{N_g}\sum_{j=1}^{N_g}\lvert i-j\rvert^2 p(i,j)$	

Table A2. Textural features included in the classification as described in [52].

Description	Formula
Inertia	$I(\delta,T) = \sum_{i=0}^{L-1}\sum_{j=0}^{L-1}(i-j)^2 s(i,j,\delta,T)$
Cluster shade	$A(\delta,T) = \sum_{i=0}^{L-1}\sum_{j=0}^{L-1}\left(i+j-\mu_i-\mu_j\right)^3 s(i,j,\delta,T)$
Cluster prominence	$B(\delta,T) = \sum_{i=0}^{L-1}\sum_{j=0}^{L-1}\left(i+j-\mu_i-\mu_j\right)^4 s(i,j,\delta,T)$

References

1. Lowder, S.K.; Skoet, J.; Singh, S. What do We Really Know about the Number and Distribution of Farms and Family Farms in the World? FAO: Rome, Italy, 2014.
2. African Development Bank, Organisation for Economic Co-operation and Development, United Nations Development Programme. *African Economic Outlook 2014: Global Value Chains and Africa's Industrialisation*; OECD Publishing: Paris, France, 2014.

3. STARS-Project. About Us—STARS Project, 2016. Available online: http://www.stars-project.org/en/about-us/ (accessed on 1 June 2016).
4. Haub, C.; Kaneda, T. World Population Data Sheet, 2013. Available online: http://auth.prb.org/Publications/Datasheets/2013/2013-world-population-data-sheet.aspx (accessed on 6 March 2017).
5. Atzberger, C. Advances in Remote Sensing of Agriculture: Context Description, Existing Operational Monitoring Systems and Major Information Needs. *Remote Sens.* **2013**, *5*, 949–981. [CrossRef]
6. Wu, B.; Meng, J.; Li, Q.; Yan, N.; Du, X.; Zhang, M. Remote sensing-based global crop monitoring: Experiences with China's CropWatch system. *Int. J. Digit. Earth* **2014**, 113–137. [CrossRef]
7. Khan, M.R. *Crops from Space: Improved Earth Observation Capacity to Map Crop Areas and to Quantify Production*; University of Twente: Enschede, The Netherlands, 2011.
8. Debats, S.R.; Luo, D.; Estes, L.D.; Fuchs, T.J.; Caylor, K.K. A generalized computer vision approach to mapping crop fields in heterogeneous agricultural landscapes. *Remote Sens. Environ.* **2016**, *179*, 210–221. [CrossRef]
9. Waldner, F.; Canto, G.S.; Defourny, P. Automated annual cropland mapping using knowledge-based temporal features. *ISPRS J. Photogramm. Remote Sens.* **2015**, *110*, 1–13. [CrossRef]
10. Foody, G.M.; Mathur, A. Toward intelligent training of supervised image classifications: Directing training data acquisition for SVM classification. *Remote Sens. Environ.* **2004**, *93*, 107–117. [CrossRef]
11. Beyer, F.; Jarmer, T.; Siegmann, B.; Fischer, P. Improved crop classification using multitemporal RapidEye data. In Proceedings of the 2015 8th International Workshop on the Analysis of Multitemporal Remote Sensing Images (Multi-Temp), Annecy, France, 22–24 July 2015; pp. 1–4.
12. Camps-Valls, G.; Gomez-Chova, L.; Calpe-Maravilla, J.; Martin-Guerrero, J.D.; Soria-Olivas, E.; Alonso-Chorda, L.; Moreno, J. Robust support vector method for hyperspectral data classification and knowledge discovery. *IEEE Trans. Geosci. Remote Sens.* **2004**, *42*, 1530–1542. [CrossRef]
13. Tatsumi, K.; Yamashiki, Y.; Torres, M.A.C.; Taipe, C.L.R. Crop classification of upland fields using Random forest of time-series Landsat 7 ETM+ data. *Comput. Electron. Agric.* **2015**, *115*, 171–179. [CrossRef]
14. Yang, C.; Everitt, J.H.; Murden, D. Evaluating high resolution SPOT 5 satellite imagery for crop identification. *Comput. Electron. Agric.* **2011**, *75*, 347–354. [CrossRef]
15. Sweeney, S.; Ruseva, T.; Estes, L.; Evans, T. Mapping Cropland in Smallholder-Dominated Savannas: Integrating Remote Sensing Techniques and Probabilistic Modeling. *Remote Sens.* **2015**, *7*, 15295–15317. [CrossRef]
16. Lu, D.; Weng, Q. A survey of image classification methods and techniques for improving classification performance. *Int. J. Remote Sens.* **2007**, *28*, 823–870. [CrossRef]
17. Waldner, F.; Li, W.; Weiss, M.; Demarez, V.; Morin, D.; Marais-Sicre, C.; Hagolle, O.; Baret, F.; Defourny, P. Land Cover and Crop Type Classification along the Season Based on Biophysical Variables Retrieved from Multi-Sensor High-Resolution Time Series. *Remote Sens.* **2015**, *7*, 10400–10424. [CrossRef]
18. Jackson, R.D.; Huete, A.R. Interpreting vegetation indices. *Prev. Vet. Med.* **1991**, *11*, 185–200. [CrossRef]
19. Peña-Barragán, J.M.; Ngugi, M.K.; Plant, R.E.; Six, J. Object-based crop identification using multiple vegetation indices, textural features and crop phenology. *Remote Sens. Environ.* **2011**, *115*, 1301–1316. [CrossRef]
20. Rao, P.V.N.; Sai, M.V.R.S.; Sreenivas, K.; Rao, M.V.K.; Rao, B.R.M.; Dwivedi, R.S.; Venkataratnam, L. Textural analysis of IRS-1D panchromatic data for land cover classification Textural analysis of IRS-1D panchromatic data for land cover classication. *Int. J. Remote Sens.* **2002**, *2317*, 3327–3345. [CrossRef]
21. Shaban, M.A.; Dikshit, O. Improvement of classification in urban areas by the use of textural features: The case study of Lucknow city, Uttar Pradesh. *Int. J. Remote Sens.* **2001**, *22*, 565–593. [CrossRef]
22. Chellasamy, M.; Zielinski, R.T.; Greve, M.H. A Multievidence Approach for Crop Discrimination Using Multitemporal WorldView-2 Imagery. *IEEE J. Sel. Top. Appl. Earth Obs. Remote Sens.* **2014**, *7*, 3491–3501. [CrossRef]
23. Hu, Q.; Wu, W.; Song, Q.; Lu, M.; Chen, D.; Yu, Q.; Tang, H. How do temporal and spectral features matter in crop classification in Heilongjiang Province, China? *J. Integr. Agric.* **2017**, *16*, 324–336. [CrossRef]
24. Misra, G.; Kumar, A.; Patel, N.R.; Zurita-Milla, R. Mapping a Specific Crop-A Temporal Approach for Sugarcane Ratoon. *J. Indian Soc. Remote Sens.* **2014**, *42*, 325–334. [CrossRef]

25. Khobragade, N.A.; Raghuwanshi, M.M. Contextual Soft Classification Approaches for Crops Identification Using Multi-sensory Remote Sensing Data: Machine Learning Perspective for Satellite Images. In *Artificial Intelligence Perspectives and Applications*; Silhavy, R., Senkerik, R., Oplatkova, Z.K., Prokopova, Z., Silhavy, P., Eds.; Springer International Publishing: Cham, Switzerland, 2015; pp. 333–346.
26. Oommen, T.; Misra, D.; Twarakavi, N.K.C.; Prakash, A.; Sahoo, B.; Bandopadhyay, S. An Objective Analysis of Support Vector Machine Based Classification for Remote Sensing. *Math Geosci.* **2008**, *40*, 409–424. [CrossRef]
27. Gómez, C.; White, J.C.; Wulder, M.A. Optical remotely sensed time series data for land cover classification: A review. *ISPRS J. Photogramm. Remote Sens.* **2016**, *116*, 55–72. [CrossRef]
28. Wozniak, M.; Graña, M.; Corchado, E. A survey of multiple classifier systems as hybrid systems. *Inf. Fusion* **2014**, *16*, 3–17. [CrossRef]
29. Kuncheva, L.I. *Combining Pattern Classifiers: Methods and Algorithms*; John Wiley & Sons, Inc.: Hoboken, NJ, USA, 2004.
30. Gopinath, B.; Shanthi, N. Development of an Automated Medical Diagnosis System for Classifying Thyroid Tumor Cells using Multiple Classifier Fusion. *Technol. Cancer Res. Treat.* **2014**, *14*, 653–662. [CrossRef] [PubMed]
31. Li, H.; Shen, F.; Shen, C.; Yang, Y.; Gao, Y. Face Recognition Using Linear Representation Ensembles. *Pattern Recognit.* **2016**, *59*, 72–87. [CrossRef]
32. Lumini, A.; Nanni, L.; Brahnam, S. Ensemble of texture descriptors and classifiers for face recognition. *Appl. Comput. Inform.* **2016**, *13*, 79–91. [CrossRef]
33. Clinton, N.; Yu, L.; Gong, P. Geographic stacking: Decision fusion to increase global land cover map accuracy. *ISPRS J. Photogramm. Remote Sens.* **2015**, *103*, 57–65. [CrossRef]
34. Lijun, D.; Chuang, L. Research on remote sensing image of land cover classification based on multiple classifier combination. *Wuhan Univ. J. Nat. Sci.* **2011**, *16*, 363–368.
35. Li, D.; Yang, F.; Wang, X. Study on Ensemble Crop Information Extraction of Remote Sensing Images Based on SVM and BPNN. *J. Indian Soc. Remote Sens.* **2016**, *45*, 229–237. [CrossRef]
36. Du, P.; Xia, J.; Zhang, W.; Tan, K.; Liu, Y.; Liu, S. Multiple classifier system for remote sensing image classification: A review. *Sensors (Basel)* **2012**, *12*, 4764–4792. [CrossRef] [PubMed]
37. Gargiulo, F.; Mazzariello, C.; Sansone, C. Multiple Classifier Systems: Theory, Applications and Tools. In *Handbook on Neural Information Processing*; Bianchini, M., Maggini, M., Jain, L.C., Eds.; Springer: Berlin/Heidelberg, Germany, 2013; Volume 49, pp. 505–525.
38. Corrales, D.C.; Figueroa, A.; Ledezma, A.; Corrales, J.C. An Empirical Multi-classifier for Coffee Rust Detection in Colombian Crops. In Proceedings of the Computational Science and Its Applications—ICCSA 2015: 15th International Conference, Banff, AB, Canada, 22–25 June 2015; Gervasi, O., Murgante, B., Misra, S., Gavrilova, L.M., Rocha, C.A.M.A., Torre, C., Taniar, D., Apduhan, O.B., Eds.; Springer International Publishing: Cham, Switzerland, 2015; pp. 60–74.
39. Song, X.; Pavel, M. Performance Advantage of Combined Classifiers in Multi-category Cases: An Analysis. In Proceedings of the 11th International Conference, ICONIP 2004, Calcutta, India, 22–25 November 2004; pp. 750–757.
40. Polikar, R. Ensemble based systems in decision making. *Circuits Syst. Mag. IEEE* **2006**, *6*, 21–45. [CrossRef]
41. Duin, R.P.W. The Combining Classifier: To Train or Not to Train? In Proceedings of the 16th International Conference on Pattern Recognition, Quebec City, QC, Canada, 11–15 August 2002.
42. Joint Experiment for Crop Assessment and Monitoring (JECAM). Mali JECAM Study Site, Mali-Koutiala—Site Description. Available online: http://www.jecam.org/?/site-description/mali (accessed on 18 April 2018).
43. Stratoulias, D.; de By, R.A.; Zurita-Milla, R.; Retsios, V.; Bijker, W.; Hasan, M.A.; Vermote, E. A Workflow for Automated Satellite Image Processing: From Raw VHSR Data to Object-Based Spectral Information for Smallholder Agriculture. *Remote Sens.* **2017**, *9*, 1048. [CrossRef]
44. Rouse, W.; Haas, R.H.; Deering, D.W. Monitoring vegetation systems in the great plains with ERTS. *Proc. Earth Resour. Technol. Satell. Symp. NASA* **1973**, *1*, 309–317.
45. Louhaichi, M.; Borman, M.M.; Johnson, D.E. Spatially located platform and aerial photography for documentation of grazing impacts on wheat. *Geocarto Int.* **2001**, *16*, 65–70. [CrossRef]
46. Huete, A.; Didan, K.; Miura, T.; Rodriguez, E.P.; Gao, X.; Ferreira, L.G. Overview of the radiometric and biophysical performance of the MODIS vegetation indices. *Remote Sens. Environ.* **2002**, *83*, 195–213. [CrossRef]

47. Huete, A.R. A Soil-Adjusted Vegetation Index (SAVI). *Remote Sens. Environ.* **1988**, *25*, 295–309. [CrossRef]
48. Qi, J.; Chehbouni, A.; Huete, A.R.; Kerr, Y.H.; Sorooshian, S. A modified soil adjusted vegetation index. *Remote Sens. Environ.* **1994**, *48*, 119–126. [CrossRef]
49. Haboudane, D.; Miller, J.R.; Tremblay, N.; Zarco-Tejada, P.J.; Dextraze, L. Integrated narrow-band vegetation indices for prediction of crop chlorophyll content for application to precision agriculture. *Remote Sens. Environ.* **2002**, *81*, 416–426. [CrossRef]
50. Gitelson, A.A.; Kaufman, Y.J.; Stark, R.; Rundquist, D. Novel algorithms for remote estimation of vegetation fraction. *Remote Sens. Environ.* **2002**, *80*, 76–87. [CrossRef]
51. Haralick, R.; Shanmugan, K.; Dinstein, I. Textural features for image classification. *IEEE Trans. Syst. Man Cybern.* **1973**, *3*, 610–621. [CrossRef]
52. Conners, R.W.; Trivedi, M.M.; Harlow, C.A. Segmentation of a high-resolution urban scene using texture operators. *Comput. Vis. Graph. Image Process.* **1984**, *25*, 273–310. [CrossRef]
53. Gorelick, N.; Hancher, M.; Dixon, M.; Ilyushchenko, S.; Thau, D.; Moore, R. Google Earth Engine: Planetary-scale geospatial analysis for everyone. *Remote Sens. Environ.* **2017**, *202*, 18–27. [CrossRef]
54. Izquierdo-Verdiguier, E.; Zurita-Milla, R.; de By, R.A. On the use of guided regularized random forests to identify crops in smallholder farm fields. In Proceedings of the 2017 9th International Workshop on the Analysis of Multitemporal Remote Sensing Images (MultiTemp), Brugge, Belgium, 27–29 June 2017; pp. 1–3.
55. Breiman, L. Random forests. *Mach. Learn.* **2001**, *45*, 5–32. [CrossRef]
56. Phillips, S.J.; Anderson, R.P.; Schapire, R.E. Maximum entropy modeling of species geographic distributions. *Ecol. Model.* **2006**, *190*, 231–259. [CrossRef]
57. Cortes, C.; Vapnik, V. Support Vector Networks. *Mach. Learn.* **1995**, *20*, 273–297. [CrossRef]
58. Duro, D.C.; Franklin, S.E.; Dubé, M.G. A comparison of pixel-based and object-based image analysis with selected machine learning algorithms for the classification of agricultural landscapes using SPOT-5 HRG imagery. *Remote Sens. Environ.* **2012**, *118*, 259–272. [CrossRef]
59. Gao, T.; Zhu, J.; Zheng, X.; Shang, G.; Huang, L.; Wu, S. Mapping spatial distribution of larch plantations from multi-seasonal landsat-8 OLI imagery and multi-scale textures using random forests. *Remote Sens.* **2015**, *7*, 1702–1720. [CrossRef]
60. Pal, M. Random forest classifier for remote sensing classification. *Int. J. Remote Sens.* **2005**, *26*, 217–222. [CrossRef]
61. Rodriguez-Galiano, V.F.; Ghimire, B.; Rogan, J.; Chica-Olmo, M.; Rigol-Sanchez, J.P. An assessment of the effectiveness of a random forest classifier for land-cover classification. *ISPRS J. Photogramm. Remote Sens.* **2012**, *67*, 93–104. [CrossRef]
62. Nitze, I.; Schulthess, U.; Asche, H. Comparison of Machine Learning Algorithms Random Forest, Artificial Neural Network and Support Vector Machine to Maximum Likelihood for Supervised Crop Type Classification. In Proceedings of the 4th international conference on Geographic Object-Based Image Analysis (GEOBIA) Conference, Rio de Janeiro, Brazil, 7–9 May 2012; pp. 35–40.
63. Akar, O.; Güngör, O. Integrating multiple texture methods and NDVI to the Random Forest classification algorithm to detect tea and hazelnut plantation areas in northeast Turkey. *Int. J. Remote Sens.* **2015**, *36*, 442–464. [CrossRef]
64. Berger, A.L.; della Pietra, S.A.; della Pietra, V.J. A Maximum Entropy Approach to Natural Language *Process. Comput. Linguist.* **1996**, *22*, 39–71.
65. Evangelista, P.H.; Stohlgren, T.J.; Morisette, J.T.; Kumar, S. Mapping Invasive Tamarisk (Tamarix): A Comparison of Single-Scene and Time-Series Analyses of Remotely Sensed Data. *Remote Sens.* **2009**, *1*, 519–533. [CrossRef]
66. Rahmati, O.; Pourghasemi, H.R.; Melesse, A.M. Application of GIS-based data driven random forest and maximum entropy models for groundwater potential mapping: A case study at Mehran Region, Iran. *Catena* **2016**, *137*, 360–372. [CrossRef]
67. Mountrakis, G.; Im, J.; Ogole, C. Support vector machines in remote sensing: A review. *ISPRS J. Photogramm. Remote Sens.* **2011**, *66*, 247–259. [CrossRef]
68. Izquierdo-Verdiguier, E.; Gómez-Chova, L.; Camps-Valls, G. Kernels for Remote Sensing Image Classification. In *Wiley Encyclopedia of Electrical and Electronics Engineering*; John Wiley & Sons, Inc.: Hoboken, NJ, USA, 2015; pp. 1–23.

69. Kohavi, R. A study of cross-validation and bootstrap for accuracy estimation and model selection. *Int. Jt. Conf. Artif. Intell.* **1995**, 1137–1143.
70. Smits, P.C. Multiple classifier systems for supervised remote sensing image classification based on dynamic classifier selection. *IEEE Trans. Geosci. Remote Sens.* **2002**, *40*, 801–813. [CrossRef]
71. Kavzoglu, T.; Colkesen, I. A kernel functions analysis for support vector machines for land cover classification. *Int. J. Appl. Earth Obs. Geoinf.* **2009**, *11*, 352–359. [CrossRef]
72. Hao, P.; Wang, L.; Niu, Z. Comparison of Hybrid Classifiers for Crop Classification Using Normalized Difference Vegetation Index Time Series: A Case Study for Major Crops in North Xinjiang, China. *PLoS ONE* **2015**, *10*, e0137748. [CrossRef] [PubMed]
73. Amici, V.; Marcantonio, M.; la Porta, N.; Rocchini, D. A multi-temporal approach in MaxEnt modelling: A new frontier for land use/land cover change detection. *Ecol. Inform. J.* **2017**, *40*, 40–49. [CrossRef]
74. Gilmore, R.P., Jr.; Millones, M. Death to Kappa: Birth of quantity disagreement and allocation disagreement for accuracy assessment. *Int. J. Remote Sens.* **2011**, *32*, 4407–4429.

Article

Nominal 30-m Cropland Extent Map of Continental Africa by Integrating Pixel-Based and Object-Based Algorithms Using Sentinel-2 and Landsat-8 Data on Google Earth Engine

Jun Xiong [1,2,*], Prasad S. Thenkabail [1], James C. Tilton [3], Murali K. Gumma [4], Pardhasaradhi Teluguntla [1,2], Adam Oliphant [1], Russell G. Congalton [5], Kamini Yadav [5] and Noel Gorelick [6]

1 Western Geographic Science Center, U. S. Geological Survey (USGS), 2255, N. Gemini Drive, Flagstaff, AZ 86001, USA; pthenkabail@usgs.gov (P.S.T.); pteluguntla@usgs.gov (P.T.); aoliphant@usgs.gov (A.O.)
2 Bay Area Environmental Research Institute (BAERI), 596 1st St West Sonoma, CA 95476, USA
3 Computational & Information Science and Technology Office, Mail Code 606.3, NASA Goddard Space Flight Center, Greenbelt, MD 20771, USA; james.c.tilton@nasa.gov
4 International Crops Research Institute for the Semi-Arid Tropics (ICRISAT), Patancheru, Hyderabad 502324, India; m.gumma@cgiar.org
5 Department of Natural Resources and the Environment, University of New Hampshire, 56 College Road, Durham, NH 03824, USA; russ.congalton@unh.edu (R.G.C.); kaminiyadav.02@gmail.com (K.Y.)
6 Google Inc., 1600 Amphitheater Parkway, Mountain View, CA 94043, USA; gorelick@google.com
* Correspondence: jun.xiong1981@gmail.com; Tel.: +1-928-556-7215

Received: 11 July 2017; Accepted: 10 October 2017; Published: 19 October 2017

Abstract: A satellite-derived cropland extent map at high spatial resolution (30-m or better) is a must for food and water security analysis. Precise and accurate global cropland extent maps, indicating cropland and non-cropland areas, are starting points to develop higher-level products such as crop watering methods (irrigated or rainfed), cropping intensities (e.g., single, double, or continuous cropping), crop types, cropland fallows, as well as for assessment of cropland productivity (productivity per unit of land), and crop water productivity (productivity per unit of water). Uncertainties associated with the cropland extent map have cascading effects on all higher-level cropland products. However, precise and accurate cropland extent maps at high spatial resolution over large areas (e.g., continents or the globe) are challenging to produce due to the small-holder dominant agricultural systems like those found in most of Africa and Asia. Cloud-based geospatial computing platforms and multi-date, multi-sensor satellite image inventories on Google Earth Engine offer opportunities for mapping croplands with precision and accuracy over large areas that satisfy the requirements of broad range of applications. Such maps are expected to provide highly significant improvements compared to existing products, which tend to be coarser in resolution, and often fail to capture fragmented small-holder farms especially in regions with high dynamic change within and across years. To overcome these limitations, in this research we present an approach for cropland extent mapping at high spatial resolution (30-m or better) using the 10-day, 10 to 20-m, Sentinel-2 data in combination with 16-day, 30-m, Landsat-8 data on Google Earth Engine (GEE). First, nominal 30-m resolution satellite imagery composites were created from 36,924 scenes of Sentinel-2 and Landsat-8 images for the entire African continent in 2015–2016. These composites were generated using a median-mosaic of five bands (blue, green, red, near-infrared, NDVI) during each of the two periods (period 1: January–June 2016 and period 2: July–December 2015) plus a 30-m slope layer derived from the Shuttle Radar Topographic Mission (SRTM) elevation dataset. Second, we selected Cropland/Non-cropland training samples (sample size = 9791) from various sources in GEE to create pixel-based classifications. As supervised classification algorithm, Random Forest (RF) was used as the primary classifier because of its efficiency, and when over-fitting issues of

RF happened due to the noise of input training data, Support Vector Machine (SVM) was applied to compensate for such defects in specific areas. Third, the Recursive Hierarchical Segmentation (RHSeg) algorithm was employed to generate an object-oriented segmentation layer based on spectral and spatial properties from the same input data. This layer was merged with the pixel-based classification to improve segmentation accuracy. Accuracies of the merged 30-m crop extent product were computed using an error matrix approach in which 1754 independent validation samples were used. In addition, a comparison was performed with other available cropland maps as well as with LULC maps to show spatial similarity. Finally, the cropland area results derived from the map were compared with UN FAO statistics. The independent accuracy assessment showed a weighted overall accuracy of 94%, with a producer's accuracy of 85.9% (or omission error of 14.1%), and user's accuracy of 68.5% (commission error of 31.5%) for the cropland class. The total net cropland area (TNCA) of Africa was estimated as 313 Mha for the nominal year 2015. The online product, referred to as the Global Food Security-support Analysis Data @ 30-m for the African Continent, Cropland Extent product (GFSAD30AFCE) is distributed through the NASA's Land Processes Distributed Active Archive Center (LP DAAC) as (available for download by 10 November 2017 or earlier): https://doi.org/10.5067/MEaSUREs/GFSAD/GFSAD30AFCE.001 and can be viewed at https://croplands.org/app/map. Causes of uncertainty and limitations within the crop extent product are discussed in detail.

Keywords: cropland mapping; cropland areas; 30-m; Landsat-8; Sentinel-2; Random Forest; Support Vector Machines; segmentation; RHSeg; Google Earth Engine; Africa

---

1. Introduction

Agricultural areas are changing rapidly over time and space across the world as a result of land cover change as well as climate variability. Mapping the geographical extent of croplands, their precise locations, and establishing areas of agricultural croplands is of great importance for managing food production systems and to study their inter-relationships with geo-political, socio-economic, health, environmental, and ecological issues [1]. In many food-insecure regions of the world, such as Africa, understanding and characterizing agricultural production remains a major challenge [2]. In addition, a primary requirement for agricultural cropland studies is a dependency on the availability of a precise map of cropland extent at high spatial resolution (30-m or better) as well as determining reliable and consistent cropland areas derived from these accurate maps [3]. The absence of such a product leads to great uncertainties in all higher-level cropland products resulting in poor assessment of global and local food security scenarios. Consequently, the demand for a baseline cropland extent product at high resolution and accuracy has been widely recognized [4]. Accuracy of higher-level cropland products such as cropping intensities, crop types, crop watering methods (e.g., irrigated or rainfed), planted or left fallow, crop health, crop productivity (productivity per unit of land, $kg \cdot m^{-2}$), and crop water productivity (productivity per unit of water or crop per drop, $kg \cdot m^{-3}$) are dependent on having a precise cropland extent product as a baseline product. In Africa, these products are particularly helpful due to the absence of high resolution cropland products that map field level details of croplands making them an invaluable baseline product for all higher-level products such as crop type, crop productivity, and crop water productivity [5,6].

Remote sensing has long been recognized as an effective tool for broad-scale crop mapping [7–9]. The two most applied remote sensing methods for land-cover mapping are manual classification based on visual interpretation [10] and digital per-pixel classification [11]. Although the human capacity for interpreting images is remarkable, visual interpretation is subjective, time-consuming, and expensive on large area. A number of cropland cover datasets on a global scale have been developed, mostly at a coarse resolution of 1-km [8,12–14]. Others have mapped cropland as one class in their land cover products at MODIS resolution [15–19]. However, all these studies suffered from inability to depict individual farm fields. Moving from existing products to high resolution (30-m or finer) greatly improves the ability to capture small and fragmented cultivated fields. On this topic, National Agricultural Statistics Service (NASS) of the US Department of Agriculture (USDA) produced Cropland Data Layers (CDLs) for the US using a decision tree approach based on millions ground samples generated from farmers' surveys over across the country as well as the National Land Cover Database [20]. However, such advanced operational approaches cannot be replicated in developing regions other than North America and Europe because of the lack of systematic collection of ground training samples. Alternate procedures have consisted of unsupervised approaches [13,21–23] and supervised methods in small regional areas with different classifiers including decision trees [12], Support Vector Machine [24,25], Random Forest [12], neural networks [26–28], data mining [29], and hybrid methods [30]. In order to improve classification results, the following issues were investigated in literature which include the selection of the dates [31], temporal windows derivation [32], input features selection [33] and automated classification methods [34]. Object-based approaches of crop identification have also been explored [35].

One issue that has confounded above cropland mapping efforts is how the term "cropland" has been defined. For instance, the U.S. Department of Agriculture (USDA) includes in its cropland definition "all areas used for the production of adapted crops for harvest", which covers both cultivated and non-cultivated areas [36]. In most global land cover products, such as Africover [37], GLC2000 [38], GlobCover [39], GLCShar [40], MODIS Land Cover [41] croplands are partly combined in mosaic or mixed classes including meadows and pastures [42], making them difficult to use in agricultural applications, either as agricultural masks or as a source for area estimates. In a previous effort to compile four existing global cropland into a 1-km global cropland extent map [43], cropland was defined as: "lands cultivated with plants harvested for food, feed, and fiber, include both seasonal crops (e.g., wheat, rice, corn, soybeans, cotton) and continuous plantations (e.g., coffee, tea, rubber, cocoa, oil palms). Cropland fallows are lands uncultivated during a season or a year but are farmlands that are equipped for cultivation, and hence included as part of croplands. From a remote-sensing perspective, a cropland in this study is a piece of land of minimum 0.09 ha (30-m × 30-m pixel) that is sowed/planted and harvest-able at least once within the 12 months after the sowing/planting date. The annual cropland produces an herbaceous cover and can sometimes be combined with some tree or woody vegetation. Some crops like sugarcane plantation and cassava crop are not necessarily planted yearly, but are still crops based on planting that had taken place during a previous year. Greenhouses, and aquaculture are part of the farmlands and have different signature from other croplands [44], but these are negligible in Africa. In a nutshell, the cropland extent includes: standing crops, cropland fallows and plantations.

In order to make continental scale classification feasible, new methods and approaches need to be adopted or developed to deal with the complex classification issues [45–48]. In this research, we propose this integrated method of pixel-based classification and object-based segmentation for large area cropland mapping. A number of earlier studies [49–52] have explored such integrated approaches. Pixel-based classification algorithms, such as the Random Forests (RF) and Support Vector Machines (SVM), are widely used due to their efficiency over large areas. These pixel-based clustering algorithms focus only on the spectral value of each pixel and often result in image speckle and overall inaccuracies when applied to high resolution imagery. Since each pixel is dealt with in isolation from its neighbors in the pixel-based paradigm, close neighbors often have different classes, despite being similar. When classification to produce discrete mapped entities is needed, an object-based segmentation approach can alleviate such problems [53]. For object-based classification, field boundaries can be derived either from a digital vector database [54] or by segmentation [55]. In landscapes with mixed agriculture and pastoral land cover classes (e.g., Sahelian countries), image segmentation methods seem to provide a considerable advantage, since these land cover types are structurally fairly dissimilar to non-cropland areas whereas they are spectrally similar [56].

The aim of this paper is to develop a 30-m crop extent map of continental Africa by integrating pixel-based and object-based approaches. The generic methodology is capable of handling high-resolution satellite imagery with support of cloud-based Google Earth Engine (GEE) computing platform [57]. First, creating two half-year period mosaics from 30-m Landsat-8 Operational Land Imager (OLI) Top of Atmosphere (TOA) and 10-m to 20-m Sentinel-2 multi-spectral instrument (MSI) L1C product from 2015 to 2016; Second, two pixel-based supervised classifiers (Random Forest and Support Vector Machines) are applied to input dataset to obtain pixel-based classification; Third, object-based segmentations from Recursive Hierarchical Image Segmentation (RHSeg) were introduced to improve the pixel-based classification. Fourth, the study will compare the cropland areas determined using 30-m cropland product for the nominal 2015 with other statistics such as from the Food and Agricultural Organization (FAO) of the United Nations (UN). This research is a part of the Global Food Security-Support Analysis Data Project at 30-m (GFSAD30).

2. Materials

2.1. Study Area

We have chosen the entire continent of Africa for the study area (Figure 1), which extends from approximately 38° N to 35° S , occupies 3037 million hectares (Mha), and has 7 distinct geologic and bio-geographic regions with varying land cover types [58]. Demographic changes of continental Africa are expected to be staggering in the 21st Century with population expected to increase from the current 1.2 billion to over 4 billion by the end of the Century [59]. Africa is endowed with a wide diversity of agro-ecological zones. These zones range from the heavy rain-forest vegetation with bi-annual rainfall to relatively sparse, dry and arid vegetation with low unimodal rainfall. This diversity is a tremendous asset, but it also poses a substantial challenge for agricultural development. On the one hand, it creates a vast potential with respect to the mix of agricultural commodities and products which can be produced and marketed in domestic and external markets. On the other hand, the diversity implies that there are no continent-wide uniform solutions to agricultural developmental problems across the continent. Thereby, a precise and accurate cropland extent map of Africa is of significant importance to study crop dynamics, water security, and food security.

2.2. Cloud-Free Satellite Imagery Composition at 30-m Resolution

Concurrent availability of 10-day Sentinel-2 and 16-day Landsat-8 data provides an unprecedented opportunity to gather high resolution data for global cropland mapping. Sentinel-2 data become available for Africa and Europe in the middle of 2015 [60] and its capabilities to map crop types and tree species have been assessed [34,61–64] and its 10-m and 20-m data provides much more details than Landsat 30-m data. The easy and simultaneous access to entire archive of Sentinel-2 and Landsat-8 products through GEE, as well as the fast and scalable computational tools that it offers, makes GEE an essential and powerful tool for this project. Creating 30-m cloud-free imagery composition for entire Africa is a challenging task because of west and east African monsoon [65], which cause constant clouds in Gulf of Guinea and east Mozambique most of the time. As a result, we established the Sentinel-2 composite during two periods (period 1: January–June 2016, and period 2: July–December 2015), each coinciding with the two main crop growing seasons [24,44,66–69] in Africa. All data were resampled to 30-m using the average value of all involved pixels. However, data-gaps still existed in small portions of the continent due to cloud and haze issues after this composition. For these gaps, Landsat multi-bands (Table 1) with similar wavelength range as Sentinel-2 MSI were used as supplementary data for gap-filling, to make sure this 30-m wall-to-wall continental mosaic was cloud-free. In the end, a total of 36,924 images (20,214 from Sentinel-2 and 16,710 from Landsat 8) were queried from GEE data-pool and was used in this study.

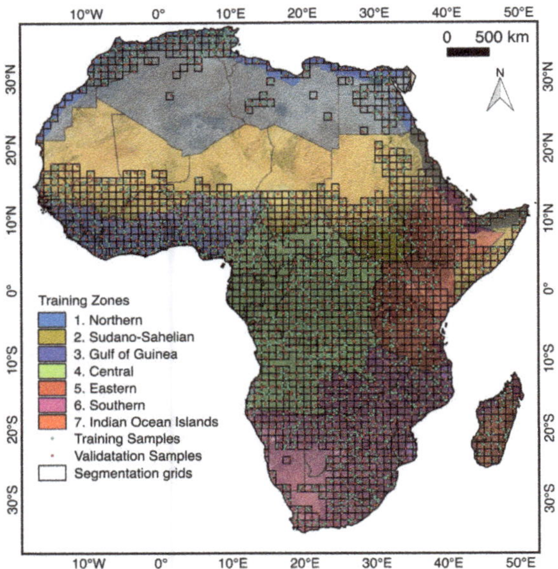

Figure 1. Map of Africa and its seven stratified zones used in the study. The pixel-based supervised classifications were run separately for each of the 7 refined agro-ecological zones (RAEZs, or simply referred to as zones: Northern, Sudano-Sahelian, Gulf of Guinea, Central, Eastern, Southern, Indian Ocean Islands). The object-based segmentation was run on each $1° \times 1°$ grids to delineate crop field boundaries. The dots represent the location of the reference training and validation samples (green indicates training and red indicates validation).

Table 1. Characteristics of Sentinel-2 MSI and Landsat-8 OLI bands used in this study. Five bands were used for each of the two periods plus slope layer from SRTM (Total 11 bands for input of classification).

Sensors	Period	Band	Use	Wavelength	Resolution	Provider
Sentinel-2 MSI Level-1C, TOA	Period 1: January – June, 2016; Period 2: July – December, 2015	B2	Blue	490 nm	10 m	ESA
		B3	Green	560 nm	10 m	
		B4	Red	665 nm	10 m	
		B8A	Near Infrared	856 nm	20 m	
		NDVI			10 m	
Landsat 8 OLI TOA	Period 1: January – June, 2016; Period 2: July – December, 2015	B2	Blue	450 – 510 nm	30 m	USGS
		B3	Green	530 – 590 nm	30 m	
		B4	Red	640 – 670 nm	30 m	
		B5	Near Infrared	850 – 880 nm	30 m	
		NDVI			30 m	
Shuttle Radar Topography Mission (SRTM) 30 m		Slope			30 m	NASA/USGS

The gap-filling of Sentinel-2 data with Landsat-8 data poses some technical challenges and requires imagery to be harmonized. The platforms and sensors differ in their orbital, spatial, and spectral configuration. As a consequence, measured physical values and radiometric attributes of the imagery are affected. For example, a root mean square error (RMSE) greater than 8% in the red band was found when comparing Sentinel and Landsat simulated data, due to the discrepancies in the nominal relative spectral response functions (RSRF) [70]. Werff compared Sentinel-2A MSI and Landsat-8 OLI Data [71], finding the correlation of their TOA reflectance products is higher than their bottom-of-atmosphere reflectance products. Besides, the combined use of multi-temporal images requires an accurate geometric registration, i.e., pixel-to-pixel correspondence for terrain-corrected products. Both systems are designed to register Level 1 products to a reference image framework. However, the Landsat-8 framework, based upon the Global Land Survey images, contains residual geolocation errors leading to an expected sensor-to-sensor misregistration of 38-m [72]. This is because although both sensor geolocation systems use parametric approaches, whereby information concerning the sensing geometry is modeled and the sensor exterior orientation parameters (altitude and position) are measured, they use different ground control and digital elevation models to refine the geolocation [72,73]. These misalignments vary geographically but should be stable for a given area. A study demonstrates that sub-pixel accuracy was achieved between 10-m resolution Sentinel-2 bands (band 3) and 15-m resolution panchromatic Landsat images (band 8) [74]. We determined that the mismatch between the geo-referencing of Landsat and Sentinel is within 30-m by comparing multiple ground control points from obvious, well-recognized locations on the land when both sensors images were available. Sentinel-2 has two NIR bands B8 (10-m) and B8A (20-m); B8 is consistently lower than B8A due to different gain settings. The B8A band was used here for 'NIR band' because it matched Landsat data better.

For each period (January–June 2016, July–December 2015) 5 bands (blue, green, red, NIR and an NDVI band (Table 1, Figure 2) were composited using Sentinel-2 and Landsat-8 combined. First, TOA reflectance values of Sentinel-2 were calculated and mosaicked using median values to create layer stacks for each of the seasons separately. Wherever "data-gaps" found, they were identified and filled using Landsat-8 data. Note that each of the 5 bands were composited over each of the two periods so we had total 10 bands from two time periods. In addition, we derived a slope surface from the Shuttle Radar Topography Mission (SRTMGL1v3) [75] digital elevation at one arc-sec (approximately 30-m) resolution dataset. These 11 bands were organized as a GEE Image Collection object, which provided a programmable way to run classification algorithms such as RF and SVM deployed on GEE.

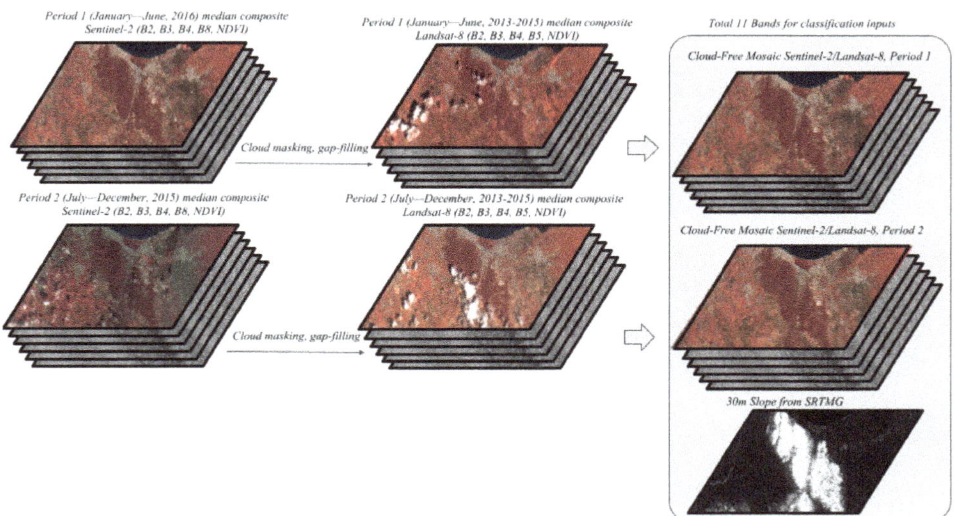

Figure 2. Illustration of the 11 input layers in this study. Five bands (Blue, Green, Red, NIR, NDVI) were composited for each of two periods (period 1: January–June 2016, and period 2: July–December 2015) for entire African continent. Total 10 bands plus the slope layer derived from SRTM elevation data were composed on GEE for pixel-based classification, 10 bands without slope layer were used for object-based segmentation.

It is noteworthy that half-yearly (January–June, July–December) composites were a measure of expediency in attaining wall-to-wall cloud-free mosaics over such a large area as Africa. Because of cloud cover, bimonthly mosaics or even trimonthly mosaics contain areas with no-data. In order to create wall-to-wall cloud-free mosaics composites of Africa, we determined through experiments that only 2 composites per year (half-yearly as mentioned above) could be reliably generated. Data availability is much higher in some areas (e.g., North Africa) than others (central Africa), which means we could set shorter composite periods in these regions, to get more composite periods within one year. However, it also makes the number of inputs bands different between regions, which increases difficulty in operational processing. Eventually, nominal 30-m cloud-free imagery were generated for the entire African continent for two main crop growing periods (January–June 2016; July–December 2015).

2.3. Reference Training Samples

We obtained reference training data (Figure 1) from following reliable sources in addition to our own collections. First, we randomly distributed 10,000 data points across land of continental Africa. We assessed these samples to ensure that they represent homogeneous cropland or non-cropland classes in a 90-m × 90-m sample frame using National Geospatial Agency (NGA) sub-meter to 5-m imagery. We removed some heterogeneous (e.g., cropland mixed with non-cropland) samples. After using 5511 randomly generated samples to train the pixel-based classification algorithm (RF and SVM), another 4280 polygons were appended which were placed by the analyst. In the end, there were 9791 training samples that were either croplands or non-croplands derived from VHRI spread across Africa. The attributes, ground photos or satellite images of these training samples are accessible through https://web.croplands.org/app/data/search.

The reference training data were then used to generate a cropland versus non-cropland knowledge-base for the algorithms. For an example of zone 1 sample sets, reflectance values of non-croplands were much higher than cropland samples (e.g., Figure 3), especially for Band 4 (red), Band 3 (green) and Band 2 (blue), while NDVI values of croplands were much higher than non-croplands. In this sample area (Figure 3), period 1 (January–June 2016) reflectance was significantly higher than period 2 (July–December 2015) due to greater intensity of cropping during this period in this sample area as well as types of crops grown during period 1 compared to period 2 (Figure 2). However, this may change in other areas depending on crop dynamics. This knowledge was used in training, classifying, and separating croplands from non-croplands in the the pixel-based supervised classification algorithms (RF, SVM).

2.4. Reference Validation Sample Polygons

Reference validation samples were also collected using similar approach as in Section 2.3. Spatial distribution of the validation data is shown in Figure 1. The validation data was hidden from the map producers and was made available only to independent accuracy assessment team. These reference datasets are publicly available for download at: https://doi.org/10.5067/MEaSUREs/GFSAD/GFSAD30AFCE.001. Accuracy error matrices were established for each of the 7 refined agro-ecological zones or RAEZs (Figure 1) separately as well as for the entire African continent. A total of 1754 validation samples were reserved and was only available to the validation team. Further, the areas computed for the 55 countries of Africa were compared with areas available from UN FAO.

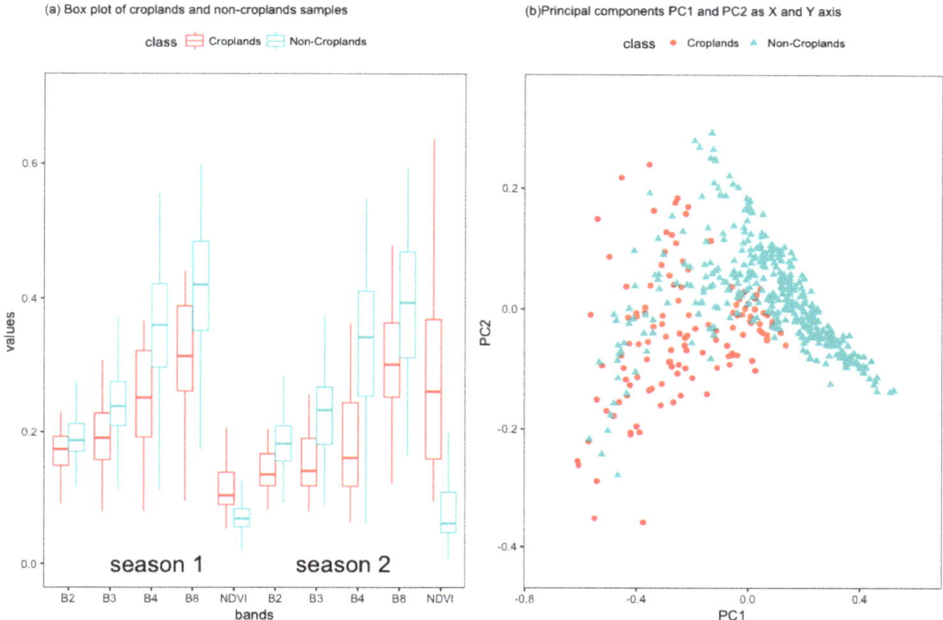

Figure 3. Creating knowledge-base to separate croplands from non-croplands involving: (**a**) waveband reflectivity and NDVI of the two seasons; and (**b**) principal component plot. The above knowledge-base is illustrated for a sub-area in one of the seven zones shown (Figure 1) of Africa.

3. Methods

There is no single classifier that is best applicable to cropland mapping [13,43] as a result of their various strengths and limitations [15], specifically for large areas [8,9]. As result, a combination of different classification techniques [47,76] were investigated. Earlier, several authors [49,50] have explored this combination of methods for land use/land cover classification over large areas, but not for cropland classification over large areas yet. In this section, an integrated methodology of pixel-based classifiers (Random Forest, Support Vector Machines) and object-based Recursive Hierarchical Segmentation is outlined in the flowchart (Figure 4) and described in the following sub-sections:

3.1. Overview of the Methodology

A comprehensive overview of the methodology is shown in Figure 4. As it is difficult to apply a single classifier over large areas (Figure 1) for cropland mapping, an ensemble of machine learning algorithms were investigated [47,76]. Specifically, the integration of pixel-based and the object-based classifiers for large area land cover mapping has been explored by several authors [49,50]. Both pixel-based and object-oriented classifiers require a large amount of reference training data (Section 2.3, Figure 1), which we established through several endeavors as discussed in Section 2 and its sub-sections.

1. 30-m mosaic (11 bands) was built using Sentinel-2 and Landsat-8 data (Section 2.2) for period 1 (January–June, 2016) and period 2 (July–December, 2015);
2. Random Forest and Support Vector Machines (Section 3.1) were used to classify input bands for croplands versus non-croplands;

3. Using same bands as inputs, recursive hierarchical segmentation (Section 3.2) was carried out in 1° by 1° grid units on NASA pleiades supercomputer;
4. The pixel-based classification was integrated with object-based segmentation into cropland extent map (Section 3.3) for further assessment (Section 3.4)
5. We compared derived cropland areas with country-wise statistics from other sources in Section 3.5 and explored the consistency between GFSAD30AFCE map and other reference maps in Section 3.6.
6. 30-m Cropland extent product is released through the NASA Land Processes Distributed Active Archive Center (LP DAAC) at: https://doi.org/10.5067/MEaSUREs/GFSAD/GFSAD30AFCE.001 and can be viewed at: https://croplands.org.

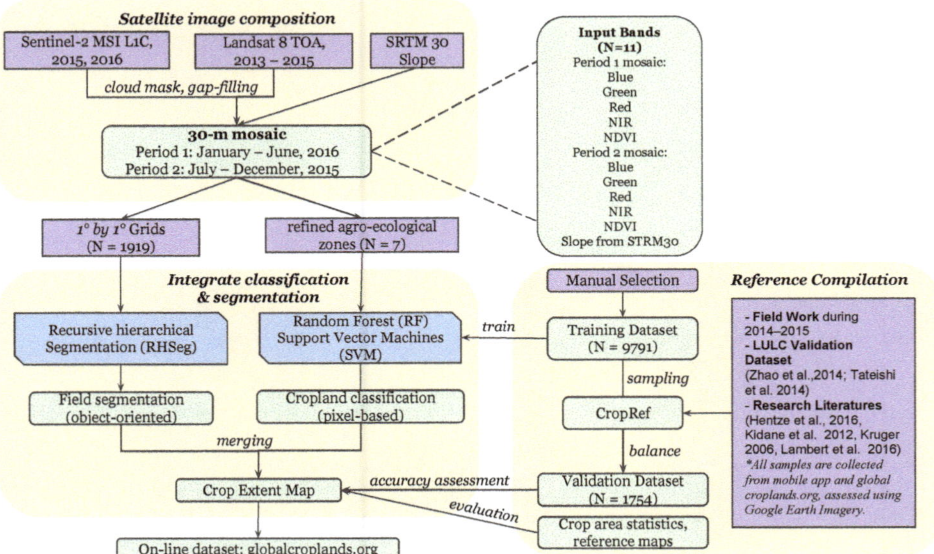

Figure 4. Overview of methodology for cropland extent mapping. The study integrates pixel-based classification involving the Random Forest (RF) and Support Vector Machines (SVM) with object-oriented Recursive Hierarchical Image Segmentation (RHSeg). The chart also shows the reference and training dataset used.

3.2. Pixel-Based Classifier: Random Forest (RF) and Support Vector Machine (SVM)

The Random Forest classifier uses bootstrap aggregating (bagging) to form an ensemble of trees by searching random subspaces from the given features and the best splitting of the nodes by minimizing the correlation between the trees. It is more robust, relatively faster in speed of classification, and easier to implement than many other classifiers [77]. Accurate land cover classification and better performance of the RF models have been described by many researchers [11,77–79].

RF was used to classify croplands in the 7 RAEZ's, as shown in Figure 1. Initially, 5511 training samples described in Section 2.3 were used for the first run of RF algorithm on GEE. In order to improve classification results, ~500-600 trees were used and the number of training samples were increased to varying degrees for each individual zone.

RF classified results were visually compared to other reference maps (GlobCover, GRIPC, GLC30, Google Earth Imagery). Based on these comparisons, the training polygon set was refined by creating additional croplands and non-croplands training polygons by drawing them in GEE editor.

After additional training polygons were added, the classification was run again. This iterative step is time-consuming, depending on the complexity of the landscape. For example, in the rainfed areas of central Africa like Tanzania, the rainfed cropland areas are mixed with natural vegetation and bare land. In such places, the training sample selection was repeated 4 times to achieve satisfactory results. For all 7 RAEZs, another 4280 polygons were added following this iterative procedure. Overall, we used 9791 training samples/polygons across the entire African continent.

Occasionally, overfitting issues happened in RF because input features are heavily correlated in specific areas; to correct for this a SVM classifier [80] with a Linear kernel was used in the problematic regions to replace RF results. SVM has also been reported to work significantly better with smaller, intelligently selected, training samples than RF in literature [81].

The pixel-based classifiers (RFs and SVMs) were run on GEE. Cloud computing offers the power of computing by linking 1000s of computers, allowing parallel processing and thus enabling the classification of individual zones with 30-m pixels in matter of hours.

3.3. Recursive Hierarchical Image Segmentation (RHSeg)

Section 3.2 discussed pixel-based classifiers which are fast and are scalable over large areas on cloud computing facilities like GEE. However, the pixel-based classification results inevitably include "salt and pepper" noise and dis-jointed farm fragments in practice. Object-based analysis can improve salt and pepper effects and increase classification accuracies over pixel-based image classification [82,83]. Image segmentation gathers several similar neighbor pixels together as objects, and categorizes or labels objects, which would be further labelled as croplands or non-croplands in the integration step with pixel-based classification in Section 3.2. Image segmentation procedures have many implementation [82,83] with very high memory and CPU requirements. GEE provides APIs related to image segmentation such as region grow [84]. However, image segmentation for entire Africa at 30-m is beyond GEE's existing capacity so we utilized NASA supercomputer facilities [85] to implement intensive segmentation over large areas.

In this study, Recursive Hierarchical Segmentation (RHSeg) software [86] was adopted to extract object information from 30-m input imagery. RHSeg is an approximation of the Hierarchical Segmentation (HSeg) algorithm that recursively subdivides large images into smaller subimages that can be effectively processed by HSeg. RHSeg then blends the results from the subimages to produce a hierarchical segmentation of the entire large image. HSeg utilizes an iterative region growing approach to produce hierarchical image segmentations. HSeg is unique in that it alternates merges of spatially adjacent and non-spatially adjacent regions to provide a simultaneous segmentation and classification. The addition of merging non-spatially adjacent regions helps stabilize the segmentation result by providing a larger sampling of each spectral class type. Other hierarchical classification strategies have been tested by several researchers with a series of per-class classifiers to minimize the effect of spectral confusion among different land cover classes [50,87].

The merging of spatially non-adjacent regions in RHSeg leads to heavy computational demands. In order to expand its capability from regional-size to continental scale, a grid scheme (Figure 1) was applied to subset the 30-m mosaic dataset created in Section 2.2, covering the non-desert areas of Africa into 1919 smaller pieces using GEE. Each scene was a 10-band image for entire Africa without the slope band (Figure 2, Table 1) at 30-m resolution, about 4000 columns by 4000 rows in size which was then used by RHSeg as input for segmentation. Generating the segmentation of these input datasets took about 74 hours using 64 CPUs on the Pleiades and Discover NASA Supercomputers under the parallel mode supported by RHSeg.

Noting that some image scenes had a large percentage of water pixels or pixels masked out due to clouds, we realized that more consistent and accurate results could be obtained by selecting results from the RHSeg segmentation hierarchy based on merge threshold instead of the number of regions. Based on the analysis of 12 representative 30-m foot-print images across Africa, we found that a merge threshold of 15.0 selected the most suitable RHSeg segmentation hierarchy layer for agricultural application.

3.4. Integration of Pixel-Based Classification and Object-Based Segmentation

Every segment in the outputs of RHSeg at the selected hierarchical level consists a group of pixels with a unique id (region label), which will be further labeled as "cropland" or "non-cropland" patch when the segment was overlaid with pixel-based classification results. In order to merge pixel-based classification and field boundary information from segmentation, we reassigned the pixel values for individual segments according to the following rules which were developed based on trial and error in 12 images under different landscapes across the continent:

If > 85% of the pixels in a segmented patch are classified as 'cropland', the whole patch was assigned to 'cropland';

If < 15% of the pixels in a segmented patch are classified as 'cropland', the whole patch was assigned to 'non-cropland';

If either condition is not met, the pixel-based classification results will be unchanged in the final crop extent map, resulting in mixed cropland and non-croplands pixels in one patch.

The example shown in Figure 5 highlights the value of the merging steps above to produce the final cropland extent map. Pixel-based classification of croplands (green) covered most highly vegetated areas, however, some cropland-pixels were missing because of cropland heterogeneity and spectral contamination among neighboring pixels (Figure 5a). In Figure 5b, the RHSeg segmentation layer is better able to determine whether pixels belong to the same field (random coloring). The results from Figure 5a,b are merged to produce a more refined and complete boundaries of cropland fields (green) (Figure 5c) which has better consistency with the true color VHRI from Google Earth (Figure 5d).

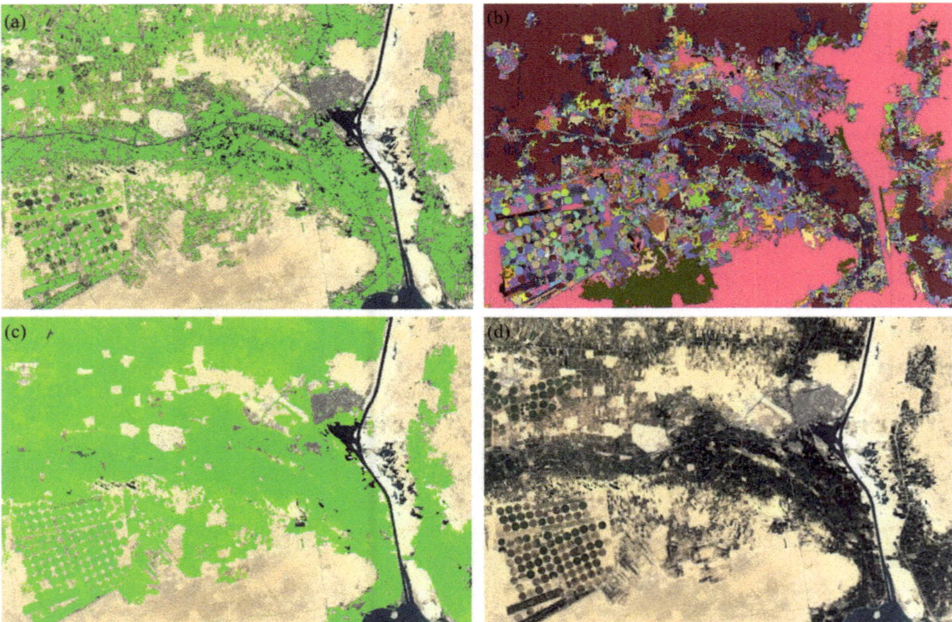

Figure 5. The example of (**a**) the pixel-based classification from random forest classifier; (**b**) the object-based RHSeg image segmentation result; (**c**) the merged results with RHSeg segmentation result with pixel-based Random Forest classification; and (**d**) a true color Google Earth Imagery for reference.

3.5. Accuracy Assessment

Map accuracy assessment is a key component of map production, especially when remote sensing data are utilized [88]. Validation exercises require high-quality reference validation data sets collected at appropriate spatial and temporal scales using random sample designs [89,90]. In addition to the accuracy analysis performed when evaluating the classification results to select the best algorithms and results, an independent validation of the product was performed. For the independent assessment, a total of 1754 samples were used to determine the accuracy of the final cropland extent map of Africa for all 7 RAEZ's (Figure 1). Error matrices were generated for each of the RAEZ's separately and also for the entire African continent providing producer's, user's, and overall accuracies. Further, the areas computed for the 55 countries of Africa were compared with areas available from UN FAO.

There are few basic important considerations that must be followed step by step in order to perform an assessment of the cropland thematic maps [89]. The process of validation usually starts with collection of a high-quality reference data independent of the training data that have already been used for mapping the same area. The reference samples have been collected from very high-resolution imagery or VHRI (sub-meter to 5-m) that were available for the entire continent through US National Geospatial Agency (NGA) though image interpretation that corresponded with the same year of mapping. It is better to adopt a continent specific sampling method to perform a meaningful assessment of global cropland products. For Africa, a stratified random sampling design [89] was used to distribute a balanced sample size using the following steps:

1. Stratified, random and balanced sampling: The African continent has been divided into 7 refined agro-ecological zones or RAEZs (Figure 1) for stratified random sampling. Due to a large crop diversity across RAEZ's (Figure 1) there is high variability in their growing periods and crop distribution. Therefore, to maintain balanced sampling for each zone, samples have been randomly distributed in each zone. The question of how many samples are sufficient to achieve statistically valid accuracy results is described in next point below.
2. Sample Size: The sample size has been chosen based on the analysis of incrementing minimum number of samples. Initially, first 50 samples were chosen as minimum number for all the 7 RAE's and then incremented in steps with another 50 more samples. A few RAEZ's in Africa have little cropland distribution so that 50 samples were enough to achieve a valid assessment. However, other RAEZ's needed up to 250 samples for their assessment. Beyond 250 samples, accuracies of all RAEZ's become asymptotic. Overall, for Africa, total 1754 samples were used from 7 RAEZ's.
3. Sample unit: The sample unit for a given validation sample must be a group of pixels (at least 3×3 pixels of 30-m resolution) in order to minimize the impact of positional accuracy [88]. This sampling unit is a 3×3 homogeneous window containing one class. If a sample at this step was recognized to be a mixed patch of cropland and non-cropland, it had to be excluded from the validation dataset in the accuracy assessment since heterogeneous windows were not considered, however excluding them is the best practical choice for accuracy assessment.
4. Sampling was balanced to keep the proportion of the cropland versus non-cropland samples close to the proportion of the cropland versus non-cropland area from the product layer to be validated.
5. Validation samples are created independently from training samples described in Section 2.3, by a different team.

The performance of the different approaches was assessed by two complementary criteria, namely the accuracy assessment and across-site robustness. Two different metrics, derived from the confusion matrix, were selected for the overall accuracy (OA) assessment. The OA evaluated the overall effectiveness of the algorithm, while the F-score measured the accuracy of a class using the precision and recall measures. For each of the 7 RAEZ's (Figure 1) of Africa, the study establishes error matrices that provides user's (UA), producer's (PA), and overall accuracies (OA) as following equations:

$$OA = \frac{S_d}{n} \times 100\% \quad (1)$$

$$UA = \frac{X_{ij}}{X_j} \times 100\% \quad (2)$$

$$PA = \frac{X_{ij}}{X_i} \times 100\% \quad (3)$$

$$F_{score} = 2 \times \frac{UA \times PA}{UA + PA} \quad (4)$$

where S_d is the total number of correctly-classified pixels, n = total number of validation pixels, X_{ij} = observation in row i column j; X_i = marginal total of row i; X_j = marginal total of column j.

3.6. Calculation of Actual Cropland Areas and Comparison with Areas from Other Sources

Generating cropland areas such as at national and sub-national levels is of great importance in food security studies. In Google Earth Engine, we convert the crop extent map to a crop area map where the pixel value represents the actual crop area converting the map to Lambert Azimuthal (equal-area) projection. In order to derived the country level cropland area statistics from the 30-m crop extent map of Africa, we used the Global Administrative Unit Layers (GAUL) from UN FAO as country boundaries to create Table 4 as well as statistics from other sources, including AquaStat [91], Mirca2000 [92], GRIPC [16] and GLC30 [45].

3.7. Consistency between GFSAD30AFCE Product and Four Existing Crop Maps

The GFSAD30AF product was also compared with other LULC/Cropland products that were published recently to establish consistency between the products. First, we remapped four existing global land cover map products according to their individual classification schemes (Table 2):

- Global Land Cover Map for 2009 (GlobCover 2009) [39]. Class 11, 14 were reclassified as "croplands" and other land cover classes were reclassified as "non-croplands";
- Global rainfed, irrigated, and paddy croplands map GRIPC [16]. All agricultural classes include rainfed, irrigated and paddy were combined as "croplands" and other classes were "non-croplands";
- 30-m global land-cover map FROM-GLC [48]. Level 1 class 10 and Level 2 Bare-cropland 94 were combined as "croplands" and other classes were "non-croplands"; and
- Global land cover GLC30 [45]. Class 10 was combined as "croplands" and other classes were "non-croplands".

To unify spatial resolution of cropland maps, the Cropland Extent Map and these four cropland maps were all resampled to 30-m resolution for comparison. In addition to visual comparison illustrations, we also evaluated statistical agreements between these cropland maps. We generated 12,627 random points across the classification extent and sampled the cropland classes from five cropland extent maps to build similarity metrics.

Table 2. Remapped land cover classes of other cropland or land use/land cover (LULC) products used to compare with the GFSAD30AFCE product of this study.

Cropland/LULC Maps	Resolution	Code	Class Name
GlobCover 2009 v2.3	300 m	11	Post-flooding or irrigated croplands
		14	Rainfed croplands
		20	Mosaic cropland/vegetation
		30	Mosaic vegetation/croplands
GRIPC	500 m	1	Rain-fed croplands
		2	Irrigated croplands
		3	Paddy croplands
FROMGC Level 2	30 m	11	Rice
		12	Greenhouse
		13	Other Crop
		94	Bare-cropland
GLC30	30 m	1	Cultivated Land

4. Results

The study produced a nominal 30-m cropland extent product (Figure 6; croplands.org) of the entire African continent using Sentinel-2 and Landsat-8 data for the year 2015. In the following sub-sections, we will discuss this product, referred to as, the Global Food Security-support Analysis Data @ 30-m of Africa, Cropland Extent (GFSAD30AFCE; Figure 6) product, its accuracies, areas derived from it, and comparison of areas with areas reported through National and sub-National statistics as reported by the Food and Agricultural Organization (FAO) of the United Nations (UN). We will also compare the GFSAD30AFCE product with other cropland and/or land use/land cover (LULC) products where cropland classes were mapped.

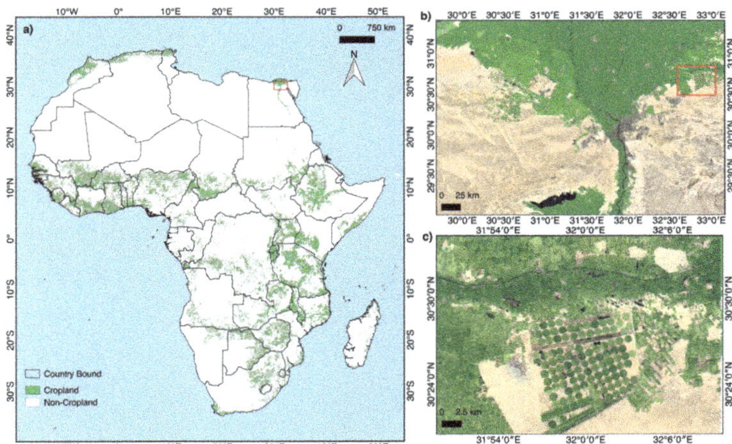

Figure 6. Global Food Security-support Analysis Data @ 30-m of Africa, Cropland Extent product (**a**). Full resolution of 30-m cropland extent can be visualized by zooming-in to specific areas as illustrated in right panel (**b**,**c**). For any area in Africa, croplands can be visualized by zooming into specific areas in croplands.org.

4.1. GFSAD30AFCE Product

The Global Food Security-support Analysis Data @ 30-m of Africa, Cropland Extent (GFSAD30AFCE; Figure 6), produced by combining the pixel-based (RF, SVM) and Object-based segmentation algorithm (RHSeg), is accessible at https://croplands.org/app/map. The data will also be soon made available for download through NASA's Land Processes Distributed Active Archive Center (LP DAAC). The product year is referred to as nominal 2015 since most Sentinel-2 images used in processing were from July 2015 to June 2016. Data users can also browse the online version of the products at croplands.org.

On the African continent, croplands are primarily dominant throughout West Africa, along the great Lakes of Africa (Lake Victoria and Lake Tanganyika), South Africa, Southern Africa, along the coasts of North Africa, and all along the Nile Basin (Figure 6). The Sahara Desert, Kalahari Desert, and overwhelming proportion of the Congo rain forests have almost no croplands (Figure 6).

4.2. GFSAD30AFCE Product Accuracies

This final cropland extent product of Africa (GFSAD30AFCE) was systematically tested for accuracies (Table 3) by independent validation datasets in each of the 7 refined agro-ecological zones or RAEZs (Figure 1). For the entire African continent, the weighted overall accuracy was 94.5% with producer's accuracy of 85.9% (errors of omissions of 14.1%) and user's accuracy of 68.5% (errors of commissions of 31.5%) for the cropland class (Table 3). When considering all 7 RAEZs, the overall accuracies varied between 90.8% and 96.8%, Producer's accuracies varied between 60.7% and 94.9%, and user's accuracies varies between 53.3% and 89.6% for cropland class (Table 3). The F score ranged between 0.65 and 0.9.

Table 3. Independent Accuracy Assessment of GFSAD30 Cropland Extent product of Africa (GFSAD30AFCE).

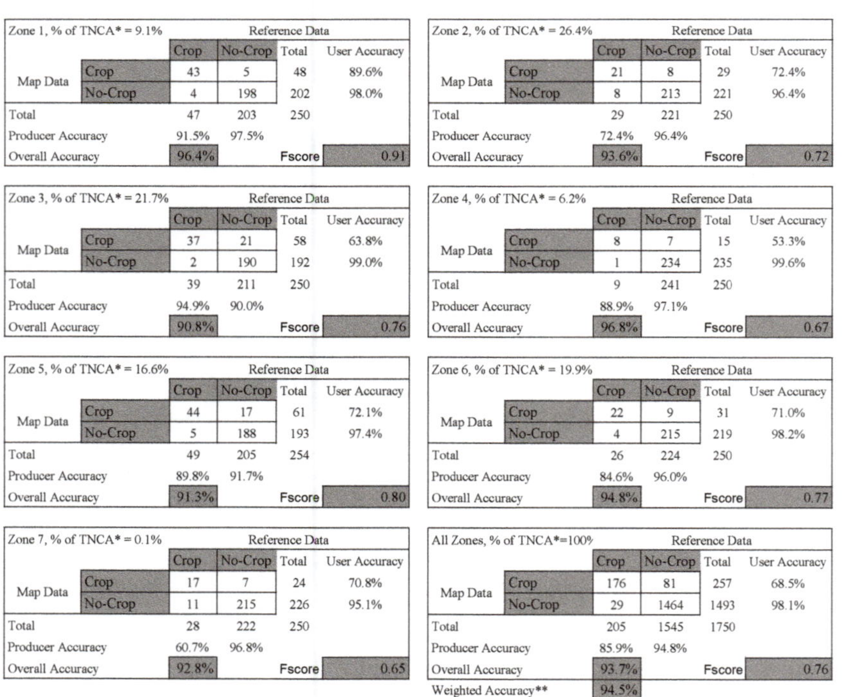

Zone 1, % of TNCA* = 9.1%		Reference Data			
		Crop	No-Crop	Total	User Accuracy
Map Data	Crop	43	5	48	89.6%
	No-Crop	4	198	202	98.0%
Total		47	203	250	
Producer Accuracy		91.5%	97.5%		
Overall Accuracy		96.4%		Fscore	0.91

Zone 2, % of TNCA* = 26.4%		Reference Data			
		Crop	No-Crop	Total	User Accuracy
Map Data	Crop	21	8	29	72.4%
	No-Crop	8	213	221	96.4%
Total		29	221	250	
Producer Accuracy		72.4%	96.4%		
Overall Accuracy		93.6%		Fscore	0.72

Zone 3, % of TNCA* = 21.7%		Reference Data			
		Crop	No-Crop	Total	User Accuracy
Map Data	Crop	37	21	58	63.8%
	No-Crop	2	190	192	99.0%
Total		39	211	250	
Producer Accuracy		94.9%	90.0%		
Overall Accuracy		90.8%		Fscore	0.76

Zone 4, % of TNCA* = 6.2%		Reference Data			
		Crop	No-Crop	Total	User Accuracy
Map Data	Crop	8	7	15	53.3%
	No-Crop	1	234	235	99.6%
Total		9	241	250	
Producer Accuracy		88.9%	97.1%		
Overall Accuracy		96.8%		Fscore	0.67

Zone 5, % of TNCA* = 16.6%		Reference Data			
		Crop	No-Crop	Total	User Accuracy
Map Data	Crop	44	17	61	72.1%
	No-Crop	5	188	193	97.4%
Total		49	205	254	
Producer Accuracy		89.8%	91.7%		
Overall Accuracy		91.3%		Fscore	0.80

Zone 6, % of TNCA* = 19.9%		Reference Data			
		Crop	No-Crop	Total	User Accuracy
Map Data	Crop	22	9	31	71.0%
	No-Crop	4	215	219	98.2%
Total		26	224	250	
Producer Accuracy		84.6%	96.0%		
Overall Accuracy		94.8%		Fscore	0.77

Zone 7, % of TNCA* = 0.1%		Reference Data			
		Crop	No-Crop	Total	User Accuracy
Map Data	Crop	17	7	24	70.8%
	No-Crop	11	215	226	95.1%
Total		28	222	250	
Producer Accuracy		60.7%	96.8%		
Overall Accuracy		92.8%		Fscore	0.65

All Zones, % of TNCA*=100%		Reference Data			
		Crop	No-Crop	Total	User Accuracy
Map Data	Crop	176	81	257	68.5%
	No-Crop	29	1464	1493	98.1%
Total		205	1545	1750	
Producer Accuracy		85.9%	94.8%		
Overall Accuracy		93.7%		Fscore	0.76
Weighted Accuracy**		94.5%			

Note: * TNCA (Total Net Croplands Area) = 313 Mha

** The all-zones Weighted Accuracy is weighted by proportion of croplands in each zone

Across RAEZs (Table 3), user's accuracies (commission errors) were significantly lower than producer's accuracies (omission errors). This was mainly because when training the random forest algorithm, we tweaked it to capture as much croplands as possible, thus ensuring high producer's accuracies (or low omission errors for the cropland class) across RAEZ's. In this process, the compromise was that some non-croplands were included as croplands, resulting in lower user's accuracies (or higher commission errors) for the cropland class. Ideally, an algorithm should optimize a classification to balance producer's and user's accuracies. However, the goal of this project is to map almost all croplands including fallow croplands. As a result, we aimed for high producer's accuracies (low errors of omissions) for the cropland class across zones (Table 3) and achieved it for most RAEZs, as evidenced by a continent-wide producer's accuracy of 85.9% (Table 3).

4.3. Cropland Areas and Comparison with Statistics from Other Sources

GFSAD30AFCE can accurately estimate cropland areas by nation or sub-national regions (e.g., state, district, county, village). Here we calculated the cropland areas by country in Africa (Table 4) for comparison with survey-based statistical area from UN FAO. Users can make use of this product to do their own computation of sub-national statistics anywhere in Africa at all kinds of administrative level and compare them to reliable reference data.

Table 4 shows a country-wise cropland area statistics of all 55 African countries generated from this study using GFSAD30AFCE product of year 2015 (Figure 6) and compared with the national census data based MIRCA2000 [92] which were updated in the year 2015 (Stefan Siebert and Portmann, personal communication). Overall, the entire African continent had total net cropland area (TNCA) of 313 million hectares (Mha). Five countries (Nigeria, Ethiopia, Sudan, Tanzania, and South Africa) constitute 40% of all cropland areas of Africa and each have 5% or more of total Africa's net cropland area (TNCA) of 313 Mha (Table 4). Nigeria is the leading cropland country in Africa with 11.4% of the 313 Mha (Table 4). Ethiopia is second with 8.21%. However, crop productivities will depend on numerous factors such as soils, whether they are irrigated or rainfed, management issues (e.g., inputs such a N, K, P), and also climate and plant genetics. Thereby, larger cropland area does not necessarily mean greater crop productivity. There are 12 countries (DR Congo, Mali, Zimbabwe, Kenya, Morocco, Algeria, Niger, Zambia, Uganda, Mozambique, Burkina Faso, Chad) which have above 2% but below 5% of Africa's TNCA of 313 Mha. The remaining 38 African countries have less than 2% of Africa's NCA. The overwhelming proportion (94%) of the cropland areas are in just 25 of the 55 countries (Table 4).

For 48 of the 55 countries (7 "outliers" countries removed) there was a strong relationship between the GFSAD30AFCE product produced cropland areas versus the MIRCA2000 produced cropland areas (Figure 7) with an R-square value of 0.78. When all 55 countries are considered, this relationship provides an R-square value of 0.65. The countries where GFSAD30AFCE under-estimated croplands include Cote d'Voire, Uganda, Cameroun, Ghana, and Tunisia (Figure 7). The countries where GFSAD30AFCE over-estimated croplands include Malawi, Kenya, Mozambique, and Egypt to mention few names (Figure 7). Causes of this variability are many. Besides uncertainties among the input data and methodology, the GFSAD30AFCE product and the national statistics differ due to these reasons:

- Different definition of "croplands" class: GFSAD30AFCE product as per definition, includes all agricultural annual standing croplands, cropland fallows, and permanent plantation crops whereas cropland areas reported in statistics may not include cropland fallows;
- Different time: GFSAD30AFCE incorporate the latest cultivated area in 2015–2016 as well as the croplands fallows whereas country reported cropland areas may happen in other years.

Table 4. Total net cropland areas (TNCA) of the African countries derived from the global food security-support analysis data @ 30-m cropland extent product (GFSAD30AFCE) and compared with other cropland area sources.

Country	Land (Mha) FAO-GAUL	Total Net Cropland Area (TNCA, Mha) MIRCA2000	This Study: GFSAD30AFCE	FAO Cultivated area (2002)	GRIPC	GFSAD250	GLC30
Resolution	--	variable	30-m	variable	500-m	250-m	30-m
Nigeria	90.56	38.62	35.67	33.00	39.30	14.05	28.20
Ethiopia	112.76	11.09	25.70	10.67	14.23	19.63	21.72
Sudan	186.88	18.40	22.74	16.65	10.43	9.09	19.94
Tanzania	93.98	5.67	22.57	5.10	4.57	28.81	18.25
South Africa	122.00	15.70	19.91	15.71	10.99	13.06	15.20
Congo DRC	232.94	9.80	16.32	7.80	11.41	22.87	4.47
Mali	125.26	4.84	12.78	4.70	9.81	10.66	5.16
Zimbabwe	39.07	3.53	12.31	3.35	0.10	10.75	8.97
Zambia	75.12	5.43	9.70	5.29	0.23	15.31	6.49
Kenya	59.34	5.35	9.23	5.16	5.74	8.91	8.38
Morocco	67.77	9.68	8.98	9.28	6.18	5.76	8.06
Algeria	231.27	5.96	8.81	8.27	3.99	3.78	7.73
Niger	118.12	14.53	8.45	4.50	1.06	0.24	6.60
Mozambique	78.57	4.82	8.42	4.44	1.30	11.76	5.80
Côte d'Ivoire	32.07	7.13	7.86	6.90	9.53	9.02	1.44
Burkina Faso	27.39	4.34	7.37	4.40	10.39	7.95	4.19
Uganda	24.13	8.58	7.19	7.20	9.37	7.69	6.24
Angola	124.71	3.67	6.32	3.30	2.17	7.94	4.29
Chad	127.09	3.72	6.31	3.63	7.13	11.07	4.80
Cameroon	46.50	7.40	5.20	7.16	4.45	5.37	1.45
Malawi	11.85	1.74	5.20	2.44	0.46	5.83	3.76
Tunisia	15.50	2.51	5.05	4.91	2.21	1.66	4.53
Egypt	98.22	4.51	4.99	3.42	3.21	4.11	4.31
Madagascar	58.98	4.34	4.63	3.55	4.02	2.55	2.07
Ghana	23.86	6.58	4.62	6.33	6.72	8.73	2.16
Senegal	19.52	2.52	4.33	2.51	5.65	2.59	3.32
Benin	11.52	2.85	3.81	2.82	2.44	2.97	2.95
Togo	5.67	2.63	2.24	2.63	1.53	1.65	1.73
Libya	161.52	0.94	2.08	2.15	0.64	0.45	1.68
Somalia	63.26	1.21	2.05	1.07	1.95	2.52	1.54
Botswana	57.84	0.87	1.90	0.38	0.02	9.75	0.84
Rwanda	2.53	1.29	1.42	1.39	1.68	1.63	1.20
South Sudan	62.43	0.00	1.26	0.00	0.00	4.13	0.00
Namibia	82.39	0.82	1.23	0.82	0.00	3.80	0.80
Central African Republic	62.03	2.06	1.02	2.02	0.61	1.18	0.69
Guinea	24.47	1.92	0.97	1.54	0.85	5.32	0.64
Lesotho	3.05	0.34	0.85	0.33	0.08	0.54	0.62
Burundi	2.72	1.08	0.85	1.35	1.26	1.80	0.67
Eritrea	12.25	0.56	0.73	0.50	0.16	0.33	0.56
Swaziland	1.74	0.24	0.69	0.19	0.19	0.38	0.52
Gambia	1.06	0.30	0.38	0.26	0.63	0.57	0.26
Congo	34.22	0.58	0.31	0.24	3.40	2.35	0.22
Guinea-Bissau	3.36	0.55	0.31	0.55	0.14	0.76	0.20
Sierra Leone	7.23	0.67	0.13	0.60	0.87	2.73	0.09
Mauritania	103.89	0.86	0.11	0.50	0.03	0.07	0.08
Liberia	9.58	0.46	0.01	0.60	0.21	2.44	0.01
Cape Verde	0.41	0.06	0.00	0.05	0.02	0.00	0.00
Gabon	26.15	0.49	0.00	0.50	0.94	1.07	0.00
Sao Tome and Principe	0.10	0.03	0.00	0.05	0.00	0.01	0.00
Equatorial Guinea	2.69	0.22	0.00	0.23	0.01	0.13	0.00
Comoros	0.17	0.08	0.00	0.13	0.04	0.00	0.00
Djibouti	2.17	0.00	0.00	0.00	0.00	0.00	0.00
Mauritius	0.20	0.16	0.00	0.11	0.12	0.00	0.05
Saint Helena	0.04	0.00	0.00	0.00	0.00	0.00	0.00
Seychelles	0.05	0.00	0.00	0.01	0.00	0.00	0.00
Total	**2988**	**232**	**313**	**211**	**202**	**296**	**223**

Note: FAO GAUL = The Food and Agricultural Organization's The Global Administrative Unit Layers (GAUL); GFSAD30AFCE = global food security support analysis data @ 30-m (this study); GFSAD250 = global food security support-analysis data @ 250-m [15]; GRIPC = Global rain-fed, irrigated, and paddy croplands [16]; MIRCA2000 = Global data set of monthly irrigated and rainfed crop areas around 2000, revised for year 2015 in this study [92]; GLC30 = Global land cover mapping at 30m resolution [45].

There are a number of other reasons for discrepancies between remote sensing and non-remote sensing sources [1,8,13,23,43]. We suggest that a detailed investigation should be conducted on this aspect to see why uncertainties exist and how to overcome them. More detailed assessment of such variability is beyond the goal of this study. On average, the GFSAD30AFCE determined about 35% higher cropland areas relative to national statistics reported by Portmann et al., and UN FAO. It is important to note that the GFSAD30AFCE of this study provided TNCA of the continent as 313 Mha, which is 5.7% higher than our earlier MODIS 250-m data based estimate of 296 Mha [15]. Other studies reported far less cropland areas for Africa, which were (Table 4): 232 Mha (MIRCA), 211 Mha (FAO), 202 Mha (GRIPC), and 223 Mha (GLC30). Therefore, these estimates were lower by about 26% to 35% relative to GFSAD30AFCE product. Various factors may contribute to such discrepancies: 1. MIRCA and FAO UN statistics were derived from a combination of national reports and their synthesis using some remote sensing, GIS and field visits. MIRCA2000 is a derived gridded dataset based on the FAOSTAT database [92]. FAO compiles the statistics reported by individual countries, which are based on national censuses, agricultural samples, questionnaire-based surveys with major agricultural producers, and independent evaluations (FAO, 2006 and The World Bank, 2010). Since each country has its own data collection mechanism, differences in data gathering, and resource limitations, the data lacks objectivity in many countries, resulting in data quality issues, particularly in Africa. For example, in 2008/09 in Malawi, cropland extent was estimated by combining household surveys with field measurements derived from a "pacing method" in which the size of crop fields is determined by the number of steps required to walk around them [93]; 2. GRIPC [16] also maps croplands but using 500-m MODIS data and using different definition and methodologies other than GFSAD30AFCE; 3. GLC30 does include croplands class [45], and its focus is land use and land cover so a lot of uncultivated, low vegetated croplands were identified as shrubs or grass instead of croplands.

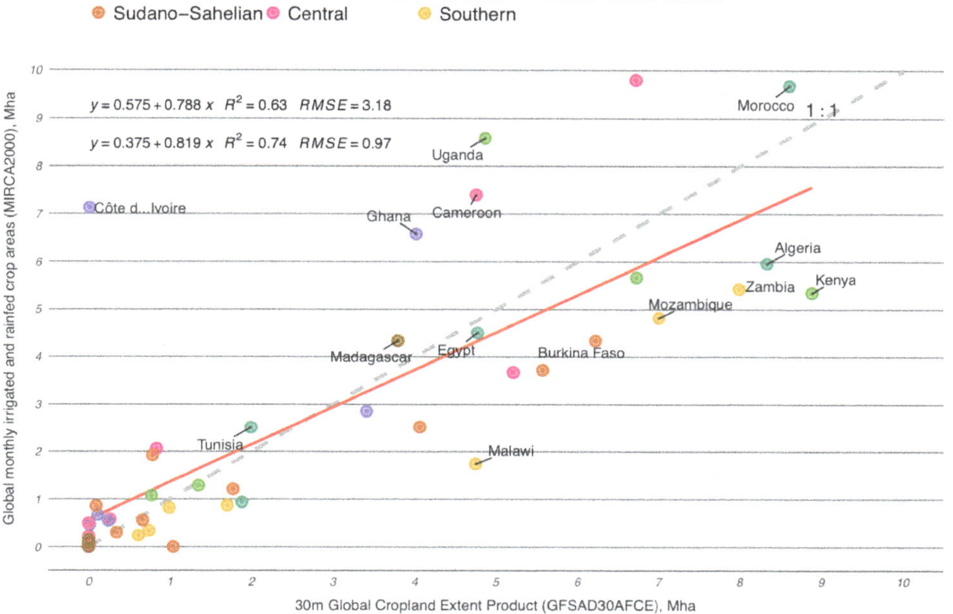

Figure 7. Scatter plot of GFSAD30AFCE derived cropland areas versus MIRCA2000 (Personal communication with Dr. Siebert and Dr. Portmann by Dr. Thenkabail) derived cropland areas country by country for Africa.

4.4. Consistency between GFSAD30AFCE Product and Four Existing Crop Maps

A similarity analysis was conducted where GFSAD30AFCE product was compared with each of the other four products (Table 2) using the 12,627 random samples spread across the African continent. The results showed that GFSAD30AFCE product matches with GLC30 product for 77.3% of the samples while matches with GRIPC500 product with 68.8% samples for the cropland class (Table 5). For the other two products (Globecover2009 and FROMGC), the cropland samples only matched 60% (Table 5). The discrepancies between the products include different reference year of the dataset, cropland definition and methodologies, resolution of the datasets, and a host of other factors; the differences between these products should be investigated further in the future.

Table 5. Similarity analysis comparing the GFSAD30AFCE product of this study with four other products using 12,627 random samples.

(a) Confusion Matrix Obtained with GlobCover 2009 (300-m, Arino et al., 2007)			
	Cropland	**Non-Cropland**	**User's Accuracy (%)**
Cropland	1125	585	65.79%
Non-Cropland	720	10,395	93.52%
Producer's Accuracy (%)	60.98%	94.67%	
(b) Confusion Matrix Obtained with GRIPC (500-m, Salmon et al., 2015)			
	Cropland	**Non-Cropland**	**User's Accuracy (%)**
Cropland	1267	443	74.09%
Non-Cropland	576	10,539	94.82%
Producer's Accuracy (%)	68.75%	95.97%	
(c) Confusion Matrix Obtained with FROMGC (30-m, Yu et al., 2013)			
	Cropland	**Non-Cropland**	**User's Accuracy (%)**
Cropland	1009	701	59.01%
Non-Cropland	676	10,439	93.92%
Producer's Accuracy (%)	59.88%	93.71%	
(d) Confusion Matrix Obtained with GLC30 (30-m, Chen et al., 2015)			
	Cropland	**Non-Cropland**	**User's Accuracy (%)**
Cropland	1409	301	82.40%
Non-Cropland	413	10,702	96.28%
Producer's Accuracy (%)	77.33%	97.26%	

Figure 8 shows a visual comparison of each product's mapping of cropland pixels (highlighted in green) using true color Google Earth Imagery as a background reference under three different landscapes: Egyptian irrigation area, South Africa irrigation area, Cote d'Ivoire mixed agricultural area. GFSADAFCE maps greatly outperformed the coarser resolution products which is significant considering these maps are frequently used in applications that monitor agricultural landscapes. Also, since images from the year 2015 were used in this study, it also covers newly cultivated farmlands which were not mapped in higher resolution product such as GLC30. Furthermore, the training/validation datasets of GFSAD will be also published with croplands maps, which means these training samples can be reused/expanded to update the latest cropland extent when necessary by using Google Earth Engine Cloud-based image composition and classification tools.

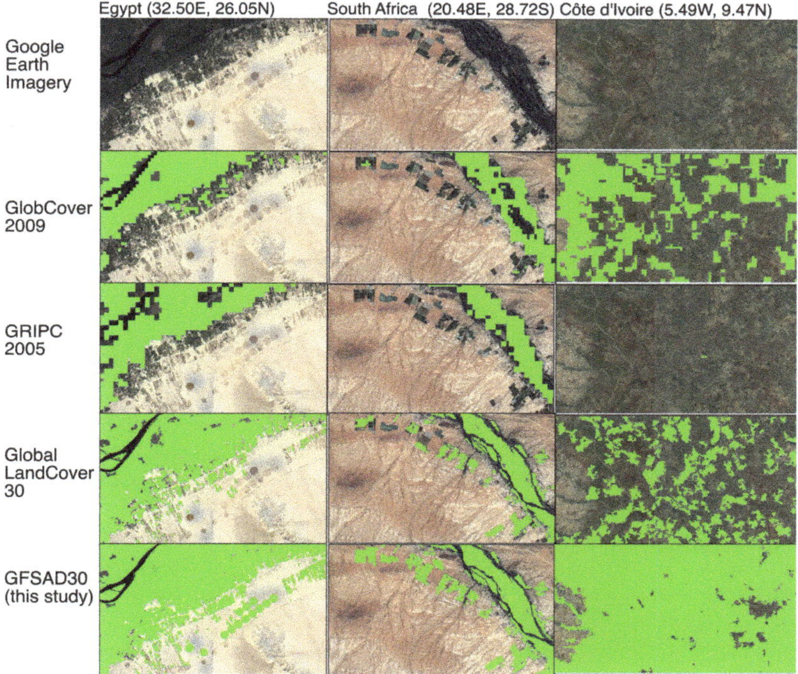

Figure 8. A visual comparison of all crop extent products (shown in green) overlaid on Google Earth Imagery.

5. Discussion

Although the value of this approach is evident, there are still problems in GFSAD30 cropland extent; some of these are discussed below. There were insufficient samples to reflect the diversity of croplands for certain regions and as a consequence, confusion between cropland and non-cropland classes exist. In Africa, the diversity of spectral properties for croplands is very high (e.g., cropland fallows in desert-margins of Sahara versus cropland fallows of the forest-margins of the rain forests). Even though we gathered a very large sample size for training and validation in this project, we still had difficulties verifying some areas. It was difficult to separate rainfed croplands from seasonal grasses in the Sahel and Northern Guinea Savanna because it is difficult to discriminate between them using VHRI imagery. In the context of phenology signatures, it is also easy to confuse croplands with bare lands and grasslands. For example, barelands, grasslands, and rainfed croplands in Sahel are very difficult to discern due to sparse vegetation of all three classes. We resolved such issues by utilizing acquisitions of quality samples from field visits, data (reference maps and ground data) from a few published articles on detailed studies in some small portions of the landscape and VHRI acquired during the exact growing seasons. Also, fallows of different ages (<1-year fallow to 5-year fallow) all have different signatures, especially in the rain forests where greater age corresponds to greater natural vegetation.

With 30-m resolution, satellite imagery is still limited for small fragmented fields in Africa, where field boundaries hardly exist and are often adjoin similar looking sparse grasslands or barrenlands. This is a specific problem throughout the Sahel and in certain places on the Northern Guinea Savannas of Africa. In such cases, RHSeg fails to identify boundaries of crop fields using 30-m imagery; however such issues can be improved by applying RHSeg to 10-m Sentinel-2 data in the future.

Another approach for improving classification accuracies is to use more refined segments as units for classification. A solution might be to integrate the FAO Farming systems map [2], which provides a finer stratification and takes into account both agro-ecological and climatic characteristics. In our earlier study using MODIS 250-m data, we did use FAO agro-ecological zones (AEZ's) for stratification [15]. However, the MODIS approach allowed monthly composites to be created, which was infeasible here when using 10 to 30-m data. By adopting more frequent (e.g., 15-day, monthly) periodic composites, we will be able to work with more detailed AEZ zones rather than the 7 RAEZ's zones as used in this study. We expect this will increase classification accuracies further.

6. Conclusions

This paper presents a practical methodology for cropland extent mapping at 30-m for the entire African continent on Google Earth Engine. Five-bands (blue, green, red, NIR, and NDVI) from 10-day time series Sentinel-2 and 16-day time-series Landsat-8 were time-composited over each of the two crop growing periods (period 1: January–June 2016; period 2: July–December 2015) along with 30-m SRTM DEM data, resulting in a 11-band stack over entire Africa. This input data was then classified using two pixel-based supervised classifiers: Random Forests (RFs) and Support Vector Machines (SVMs) which were merged with RHSeg, an object-based segmentation algorithm. A total of 9791 training samples/polygons were used to train the supervised classifiers. A total of 1754 validation samples were used for assessing accuracies, errors, and uncertainties.

The study produced the first cropland extent map of Africa at nominal 30-m resolution for the nominal year 2015. The product is referred to as the Global Food Security-support Analysis Data@ 30-m of Africa, Cropland Extent (GFSAD30AFCE; Figure 6). The weighted overall accuracy of the entire Africa continent was 94.5% with producer's accuracy of 85.9% (errors of omissions of 14.1%) and user's accuracy of 68.5% (errors of commissions of 31.5%). Across the 7 zones of Africa, accuracies vary with: 90.8% –96.8% overall accuracies, 60.7% –94.9% producer's accuracies, and 53.3% –89.6% user's accuracies. The F-score ranged between 0.65 and 0.90 across all 7 zones.

Derived from GFSAD30AFCE, total net croplands areas (TNCA's) was 313 million hectares for the African continent for the year 2015. In comparison to other cropland products of the past for African continent these area estimates were 26% to 35% higher. Five countries constitute 40% of all cropland areas of Africa, Nigeria (11.4%), Ethiopia (8.2%), Sudan (7.3%), Tanzania (7.2%) and South Africa (6.4%). There are 12 countries (DR Congo, Mali, Zimbabwe, Kenya, Morocco, Algeria, Niger, Zambia, Uganda, Mozambique, Burkina Faso, Chad) which each have above 2% but below 5% of Africa's TNCA. The remaining 38 African countries each have less than 2% of Africa's TNCA. The GFSAD30AFCE cropland areas explained 65% –78% percent of variability in UN FAO country-wise cropland areas.

The GFSAD30AFCE products are viewable at: https://croplands.org/app/map. The GFSAD30AFCE (https://doi.org/10.5067/MEaSUREs/GFSAD/GFSAD30AFCE.001) is released through NASA's Land Processes Distributed Active Archive Center (LP DAAC) for download by user community.

Cloud-based computing platforms such as Google Earth Engine and new earth-observing satellites like the Sentinel-2 constellation have brought significant paradigm-shifts in LULC mapping and agricultural cropland mapping and monitoring. The production of standard static maps will be replaced by the dynamic creation of maps from big data using crowd sourced training samples, and cloud computing which will better serve land managers, NGO's and the scientific community.

Acknowledgments: The research is funded by NASA MEaSUREs (Making Earth System Data Records for Use in Research Environments). The United States Geological Survey (USGS) provided supplemental funding as well as numerous other direct and indirect support through its Land Change Science (LCS), Land Remote Sensing (LRS) programs, and Climate and Land Use Change Mission Area. The NASA MEaSUREs project grant number: NNH13AV82I, the USGS Sales Order number: 29039. The authors would like to thank following persons for their support: Felix T. Portman and Stefan Siebert for providing statistics of MIRCA2000 data; Peng Gong for sharing of FROMGLC Validation Dataset (Zhao et al., 2014); Ryutaro Tateishi for sharing of CEReS Gaia validation data (Tateishi et al., 2014); Friedl Mark for sharing GRIPC500 dataset for inter-comparison, and Fabio Grita and Michela

Marinelli's help from FAO/CountrySTAT team. Special thanks to Jennifer L. Dungan and Mutlu Ozdogan for their suggestion to the manuscript. We would like to thanks the anonymous reviewers that helped improve this paper with their comments.

Author Contributions: Jun Xiong and Prasad Thenkabail conceived and designed the methodology; Adam Oliphant contributed Google Earth Engine scripting; Russell Congalton and Kamini Yadav performed validation; Murali K. Gumma and Pardhasaradhi Teluguntla helped visual assessment; James Tilton contributed to data pipeline; Jun Xiong wrote the paper.

Conflicts of Interest: The authors declare no conflict of interest.

Abbreviations

The following abbreviations are used in this manuscript:

GFSAD30: Global Food Security Analysis-Support Data Project
GEE: Google Earth Engine

References

1. Thenkabail, P.S.; Hanjra, M.A.; Dheeravath, V.; Gumma, M. A Holistic View of Global Croplands and Their Water Use for Ensuring Global Food Security in the 21st Century through Advanced Remote Sensing and Non-remote Sensing Approaches. *Remote Sens.* **2010**, *2*, 211–261.
2. Fritz, S.; See, L.; Rembold, F. Comparison of global and regional land cover maps with statistical information for the agricultural domain in Africa. *Int. J. Remote Sens.* **2010**, *31*, 2237–2256.
3. Herold, M.; Mayaux, P.; Woodcock, C.E.; Baccini, A.; Schmullius, C. Some challenges in global land cover mapping: An assessment of agreement and accuracy in existing 1 km datasets. *Remote Sens. Environ.* **2008**, *112*, 2538–2556.
4. See, L.; Fritz, S.; You, L.; Ramankutty, N.; Herrero, M.; Justice, C.; Becker-Reshef, I.; Thornton, P.; Erb, K.; Gong, P.; et al. Improved global cropland data as an essential ingredient for food security. *Glob. Food Secur.* **2015**, *4*, 37–45.
5. Delrue, J.; Bydekerke, L.; Eerens, H.; Gilliams, S.; Piccard, I.; Swinnen, E. Crop mapping in countries with small-scale farming: A case study for West Shewa, Ethiopia. *Int. J. Remote Sens.* **2012**, *34*, 2566–2582.
6. Hannerz, F.; Lotsch, A. Assessment of land use and cropland inventories for Africa. In *CEEPA Discussion Papers*; University of Pretoria: Pretoria, South Africa, 2006; Volume 22.
7. Gallego, F.J.; Kussul, N.; Skakun, S.; Kravchenko, O.; Shelestov, A.; Kussul, O. Efficiency assessment of using satellite data for crop area estimation in Ukraine. *Int. J. Appl. Earth Obs. Geoinf.* **2014**, *29*, 22–30.
8. Thenkabail, P.S.; Wu, Z. An Automated Cropland Classification Algorithm (ACCA) for Tajikistan by Combining Landsat, MODIS, and Secondary Data. *Remote Sens.* **2012**, *4*, 2890–2918.
9. Wu, W.; Shibasaki, R.; Yang, P.; Zhou, Q.; Tang, H. Remotely sensed estimation of cropland in China: A comparison of the maps derived from four global land cover datasets. *Can. J. Remote Sens.* **2014**, *34*, 467–479.
10. Büttner, G. CORINE Land Cover and Land Cover Change Products. In *Land Use and Land Cover Mapping in Europe*; Springer: Dordrecht, The Netherlands, 2014; pp. 55–74.
11. Tian, S.; Zhang, X.; Tian, J.; Sun, Q. Random Forest Classification of Wetland Landcovers from Multi-Sensor Data in the Arid Region of Xinjiang, China. *Remote Sens.* **2016**, *8*, 954, doi:10.3390/rs8110954.
12. Pittman, K.; Hansen, M.C.; Becker-Reshef, I.; Potapov, P.V.; Justice, C.O. Estimating Global Cropland Extent with Multi-year MODIS Data. *Remote Sens.* **2010**, *2*, 1844–1863.
13. Thenkabail, P.S.; Biradar, C.M.; Noojipady, P.; Dheeravath, V.; Li, Y.; Velpuri, M.; Gumma, M.; Gangalakunta, O.R.P.; Turral, H.; Cai, X.; et al. Global irrigated area map (GIAM), derived from remote sensing, for the end of the last millennium. *Int. J. Remote Sens.* **2009**, *30*, 3679–3733.

14. Teluguntla, P.; Thenkabail, P.S.; Xiong, J.; Gumma, M.K.; Congalton, R.G.; Oliphant, A.; Poehnelt, J.; Yadav, K.; Rao, M.N.; Massey, R. Spectral Matching Techniques (SMTs) and Automated Cropland Classification Algorithms (ACCAs) for Mapping Croplands of Australia using MODIS 250-m Time-series (2000–2015) Data. *Int. J. Digit. Earth* **2017**, 944–977.
15. Xiong, J.; Thenkabail, P.S.; Gumma, M.K.; Teluguntla, P.; Poehnelt, J.; Congalton, R.G.; Yadav, K.; Thau, D. Automated cropland mapping of continental Africa using Google Earth Engine cloud computing. *ISPRS J. Photogramm. Remote Sens.* **2017**, *126*, 225–244.
16. Salmon, J.M.; Friedl, M.A.; Frolking, S.; Wisser, D.; Douglas, E.M. Global rain-fed, irrigated, and paddy croplands: A new high resolution map derived from remote sensing, crop inventories and climate data. *Int. J. Appl. Earth Obs. Geoinf.* **2015**, *38*, 321–334.
17. Alcántara, C.; Kuemmerle, T.; Prishchepov, A.V.; Radeloff, V.C. Mapping abandoned agriculture with multi-temporal MODIS satellite data. *Remote Sens. Environ.* **2012**, *124*, 334–347.
18. Estel, S.; Kuemmerle, T.; Alcántara, C.; Levers, C.; Prishchepov, A.; Hostert, P. Mapping farmland abandonment and recultivation across Europe using MODIS NDVI time series. *Remote Sens. Environ.* **2015**, *163*, 312–325.
19. Friedl, M.A.; Sulla-Menashe, D.; Tan, B.; Schneider, A.; Ramankutty, N.; Sibley, A.; Huang, X. MODIS Collection 5 global land cover: Algorithm refinements and characterization of new datasets. *Remote Sens. Environ.* **2010**, *114*, 168–182.
20. Boryan, C.; Yang, Z.; Mueller, R.; Craig, M. Monitoring US agriculture: The US Department of Agriculture, National Agricultural Statistics Service, Cropland Data Layer Program. *Geocarto Int.* **2011**, *26*, 341–358.
21. Vintrou, E.; Desbrosse, A.; Bégué, A.; Traoré, S. Crop area mapping in West Africa using landscape stratification of MODIS time series and comparison with existing global land products. *Int. J. Appl. Earth Obs. Geoinf.* **2012**, *14*, 83–93.
22. Dheeravath, V.; Thenkabail, P.S.; Thenkabail, P.S.; Noojipady, P.; Chandrakantha, G.; Reddy, G.P.O.; Gumma, M.K.; Biradar, C.M.; Velpuri, M.; Gumma, M.K. Irrigated areas of India derived using MODIS 500 m time series for the years 2001–2003. *ISPRS J. Photogramm. Remote Sens.* **2010**, *65*, doi:10.1016/j.isprsjprs.2009.08.004.
23. Biradar, C.M.; Thenkabail, P.S.; Noojipady, P.; Li, Y.; Dheeravath, V.; Turral, H.; Velpuri, M.; Gumma, M.K.; Gangalakunta, O.R.P.; Cai, X.L.; et al. A global map of rainfed cropland areas (GMRCA) at the end of last millennium using remote sensing. *Int. J. Appl. Earth Obs. Geoinf.* **2009**, *11*, 114–129.
24. Lambert, M.J.; Waldner, F.; Defourny, P. Cropland Mapping over Sahelian and Sudanian Agrosystems: A Knowledge-Based Approach Using PROBA-V Time Series at 100-m. *Remote Sens.* **2016**, *8*, doi:10.3390/rs8030232.
25. Shao, Y.; Lunetta, R.S. Comparison of support vector machine, neural network, and CART algorithms for the land-cover classification using limited training data points. *ISPRS J. Photogramm. Remote Sens.* **2012**, *70*, 78–87.
26. Kussul, N.; Skakun, S.; Shelestov, A. Regional scale crop mapping using multi-temporal satellite imagery. *Int. Arch. Photogramm. Remote Sens. Spat. Inf. Sci.* **2015**, *40*, 45–52.
27. Skakun, S.; Kussul, N.; Shelestov, A.Y.; Lavreniuk, M.; Kussul, O. Efficiency Assessment of Multitemporal C-Band Radarsat-2 Intensity and Landsat-8 Surface Reflectance Satellite Imagery for Crop Classification in Ukraine. *IEEE J. Sel. Top. Appl. Earth Obs. Remote Sens.* **2015**, *9*, 3712–3719.
28. Huang, C.; Davis, L.S.; Townshend, J.R.G. An assessment of support vector machines for land cover classification. *Int. J. Remote Sens.* **2010**, *23*, 725–749.
29. Vintrou, E.; Ienco, D.; Bégué, A. Data mining, a promising tool for large-area cropland mapping. *IEEE J. Sel. Top. Appl. Earth Obs. Remote Sens.* **2013**, *6*, 2132–2138.
30. Pan, Y.; Hu, T.; Zhu, X.; Zhang, J. Mapping cropland distributions using a hard and soft classification model. *IEEE Trans. Geosci. Remote Sens.* **2012**, *50*, 4301–4312.
31. Van Niel, T.G.; McVicar, T.R. Determining temporal windows for crop discrimination with remote sensing: A case study in south-eastern Australia. *Comput. Electron. Agric.* **2004**, *45*, 91–108.
32. Conrad, C.; Dech, S.; Dubovyk, O.; Fritsch, S.; Klein, D.; Löw, F.; Schorcht, G.; Zeidler, J. Derivation of temporal windows for accurate crop discrimination in heterogeneous croplands of Uzbekistan using multitemporal RapidEye images. *Comput. Electron. Agric.* **2014**, *103*, 63–74.

33. Löw, F.; Michel, U.; Dech, S.; Conrad, C. Impact of feature selection on the accuracy and spatial uncertainty of per-field crop classification using Support Vector Machines. *ISPRS J. Photogramm. Remote Sens.* **2013**, *85*, 102–119.
34. Matton, N.; Canto, G.; Waldner, F.; Valero, S.; Morin, D.; Inglada, J.; Arias, M.; Bontemps, S.; Koetz, B.; Defourny, P. An Automated Method for Annual Cropland Mapping along the Season for Various Globally-Distributed Agrosystems Using High Spatial and Temporal Resolution Time Series. *Remote Sens.* **2015**, *7*, 13208–13232.
35. Peña-Barragán, J.M.; Ngugi, M.K.; Plant, R.E.; Six, J. Object-based crop identification using multiple vegetation indices, textural features and crop phenology. *Remote Sens. Environ.* **2011**, *115*, 1301–1316.
36. Johnson, D.M.; Mueller, R. The 2009 Cropland Data Layer. *Photogramm. Eng. Remote Sens.* **2010**, *76*, 1201–1205.
37. Kalensky, Z.D. AFRICOVER Land Cover Database and Map of Africa. *Can. J. Remote Sens.* **1998**, *24*, 292–297.
38. Bartholomé, E.; Belward, A.S. GLC2000: A new approach to global land cover mapping from Earth observation data. *Int. J. Remote Sens.* **2005**, *26*, 1959–1977.
39. Arino, O.; Gross, D.; Ranera, F.; Leroy, M.; Bicheron, P.; Brockman, C.; Defourny, P.; Vancutsem, C.; Achard, F.; Durieux, L.; et al. GlobCover: ESA service for global land cover from MERIS. In Proceedings of the IEEE International Geoscience and Remote Sensing Symposium, Barcelona, Spain, 23–27 July 2007; pp. 2412–2415.
40. Latham, J.; Cumani, R.; Rosati, I.; Bloise, M. *Global Land Cover Share (GLC-SHARE) Database Beta-Release Version 1.0-2014*; FAO: Rome, Italy, 2014. Available online: http://csdms.colorado.edu/wiki/Data:GLC-SHARE (accessed on 31 May 2017).
41. Friedl, M.A.; McIver, D.K.; Hodges, J.C.F.; Zhang, X.Y.; Muchoney, D.; Strahler, A.H.; Woodcock, C.E.; Gopal, S.; Schneider, A.; Cooper, A.; et al. Global land cover mapping from MODIS: algorithms and early results. *Remote Sens. Environ.* **2002**, *83*, 287–302.
42. Waldner, F.; Fritz, S.; Di Gregorio, A.; Defourny, P. Mapping Priorities to Focus Cropland Mapping Activities: Fitness Assessment of Existing Global, Regional and National Cropland Maps. *Remote Sens.* **2015**, *7*, 7959–7986.
43. Teluguntla, P.; Thenkabail, P.; Xiong, J.; Gumma, M.K.; Giri, C.; Milesi, C.; Ozdogan, M.; Congalton, R.; Yadav, K. CHAPTER 6—Global Food Security Support Analysis Data at Nominal 1 km (GFSAD1 km) Derived from Remote Sensing in Support of Food Security in the Twenty-First Century: Current Achievements and Future Possibilities. In *Remote Sensing Handbook (Volume II): Land Resources Monitoring, Modeling, and Mapping with Remote Sensing*; Thenkabail, P.S., Ed.; CRC Press: Boca Raton, FL, USA; London, UK; New York, NY, USA, 2015; pp. 131–160.
44. Waldner, F.; De Abelleyra, D.; Verón, S.R.; Zhang, M.; Wu, B.; Plotnikov, D.; Bartalev, S.; Lavreniuk, M.; Skakun, S.; Kussul, N.; et al. Towards a set of agrosystem-specific cropland mapping methods to address the global cropland diversity. *Int. J. Remote Sens.* **2016**, *37*, 3196–3231.
45. Chen, J.; Chen, J.; Liao, A.; Cao, X.; Chen, L.; Chen, X.; He, C.; Han, G.; Peng, S.; Zhang, W.; et al. Global land cover mapping at 30 m resolution: A POK-based operational approach. *ISPRS J. Photogramm. Remote Sens.* **2015**, *103*, 7–27.
46. Gong, P.; Wang, J.; Yu, L.; Zhao, Y.; Zhao, Y.; Liang, L.; Niu, Z.; Huang, X.; Fu, H.; Liu, S.; et al. Finer resolution observation and monitoring of global land cover: first mapping results with Landsat TM and ETM+ data. *Int. J. Remote Sens.* **2013**, *34*, 2607–2654.
47. Hansen, M.C.; Loveland, T.R. A review of large area monitoring of land cover change using Landsat data. *Remote Sens. Environ.* **2012**, *122*, 66–74.
48. Yu, L.; Wang, J.; Clinton, N.; Xin, Q.; Chen, Y.; Zhong, L.; Gong, P. FROM-GC: 30 m global cropland extent derived through multisource data integration. *Int. J. Digit. Earth* **2013**, *6*, 521–533, doi:10.1080/17538947.2013.822574.
49. Costa, H.; Carrão, H.; Bação, F.; Caetano, M. Combining per-pixel and object-based classifications for mapping land cover over large areas. *Int. J. Remote Sens.* **2014**, *35*, 738–753.
50. Myint, S.W.; Gober, P.; Brazel, A.; Gober, P.; Brazel, A.; Grossman-Clarke, S.; Grossman-Clarke, S.; Weng, Q. Per-pixel vs. object-based classification of urban land cover extraction using high spatial resolution imagery. *Remote Sens. Environ.* **2011**, *115*, 1145–1161.

51. Malinverni, E.S.; Tassetti, A.N.; Mancini, A.; Zingaretti, P.; Frontoni, E.; Bernardini, A. Hybrid object-based approach for land use/land cover mapping using high spatial resolution imagery. *Int. J. Geogr. Inf. Sci.* **2011**, *25*, 1025–1043.
52. Dingle Robertson, L.; King, D.J. Comparison of pixel- and object-based classification in land cover change mapping. *Int. J. Remote Sens.* **2011**, *32*, 1505–1529.
53. Ok, A.O.; Akar, O.; Gungor, O. Evaluation of random forest method for agricultural crop classification. *Eur. J. Remote Sens.* **2012**, *45*, doi:10.5721/EuJRS20124535.
54. De Wit, A.J.W.; Clevers, J.G.P.W. Efficiency and accuracy of per-field classification for operational crop mapping. *Int. J. Remote Sens.* **2010**, *25*, 4091–4112.
55. Castillejo-González, I.L.; López-Granados, F.; García-Ferrer, A.; Peña-Barragán, J.M.; Jurado-Expósito, M.; de la Orden, M.S.; González-Audicana, M. Object- and pixel-based analysis for mapping crops and their agro-environmental associated measures using QuickBird imagery. *Comput. Electron. Agric.* **2009**, *68*, 207–215.
56. Marshall, M.T.; Husak, G.J.; Michaelsen, J.; Funk, C.; Pedreros, D.; Adoum, A. Testing a high-resolution satellite interpretation technique for crop area monitoring in developing countries. *Int. J. Remote Sens.* **2011**, *32*, 7997–8012.
57. Gorelick, N.; Hancher, M.; Dixon, M.; Ilyushchenko, S.; Thau, D.; Moore, R. Google Earth Engine: Planetary-scale geospatial analysis for everyone. *Remote Sens. Environ.* **2017**, doi:10.1016/j.rse.2017.06.031.
58. Lupien, J.R. *Agriculture Food and Nutrition for Africa–A Resource Book for Teachers of Agriculture*; FAO: Rome, Italy, 1997. Available online: http://www.fao.org/docrep/w0078e/w0078e00.htm (accessed on 12 October 2016).
59. Gerland, P.; Raftery, A.E.; ikova, H.E.; Li, N.; Gu, D.; Spoorenberg, T.; Alkema, L.; Fosdick, B.K.; Chunn, J.; Lalic, N.; et al. World population stabilization unlikely this century. *Science* **2014**, *346*, 234–237.
60. Drusch, M.; Del Bello, U.; Carlier, S.; Colin, O.; Fernandez, V.; Gascon, F.; Hoersch, B.; Isola, C.; Laberinti, P.; Martimort, P.; et al. Sentinel-2: ESA's Optical High-Resolution Mission for GMES Operational Services. *Remote Sens. Environ.* **2012**, *120*, 25–36.
61. Immitzer, M.; Vuolo, F.; Atzberger, C. First Experience with Sentinel-2 Data for Crop and Tree Species Classifications in Central Europe. *Remote Sens.* **2016**, *8*, 166, doi:10.3390/rs8030166.
62. Battude, M.; Al Bitar, A.; Morin, D.; Cros, J.; Huc, M.; Sicre, C.M.; Le Dantec, V.; Demarez, V. Estimating maize biomass and yield over large areas using high spatial and temporal resolution Sentinel-2 like remote sensing data. *Remote Sens. Environ.* **2016**, *184*, 668–681.
63. Inglada, J.; Arias, M.; Tardy, B.; Hagolle, O.; Valero, S.; Morin, D.; Dedieu, G.; Sepulcre, G.; Bontemps, S.; Defourny, P.; et al. Assessment of an Operational System for Crop Type Map Production Using High Temporal and Spatial Resolution Satellite Optical Imagery. *Remote Sens.* **2015**, *7*, 12356–12379.
64. Valero, S.; Morin, D.; Inglada, J.; Sepulcre, G.; Arias, M.; Hagolle, O.; Dedieu, G.; Bontemps, S.; Defourny, P.; Koetz, B. Production of a Dynamic Cropland Mask by Processing Remote Sensing Image Series at High Temporal and Spatial Resolutions. *Remote Sens.* **2016**, *8*, 55, doi:10.3390/rs8010055.
65. Lohou, F.; Kergoat, L.; Guichard, F.; Boone, A.; Cappelaere, B.; Cohard, J.M.; Demarty, J.; Galle, S.; Grippa, M.; Peugeot, C.; et al. Surface response to rain events throughout the West African monsoon. *Atmos. Chem. Phys.* **2014**, *14*, 3883–3898.
66. Hentze, K.; Thonfeld, F.; Menz, G. Evaluating Crop Area Mapping from MODIS Time-Series as an Assessment Tool for Zimbabwe's "Fast Track Land Reform Programme". *PLoS ONE* **2016**, *11*, doi:10.1371/journal.pone.0156630.
67. Kidane, Y.; Stahlmann, R.; Beierkuhnlein, C. Vegetation dynamics, and land use and land cover change in the Bale Mountains, Ethiopia. *Environ. Monit. Assess.* **2012**, *184*, 7473–7489.
68. Kruger, A.C. Observed trends in daily precipitation indices in South Africa: 1910–2004. *Int. J. Climatol.* **2006**, *26*, 2275–2285.
69. Motha, R.P.; Leduc, S.K.; Steyaert, L.T.; Sakamoto, C.M.; Strommen, N.D.; Motha, R.P.; Leduc, S.K.; Steyaert, L.T.; Sakamoto, C.M.; Strommen, N.D. Precipitation Patterns in West Africa. *Mon. Weather Rev.* **1980**, *108*, 1567–1578.
70. D'Odorico, P.; Gonsamo, A.; Damm, A.; Schaepman, M.E. Experimental Evaluation of Sentinel-2 Spectral Response Functions for NDVI Time-Series Continuity. *IEEE Trans. Geosci. Remote Sens.* **2013**, *51*, 1336–1348.

71. Van der Werff, H.; van der Meer, F. Sentinel-2A MSI and Landsat 8 OLI Provide Data Continuity for Geological Remote Sensing. *Remote Sens.* **2016**, *8*, 883, doi:10.3390/rs8110883.
72. Storey, J.; Roy, D.P.; Masek, J.; Gascon, F.; Dwyer, J.; Choate, M. A note on the temporary misregistration of Landsat-8 Operational Land Imager (OLI) and Sentinel-2 Multi Spectral Instrument (MSI) imagery. *Remote Sens. Environ.* **2016**, *186*, 121–122.
73. Languille, F.; Déchoz, C.; Gaudel, A.; Greslou, D.; de Lussy, F.; Trémas, T.; Poulain, V. Sentinel-2 geometric image quality commissioning: First results. *Proc. SPIE* **2015**, *9643*, 964306, doi:10.1117/12.2194339.
74. Barazzetti, L.; Cuca, B.; Previtali, M. Evaluation of registration accuracy between Sentinel-2 and Landsat 8. *Proc. SPIE* **2016**, doi:10.1117/12.2241765.
75. Farr, T.G.; Rosen, P.A.; Caro, E.; Crippen, R.; Duren, R.; Hensley, S.; Kobrick, M.; Paller, M.; Rodriguez, E.; Roth, L.; et al. The Shuttle Radar Topography Mission. *Rev. Geophys.* **2007**, *45*, doi:10.1029/2005RG000183.
76. Aitkenhead, M.J.; Aalders, I.H. Automating land cover mapping of Scotland using expert system and knowledge integration methods. *Remote Sens. Environ.* **2011**, *115*, 1285–1295.
77. Pelletier, C.; Valero, S.; Inglada, J.; Champion, N.; Dedieu, G. Assessing the robustness of Random Forests to map land cover with high resolution satellite image time series over large areas. *Remote Sens. Environ.* **2016**, *187*, 156–168.
78. Sharma, R.; Tateishi, R.; Hara, K.; Iizuka, K. Production of the Japan 30-m Land Cover Map of 2013–2015 Using a Random Forests-Based Feature Optimization Approach. *Remote Sens.* **2016**, *8*, 429, doi:10.3390/rs8050429.
79. Wessels, K.; van den Bergh, F.; Roy, D.; Salmon, B.; Steenkamp, K.; MacAlister, B.; Swanepoel, D.; Jewitt, D. Rapid Land Cover Map Updates Using Change Detection and Robust Random Forest Classifiers. *Remote Sens.* **2016**, *8*, 888.
80. Vapnik, V.N.; Vapnik, V. *Statistical Learning Theory*; Wiley: New York, NY, USA, 1998.
81. Shi, D.; Yang, X. Support Vector Machines for Land Cover Mapping from Remote Sensor Imagery. In *Monitoring and Modeling of Global Changes: A Geomatics Perspective*; Springer: Dordrecht, The Netherlands, 2015; pp. 265–279.
82. Im, J.; Jensen, J.R.; Tullis, J.A. Object-based change detection using correlation image analysis and image segmentation. *Int. J. Remote Sens.* **2008**, *29*, 399–423.
83. Stow, D.; Hamada, Y.; Coulter, L.; Anguelova, Z. Monitoring shrubland habitat changes through object-based change identification with airborne multispectral imagery. *Remote Sens. Environ.* **2008**, *112*, 1051–1061.
84. Espindola, G.; Câmara, G.; Reis, I.; Bins, L.; Monteiro, A. Parameter selection for region-growing image segmentation algorithms using spatial autocorrelation. *Int. J. Remote Sens.* **2006**, *27*, 3035–3040.
85. Nemani, R.; Votava, P.; Michaelis, A.; Melton, F.; Milesi, C. Collaborative supercomputing for global change science. *Eos Trans. Am. Geophys. Union* **2011**, *92*, 109–110.
86. Tilton, J.C.; Tarabalka, Y.; Montesano, P.M.; Gofman, E. Best Merge Region-Growing Segmentation with Integrated Nonadjacent Region Object Aggregation. *IEEE Trans. Geosci. Remote Sens.* **2012**, *50*, 4454–4467.
87. Sulla-Menashe, D.; Friedl, M.A.; Krankina, O.N.; Baccini, A.; Woodcock, C.E.; Sibley, A.; Sun, G.; Kharuk, V.; Elsakov, V. Hierarchical mapping of Northern Eurasian land cover using MODIS data. *Remote Sens. Environ.* **2011**, *115*, 392–403.
88. Congalton, R.G.; Green, K. *Assessing the Accuracy of Remotely Sensed Data: Principles and Practice*, 2nd ed.; CRC/Taylor & Francis: Boca Raton, FL, USA, 2009; p. 183.
89. Congalton, R.G. Assessing Positional and Thematic Accuracies of Maps Generated from Remotely Sensed Data. In *"Remote Sensing Handbook" (Volume I): Remotely Sensed Data Characterization, Classification, and Accuracies*; Thenkabail, P.S., Ed.; CRC Press: Boca Raton, FL, USA; London, UK; New York, NY, USA, 2015; pp. 583–602.
90. Thenkabail, P.S.; Knox, J.W.; Ozdogan, M.; Gumma, M.K.; Congalton, R.G.; Wu, Z.; Milesi, C.; Finkral, A.; Marshall, M.; Mariotto, I.; et al. Assessing Future Risks to Agricultural Productivity, Water Resources and Food Security: How Can Remote Sensing Help? *Photogramm. Eng. Remote Sens.* **2012**, *78*, 773–782.
91. Chapagain, A.K.; Hoekstra, A.Y. The global component of freshwater demand and supply: An assessment of virtual water flows between nations as a result of trade in agricultural and industrial products. *Water Int.* **2008**, *33*, 19–32.

92. Portmann, F.T.; Siebert, S.; Döll, P. MIRCA2000—Global monthly irrigated and rainfed crop areas around the year 2000: A new high-resolution data set for agricultural and hydrological modeling. *Glob. Biogeochem. Cycles* **2010**, *24*, 1–24.
93. Dorward, A.; Chirwa, E. A Review of Methods for Estimating Yield and Production Impacts. 2010. Available online: http://eprints.soas.ac.uk/16731/1/FISP%20Production%20Methodologies%20review%20Dec%20Final.pdf (accessed on 10 August 2016).

Article

SnowCloudHydro—A New Framework for Forecasting Streamflow in Snowy, Data-Scarce Regions

Eric A. Sproles [1,2,*], Ryan L. Crumley [3], Anne W. Nolin [1,4], Eugene Mar [1] and Juan Ignacio Lopez Moreno [5]

1 College of Earth, Ocean, and Atmospheric Sciences, Oregon State University, Corvallis, OR 97331, USA; anne.nolin@gmail.com (A.W.N.); gmar31@comcast.net (E.M.)
2 Department of Earth Sciences, Montana State University, Bozeman, MT 59717, USA
3 Water Resources Science, Oregon State University, Corvallis, OR 77331, USA; ryanlcrumley@gmail.com
4 Department of Geography, University of Nevada, Reno, NV 97331, USA
5 Pyrenean Institute of Ecology, CSIC, 50059 Zaragoza, Spain; nlopez@ipe.csic.es
* Correspondence: eric.sproles@montana.edu

Received: 19 June 2018; Accepted: 1 August 2018; Published: 13 August 2018

Abstract: We tested the efficacy and skill of SnowCloud, a prototype web-based, cloud-computing framework for snow mapping and hydrologic modeling. SnowCloud is the overarching framework that functions within the Google Earth Engine cloud-computing environment. SnowCloudMetrics is a sub-component of SnowCloud that provides users with spatially and temporally composited snow cover information in an easy-to-use format. SnowCloudHydro is a simple spreadsheet-based model that uses Snow Cover Frequency (SCF) output from SnowCloudMetrics as a key model input. In this application, SnowCloudMetrics rapidly converts NASA's Moderate Resolution Imaging Spectroradiometer (MODIS) daily snow cover product (MOD10A1) into a monthly snow cover frequency for a user-specified watershed area. SnowCloudHydro uses SCF and prior monthly streamflow to forecast streamflow for the subsequent month. We tested the skill of SnowCloudHydro in three snow-dominated headwaters that represent a range of precipitation/snowmelt runoff categories: the Río Elqui in Northern Chile; the John Day River, in the Northwestern United States; and the Río Aragón in Northern Spain. The skill of the SnowCloudHydro model directly corresponded to snowpack contributions to streamflow. Watersheds with proportionately more snowmelt than rain provided better results (R^2 values: 0.88, 0.52, and 0.22, respectively). To test the user experience of SnowCloud, we provided the tools and tutorials in English and Spanish to water resource managers in Chile, Spain, and the United States. Participants assessed their user experience, which was generally very positive. While these initial results focus on SnowCloud, they outline methods for developing cloud-based tools that can function effectively across cultures and languages. Our approach also addresses the primary challenges of science-based computing; human resource limitations, infrastructure costs, and expensive proprietary software. These challenges are particularly problematic in countries where scientific and computational resources are underdeveloped.

Keywords: cloud computing; remote sensing; snow hydrology; water resources; Google Earth Engine; user assessment; MODIS; snow cover

---

1. Introduction

Mountain snowpack collects, stores, and releases water that fills streams and recharges aquifers, functioning as an essential water resource for people, economies, and ecosystems. This annual cycle of accumulation and melt represents one of the most profound seasonal changes on the surface of the Earth [1] (Figure 1). Globally, however, warmer conditions have limited the accumulation of mountain snowpack and hastened its melt, leading to earlier spring snowmelt runoff [2–8]. These same shifts are expected to negatively affect groundwater recharge in mountainous regions [9], and in turn groundwater levels. Currently over one billion people rely on glaciers and seasonal snowpack as their water supply [10], and the global demand for water is projected to increase with growing populations and changing global economies [11]. These shifts in demand are coupled with supply becoming increasingly uncertain in the face of current climate trends [10,12].

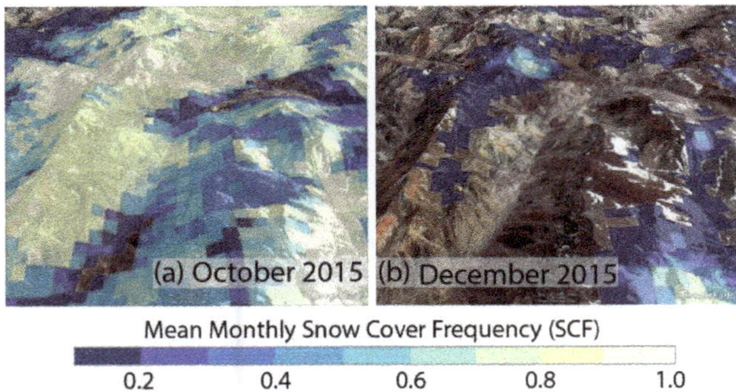

Figure 1. Maps of monthly snow cover frequency in the Chilean Andes (30° S, ranging in elevation from 3135 to 6200 m) for October ((**a**); spring) and December ((**b**); early summer) showing the extreme changes in seasonal snow.

Despite its importance, measurements of mountain snowpacks are sparse, and even when available they rely on monitoring networks that are commonly not representative of general topographic conditions (i.e., they are situated on flat ground in mountainous landscapes and in areas where the forest is has been cleared) [1,13]. Additionally, these are point-based measurements at stationary locations—functioning in a non-stationary climate. Since the availability and scarcity of water vary in time and space [14], the ability to better understand and quantify the accumulation and melt of mountain snowpack would advance science and improve the capacity for better-informed water resource management.

Remotely-sensed data capture the variability of snow across rugged mountain topography and bridge sparse monitoring networks [1,15]. For example, NASA's Moderate Resolution Imaging Spectroradiometer (MODIS) provides daily snowcover data (MOD10A1) with global coverage. MODIS and other remotely-sensed snow (RSS) products provide daily global-to-local scale coverage of changing mountain snowcover [15,16]. Since RSS measurements provide near-real time data, global coverage, and a consistent historical data record [17], they are pivotal to better understanding snow hydrology, climate change, and related socio-environmental systems [18]. With broad spatial and temporal coverage, RSS products support new snow metrics and novel insights into the spatial and temporal connections between mountain snowpack and downstream water resources that support both basic and applied scientific insights [19]. Additionally, RSS products provide key input data for streamflow prediction models, and are especially valuable in data sparse watersheds [20–22].

While geographically versatile, RSS data do not readily transfer across institutions and user groups. Logistical and computational challenges abound in transitioning RSS data to actionable water resources information [23]. Traditionally, working with these massive datasets would require high bandwidth Internet, large digital storage capacity, and expensive hardware and software to download, process, and analyze. Such processing also typically requires a high level of technical expertise. These computational and human resource burdens limit these types of data from being readily implemented by managers and researchers [24].

These challenges limit the use of RSS data in both developed and developing countries, even though there is a critical global need for accessible snowcover information by resource managers and decision-makers. This demand and underutilization of RSS echoes similar concerns voiced by Peter H. Raven regarding the successful dissemination of scientific information. In his presidential address to the American Association for the Advancement of Science in 2002, Raven stated that disseminating scientific information will be fulfilled by approaches that combine advances in understanding, social capacity, and technology [25]. While progress has been made in these efforts, resource managers remain underserved in access to actionable scientific information needed to form decisions for complex, long-term challenges [26].

Cloud computing provides valuable opportunities to address issues of data access, human resource capacity, and computational intensity associated with geospatial information [27,28]. The web-based access and efficiency of cloud computing relieves the computational, budgetary, and logistical challenges of RSS products, provides a low barrier to users implementing this technology, and allows stakeholders and scientists to collaborate on timely and informed decisions in addition to basic and applied scientific discovery [27–29]. These advances speak to the long-term relevance of cloud computing to provide a reliable framework for RSS products and to the democratization of data [30,31]. However, simply developing cloud-based tools will not disseminate scientific information. The success of this new cloud-based paradigm will require advances in technological and social capacity that connect data providers to scientists and resource managers to create an end-to-end information system.

We present SnowCloud, an end-to-end cloud-computing framework comprised of (i) SnowCloudMetrics, cloud-based tools that transition RSS products into actionable snow metrics [19]; and (ii) SnowCloudHydro, a simple hydrologic model for snow dominated watersheds that relies solely on monthly Snow Cover Frequency (SCF) and previous streamflow to forecast monthly streamflow with a one-month lead-time. We also present the insights of water resource professionals collected from an anonymous survey that reflect their perspectives on the value of SnowCloud and cloud-based computing for improving water resource management. The intent of this paper is to demonstrate the prototype SnowCloud framework, showing snow mapping and hydrologic forecasting results from three case study watersheds. We also present ideas and methods on how to better integrate and manage large datasets for hydrological forecasts using an interactive, cloud-based approach.

2. Materials and Methods

The SnowCloud framework was tested in three snow-dominated watersheds: La Laguna (the headwaters of the Río Elqui in semi-arid, Northern Central Chile), the John Day River (a semi-arid watershed in Eastern Oregon, Northwestern USA), and the Río Aragón (in the Spanish Pyrenees) (Table 1). These watersheds were chosen to test the model framework where snowmelt is a major water resource, but in distinctly different climates. The watersheds were delineated from each respective stream gage, and streamflow data were provided by the respective managing agency [32–34]. Table 1 provides brief descriptions of each watershed's topography, climate, mean annual SCF (2002–2016), and the runoff ratio (long-term average streamflow, Q, to long-term average precipitation, P) [35].

Figure 2 provides a conceptual overview of the SnowCloud framework for the reader, which is explained in greater detail in the Methods section.

Table 1. The topographic and climatic characteristics of the three study watersheds. Mean temperature (T) and Snow Cover Frequency (SCF) corresponds to the four primary winter months (May–September in the southern hemisphere, and November–March in the northern hemisphere).

	La Laguna	John Day	Aragón
Latitude	30°S	44°N	42°N
Mean T (°C)	−4.9	−1.3	9.5
Min Elevation (m)	3135	933	794
Max Elevation (m)	6200	2733	2858
Average Precipitation (mm year^{-1})	250	600	800
Mean SCF	0.33	0.27	0.27
Runoff Ratio (mm/mm)	0.41	0.26	0.77
Area (km^2)	568	1036	242

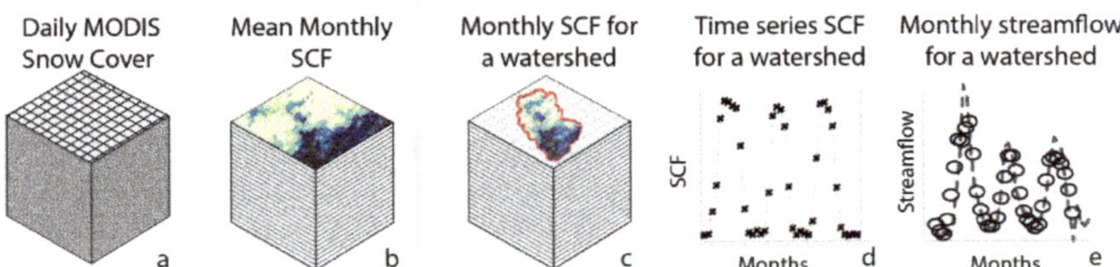

Figure 2. A conceptual flowchart of the SnowCloud framework. (**a**–**c**) SnowCloudMetrics calculating Snow Cover Frequency (SCF) for a watershed using Moderate Resolution Imaging Spectroradiometer (MODIS) satellite data and is (**d**,**e**) transitioned into a streamflow forecast using SnowCloudHydro. (**a**) Data arrays and tiling of satellite images are used to (**b**) calculate SCF for each month. (**c**) These monthly data are then spatially subset and averaged by study watershed, and (**d**) converted into a time series for the MODIS record (three years of data shown in this figure). (**e**) The SCF data are then implemented into a streamflow forecast model for the watershed with a one-month lead-time.

Here we used SnowCloudMetrics to compute the monthly SCF from the daily MOD10A1 [36] snow cover product at 500-m resolution. For each MODIS pixel, SCF represents the ratio of the number of days in a month that the pixel is snow covered:

$$SCF_{monthly} = \frac{\text{\# of snow observations}}{\text{\# of valid observations}} \quad (1)$$

Since cloud cover commonly obscures satellite observations of snow, a per-pixel cloud correction was implemented based upon the following rules:

(i) If consecutive cloudy days occur between two snow-covered days, then the cloudy days are interpreted as snow-covered;
(ii) If consecutive cloudy days occur between two non-snow-covered days, then the cloudy days are interpreted as non-snow-covered;
(iii) If consecutive cloudy days occur between an antecedent snow-covered day and a subsequent non-snow-covered day, then the cloudy days are interpreted as non-snow-covered; and
(iv) If consecutive cloudy days occur between an antecedent non-snow-covered day and a subsequent snow-covered day, then the cloudy days are interpreted as snow-covered.

The two corollaries to the four conditions listed above are (i) if cloudy days are succeeded by a snow day, then the cloudy days are considered as snow days; and (ii) if cloudy days are succeeded by a non-snow-covered day, then the cloudy days are considered non-snow-covered.

If the end of the analysis period has consecutive cloudy days, an additional test is performed. For this special case, the underlying snow condition for cloud pixels is estimated by including an additional 30 days at the end of the analysis period, then stepping backwards, applying the two corollaries to identify the snow or no-snow land condition for each pixel. When analysis of that 30-day period is complete, the snow/no-snow condition is accordingly assigned to any cloudy pixels at the end of the analysis period.

After the monthly calculations have been completed for each pixel, they are averaged over the watershed to provide a spatially aggregated fraction of days in a month with snow cover.

SCF values are computed in the Google Earth Engine (GEE) Code Editor, an integrated platform that provides planetary-scale geospatial analysis in a cloud computing environment [24]. Google Earth Engine allows users to access and analyze data from a multi-petabyte data catalog (including MODIS). Developers can create fast computing tools using code written in either JavaScript or Python in the Code Editor. The code runs on the GEE cloud and results are displayed as map-based visualizations [24]. The results of the SCF calculations are then exported to a comma-separated value file that can be readily input to any spreadsheet application.

These SCF calculations were used as input for SnowCloudHydro, a runoff model developed to forecast streamflow for snow-dominated watersheds in data-scarce northern Chile [21]. The structure of the model requires only monthly SCF and previous monthly streamflows to forecast monthly streamflow (Q_m) with one-month lead time based upon the following algorithm:

$$Q_m = a\overline{SCF}_{m-1}^{b} + c\overline{Q}_{m-1}^{*} + d\overline{Q}_{m-1}^{\prime} \tag{2}$$

where $\overline{SCF}_{m-1}$ represents the six-month running average of snow cover frequency, encapsulating annual accumulation and melt cycles. This approach also allows winters with more extensive and prolonged snowpack to provide greater melt contributions later into the spring and early summer. $\overline{Q}_{m-1}^{*}$ is the two-month moving average of streamflow that represents seasonal changes and conditions, and whether streamflows are increasing or decreasing. $\overline{Q}_{m-1}^{\prime}$ is the twelve-month moving average of streamflow that represents annual base flow contributions [21]. The parameters a, b, c, and d are scaling coefficients.

The initial model was calibrated using data from 36 of 144 months. The calibration implemented code that executed the GLUE methodology [37] to find solutions for the scaling parameters (a, b, c, and d) that optimize the model. GLUE incorporates a series of dotty plots comprised of 5000 Monte Carlo simulations per iteration. In the optimization process, each scaling parameter was individually optimized through visual inspection using dotty plots (which show objective function values as a function of model parameters) and computational metrics (Nash-Sutcliffe Efficiency (NSE) [38,39] and R^2). The model was then validated using the remaining 108 months of data. This rigorous calibration and validation helped ensure that the model was getting the right answers for the right reasons [40]. The original SCF-Runoff model displayed a high level of skill in predicting streamflow with NSE and R^2 values of 0.83. For a more detailed description of the model, its uncertainties, and its results please refer to Sproles, et al. [21].

One of the primary goals of the SnowCloudHydro framework is to alleviate the computational and software demands required when using a robust methodology like the GLUE method. GLUE requires coding expertise and access to advanced computing resources. The same SnowCloudHydro model was tested in a cloud-based spreadsheet application (Google Sheets) by implementing the Solver extension and the non-linear least squares method to solve for the a, b, c, and d parameters.

Spreadsheets for each of the three watersheds were developed that included SCF calculations and monthly streamflow data. Optimized solutions for the a, b, c, and d parameters were solved in the spreadsheets for each watershed using only a calibration process. To test the efficacy of the optimized non-linear least squares solutions in Solver, the GLUE methodology was also completed for each watershed using a calibration and validation process within MATLAB® scientific computing software [41].

Since the intent of SnowCloud is provide end users with actionable scientific information, we introduced water resource managers to the cloud-based decision support tools through a series of web-based tutorials in English and Spanish. To assess the efficacy of SnowCloudHydro and the user's experiences, a dual-language survey was developed and sent to water resource managers in the three participating countries (Appendix A). The survey provided anonymity for participants, and the only distinguishing component was whether or not it was conducted in English or Spanish. The distribution of the web-based tutorials and surveys were sent by email to participants using an address list curated by the authors. Additionally, the same introductory email and request for participation was emailed through listservs.

This qualitative component of the project was designed and implemented in accordance and compliance with the Human Research Protection Program and the Internal Review Board at Oregon State University (Study ID 8161).

3. Results

3.1. Snow Cover Frequency Calculations Using SnowCloudMetrics

For each of the three case study watersheds, we computed SCF within SnowCloudMetrics. This was done for the period 24 February 2000 to 31 December 2016 (6156 days). The global pre-processing (Figure 2a; creating the data arrays and tiling the daily MODIS datasets) for SnowCloudMetrics required 3.75 min to compile a global MOD10A1 array. Calculations of the mean SCF for an individual month required between 5–10 s for each watershed (Figures 2b,c and 3a–f). The calculations of mean monthly SCF for February 2000–December 2016 required between 4–5 min on average (Figure 2d). The calculations in the cloud also avoided downloading and organizing the requisite MODIS data, which, in total, would have been around 30–40 GB per watershed.

3.2. SnowCloudHydro

The model's skill in predicting streamflow one month in advance varied across watersheds. In the higher elevation La Laguna watershed (mean elev. 4300 m) the simulations provided a high-degree of skill in forecasting streamflow (NSE = 0.87, Figure 3g). In the lower elevation John Day River (mean elev. 1514 m) and Río Aragón (mean elev. 1600 m) watersheds, model skill was lower (R^2 = 0.52 and R^2 = 0.21, Figure 3h,i, respectively). The streamflow simulations for the John Day River do not capture peak streamflow events during the winter, presumably because of rain contributions during these events. Similarly, in the Río Aragón watershed the model does not capture numerous peaks in streamflow, occurring primarily from late spring to mid fall, mainly associated with rain events. These results indicate a need for more detailed analysis of the model's seasonal indicators, however this analysis lies outside the scope of this paper.

The optimized parameters for the GLUE and the non-linear least squares calibration were similar except for the a parameter in the Río Aragón, which was much higher in the non-linear least squares version (Table 2). In all three watersheds the non-linear least squares approach performed slightly better than the more robust GLUE and dotty plot method (Table 2).

Table 2. Values for the different model parameters (*a*–*d*) from the GLUE and non-linear least squares methods from the three study watersheds. The R^2 values associated with method are also provided.

Watershed	Parameter	GLUE	Non-Linear Least Squares
La Laguna	a	5.46	4.21
	b	3.98	2.96
	c	0.75	0.74
	d	0.09	0.05
	NSE	0.83	0.87
	R^2	0.83	0.88
John Day River	a	21.94	22.18
	b	1.74	1.99
	c	0.13	0.11
	d	0.25	0.40
	NSE	0.50	0.52
	R^2	0.50	0.52
Río Aragón	a	38.42	111.26
	b	3.64	5.45
	c	0.12	0.13
	d	0.59	0.61
	NSE	0.18	0.21
	R^2	0.19	0.22

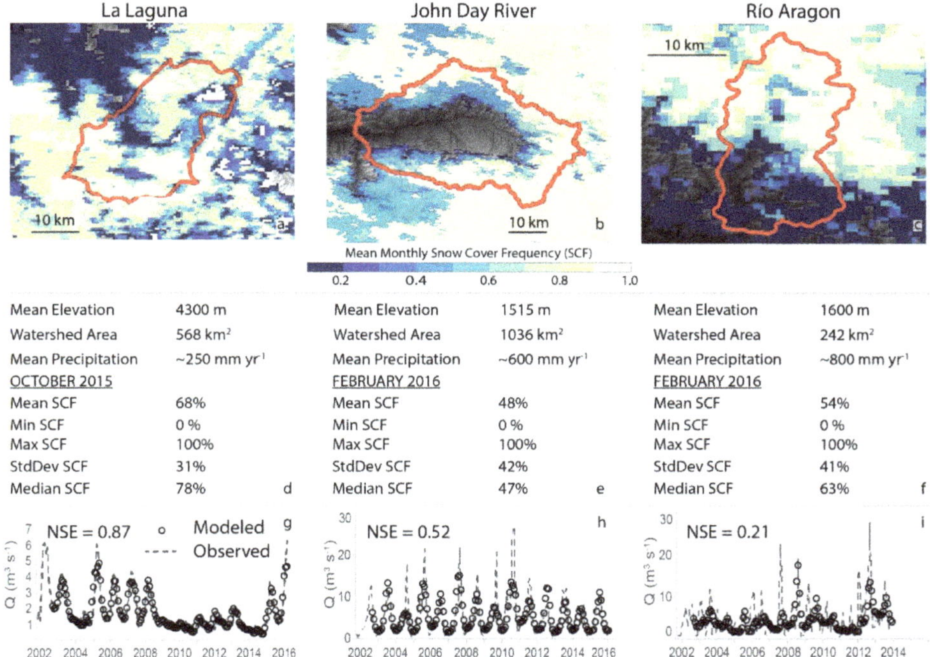

Figure 3. (**a**–**c**) Maps of monthly SCF calculated in SnowCloudMetrics for the three watersheds. (**d**–**f**) Topographic, climatic, and SCF metrics for the three watersheds. (**g**–**i**) The SnowCloudHydro results for the three watersheds. NSE refers to the Nash-Sutcliffe Efficiency coefficient, a standard metric applied to assess the predictive skill of hydrologic models [38,39].

3.3. Users' Assessment of SnowCloud

The participants' responses from the qualitative survey provided an overall positive assessment of the SnowCloud framework (SnowCloudMetrics and SnowCloudHydro) (Figure 4a–j). The professional background of the users was comprised of more experienced water resource professionals (Figure 4a; 9.5 years for English speakers, and 12.6 years for Spanish speakers). The participants' background in cloud computing was varied, ranging from beginners to experienced users (Figure 4b). Following this same trend, the participants used both downloaded and cloud-based data (Figure 4c).

Participants generally perceived the SnowCloud framework and its tools as moderately or extremely useful in calculating SCF and streamflow (Figure 4d,e), and only one English speaker found these tools moderately useless. To better understand which components of the web-interface were most useful, participants were provided an image of the Code Editor web-interface and were asked to identify which components were useful or not useful (Figure 4f and Appendix A). If the area was not selected it was considered neutral. The *Map* and the *Chart* that display the results in SnowCloudMetrics were considered the most useful by participants (66% and 63%, respectively). The *Statistics* (outputs similar to those in Figure 3d–f) and the ability to change *User Inputs* (select different watersheds) were also evaluated as useful, but to a lesser degree. While each of the components of the web-interface were classified as not useful by at least one participant, no one single component stood out as having less utility than the other components (all were below 10% of respondents).

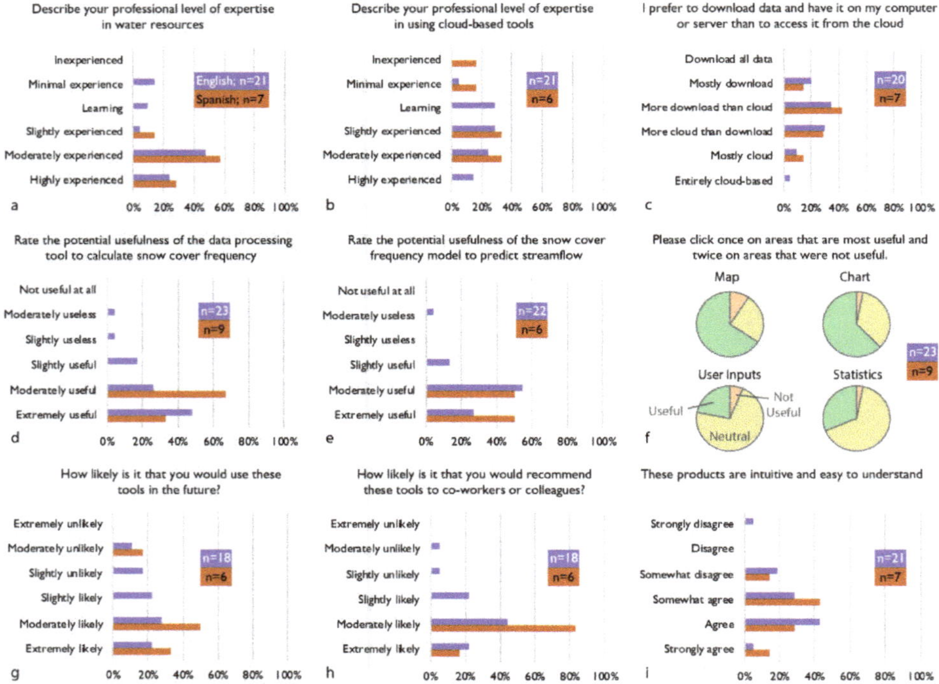

Figure 4. The questions and responses (**a–j**) from the qualitative assessments completed by participants. Each question provided to the participants is above its respective sub-figure. In all sub-figures the number of responses (n) are represented in purple for English-speaking participants and orange for Spanish-speaking participants.

Participants also indicated that they were moderately-to-extremely likely to use these tools in the future or would recommend them to a co-worker (Figure 4g,h). Despite these positive perceptions regarding the overarching SnowCloud framework, there was only modest agreement that these tools were intuitive to users (Figure 4i). This highlights the need to have clear directions and guidelines for the use of these tools.

The goal of this survey was to obtain user feedback on the prototype version of SnowCloud developed in the GEE Code Editor. The number of responses for each question is noted in each of the sub-figures. As such these results should be interpreted as indicative of user impressions, but do not reflect the opinions of the full water resources professional community.

4. Discussion

The case studies presented in this paper tested the ability of cloud computing frameworks to transition RSS data into actionable water resources information, a new paradigm for information delivery for snow hydrology and natural resource management. The initial inspiration for this project was based in the authors' frustrations regarding the amount of computational and human resources required for monthly updates to the original SCF-runoff model in Chile [21]. Prior to SnowCloud, monthly streamflow forecasts of the La Laguna watershed required 2–3 days to complete. That earlier process required a technician to download MODIS snow cover data (with variable broadband rates), process and re-project the spatial data, compute SCF for the watershed, and finally to compute the streamflow. Requiring only a Google account and a web browser, SnowCloud performs these same spatial calculations in less than 10 min, and with a similar level of forecast skill. Through GEE, the SnowCloud model framework provides access to the data, processing, models, and visualizations that complete the detailed analyses of watersheds in snowy regions.

These case studies also identified the deficiencies of the snow-only framework of the SCF-Runoff model. Model skill directly corresponded to the snowpack contribution to streamflow, decreasing as the ratio of rain/snow increases at the basin scale. In the high-elevation La Laguna watershed, winter conditions consistently remain below 0 °C and almost all precipitation occurs during the winter months, arriving almost entirely as snow and covering the entire watershed (Figure 3a). Here the SnowCloudHydro model works with a high level of forecast skill (NSE = 0.87). By comparison in the lower elevation and warmer John Day River and Río Aragón watersheds (Table 1), the model skill drops considerably (NSE = 0.52 and NSE = 0.21, respectively). Average winter temperatures in the John Day River and Río Aragón watersheds are at, or slightly above, 0 °C in the middle and lower elevations of the watersheds. This 0 °C isotherm infers a higher likelihood that winter precipitation events will have both rain and snow during the course of the winter. As a result, the mid- to lower-elevation portions of these watersheds are only partially snow covered (Figure 3b,c). Further impacting model skill is summer precipitation in the John Day and Río Aragón watersheds, as 40% of annual precipitation occurs during May-September when precipitation falls as rain. These events are not captured in the current SnowCloudHydro model structure. The relative lower skill of the Río Aragón can be attributed to a watershed that is more responsive to rain events and/or has more intense summer precipitation events as evidenced in the peaks of measured streamflow during the summer (Figure 3j) [34].

While the model does work well in high altitude, snow-dominated regions like the headwaters of the La Laguna sub-basin, in regions where there is a higher contribution to streamflow from rain (like the John Day River and Río Aragón), the model underperforms. The authors recognize that improvements to this prototype model structure should enhance forecast skill in regions where rain has a pronounced role in streamflow generation. The current limitations will be addressed into subsequent versions of the SnowCloud model framework. We anticipate integrating precipitation data from NASA's Global Precipitation Measurement (GPM) that are readily available in the GEE data library. Successful implementation of these precipitation forcings would greatly enhance the model structure, and potentially alleviate a priori streamflow input requirements. This enhanced model would thus allow cloud-based streamflow forecasts in ungauged basins that are influenced by both snow and rain. Increasing the temporal frequency of model predictions (monthly to weekly or biweekly) would require approximately four times as many calculations (ca. four weeks in a month), but would be able to provide more frequent insights into the water resources within a basin.

Temporally, the coarse resolution of monthly forecasts also provides a means to improve the current model structure. While informative, the monthly forecasts provide a general overview of future streamflow. Transitioning the model to weekly streamflow forecasts would provide managers with more timely information and the opportunity for more proactive resource management.

Another potential means to improve SnowCloudHydro would be to validate the SCF maps using high-resolution remote sensing imagery or in situ observations. This component was not included in this study because (1) the goal was to present the prototype framework, and (2) SnowCloudHydro is designed specifically for data scarce regions, which would limit these types of data.

Comparison of the optimized parameter solutions and model results for the two modeling approaches (GLUE and non-linear least squares) presents several questions regarding hydrologic models that are more broad in scope. The more rigorous GLUE methodology incorporates a calibration and validation period using Monte Carlo simulations. This calibration and validation approach is more common in research-based applications and generally requires more sophisticated software and programming and is not embedded in SnowCloudHydro, however, we do provide the source code for replicability [42]. Despite its complexity, the GLUE-based model performed with slightly less skill than the more simplistic non-linear least squares approach. These results suggest that model complexity does not always ensure better model results especially in data scarce regions, and that with straightforward models (like SnowCloudHydro) simplistic approaches can function well. Since the non-linear least squares approach was successfully implemented using a cloud-based spreadsheet, this technique provides a lower barrier to hydrological forecasting as compared to research-based computing and software associated with the GLUE methodology. This question of model complexity is relevant for researchers as they begin the initial stages of a research project, when they begin to conceptualize and develop analysis tools. During these initial stages it is important to have the end user in mind, identify the goals of the project, and understand the computational and human resources that are available for potential users.

The computation times for GEE will vary depending on the complexity of the processing task and the amount of computational resources available on the GEE servers. The times provided in the results are intended to be relative (minutes, hours, or days), and not exact (minutes: seconds). The important message is that these cloud-based calculations collapse the amount of time needed to perform complex processing and data generation from days to minutes. This represents a new paradigm for information creation and delivery.

To better connect these and subsequent efforts to the end user, our qualitative survey provides initial insights into the needs, interests, and expertise of end users across broad geographies. A majority of participants (water resource managers) classified SnowCloudMetrics and SnowCloudHydro as potentially useful, even though there was a wide-range of user experience with regards to cloud computing. Participants indicated that these types of tools would be used or recommended in the future. Regardless of their level of expertise with cloud computing, end users imply a willingness willing to use an end-to end, decision support framework. The perception that the framework is only moderately intuitive encourages us (and subsequent developers) to develop better guidelines for our product, facilitating its use for a wide user base. These survey results will help guide the conversion of the current SnowCloudMetrics and more robust versions of SnowCloudHydro into a freely accessible application (SnowCloud.app). Facilitating the use of SnowCloud is important as participants had a positive perspective of the SnowCloud framework that persisted across languages and continents, suggesting that similar multi-lingual platforms could also be well-received in other cultures and languages. This leads to a potentially overlooked benefit of cloud computing—its ability for data providers to expand their user base globally.

While the SnowCloud framework was developed using only MODIS data, it could readily be adapted to incorporate other types of spatial data. For example, other space-borne images with higher spatial resolution could be implemented as combined reanalysis products (temperature and precipitation) in areas where observational data is extremely sparse topographically complex. In these regions the 500 m resolution of MODIS may not effectively capture the complexity of the local mountain hydro-climatology.

5. Conclusions

SnowCloud, represents a prototype end-to-end framework comprised of cloud computing tools specifically designed to calculate monthly snow cover frequency and predict streamflow one month in advance. SnowCloud and the qualitative research that accompanies this paper are not intended as an end product, but rather as initial insights into this new paradigm that provides opportunities for snow scientists and hydrologists to accelerate basic and applied research in data sparse, snowy watersheds. These advances align well with the goals of applied science to support the implementation of Earth observations into better informed management of water resources [43].

Whether using GEE or other platforms, cloud-based computing lowers the computational, budgetary, and logistical barriers to implement RSS data that previously existed when working with large RSS datasets. Avoiding the requirement of downloading and pre-processing data in itself alleviates considerable technical and human capacity barriers that have existed previously. The geographic and cultural breadth of the participants and their overall positive assessment of SnowCloud speak to the potential opportunities that cloud computing can provide natural resource managers regardless of their location or level of expertise. This democratization of science and data [30,31] offers exciting new possibilities for basic and applied science beyond water resources by reducing the complexity around data management and processing. These efficiencies allow the research manager to focus time and resources on basic and applied scientific discovery, and away from computational barriers. The evolution of cloud computing and the accompanying science support the Raven's [25] perspective that disseminating scientific information will be fulfilled by approaches that combine advances in understanding, social capacity, and technology. The resource management and research communities now have access to computational power that was previously limited to supercomputers. It is now up to these same communities to make use of it.

Author Contributions: Conceptualization: E.S., R.C., and A.N.; methodology: E.S., R.C., A.N., and E.M.; software: E.S., R.C., and E.M.; validation: E.S., R.C., and E.M.; formal analysis: E.S., R.C., A.N., and I.M.; investigation: E.S., R.C., A.N., and I.M.; resources: E.S., R.C., A.N., and I.M.; data curation: E.S. and I.M.; writing—original draft preparation: E.S.; writing—review and editing: E.S., R.C., A.N., E.M., and I.M.; visualization: E.S. and E.M.; supervision: E.S., and A.N.; project administration: E.S., and A.N.; funding acquisition: E.S., and A.N.

Funding: This research was funded by an incubator grant from the Earth Science Information Partners (ESIP) and by NASA grant No. NNX16AG35G.

Acknowledgments: The authors would like to thank the water resource professionals that participated in this project and provided anonymous assessments of SnowCloud. E.S. would also like to thank the Department of Geography at the University of Oregon for providing office space to facilitate the completion of this manuscript. Additionally, the authors would like to thank the four anonymous reviewers for their expertise and perspectives that helped improve the quality of the final manuscript.

Conflicts of Interest: The authors declare no conflict of interest. Any brand name or companies in the paper are not endorsements. The founding sponsors had no role in the design of the study; in the collection, analyses, or interpretation of data; in the writing of the manuscript, and in the decision to publish the results.

Appendix A

Questions provided in the online assessment (English version)

Read through the first page on the website, and then follow the instructions at the bottom of each page. There are four short pages in total.

Clicking on the "Agree" button indicates that:

- You have read the above information
- You voluntarily agree to participate
- You understand that participation is voluntary, and you can stop at any time
- You understand there is no monetary compensation for your participation

Thanks! Team SnowCloud
Agree (1)

(1) Rate the potential usefulness of the data processing tool to calculate snow cover frequency.

- Extremely useful (1)
- Moderately useful (2)
- Slightly useful (3)
- Slightly useless (4)
- Moderately useless (5)
- Not useful at all (6)

(2) Rate the potential usefulness of the snow cover frequency model to predict streamflow.

- Extremely useful (1)
- Moderately useful (2)
- Slightly useful (3)
- Slightly useless (4)
- Moderately useless (5)
- Not useful at all (6)

(3) Please click once on areas that are most useful (green) and twice on areas that were not useful (red).

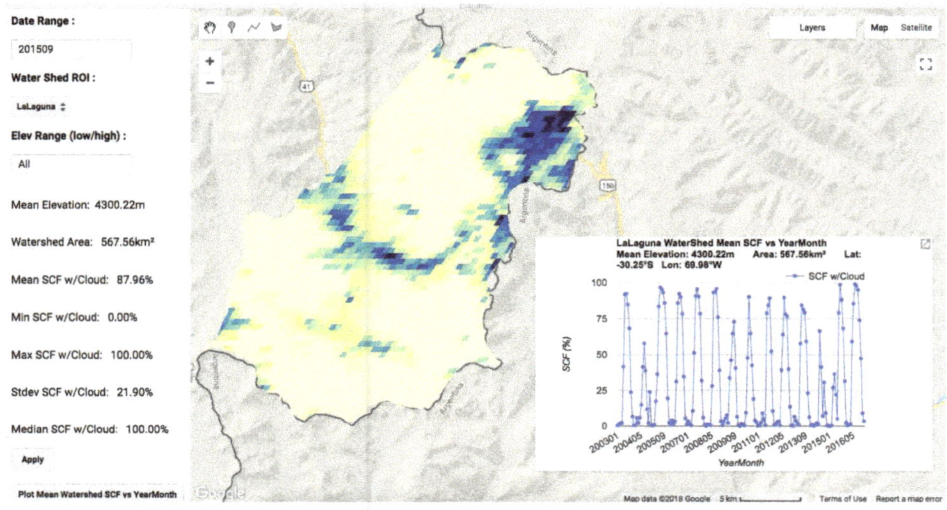

(4) Describe your professional level of expertise in using cloud-based tools.

- Highly experienced (1)
- Moderately experienced (2)
- Slightly experienced (3)

- Learning (4)
- Minimal experience (5)
- Inexperienced (6)

(5) Describe your professional level of expertise in water resources.

- Highly experienced (1)
- Moderately experienced (2)
- Slightly experienced (3)
- Learning (4)
- Minimal experience (5)
- Inexperienced (6)

(6) I prefer to download data and have it on my computer or server than to access it from the cloud.

- Entirely cloud-based (1)
- Mostly cloud (2)
- More cloud than download (3)
- More download than cloud (4)
- Mostly download (5)
- Download all data (6)

(7) These products are intuitive and easy to understand.

- Strongly agree (1)
- Agree (2)
- Somewhat agree (3)
- Somewhat disagree (4)
- Disagree (5)
- Strongly disagree (6)

(8) How likely is it that you would use these tools in the future?

- Extremely likely (1)
- Moderately likely (2)
- Slightly likely (3)
- Slightly unlikely (4)
- Moderately unlikely (5)
- Extremely unlikely (6)

(9) How likely is it that you would recommend these tools to co-workers or colleagues?

- Extremely likely (1)
- Moderately likely (2)
- Slightly likely (3)
- Slightly unlikely (4)
- Moderately unlikely (5
- Extremely unlikely (6)

(10) Please provide the number of years that you have worked in your current professional capacity.

References

1. Dozier, J. Mountain hydrology, snow color, and the fourth paradigm. *Eos Trans. Am. Geophys. Union* **2011**, *92*, 373. [CrossRef]
2. Fritze, H.; Stewart, I.T.; Pebesma, E. Shifts in Western North American Snowmelt Runoff Regimes for the Recent Warm Decades. *J. Hydrometeorol.* **2011**, *12*, 989–1006. [CrossRef]
3. Yucel, I.; Güventürk, A.; Sen, O.L. Climate change impacts on snowmelt runoff for mountainous transboundary basins in eastern Turkey. *Int. J. Climatol.* **2014**. [CrossRef]
4. Nogués-Bravo, D.; Araújo, M.B.; Errea, M.P.; Martínez-Rica, J.P. Exposure of global mountain systems to climate warming during the 21st Century. *Glob. Environ. Chang.* **2007**, *17*, 420–428. [CrossRef]
5. Stewart, I.T. Changes in snowpack and snowmelt runoff for key mountain regions. *Hydrol. Process.* **2009**, *23*, 78–94. [CrossRef]
6. Stewart, I.T.; Cayan, D.R.; Dettinger, M.D. Changes toward earlier streamflow timing across western North America. *J. Clim.* **2005**, *18*, 1136–1155. [CrossRef]
7. Smerdon, B.D.; Allen, D.M.; Grasby, S.E.; Berg, M.A. An approach for predicting groundwater recharge in mountainous watersheds. *J. Hydrol.* **2009**, *365*, 156–172. [CrossRef]
8. Berghuijs, W.R.; Woods, R.A.; Hrachowitz, M. A precipitation shift from snow towards rain leads to a decrease in streamflow. *Nat. Clim. Chang.* **2014**, *4*, 583–586. [CrossRef]
9. Meixner, T.; Manning, A.H.; Stonestrom, D.A.; Allen, D.M.; Ajami, H.; Blasch, K.W.; Brookfield, A.E.; Castro, C.L.; Clark, J.F.; Gochis, D.J.; et al. Implications of projected climate change for groundwater recharge in the western United States. *J. Hydrol.* **2016**, *534*, 124–138. [CrossRef]
10. Barnett, T.P.; Adam, J.C.; Lettenmaier, D.P. Potential impacts of a warming climate on water availability in snow-dominated regions. *Nature* **2005**, *438*, 303–309. [CrossRef] [PubMed]
11. Wada, Y.; Bierkens, M.F.P. Sustainability of global water use: Past reconstruction and future projections. *Environ. Res. Lett.* **2014**, *9*, 104003. [CrossRef]
12. Gosling, S.N.; Arnell, N.W. A global assessment of the impact of climate change on water scarcity. *Clim. Chang.* **2016**, *134*, 371–385. [CrossRef]
13. Nolin, A.W. Perspectives on Climate Change, Mountain Hydrology, and Water Resources in the Oregon Cascades, USA. *Mt. Res. Dev.* **2012**, *32*, S35–S46. [CrossRef]
14. Jaeger, W.K.; Plantinga, A.J.; Chang, H.; Dello, K.; Grant, G.; Hulse, D.; McDonnell, J.J.; Lancaster, S.; Moradkhani, H.; Morzillo, A.T.; et al. Toward a formal definition of water scarcity in natural-human systems. *Water Resour. Res.* **2013**, *49*, 4506–4517. [CrossRef]
15. Bales, R.C.; Molotch, N.P.; Painter, T.H.; Dettinger, M.D.; Rice, R.; Dozier, J. Mountain hydrology of the western United States. *Water Resour. Res.* **2006**, *42*, W08432. [CrossRef]
16. Yu, J.; Zhang, G.; Yao, T.; Xie, H.; Zhang, H.; Ke, C.; Yao, R. Developing daily cloud-free snow composite products from MODIS Terra–Aqua and IMS for the Tibetan plateau. *IEEE Trans. Geosci. Remote Sens.* **2016**, *54*, 2171–2180. [CrossRef]
17. AghaKouchak, A.; Farahmand, A.; Melton, F.S.; Teixeira, J.; Anderson, M.C.; Wardlow, B.D.; Hain, C.R. Remote sensing of drought: Progress, challenges and opportunities. *Rev. Geophys.* **2015**, *53*, 452–480. [CrossRef]
18. Wang, J.; Li, H.; Hao, X.; Huang, X.; Hou, J.; Che, T.; Dai, L.; Liang, T.; Huang, C.; Li, H. Remote sensing for snow hydrology in China: challenges and perspectives. *J. Appl. Remote Sens.* **2014**, *8*, 84687. [CrossRef]
19. Nolin, A.W.; Sproles, E.A.; Crumley, R.L.; Wilson, A.; Mar, E.; van de Kerk, M.; Prugh, L. Cloud-based computing and applications of new snow metrics for societal benefit. In *Proceedings of the American Geophysical Fall Meeting*; American Geophysical Union: New Orleans, LA, USA, 2017.
20. Xu, X.; Li, J.; Tolson, B.A. Progress in integrating remote sensing data and hydrologic modeling. *Prog. Phys. Geogr.* **2014**, *38*, 464–498. [CrossRef]
21. Sproles, E.A.; Kerr, T.; Nelson, C.O.; Aspe, D.L. Developing a Snowmelt Forecast Model in the Absence of Field Data. *Water Resour. Manag.* **2016**, *30*. [CrossRef]
22. Huang, X.; Deng, J.; Wang, W.; Feng, Q.; Liang, T. Impact of climate and elevation on snow cover using integrated remote sensing snow products in Tibetan Plateau. *Remote Sens. Environ.* **2017**, *190*, 274–288. [CrossRef]

23. Ma, Y.; Wu, H.; Wang, L.; Huang, B.; Ranjan, R.; Zomaya, A.; Jie, W. Remote sensing big data computing: Challenges and opportunities. *Futur. Gener. Comput. Syst.* **2015**, *51*, 47–60. [CrossRef]
24. Gorelick, N.; Hancher, M.; Dixon, M.; Ilyushchenko, S.; Thau, D.; Moore, R. Google Earth Engine: Planetary-scale geospatial analysis for everyone. *Remote Sens. Environ.* **2017**, *202*, 18–27. [CrossRef]
25. Raven, P.H. Science, Sustainability, and the Human Prospect. *Science* **2002**, *297*, 954–958. [CrossRef] [PubMed]
26. Beier, P.; Hansen, L.J.; Helbrecht, L.; Behar, D. A How-to Guide for Coproduction of Actionable Science. *Conserv. Lett.* **2016**. [CrossRef]
27. Yang, C.; Goodchild, M.; Huang, Q.; Nebert, D.; Raskin, R.; Xu, Y.; Bambacus, M.; Fay, D. Spatial cloud computing: How can the geospatial sciences use and help shape cloud computing? *Int. J. Digit. Earth* **2011**, *4*, 305–329. [CrossRef]
28. Granell, C.; Havlik, D.; Schade, S.; Sabeur, Z.; Delaney, C.; Pielorz, J.; Usländer, T.; Mazzetti, P.; Schleidt, K.; Kobernus, M. Future Internet technologies for environmental applications. *Environ. Model. Softw.* **2016**, *78*, 1–15. [CrossRef]
29. Nativi, S.; Mazzetti, P.; Santoro, M.; Papeschi, F.; Craglia, M.; Ochiai, O. Big Data challenges in building the Global Earth Observation System of Systems. *Environ. Model. Softw.* **2015**, *68*, 1–26. [CrossRef]
30. Foster, I. Globus Online: Accelerating and democratizing science through cloud-based services. *IEEE Internet Comput.* **2011**, *15*, 70–73. [CrossRef]
31. Sultan, N. Cloud computing: A democratizing force? *Int. J. Inf. Manag.* **2013**, *33*, 810–815. [CrossRef]
32. Junta de Vigilancia del Rio Elqui y sus Afluentes. *Estimated Inflows in La Laguna Reservoir*; Junta de Vigilancia del Rio Elqui y sus Afluentes: La Serena, Chile, 2016.
33. United States Geological Survey US Geological Survey National Water Information System. Available online: http://waterdata.usgs.gov/nwis/ (accessed on 10 January 2017).
34. Sanmiguel-Vallelado, A.; Morán-Tejeda, E.; Alonso-González, E.; López-Moreno, J.I. Effect of snow on mountain river regimes: an example from the Pyrenees. *Front. Earth Sci.* **2017**, *11*, 515–530. [CrossRef]
35. Sawicz, K.; Wagener, T.; Sivapalan, M.; Troch, P.A.; Carrillo, G. Catchment classification: empirical analysis of hydrologic similarity based on catchment function in the eastern USA. *Hydrol. Earth Syst. Sci.* **2011**, *15*, 2895–2911. [CrossRef]
36. Hall, D.K.; Riggs, G.A. *MODIS/Terra Snow Cover Daily L3 Global 500m Grid, Version 6*; NASA National Snow and Ice Data Center Distributed Archive Center: Boulder, CO, USA, 2016. Available online: https://nsidc.org/data/mod10a1 (accessed on 22 May 2018). [CrossRef]
37. Beven, K.; Binley, A. The Future of Distributed Models: Model Calibration and Uncertainty Prediction. *Hydrol. Process.* **1992**, *6*, 279–298. [CrossRef]
38. Nash, J.E.; Sutcliffe, J.V. River flow forecasting through conceptual models part I: a discussion of principles. *J. Hydrol.* **1970**, *10*, 282–290. [CrossRef]
39. Dingman, S.L. *Physical Hydrology*; Prentice Hall: Upper Saddle River, NJ, USA, 2002; ISBN 0130996955 9780130996954.
40. Kirchner, J.W. Getting the right answers for the right reasons: Linking measurements, analyses, and models to advance the science of hydrology. *Water Resour. Res.* **2006**, *42*. [CrossRef]
41. MathWorks. *The MathWorks MATLAB and Statistics Toolbox Release 2013a*; MathWorks: Natick, MA, USA, 2013.
42. Sproles, E. *SnowCloudHydro*; Github Repository: Corvallis, OR, USA, 2017. Available online: https://github.com/MountainHydroClimate/SnowCloudHydro (accessed on 22 May 2018). [CrossRef]
43. NASA—Applied Sciences Program Water Resources Program | Applied Sciences Website. Available online: https://appliedsciences.nasa.gov/programs/water-resources-program (accessed on 22 May 2018).

Review

Flood Prevention and Emergency Response System Powered by Google Earth Engine

Cheng-Chien Liu [1,2,*], Ming-Chang Shieh [3,4], Ming-Syun Ke [2] and Kung-Hwa Wang [2]

1 Department of Earth Sciences, National Cheng Kung University, Tainan 70101, Taiwan
2 Global Earth Observation and Data Analysis Center, National Cheng Kung University, Tainan 70101, Taiwan; take999kimo@gmail.com (M.-S.K.); kh.wang.peter@gmail.com (K.-H.W.)
3 Water Hazard Mitigation Center, Water Resource Agency, Taipei 10651, Taiwan; mcshieh59@gmail.com
4 Now at the 10th River Management Office, Water Resource Agency, New Taipei 22061, Taiwan
* Correspondence: ccliu88@mail.ncku.edu.tw; Tel.: +886-6-275-7575 (ext. 65422)

Received: 4 July 2018; Accepted: 10 August 2018; Published: 14 August 2018

Abstract: This paper reviews the efforts made and experiences gained in developing the Flood Prevention and Emergency Response System (FPERS) powered by Google Earth Engine, focusing on its applications at the three stages of floods. At the post-flood stage, FPERS integrates various remote sensing imageries, including Formosat-2 optical imagery to detect and monitor barrier lakes, synthetic aperture radar imagery to derive an inundation map, and high-spatial-resolution photographs taken by unmanned aerial vehicles to evaluate damage to river channels and structures. At the pre-flood stage, a huge amount of geospatial data are integrated in FPERS and are categorized as typhoon forecast and archive, disaster prevention and warning, disaster events and analysis, or basic data and layers. At the during-flood stage, three strategies are implemented to facilitate the access of the real-time data: presenting the key information, making a sound recommendation, and supporting the decision-making. The example of Typhoon Soudelor in August of 2015 is used to demonstrate how FPERS was employed to support the work of flood prevention and emergency response from 2013 to 2016. The capability of switching among different topographic models and the flexibility of managing and searching data through a geospatial database are also explained, and suggestions are made for future works.

Keywords: flood; disaster prevention; emergency response; decision making; Google Earth Engine

---

1. Introduction

Like the other islands between Japan and the Philippines off the eastern and southeastern coasts of Asia, Taiwan is visited by three to four typhoons per year, on average. Some extreme cases of torrential and sustained rainfall brought by typhoons have caused flooding, severe damage, and significant loss of lives and properties in the past two decades, such as the flooding in Taipei caused by severe Typhoon Winnie (18 August 1997); the flooding in Kaohsiung and Pingtung caused by Tropical Storm Trami (11 July 2001); the flooding in Taipei and Keelung caused by Typhoon Nari (17 September 2001); the flooding in Kaohsiung and Pingtung caused by Typhoon Mindulle (2–4 July 2004); the flooding in middle and southern Taiwan caused by Typhoon Kalmaegi (17–18 July 2008), and the flooding in middle, eastern, and southern Taiwan caused by Typhoon Morakot (6–10 August 2009). To improve our knowledge of flood prevention and emergency response requires a sound collection and analysis of information that is transformed from a huge amount of data. Apart from the historical, real-time, and forecast data in meteorology, the status of pumping stations and mobile machines, the available resources and facilities, as well as many other data, are both crucial and useful to the government and general public. The data must be identified, collected, integrated, processed, analyzed, distributed, and visualized rapidly through the Internet, in order to support a variety of management decisions, particularly a timely response to an urgent event. This was not possible until the Google Earth application programming interface (API) was released on 28 May 2008. After the devastating Typhoon Morakot hit Taiwan in August 2009, the Google Earth API was employed to develop the Formosat-2 pre- and post-Morakot image-comparison system [1] that succeeded in rapid distribution of geospatial information to the general public and government agencies (http://research.ncku.edu.tw/re/articles/e/20110429/1.html). In light of the great potential revealed by the system, the Water Resource Agency (WRA) of Taiwan initiated a multi-year project from 2012 to 2016, which led to the development of the Flood Prevention and Emergency Response System (FPERS) powered by Google Earth Engine (GEE).

This paper reviews the efforts made and experiences gained in developing FPERS using GEE, with focus on its applications at different stages and some examples of success. Motivated by the demand for processing a large amount of Formosat-2 imagery (2 m resolution) to support a rapid response to disasters caused by Typhoon Morakot, a new super-overlay tool for creating image tiles was developed [1] and incorporated into the Formosat-2 Automatic Image Processing System (F-2 AIPS) [2]. In the aftermath of Japan's earthquake and tsunami on March 11, 2011, we demonstrated that Formosat-2 imagery can be rapidly acquired, processed, and distributed through the Internet to global users by deploying the system through cloud servers [3]. Therefore, FPERS was developed with an intention to collect and display the huge amount of relevant geospatial imagery, including Formosat-2 pre- and post-flood imagery (2-m resolution) used to detect and monitor barrier lakes, the synthetic aperture radar (SAR) imagery used to derive an inundation map [4], and the high-spatial-resolution photos taken by unmanned aerial vehicles (UAV) to evaluate the damage to river channels and structures due to a debris flow [5]. In spite of these successes, these data were mainly used for disaster assessment at the post-flood stage.

To expand the application of FPERS to the pre-flood stage, numerous hydrology and meteorology observation data and all the related geospatial data were identified and integrated in FPERS. The challenge was to host and share this tremendous amount of data with the flexible functions provided by GEE. These efforts marked the application of FPERS to the pre-flood stage. To support decision-making during typhoon events, a huge amount of real-time monitoring data from various sensors, such as the videos of water gates and bridge pillars captured by closed-circuit televisions (CCTVs), was also linked to FPERS. A special outsourcing function was developed that enabled users to upload photographs, video clips, or a paragraph of text to report the disaster areas. With such an unprecedented dataset integrated using FPERS, we proposed a set of standard operating procedures (SOPs) in which all related data are automatically loaded and displayed at different stages of a flood. Rapid data processing and analyses were also conducted to make suggestions of action to the decision makers. FPERS was successfully employed by the Water Hazard Mitigation Center (WHMC) of WRA to support the flood prevention and emergency response for Typhoon Soudelor in August 2015.

The successful development and operation of FPERS powered by GEE demonstrated that a huge amount of geospatial data can be effectively integrated and distributed through the Internet to support a variety of demands at different stages of response to floods. The capability of switching among different topographic models, as well as the flexibility of managing and searching for data through a geospatial database system, are also explained in this paper, and suggestions are made for future works.

2. Flood Prevention and Emergency Response System

The prototype of FPERS was the web-based Google Earth that enabled users to select image-based or vector-based datasets from a preset list and display them on a map as overlaid layers. With the backbone of GEE, FPERS not only had an easily accessible and user-friendly front-end, but also a powerful back-end with petabyte-scale data prepared and updated by Google for scientific analysis and visualization. We were able to add and curate more data and collections to FPERS, as long as the data was preprocessed and stored as standard tiles, following the rules specified by the Google

Earth API. Displaying huge amounts of geospatial data in a 3D fashion, especially through the Internet, is supposed to be a resource-demanding task that required tedious amounts of coding and a large network bandwidth. Google would undertake all the required processing, whether a few or a multitude of users were using FPERS simultaneously from anywhere all over the world through the Internet. This capacity allowed this study to bypass purchasing the large processing power in the form of the latest computers or the latest software and allowed us to focus on the development and application of FPERS in three stages: post-, pre-, and during-flood.

2.1. Post-Flood Stage

In light of the successful case of rapid response to Typhoon Morakot using Formosat-2 imagery [1], acquiring, processing, and sharing a huge amount of Formosat-2 imagery to the general public was set up as FPERS' primary function. Although optical images obtained from spaceborne platforms are limited by weather in terms of mapping the inundated regions, they are advantageous for monitoring existing barrier lakes and detecting emergent ones that are formed in mountainous areas after a major typhoon or earthquake. This function is crucial because a barrier lake might have a large water capacity, and the catastrophic burst of a dam would result in significant casualties [6–8].

2.1.1. Optical Satellite Imagery

When a natural disaster occurs, Formosat-2 is scheduled to be used to acquire images as soon as possible. All images are processed by F-2 AIPS, which is able to digest the Gerald format of the raw data; apply the basic radiometric and geometric corrections; output the level-1A product; and rigorously conduct band-to-band co-registration [9], automatic orthorectification [10], multi-temporal image geometrical registration [11], multi-temporal image radiometric normalization [12], spectral summation intensity modulation pan-sharpening [9], the absolute radiometric calibration [13], and super-overlay processing [1]. For each image, all tiles are uploaded to a cloud machine. The link to the KML (Keyhole Markup Language) files is added to FPERS, so users can select and browse through a particular image in FPERS at different level-of-detail (LOD) in a 3D fashion. Both the pre- and post-disaster images of the same region are displayed synchronously in the dual window mode as well.

A total of 17 barrier lakes were formed after Typhoon Morakot in 2009. They have been continuously monitored by the Forest Bureau using various approaches and instruments ever since. Whenever there was a typhoon or torrential rain event, the most up-to-date status of these lakes were checked one by one via FPERS, using the pre- and post-event images in the dual window mode. As an example, Figure 1 shows a barrier lake located upstream of the Qishan River that is inaccessible by vehicle. Its formation can be confirmed by comparing the Formosat-2 images acquired before (Figure 1a) and after (Figure 1b) Typhoon Morakot in 2009, while its destruction can also be confirmed by comparing the Formosat-2 images acquired before (Figure 1c) and after (Figure 1d) the torrential rain on 10 June 2012. All figures with their original resolutions and sizes are available in the Supplementary Materials.

Apart from monitoring and detecting barrier lakes, comparing the pre- and post-event images acquired by Formosat-2 is also advantageous in terms of examining water conservancy facilities, including dikes, revetments, water intake and drainage facilities, canals, culverts, reservoirs, dams, and affiliated structures. Some examples include mapping the coverage of driftwood in Shimen Reservoir caused by Typhoon Soulik, tracking the spatiotemporal variation of the Gaoping River plume [14], and monitoring reservoir water quality [12]. However, when it comes to the original goal of FPERS—flood prevention and management—the optical imagery acquired from the spaceborne platforms tends to be limited by clouds. Therefore, the second function of FPERS is to photograph inundated areas using unmanned aerial vehicles (UAVs).

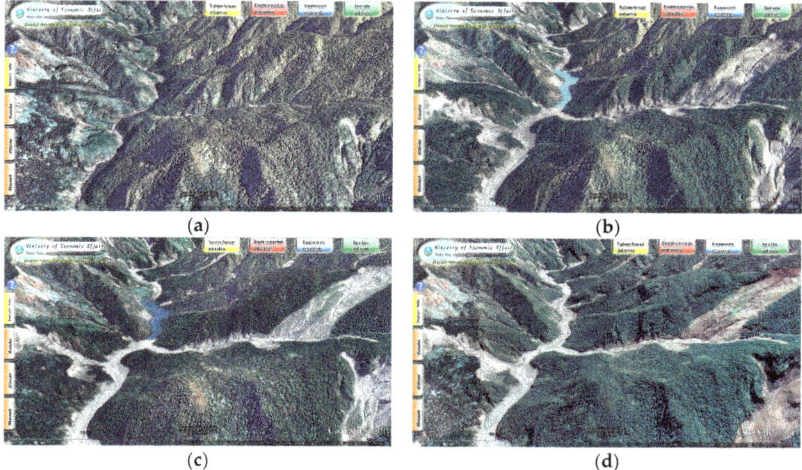

Figure 1. The formation and destruction of a barrier lake located upstream of the Qishan River, by comparing the Formosat-2 images acquired on (a) 15 November 2008; (b) 23 February 2010; (c) 16 April 2011; and (d) 28 June 2012.

2.1.2. Unmanned Aerial Vehicle Photographs

UAVs provide an alternative approach to remote sensing that is much cheaper, safer, and more flexible for deployment in small areas. An automatic mission-planning and image-processing system was developed to plan a flight mission that generated three levels of georeferenced products. It was successfully employed to rapidly respond to several disaster events involving landslides or debris flows in Taiwan [5]. To deploy a UAV to map the inundated areas, however, would be much more challenging because a flood usually rises and subsides in a very short period of time. The operation of UAVs is also restricted by rain and wind conditions, which are usually poor during a flood event. To tackle this challenge, we sent out a team to standby near the flooded region when the forecast and warning were received. Once the wind subsided, and the rain stopped, the UAV was deployed and operated in the first person view, with the intention to fly to the boundary of the inundated area. Figure 2a shows one of the 74 high-spatial-resolution photographs taken by a UAV in Minxiong, Chiayi County of Taiwan, caused by Typhoon Kong-Rey on 29 August 2013. The boundary of the inundated area can be clearly delineated from these photographs. All photographs were processed to one seamless, color-balanced, and georeferenced mosaic to publish on FPERS in dual-window mode (left: before-flood; right: after-flood) within 24 h, as shown in Figure 2b.

2.1.3. Synthetic Aperture Radar Imagery

The feasibility of mapping the inundated areas with a low-cost UAV was proven by the Minxiong mission. However, the limitations of deploying a UAV under severe weather conditions to cover a large area were also highlighted by this mission. To remove those limitations, we were motivated to seek another type of imagery from SAR to derive an inundation map [4]. Chung et al. reported a successful case of rapid response with a map of inundated areas derived from COSMO-SkyMed 1 radar satellite imagery during the July 2013 flood in I-Lan County in Taiwan, which was caused by Typhoon Soulik [4]. An SOP was established to identify the inundated area on an SAR image with ancillary information from Formosat-2, and the flood depth was inferred based on a comparison between the flood extent map and different inundation potential maps. This procedure was adopted again after Typhoon Soudelor (August 2015), and the results were shared by FPERS in dual-window mode within 24 h after the SAR image was received, as shown in Figure 3. The left window shows the SAR image of I-Lan County taken by COSMO-SkyMed 1 on 8 August 2015 (Typhoon Soudelor), overlaid with the inundated areas (red polygons), while the right window shows the same SAR image overlaid with the inferred flood depths (color polygons).

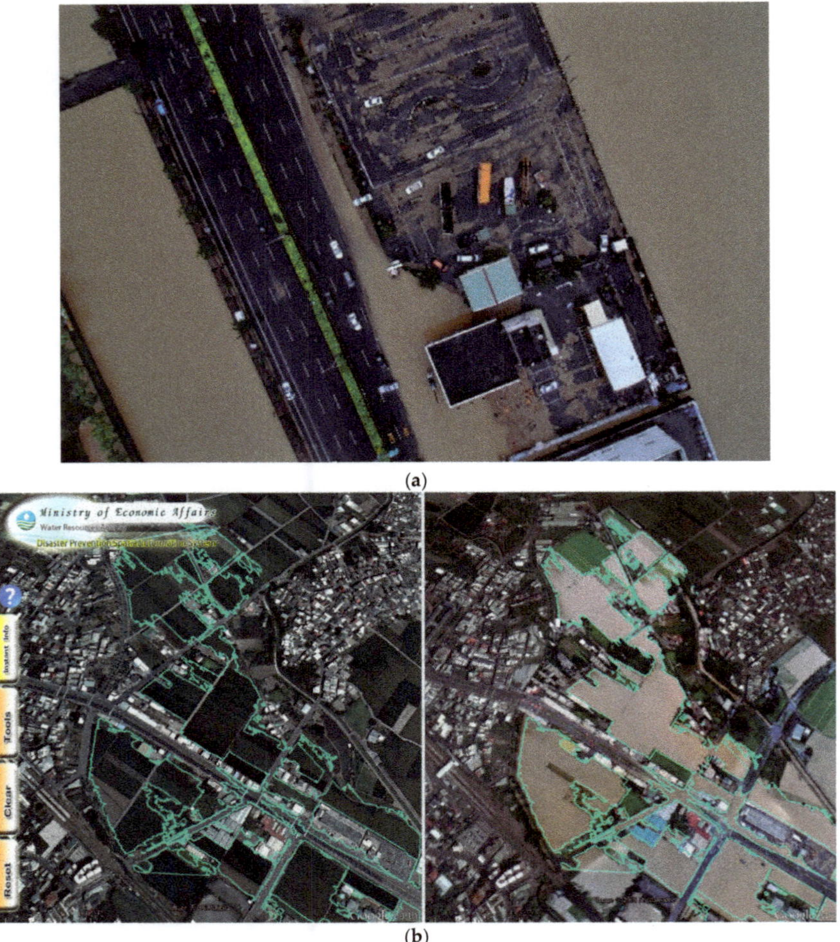

Figure 2. Illustration of the feasibility of mapping the inundated areas with a low-cost unmanned aerial vehicle (UAV). (**a**) One of the 74 high-spatial-resolution photographs taken by an UAV in Minxiong, Chiayi County in Taiwan on 29 August 2013. The boundary of the inundated area caused by Typhoon Kong-Rey can be clearly delineated from these photographs; and (**b**) the seamless, color-balanced, and georeferenced mosaic was published on Flood Prevention and Emergency Response System (FPERS) in dual-window mode (left: before-flood; right: after-flood) within 24 h.

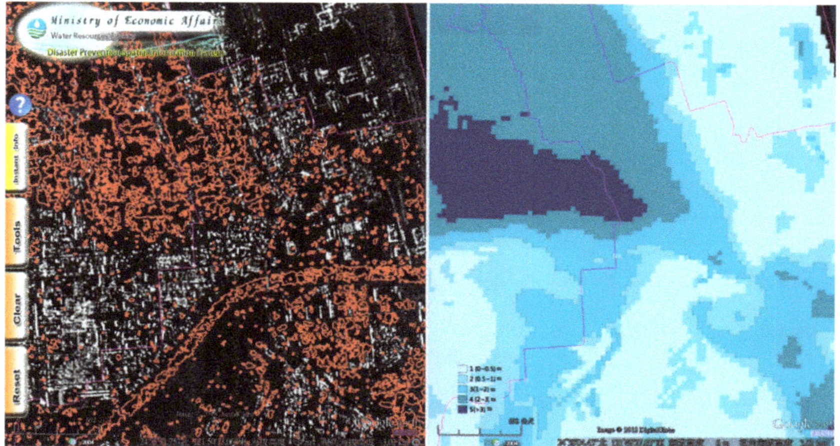

Figure 3. Synthetic aperture radar (SAR) image of I-Lan County taken by COSMO-SkyMed 1 on 8 August 2015 (Typhoon Soudelor) publish on FPERS in dual-window mode. Left window: overlaid with the inundated areas (red polygons); right window: overlaid with the inferred flood depth based on a comparison between the flood extent map and different inundation potential maps. The unit shown in legend is meter.

2.2. Pre-Flood Stage

The successful application of FPERS at the post-flood stage demonstrated its advantages related to rapid distribution of geospatial information, which also encouraged us to expand the FPERS application to the pre-flood stage. This involved identifying, collecting, integrating, processing, analyzing, distributing, and visualizing data rapidly through the Internet. We put the hydrology and meteorology data, as well as all the related geospatial data into the following four categories, because they are logical and convenient to store and search these data.

2.2.1. Typhoon Forecast and Archive

When a typhoon is approaching, and the warning is issued, the most updated forecasts of the typhoon path from various agencies always receive a lot of attention. Apart from the official forecast from the Central Weather Bureau (CWB) of Taiwan, the forecasts provided by the other agencies are all of great values as a reference, including the Japan Meteorological Agency (JMA), Hong Kong Observatory (HKO), the Joint Typhoon Warning Center (JTWC), the National Meteorological Center (NMC) of China Meteorological Administration (CMA), the Korea Meteorological Administration (KMA), the Philippine Atmospheric Geophysical and Astronomical Services Administration (PAGASA), and the Macao Meteorological and Geophysical Bureau (SMG). Although users could access each forecast from its original provider, FPERS integrated all forecasts in one place and allowed the users to overlay them with other useful geospatial layers, such as sea surface temperature and atmospheric pressure. This combination produced guidelines for quick evaluations of the forecasted typhoon path.

Another advantage of FPERS was that a comprehensive archive of typhoons provided by the CWB had also been integrated into the system, which traced back to 1958. To make the best use of this archive, we developed a function to compare the forecast of a typhoon path to all historical paths in the archive. When a typhoon is approaching Taiwan, 24 to 36 h before a typhoon warning is issued, the angle between each historical typhoon track and the CWB forecast typhoon track is calculated automatically. Five historical typhoons with the least values of angle will be listed in order.

Similar historical typhoons obtained under the above conditions will vary with the actual location of current typhoons; therefore, the information users see is consistent with the latest and most immediate calculation result. Figure 4 illustrates the application of the typhoon forecast and archive provided by FPERS for Typhoon Nepartak (8 July 2015). Based on the official forecast from the CWB at 14:00 on 7 July 2015 (Figure 4a), the decision makers could evaluate this forecast by comparing it with the forecasts provided by the other agencies (Figure 4b). In the meantime, FPERS compared this forecast with all historical paths in the archive and indicated that Typhoon Bilis (21 August 2000) had the closest path (Figure 4c). The decision makers could refer to the historical typhoon and evaluate the possible disaster regions (Figure 4d).

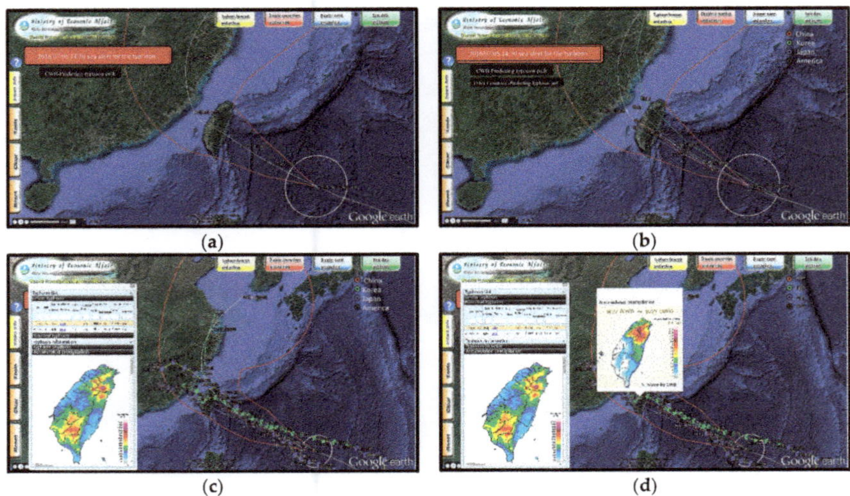

Figure 4. Application of typhoon forecasts and archives provided by FPERS for Typhoon Nepartak (8 July 2015). (**a**) The official forecast provided by the Central Weather Bureau of Taiwan; (**b**) the forecasts provided by National Meteorological Center (NMC) of China Meteorological Administration (CMA) (red), Korea Meteorological Administration (KMA) (green), Japan Meteorological Agency (JMA) (blue), and Joint Typhoon Warning Center (JTWC) (grey); (**c**) FPERS compared this forecast with all historical paths in the archive, which indicates that Typhoon Bilis (21 August 2000) had the closest path; (**d**) the disasters caused by Typhoon Bilis in the past served as a good reference from which to evaluate the possible disaster regions that would potentially be caused by Typhoon Nepartak.

2.2.2. Disaster Prevention and Warning

Apart from typhoon forecast and archive, all other data related to disaster prevention and warning were identified, collected, and integrated in FPERS to support the decision-making and timely response to a flood event. The type of data used includes meteorology data, hydrology data, auxiliary data, or disaster alert.

(1) Meteorology Data: The rainfall distribution in Taiwan is extremely uneven as a result of the four major mountain ranges in the central region that include more than 200 peaks rising higher than 3000 m. For the same reason, the locations of the 916 rainfall stations in Taiwan are also restricted by topography and transportation (Figure 5a), resulting in sparse observations of rainfall in mountainous areas. To tackle this problem, we followed the same Kriging interpolation approach that the CWB employs to generate the gridded precipitation from the real-time rainfall data transmitted from the 916 rainfall stations at intervals of 10 min. This gridded precipitation was further processed as basin, watershed, and administrative area rainfall, based on the polygons

of integration in space (Figure 5b). These products could be further processed to generate the one-hour, three-hour, six-hour, twelve-hour, twenty-four-hour, and daily accumulative rainfall, based on the polygons of integration in time (Figure 5c). All these totals were automatically calculated, stored, and managed through a Microsoft SQL (Structured Query Language) database as the real-time rainfall data were transmitted from the 916 rainfall stations every 10 min. A user-friendly function was developed and implemented in FPERS to facilitate user searches, queries, comparisons, and sorting of all meteorological products (Figure 5d). This function can help decision-makers convert the point precipitation from the 916 rainfall stations into colorful polygons of accumulative rainfall, based on the specified basin, watershed, or administrative area.

(2) Hydrology Data: The WRA is in charge of collecting and archiving most of Taiwan's hydrology data on a systematic basis. Among these data, the water levels from reservoirs, river stations, and tide stations provide valuable information for flood prevention and emergency response. The intensive precipitation brought by typhoons in the mountainous areas not only raises the water level in reservoirs, but also propagates the peak flow downstream, resulting in the rise of water levels at river stations one by one along the river. The time at which the peak flow arrives at the plain and river mouth can be calculated by simply comparing the hydrographs of different river stations, as shown in the Figure 6b,c. On the other hand, the water level at a tide station is mainly dominated by tides that are predictable, as shown in Figure 6c. Flooding becomes a certainty if the high tide coincides with the peak flow. To make the best use of the hydrology data, a user-friendly function was developed and implemented in FPERS to facilitate the ability of users to display hydrographs by clicking any reservoirs, river stations, or tide stations.

(3) Auxiliary Data: There are many valuable, yet sensitive data that are crucial for disaster prevention and decision-making, such as the most updated map of the status of pumping stations and mobile machines, the available resources and facilities, flood defense materials, the gaps and breaks of levees, flood potential maps, the streaming of real-time video captured and transmitted by CCTVs, and so on. Because of security and privacy issues, we were not able to include these data in FPERS at first, until the WRA purchased the Google Earth Enterprise product that allowed developers to create maps and 3D globes for private use. All sensitive data are hosted by this commercial product and linked to FPERS. To access these data, users must register and login to FPERS; then, another list will be available for selecting and viewing the data. In this manner, the issue of security and privacy was resolved by increasing the amount of auxiliary data that can be integrated in FPERS. Figure 7 presents one example of displaying the streaming of a real-time CCTV video in FPERS in single-window mode and multi-window mode, respectively.

(4) Disaster Alert: Different government agencies issue various disaster alerts, such as the yellow or red alerts for debris flow issued by the Soil and Water Conservation Bureau (SWCB), notices of road and bridge closures issued by the Directorate General of Highways (DGH), heavy or torrential rain alerts issued by the CWB, and river water level warnings and reservoir discharge warnings issued by the WRA. Thanks to the wide applications of GEE and the standard KML/KMZ (Zipped KML) format, each warning can be dynamically integrated into FPERS by linking to specific KML/KMZ files. As long as the KML/KMZ file is maintained and updated by the provider of the warning, the most updated and complete warnings are accessible in FPERS. Figure 8 presents an example of accessing various disaster alerts in FPERS in the event of Typhoon Nepartak (8 July 2015).

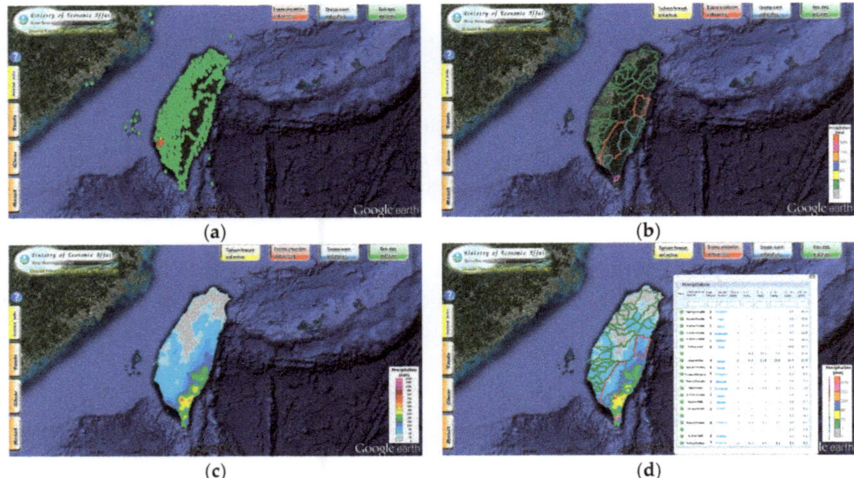

Figure 5. Meteorology data integrated using FPERS for Typhoon Nepartak (8 July 2015). (**a**) Real-time rainfall data transmitted from 916 rainfall stations at 10-min intervals; (**b**) the totals for basin, watershed, and administrative area rainfall; (**c**) the one-hour, three-hour, six-hour, twelve-hour, twenty-four-hour, and daily accumulative rainfall; (**d**) a user-friendly function to facilitate searches, queries, comparisons, and sorting with all meteorological products.

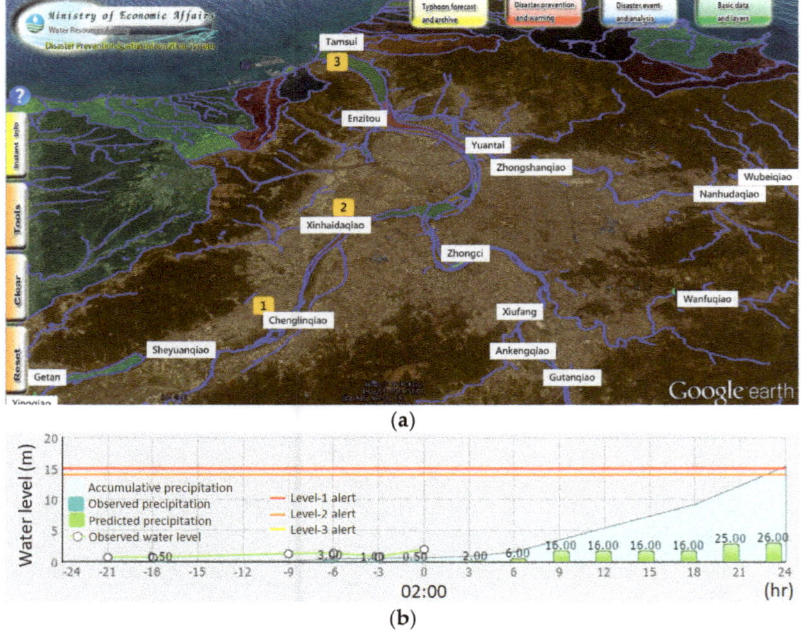

Figure 6. *Cont.*

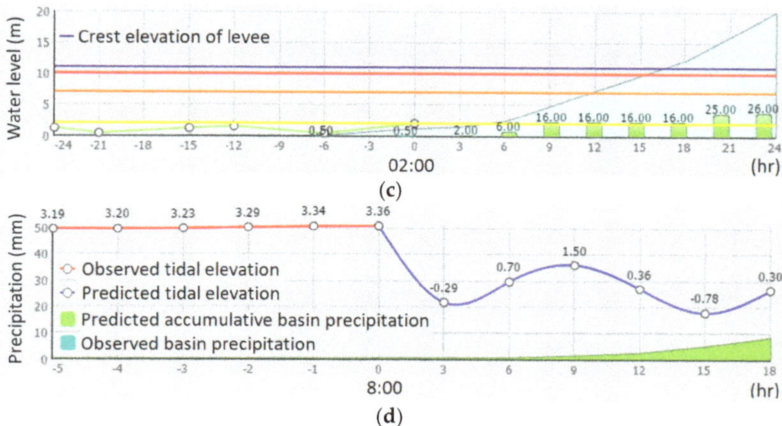

Figure 6. Hydrology data integrated by FPERS for Typhoon Nepartak (8 July 2015). (**a**) Zoom into Taipei; (**b**) hydrograph of water level at the Chen-Lin bridge river station (Label 1); (**c**) hydrograph of water level at the Hsin-Hai bridge river station (Label 2); (**d**) hydrographs of precipitation at the Damsui tide station (Label 3).

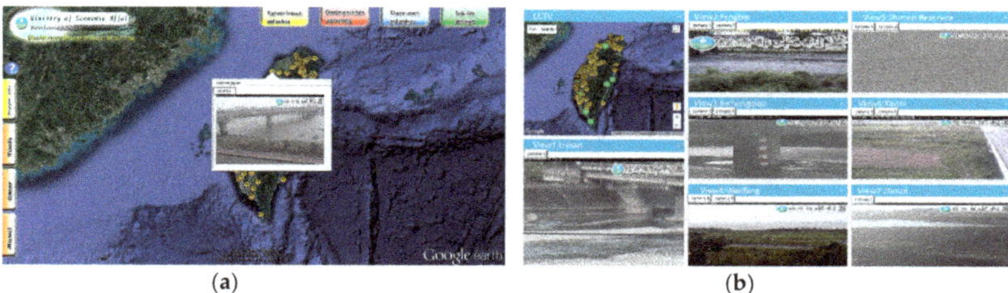

Figure 7. Auxiliary CCTV data integrated using FPERS for Typhoon Nepartak (8 July 2015). (**a**) Single window mode and (**b**) multi-window mode.

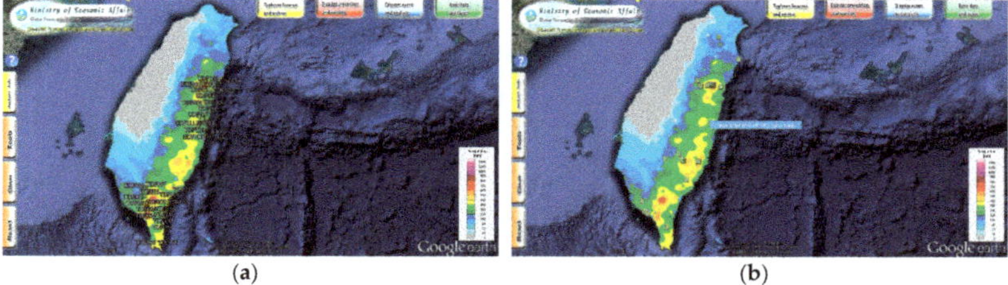

Figure 8. *Cont.*

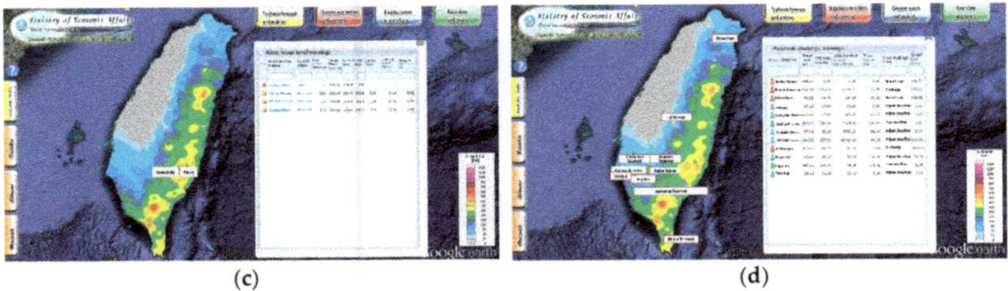

(c) (d)

Figure 8. Disaster alerts integrated by FPERS for Typhoon Nepartak (8 July 2015). (**a**) The yellow or red alerts for debris flow issued by the Soil and Water Conservation Bureau (SWCB); (**b**) the notice of road and bridge closures issued by the Directorate General of Highways (DGH); (**c**) the river water level warnings issued by the Water Resource Agency (WRA); (**d**) the reservoir discharge warnings issued by the WRA.

2.2.3. Disaster Events and Analysis

After FPERS was put online in 2013, a considerable amount of data related to a total of 19 typhoons or torrential rain events and the associated analysis results were collected and integrated into FPERS, as listed in Table 1. FPERS served not only as a comprehensive database that archived all information in a systematic way, but also as a powerful form of media that displayed various data in a flexible fashion. The time slider function is particularly useful for reflecting on the decision-making process at different stages. Users can pause at any stage and bring up more data, images, or analyses to see if there are any differences. It should be noted that FPERS is used as a database to review disaster events rather than as a model to simulate disaster scenarios. A short video review was made and posted online for each typhoon or torrential rain from 2013 to 2016.

Table 1. Disaster events and analysis integrated in Flood Prevention and Emergency Response System (FPERS) from 2013 to 2016.

Event	Start	End
Typhoon Soulik	Sea alert: 11 Jul 2013 08:30 Land warning: 11 Jul 2013 20:30	Land warning: 13 Jul 2013 23:30 Sea alert: 13 Jul 2013 23:30
Typhoon Trami	Sea alert: 20 Aug 2013 11:30 Land warning: 20 Aug 2013 20:30	Land warning: 22 Aug 2013 08:30 Sea alert: 22 Aug 2013 08:30
Tropical Storm Kong-Rey	Sea alert: 27 Aug 2013 11:30 Land warning: 28 Aug 2013 11:30	Land warning: 29 Aug 2013 17:30 Sea alert: 29 Aug 2013 20:30
Typhoon Matmo	Sea alert: 21 Jul 2014 17:30 Land warning: 22 Jul 2014 02:30	Land warning: 23 Jul 2014 23:30 Sea alert: 23 Jul 2014 23:30
7 August 2014 torrential rain	7 Aug 2014	15 Aug 2014
Tropical Storm Fung-Wong	Sea alert: 19 Sep 2014 08:30 Land warning: 19 Sep 2014 20:30	Land warning: 22 Sep 2014 05:30 Sea alert: 22 Sep 2014 08:30
Typhoon Noul	Sea alert: 10 May 2015 08:30	Sea alert: 2015-05-11 20:30
20 May 2015 torrential rain	19 May 2015	27 May 2015
Tropical Storm Linfa	Sea alert: 6 Jul 2015 08:30	Sea alert: 9 Jul 2015 05:30
Typhoon Chan-Hom	Sea alert: 9 Jul 2015 05:30 Land warning: 9 Jul 2015 20:30	Land warning: 10 Jul 2015 23:30 Sea alert: 11 Jul 2015 11:30
Typhoon Soudelor	Sea alert: 6 Aug 2015 11:30 Land warning: 6 Aug 2015 20:30	Land warning: 9 Aug 2015 08:30 Sea alert: 9 Aug 2015 08:30
Typhoon Goni	Sea alert: 20 Aug 2015 17:30	Sea alert: 23 Aug 2015 20:30
Typhoon Dujuan	Sea alert: 27 Sep 2015 08:30 Land warning: 27 Sep 2015 17:30	Land warning: 29 Sep 2015 17:30 Sea alert: 29 Sep 2015 17:30
11 June 2016 torrential rain	11 Jun 2016 16:00	14 Jun 2016 20:00
Typhoon Nepartak	Sea alert: 6 Jul 2016 14:30 Land warning: 6 Jul 2016 20:30	Land warning: 9 Jul 2016 14:30 Sea alert: 9 Jul 2016 14:30
Typhoon Meranti	Sea alert: 12 Sep 2016 23:30 Land warning: 13 Sep 2016 08:30	Land warning: 15 Sep 2016 11:30 Sea alert: 15 Sep 2016 11:30
Typhoon Malakas	Sea alert: 15 Sep 2016 23:30 Land warning: 16 Sep 2016 08:30	Land warning: 18 Sep 2016 02:30 Sea alert: 18 Sep 2016 08:30
Typhoon Megi	Sea alert: 25 Sep 2016 23:30 Land warning: 26 Sep 2016 11:30	Land warning: 28 Sep 2016 17:30 Sea alert: 28 Sep 2016 17:30
Tropical Storm Aere	Sea alert: 5 Oct 2016 11:30	Sea alert: 6 Oct 2016 14:30

2.2.4. Basic Data and Layers

After the security and privacy issue was resolved by linking to the private globes provided by Google Earth Enterprise, more basic data and layers hosted by other agencies were added to FPERS, as listed in Table 2. Most of them were in a vector format with various attributes. Although GEE allowed FPERS to load several large-sized vector layers all at once, transmitting and displaying huge amounts of data simultaneously would inevitably affect its performance. Thus, manipulating more than five vector layers at the same time was not practical. This motivated us to consider not just adding more data, but having the right data.

Table 2. Basic data and layers integrated in FPERS. WRA— Water Resource Agency; CWB—Central Weather Bureau; NLSC— National Land Surveying and Mapping Center.

Category	Data/layers	Source	Type
Drainage and flood control	Locations of water gate	WRA	Vector
	Locations of water pumping station	WRA	Vector
	Sites for flood protection	WRA	Vector
	Locations of levees	WRA	Vector
	Locations of dikes/revetments	WRA	Vector
Reservoir conservation	Locations of the Water Resources Agency and sub-units	WRA	Vector
	Locations of reservoir dams	WRA	Vector
	Reservoir storage areas	WRA	Vector
	Water quality and quantity protection areas	WRA	Vector
	Water source districts	WRA	Vector
	Water Resources Office jurisdiction	WRA	Vector
	Water resources zoning map	WRA	Vector
River hydrology	Locations of river stage observation stations	WRA	Vector
	Locations of river flow observation stations	WRA	Vector
	Locations of sand content observation stations	WRA	Vector
	Locations of coastal ocean tide stations	WRA	Vector
	Distribution of Water Resources Agency coastal data buoys	WRA	Vector
	Locations of coastal ocean weather observation stations	WRA	Vector
	Rivers (subsidiary basin)	WRA	Vector
	Rivers (course basin)	WRA	Vector
	River basin range	WRA	Vector
Weather observation	Ensemble Typhoon Quantitative Precipitation Forecast (ETQPF)	CWB	Numerical
	CWB quantitative precipitation forecast (CWBQPF)	CWB	Numerical
	Himawari-8 satellite images	CWB	Raster
	Radar echo charts	CWB	Raster
Other data/layers	Second generation of inundation potential maps	WRA	Raster
	Flooding survey reports	WRA	Document
	Topographical maps	NLSC	Raster

2.3. *During-Flood Stage*

The most challenging stage for the use of FPERS was the during-flood stage. The during-flood state refers to the time between the issuance and lifting of a typhoon sea alert. The major considerations for this stage are facilitating access to real-time monitoring data, presenting key information, making sound recommendations, and supporting decision-making. Three strategies were implemented as described below.

2.3.1. Real-Time Monitoring Data

Real-time monitoring of data from various sensors is crucial at the during-flood stage, such as videos of water gates and bridge piers captured by CCTVs, as well as water level readings recorded at reservoir stations, river stations, and tide stations. In the study, the data had already been identified and linked to FPERS for the pre-flood stage. To quickly grasp the changes in the during-flood stage, all data were automatically sorted in high-to-low order and displayed in the form of a slideshow window. The data were highlighted in red if a preset threshold was reached. A special outsourcing function was also developed to enable users to upload photographs, video clips, or a paragraph of text to report the disaster areas. All information was integrated into one disaster map published and updated by FPERS, as shown in Figure 9.

2.3.2. User-Friendly Functions

To facilitate access to the huge amount of data integrated in FPERS and quick acquisition of key information on a limited screen size, four user-friendly functions were designed and placed on the left-hand side of the FPERS screen display.

(1) "Instant Info" would bring up the slideshow window with selected meteorology data, hydrology data, geospatial information data, and disaster alerts.
(2) "Tools" would toggle between the single- and dual-window mode; adjust the transparency of each layer; measure the perimeter or area by drawing a polyline or a polygon on the screen; and pinpoint the location with inputs of coordinates, address, place name, or some key words.
(3) "Clear" would unselect all layers that have been displayed, but keep them in the list. Users can reselect the layers they want in order to avoid too much data overcrowding the screen.
(4) "Reset" would return the view to the entire Taiwan, focalized on its center. The list of all accessed data and layers would be emptied.

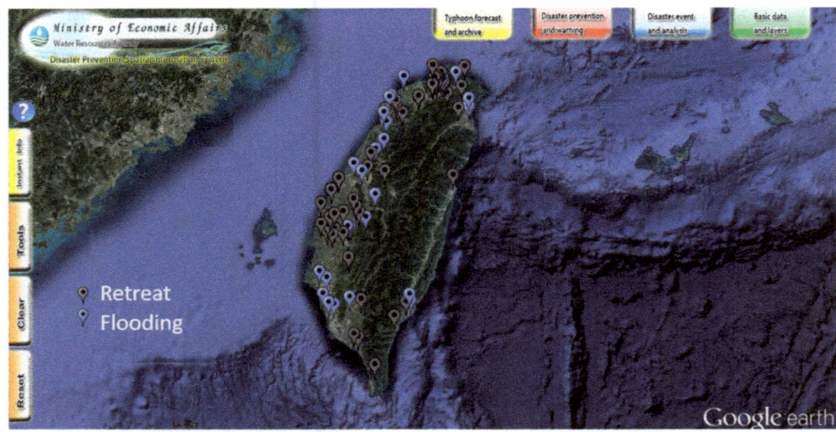

Figure 9. Disaster map published and updated by FPERS at the during-flood stage of Typhoon Nepartak on 8 July 2015.

2.3.3. Standard Operation Procedure

The time for making decisions in the during-flood stage is fleeting, and the selection of data can be very versatile and time-consuming. To ensure all important data were being properly displayed and reviewed as time progressed, we proposed an SOP that comprised eight steps, as listed in Table 3. The purpose and intention of each step are outlined below. A special button was designed to bring up all required data and layers that should be displayed and reviewed with merely one click at each step. Together with this SOP, FPERS was employed by the WHMC of the WRA to support flood prevention and emergency responses from 2013 to 2016. The application of FPERS during Typhoon Soudelor in August of 2015 is used as an example and is described in detail below.

Table 3. Standard operation procedure at during-flood stage.

Step	Purpose/Intention	Data/Layers
Typhoon formation	Determine whether a typhoon will invade Taiwan	• Typhoon track forecast issued by the CWB • Typhoon track forecast issued by other countries
Sea alert is issued	Resource dispatching	• Typhoon track forecast issued by the CWB • Typhoon track forecast issued by other countries • Distribution of pumping machines • Flood protection materials
Land warning is issued	Determine the rainfall distribution	• Typhoon track forecast issued by the CWB • Typhoon track forecast issued by other countries • Satellite cloud image • Radar echo chart • 3-h accumulated rainfall chart • ETQPF
12 h before landfall	Determine the rainfall distribution	• Typhoon track forecast issued by the CWB • Microwave imagery • Satellite cloud image • Radar echo chart • 3-h accumulated rainfall chart • ETQPF • Alert information
Landfall	Latest warnings and information	• Typhoon track forecast issued by the CWB • Microwave imagery • Satellite cloud image • Radar echo chart • 3-h accumulated rainfall chart • ETQPF • Alert information
Real-time disaster report	Emergency response	• Flooding situation • Water resource facility
Sea alert and land warning are Lifted	Monitoring the rainfall situation after a typhoon	• Typhoon track forecast issued by the CWB • Satellite cloud image • Radar echo chart

3. Application of FPERS in Typhoon Soudelor

Typhoon Soudelor was the third most intense tropical cyclone worldwide in 2015. It formed as a tropical depression on 29 July, made landfall on Saipan on 2 August, and developed into a Category 5-equivalent super typhoon under favorable environmental conditions later on 3 August. Late on 7 August, Soudelor made landfall over Hualien, Taiwan, and moved out to the Taiwan Strait early the next day. The torrential rains and destructive winds caused widespread damage and disruptions, resulting in eight deaths, and 420 people sustained injuries. Taiping Mountain recorded the heaviest rains during this event, with accumulations peaking at 1334 mm, which caused floods in I-Lan County. Rainfall in the Wulai District reached 722 mm in 24 h and triggered a large landslide.

The case of Typhoon Soudelor is used as an example to demonstrate how FPERS was employed to support flood prevention and emergency responses. All related data and layers collected and employed at different stages were recorded in a tour file, which is ideal for reviewing this event by replaying the scenarios in chronological order. Users could even pause or reverse the tour, and they could also add more data or layers at any stage. This tour file is available in the Supplementary Materials. From the file, eight scenarios were captured at different stages, as shown in Figure 10. These are explained in detail as follows.

When Typhoon Soudelor formed on 29 July, the forecasts of the typhoon path made by the CWB of Taiwan and other countries were acquired and displayed in FPERS (Figure 10a), with an intention to see whether this typhoon was coming towards Taiwan or not. Requests for urgent SAR image acquisition were also sent to a few distributors to evaluate the suitable slots and coverages from different satellites. As the order had to be placed at least 24 h before the image acquisition, the decision was made when CWB issued the Typhoon Soudelor sea alert at 11:00 on 6 August. At the same time, WHMC of WRA also initiated the SOP to respond to this event. The most updated forecasts of typhoon paths were overlaid on the map of pumping machines and flood protection materials (Figure 10b), which could be used to determine whether the dispatch of resources was appropriate or not. Both the wind and rain intensified as the typhoon was approaching Taiwan. Later, at 20:30 that day, the land warning was issued by the CWB, and the major concern at this stage was the possible amount and distribution of precipitation. Therefore, the most updated forecasts of typhoon paths were overlaid on the satellite cloud images, the radar echo chart, the accumulated rainfall chart, as well as the Ensemble Typhoon Quantitative Precipitation Forecast (ETQPF) (Figure 10c), to identify the possible sites of floods and to reallocate pumping machines and flood protection materials if necessary. The data and layers were kept updated, and the same assessment procedure was repeated until 12 h before the typhoon landed. All kinds of disaster alerts were annotated on the map (Figure 10d), including the yellow or red debris flow alerts issued by the SWCB, the notice of road and bridge closures issued by the DGH, the heavy or torrential rain alerts issued by the CWB, and the reservoir discharge warning issued by the WRA. Based on these alerts, people staying in regions with higher risks were asked to make necessary and compulsory evacuations.

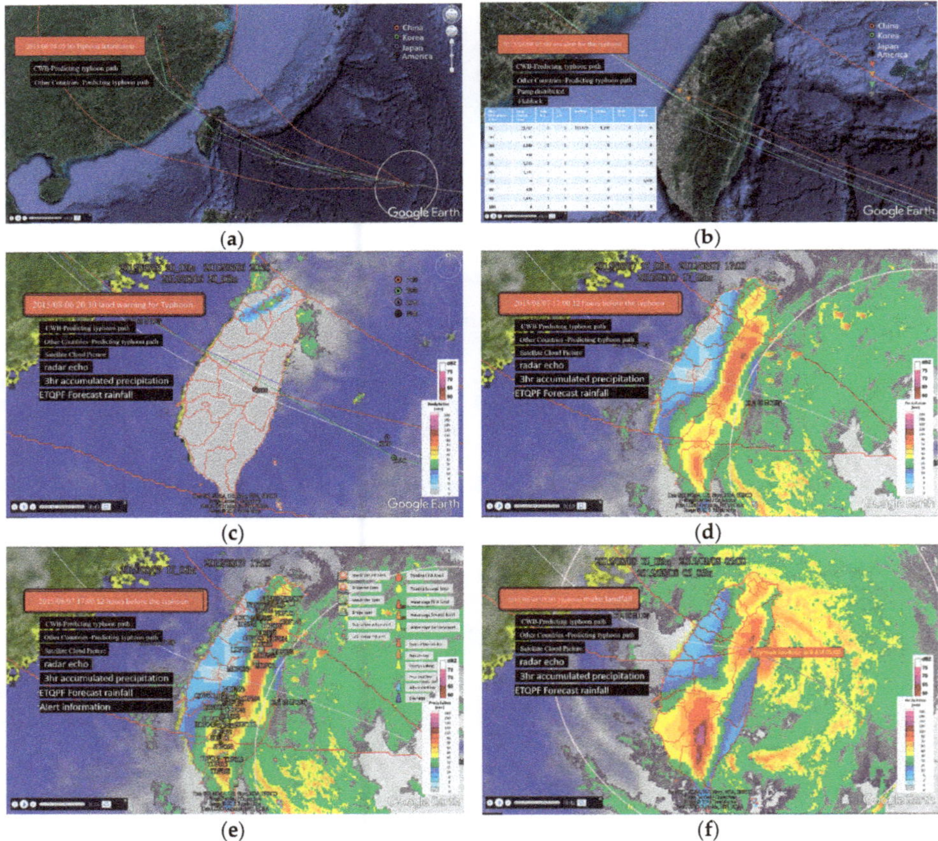

Figure 10. *Cont.*

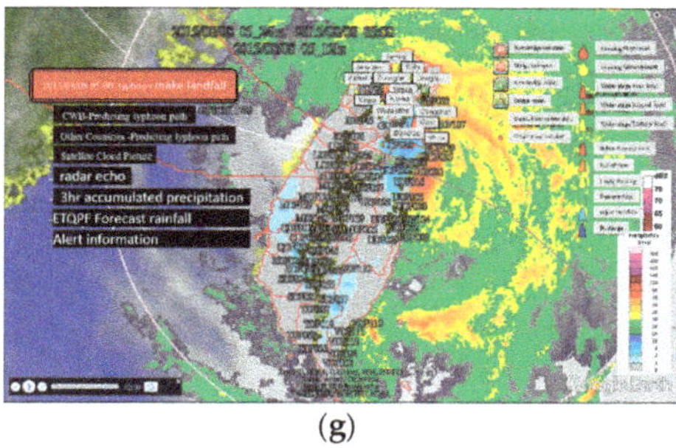

(**g**)

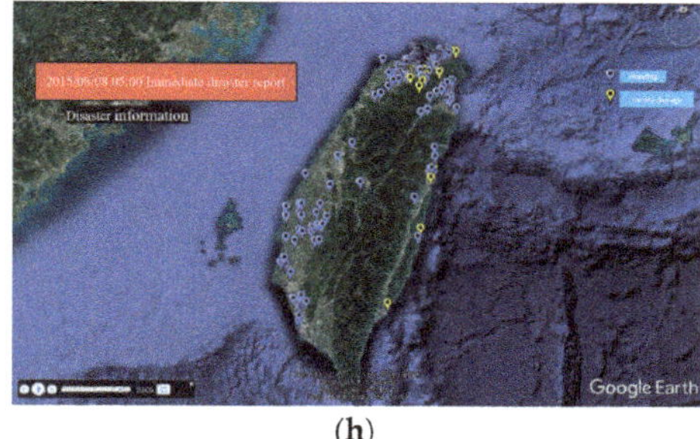

(**h**)

Figure 10. Demonstration of how FPERS was employed to support flood prevention and emergency responses for Typhoon Soudelor. (**a**) The forecasts of the typhoon path made by the Central Weather Bureau (CWB) of Taiwan and other countries when Typhoon Soudelor formed; (**b**) the most updated forecasts of typhoon paths overlaid on the map of pumping machines and flood protection materials when the CWB issued the sea alert. The most updated forecasts of typhoon paths overlaid on the satellite cloud images, the radar echo chart, the accumulated rainfall chart; as well as the Ensemble Typhoon Quantitative Precipitation Forecast (ETQPF) (**c**) when the CWB issued the land warning; and (**d**) 12 h before the typhoon landed; (**e**) SAR image of the I-Lan area acquired by COSMO-SkyMed 3 at 05:56 on August 8. Integration of a large number of geospatial data/layers and real-time monitoring data after landfall; including (**f**) ETQPF; (**g**) alert information; and (**h**) real-time disaster reports.

The center of Typhoon Soudelor eventually made landfall on the east coast of Taiwan at 04:40 on 8 August, and both the wind and rain reached their maximum values. This was also the time when the runoff was accumulating and when most of the floods were triggered. Even though the clouds and rains were overcasting the northern part of Taiwan, the active radar signals could penetrate the storm, and one SAR image of I-Lan area was successfully acquired at 05:56 on 8 August. This image was processed using the approach described in the literature [4] and was posted in FPERS (Figure 10e). As Soudelor was moving across the northern part of Taiwan, multiple disasters were reported to the local governments. This was the busiest and most stressful time for the WHMC of the WRA, because all information and requests related to floods were forwarded and integrated there. The WHMC had to coordinate various departments and agencies, make recommendations according to the available resources, and follow each case of reported disasters. Thanks to the user-friendly functions implemented in FPERS, a large number of geospatial data/layers and real-time monitoring data were integrated after landfall, including ETQPF (Figure 10f), alert information (Figure 10g), and real-time disaster reports (Figure 10h). The key information was quickly acquired from the screen using the tools developed for FPERS. Note that the hot spots at this stage were the water resource facilities, such as water gates, levees, dams, and so on. They were monitored by FPERS conveniently because not only were the real-time data of water levels linked, but also the streaming of real-time CCTV video was displayed in either the single-window or the multi-window mode. The inundated areas derived from the SAR image of I-Lan were completed and posted on FPERS at about 17:00 on 8 August, when the sea alert had not yet been lifted.

Typhoon Soudelor moved out to sea at about 11:00 and kept moving northwest. Later that day at about 22:00, it made landfall in Fujian, China. Both the land warning and sea alert were lifted at 08:30 the next day. However, the WHMC still kept an eye on the southwesterly airstream induced by the passage of Soudelor, for it could bring downpours even if the typhoon had left. For the case of Typhoon Soudelor, several landslides were triggered in Wulai, and the water of Nanshi River was contaminated and soon mixed with the primary water source in the Taipei-Keelung metropolitan area. As a result, the regional purification system was overwhelmed, and the water supply to Taipei was polluted for the next few days. To gain an overview of the entire Wulai area, the Formosat-2 images acquired before and after Typhoon Soudelor were compared, and the total areas and distribution of the landslides were mapped and posted on FPERS. The animation of virtual flight with a synchronous comparison of Formosat-2 before- and after-Soudelor images is available in the Supplementary Materials.

4. Discussion

FPERS powered by GEE provided an innovative platform that enabled us to integrate a huge amount of geospatial data, resolving the data accessibility issue. From a practical point of view, however, manipulating more than five layers simultaneously is not very helpful for decision making. Instead, referring to the right data at the right time is much more important, particularly for timely responses to an urgent event. Determining the right data and the right time requires domain knowledge for different types of disasters. Therefore, in this work, a set of SOPs was proposed for the pre-, during-, and post-flood stages to ensure that all important data were properly displayed and reviewed as time progressed. At the pre-flood stage, the huge amount of geospatial data were categorized as typhoon forecast and archive, disaster prevention and warning, disaster events and analysis, and basic data and layer. At the during-flood stage, three strategies were implemented to facilitate access to real-time data, present key information, make sound recommendations, and support decision-making. At the post-flood stage, various remote sensing imageries were integrated, including Formosat-2 optical imagery to detect and monitor barrier lakes, the synthetic aperture radar imagery to derive an inundation map, and the high-spatial-resolution photographs taken by unmanned aerial vehicles to evaluate the damage to river channels and structures. The prevention and urgent response experiences gained from FPERS can be applied to other types of urgent events, such as sediment-related or earthquake-related disasters.

One advantage of GEE is that petabyte-scale data (mostly images) are prepared and updated by Google for scientific analysis and visualization. Apart from the high-spatial-resolution of aerial photographs and satellite images for densely populated areas, the latest Landsat-8 images are included for the less populated areas, but only at a frequency of about half a year using cloudless scenes. Therefore, there might have been a lag of a few months on the Landsat-8 images provided by GEE. Thanks to the success of the Sentinel-2 mission and the Copernicus program adopting a free and open data policy, all Sentinel-2 high-spatial-resolution (10 m) data are available for free to all registered users on a five-day basis. They serve as an ideal, cost-effective data source to be included by GEE. For Taiwan, the standard Level-1C products of Sentinel-2 imagery processed and provided by the Copernicus Open Access Hub have been mosaicked, contrast-enhanced, converted to a set of pyramid image tiles, and distributed to the general public through the Open Access Satellite Image Service (OASIS, http://oasis.ncku.edu.tw) [15]. They can be integrated into FPERS as alternatives for base images, but this raises the issue of archiving and analyzing a very large number of satellite images. As the main concern is the changes among images rather than the images themselves, an efficient solution would be to archive and display only the changes, which would require a reliable, automatic approach to change detection.

Another advantage of GEE is that the global topographic data has already been incorporated, and all geospatial data can be displayed in 3D through the Internet. Google uses a range of digital elevation model (DEM) data sources to derive the terrain layer, such as the USGS National Elevation Dataset, Shuttle Radar Topography Mission data, and the NOAA Global Land One-km Base Elevation Project dataset. However, the default 3D GEE globe is not replaceable. Even though the Google Earth Enterprise product allows developers to create maps and 3D globes for private use, that cannot be done on the fly. Considering the fact that variations in relief features are always the primary interest and that there are many ways to collect the most updated topographic data, it is preferable to have the capability of switching among different topographic models. For practical application to flood prevention and emergency responses, for example, a quick way to assess the status of water a resource facility is to compare the regional DEMs collected before- and after-flood using UAV. These valuable data are available, but not able to be displayed in the current version of FPERS.

The lack of flexibility in managing or searching for a huge geospatial database by GEE has become increasingly more apparent as more geospatial data are being integrated in FPERS. Taking Google Earth as an example, users have to select the data/layer first and then display or analyze the selected ones. This implies the users must know the existence, content, or type of the data/layer, and make the selections one by one on their own. This is not practical for a system like FPERS, which has integrated a huge geospatial database. The logical and reasonable order is to narrow down the entire database first with some specified conditions, such as the time frame, spatial extent, and data type. Then, the final selection can be made accordingly. This method requires a powerful database with a special consideration of geospatial attributes. All data and collections for FPERS should be added and curated through this geospatial database.

After the release of Google Earth API in 2008, FPERS and many other web-based geospatial systems were developed and powered by GEE to support a variety of management decisions [16–18]. As experiences were gained and systems were improved gradually, innovative and reliable services were created and implemented. However, the Google Earth API was depreciated as of 15 December 2014 and remained supported until 15 December 2015. Google Chrome also ended support for the Netscape Plugin API (which the Google Earth API relies on) at the end of 2016. As a result, all systems, including FPERS, were forced to be depreciated or downgraded to use Google Map API, which is a 2D platform that does not support 3D displays. Although Google decided to make all codes behind the Google Earth Enterprise available on GitHub in March 2017, Google has not open-sourced the implementations for Google Earth Enterprise Client, Google Maps JavaScript API, and Google Earth API. The depreciation of Google Earth Enterprise and Google Earth API in 2014 inevitably affected government agencies' confidence and their willingness to rely on only one commercial platform.

As a result of these changes, the WRA decided to terminate the development and maintenance of FPERS in late 2015; the online service of FPERS was also removed. The old version of Google Chrome V4.6 (Google LLC, Mountain View, California, U.S.) that supported the Netscape Plugin API and Google Earth API, together with the entire FPERS, has been saved in one compressed file. This file can be downloaded to the user's personal computer and executed to launch the old version of Google Chrome to run FPERS with the full set of functions. On the other hand, some similar, yet fully open solutions, such as OpenLayers, have also become available. Despite the fact that more effort is required to put everything together in these open solutions compared with GEE, their capability and potential have been proven. Based on the experiences gained in developing FPERS, a new version of FPERS powered by OpenLayers is being planned.

5. Conclusions

Flooding caused by typhoons and torrential rainfalls is one of the major disasters that leads to severe damage and significant loss of lives and property. Supporting a variety of management decisions, particularly a timely response to an urgent event, requires a sound approach to identify, collect, integrate, process, analyze, distribute, and visualize a huge amount of data rapidly through the Internet. This was not possible until the Google Earth API was released on 28 May 2008. Powered by GEE, FPERS was successfully developed and employed to support flood prevention and emergency responses in 19 typhoons or torrential rain events from 2013 to 2016, including the example of Typhoon Soudelor presented in this paper. This work proposes a set of SOPs for the pre-, during-, and post-flood stages to ensure all important data are properly displayed and reviewed over time. Apart from integrating a huge amount of geospatial data, FPERS enabled us to refer to the right data at the right time. The capability of switching among different topographic models and the flexibility of managing and searching the data through a geospatial database were also shown to be main features of the system. Experiences gained from FPERS would benefit the application of prevention work and urgent response to other types of catastrophic events, such as sediment-related or earthquake-related disasters.

Supplementary Materials: The following are available online at http://www.mdpi.com/2072-4292/10/8/1283/s1: Video S1: How FPERS was employed to support flood prevention and emergency responses related to Typhoon Soudelor. Video S2: Virtual flight with synchronous comparison of Formosat-2 before- and after-Soudelor images. KML S1: Tour file to demonstrate how FPERS was employed to support the work of flood prevention and emergency responses to Typhoon Soudelor. Figure S1: All figures with their original resolutions and sizes.

Author Contributions: C.-C.L. organized the research and wrote the paper; C.-C.L. and M.-C.S. designed and proposed the main structure of this research; M.-C.S. proposed the concept of pre-, during-, and post-flood stages, and suggested and provided the required data; C.-C.L. and M.-S.K. developed the system and contributed the required tools for data processing; K.-H.W. processed and analyzed the metrological data.

Funding: This research was supported by the Water Resource Agency of Taiwan under Contract No MOEA WRA1050069., as well as the Ministry of Science and Technology of Taiwan under Contract No. MoST 2017-2611-M-006-002.

Acknowledgments: The authors acknowledge the assistance received from the Central Weather Bureau (CWB) and the National Science and Technology Center for Disaster Reduction (NCDR) of Taiwan.

Conflicts of Interest: The authors declare no conflicts of interest.

References

1. Liu, C.-C.; Chang, C.-H. Searching the strategy of responding to the extreme weather events from the archive of Formosat-2 remote sensing imagery. *Geology* **2009**, *28*, 50–54. (In Chinese)
2. Liu, C.-C. Processing of Formosat-2 daily revisit imagery for site surveillance. *IEEE Trans. Geosci. Remote Sens.* **2006**, *44*, 3206–3214. [CrossRef]
3. Liu, C.-C.; Chen, N.-Y. Responding to natural disasters with satellite imagery. *SPIE Newsroom* **2011**. [CrossRef]
4. Chung, H.-W.; Liu, C.-C.; Cheng, I.-F.; Lee, Y.-R.; Shieh, M.-C. Rapid response to a typhoon-induced flood with an sar-derived map of inundated areas: Case study and validation. *Remote Sens.* **2015**, *7*, 11954–11973. [CrossRef]
5. Liu, C.-C.; Chen, P.-L.; Matsuo, T.; Chen, C.-Y. Rapidly responding to landslides and debris flow events using a low-cost unmanned aerial vehicle. *J. Appl. Remote Sens.* **2015**, *9*, 096016. [CrossRef]
6. Lee, C.-Y.; Lai, W.-C.; Chen, C.-Y.; Huang, H.-Y.; Kuo, L.-H. The reconstruction of the processes of catastrophic disasters caused by the 2009 Typhoon Morakot. *J. Chin. Soil W. Conserv.* **2011**, *42*, 313–324.
7. Liu, L.; Wu, Y.; Zuo, Z.; Chen, Z.; Wang, X.; Zhang, W. Monitoring and assessment of barrier lakes formed after the Wenchuan earthquake based on multitemporal remote sensing data. *J. Appl. Remote Sens.* **2009**, *3*, 031665. [CrossRef]
8. Zou, Q.; Su, Z.M.; Zhu, X.H. Mechanism of landslide-debris flow-barrier lake disaster chain after the Wenchuan earthquake. In *Earthquake-Induced Landslides: Proceedings of the International Symposium on Earthquake-Induced Landslides, Kiryu, Japan, 2012*; Ugai, K., Yagi, H., Wakai, A., Eds.; Springer: Berlin/Heidelberg, Germany, 2013; pp. 917–924.
9. Liu, C.-C.; Liu, J.-G.; Lin, C.-W.; Wu, A.-M.; Liu, S.-H.; Shieh, C.-L. Image processing of Formosat-2 data for monitoring south asia tsunami. *Int. J. Remote Sens.* **2007**, *28*, 3093–3111. [CrossRef]
10. Liu, C.-C.; Chen, P.-L. Automatic extraction of ground control regions and orthorectification of Formosat-2 imagery. *Opt. Express* **2009**, *17*, 7970–7984. [CrossRef] [PubMed]
11. Liu, C.-C.; Shieh, C.-L.; Wu, C.-A.; Shieh, M.-L. Change detection of gravel mining on riverbeds from the multi-temporal and high-spatial-resolution Formosat-2 imagery. *River Res. Appl.* **2009**, *25*, 1136–1152. [CrossRef]
12. Chang, C.-H.; Liu, C.-C.; Wen, C.-G.; Cheng, I.-F.; Tam, C.-K.; Huang, C.-S. Monitoring reservoir water quality with Formosat-2 high spatiotemporal imagery. *J. Environ. Monit.* **2009**, *11*, 1982–1992. [CrossRef] [PubMed]
13. Liu, C.-C.; Kamei, A.; Hsu, K.H.; Tsuchida, S.; Huang, H.M.; Kato, S.; Nakamura, R.; Wu, A.M. Vicarious calibration of the Formosat-2 remote sensing instrument. *IEEE Trans. Geosci. Remote Sens.* **2010**, *48*, 2162–2169.
14. Chung, H.; Liu, C.; Chiu, Y.; Liu, J. Spatiotemporal variation of gaoping river plume observed by Formosat-2 high resolution imagery. *J. Mar. Syst.* **2014**, *132*, 28–37. [CrossRef]
15. Liu, C.-C. Towards an automatic change detection system for land use and land cover. In Proceedings of the 2nd Sirindhorn Conference on Geoinformatics 2018, Chaeng Watthana, Bangkok, Thailand, 1–2 February 2018; p. 155.

16. Goldblatt, R.; Rivera Ballesteros, A.; Burney, J. High spatial resolution visual band imagery outperforms medium resolution spectral imagery for ecosystem assessment in the semi-arid brazilian sertão. *Remote Sens.* **2017**, *9*, 1336. [CrossRef]
17. Markert, K.; Schmidt, C.; Griffin, R.; Flores, A.; Poortinga, A.; Saah, D.; Muench, R.; Clinton, N.; Chishtie, F.; Kityuttachai, K.; et al. Historical and operational monitoring of surface sediments in the lower mekong basin using landsat and google earth engine cloud computing. *Remote Sens.* **2018**, *10*, 909. [CrossRef]
18. Xiong, J.; Thenkabail, P.; Tilton, J.; Gumma, M.; Teluguntla, P.; Oliphant, A.; Congalton, R.; Yadav, K.; Gorelick, N. Nominal 30-m cropland extent map of continental africa by integrating pixel-based and object-based algorithms using sentinel-2 and landsat-8 data on google earth engine. *Remote Sens.* **2017**, *9*, 1065. [CrossRef]

Article

Leveraging the Google Earth Engine for Drought Assessment Using Global Soil Moisture Data

Nazmus Sazib [1,2,*], Iliana Mladenova [1,3] and John Bolten [1]

1 Hydrological Sciences Branch, NASA Goddard Space Flight Center, Greenbelt, MD 20706, USA; iliana.e.mladenova@nasa.gov (I.M.); john.bolten@nasa.gov (J.B.)

2 Science Application International Corporation (SAIC), Lanham, MD 20706, USA

3 Earth System Science Interdisciplinary Center, University of Maryland, College Park, MD 20742, USA

* Correspondence: nazmus.s.sazib@nasa.gov; Tel.: +1-301-614-6384

Received: 29 June 2018; Accepted: 8 August 2018; Published: 11 August 2018

Abstract: Soil moisture is considered to be a key variable to assess crop and drought conditions. However, readily available soil moisture datasets developed for monitoring agricultural drought conditions are uncommon. The aim of this work is to examine two global soil moisture datasets and a set of soil moisture web-based processing tools developed to demonstrate the value of the soil moisture data for drought monitoring and crop forecasting using the Google Earth Engine (GEE). The two global soil moisture datasets discussed in the paper are generated by integrating the Soil Moisture Ocean Salinity (SMOS) and Soil Moisture Active Passive (SMAP) missions' satellite-derived observations into a modified two-layer Palmer model using a one-dimensional (1D) ensemble Kalman filter (EnKF) data assimilation approach. The web-based tools are designed to explore soil moisture variability as a function of land cover change and to easily estimate drought characteristics such as drought duration and intensity using soil moisture anomalies and to intercompare them against alternative drought indicators. To demonstrate the utility of these tools for agricultural drought monitoring, the soil moisture products and vegetation- and precipitation-based products were assessed over drought-prone regions in South Africa and Ethiopia. Overall, the 3-month scale Standardized Precipitation Index (SPI) and Normalized Difference Vegetation Index (NDVI) showed higher agreement with the root zone soil moisture anomalies. Soil moisture anomalies exhibited lower drought duration, but higher intensity compared with SPIs. Inclusion of the global soil moisture data into the GEE data catalog and the development of the web-based tools described in the paper enable a vast diversity of users to quickly and easily assess the impact of drought and improve planning related to drought risk assessment and early warning. The GEE also improves the accessibility and usability of the earth observation data and related tools by making them available to a wide range of researchers and the public. In particular, the cloud-based nature of the GEE is useful for providing access to the soil moisture data and scripts to users in developing countries that lack adequate observational soil moisture data or the necessary computational resources required to develop them.

Keywords: soil moisture; Soil Moisture Ocean Salinity; Soil Moisture Active Passive; Google Earth Engine; drought

---

1. Introduction

The remote sensing advances made over the past three decades have radically improved our ability to obtain routine, global information about the amount of water present in the soil [1–3]. Several well-evaluated soil moisture (SM) datasets have proven useful for a wide range of applications, including weather and climate forecasting, monitoring of drought and wildfires, tracking floods and landslides, and enhanced agricultural productivity [4–7]. Despite all the advantages satellite-based soil moisture monitoring offers, such as global coverage, high accuracy, and high temporal frequency essential for operational applications, the shallow penetration depth of the microwave frequencies used for soil moisture monitoring remains a limiting factor [3]. In fact, many hydrological processes and forecasting and decision-making activities linked to soil moisture require knowledge of the root zone soil moisture (RZSM) information. The latter is typically estimated using hydrologic land surface models, which are traditionally driven by some weather-related observations such as precipitation, temperature, and so forth. The credibility of the modeled RZSM estimates is strongly dependent on the quality of the forcing data [8,9].

Hydrologic data assimilation (DA) offers a way to reduce the impact of precipitation-related errors and enhance the quality of the modeled RZSM data by integrating satellite-based observations into the model. It is essentially an optimal merging technique that results in an enhanced analysis product with improved accuracy over either of the parent products alone (i.e., the model data and the satellite observations). This paper focuses on the value of a combined model-satellite soil moisture dataset for near-real time drought and crop condition monitoring, developed in an effort to improve the RZSM information used by the United States Department of Agriculture-Foreign Agricultural Services (USDA-FAS).

One of the operational objectives of the USDA-FAS is the development of global crop yield forecasts, which heavily employ modeled soil moisture to force crop yield models. These forecasts are generated utilizing information from the Crop Condition Data Retrieval and Evaluation database management system, where soil moisture is an essential indicator used to assess crop health and monitor drought conditions and evaluate their impact on expected end of season yields. The baseline USDA-FAS RZSM information is generated using the modified Palmer model (PM), which is a relatively simple two-layer water balance model driven by daily estimates of precipitation and temperature [10,11]. The PM is highly susceptible to the quality of the precipitation forcing data, which are more error-prone over poorly instrumented areas [8]. Over a decade of research conducted in collaboration with USDA-FAS demonstrated that the agency's model-based RZSM estimates produced by the PM can be improved by assimilating satellite-derived observations [11,12]. Here, we focus on the operational implementation of the DA-enhanced PM using soil moisture retrievals from two passive microwave missions, the European Space Agency (ESA)'s Soil Moisture and Ocean Salinity (SMOS) [13] mission and the National Aeronautics Space Agency (NASA)'s Soil Moisture Active Passive (SMAP) mission [14] which launched in 2009 and 2015 respectively, specifically designed to monitor near-surface soil moisture at a global scale using L-band frequency.

The goal of this paper is to announce the availability of these global soil moisture datasets and to demonstrate their value for drought monitoring using the Google Earth Engine (GEE). The GEE is a web-based service that stores a petabyte archive of Earth observations and related data and provides an efficient processing software, which enables users to develop complex geospatial analyses and visualizations utilizing high-performance computing resources. The GEE capabilities have been utilized for a range of applications, including soil mapping, malaria risk assessment, and automated cropland mapping [15–18]. In this study, we demonstrate the value of the SMOS- and SMAP-datasets and the web-based tools utilizing the global soil moisture dataset generated using the satellite-enhanced PM available in the GEE data catalog. The GEE and the available tools enable users to acquire, process, analyze, and visualize Earth observing data rapidly for any user-specified region across the globe without downloading and processing a large volume of data on the user's desktop. The web-based drought assessment tools alleviate the need for users to install and work with desktop data managing and processing software, which are often labor intensive, time consuming, and difficult to reproduce, thereby overcoming compatibility limitations and enhancing usability and reproducibility of the analyses and results.

The paper is organized as follows: Section 2 provides a detailed description of the soil moisture data, modeling approach, and preparation steps for integrating the data into the GEE platform; Section 3 focuses on the functionality of the GEE tools developed for drought assessment using the satellite-enhanced PM global soil moisture data; Section 4 describes the application of the GEE tools over South Africa and Ethiopia; and Sections 5 and 6 provide some discussion of the results and conclusions, respectively.

2. Data Processing for the Google Earth Engine Platform

An overview of the major methodological steps applied in this study is provided in Figure 1. First, we processed satellite-based soil moisture datasets to estimate surface and RZSM and their anomalies. Then, we used RZSM and precipitation data to explore their spatial and temporal variability with different land cover types. Next, we estimated drought characteristics from RZSM anomalies and compare against other alternative drought indices. Details about these datasets are provided in Table 1. One of the primary goals of this study is to introduce global soil moisture data sets in the GEE; hence we provide detailed descriptions of the soil moisture datasets in the following sub-section.

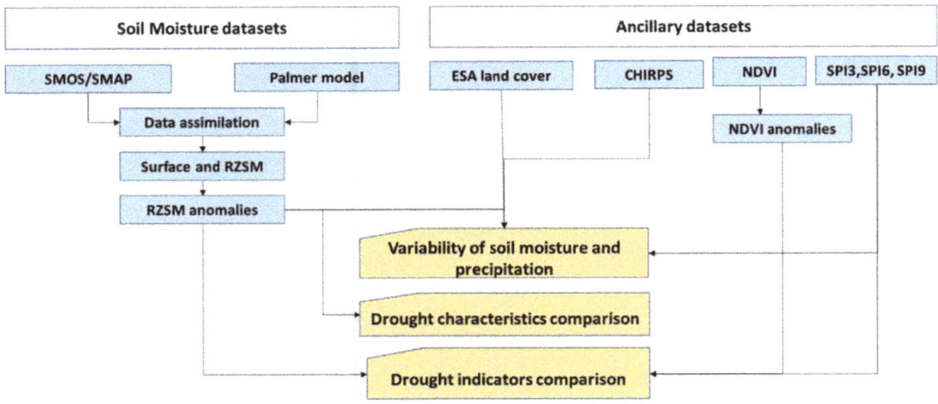

Figure 1. Schematic overview of the methodological approach. SMOS, Soil Moisture Ocean Salinity mission; SMAP, Soil Moisture Active Passive mission; ESA, European Space Agency; CHIRPS; Climate Hazards Group Infrared Precipitation with Station; NDVI, Normalized Difference Vegetation Index; SPI, Standardized Precipitation Index; and RZSM, root zone soil moisture.

Table 1. Datasets used in this study.

No	Name	Spatial Resolution	Temporal Resolution	URL
1	NASA-USDA Global Soil Moisture Data	0.25°	3 days	[19]
2	NASA-USDA SMAP Global Soil Moisture Data	0.25°	3 days	[20]
3	SPI	1°	monthly	[21]
4	NDVI	0.00225°	8 days	[22]
5	CHIRPS Pentad: Climate Hazards Group InfraRed Precipitation with Station Data	0.05°	5 days	[23]
6	Global Land Cover Map	0.00275°		[24]

NASA, National Aeronautics Space Agency; USDA, United States Department of Agriculture.

2.1. Soil Moisture Data Set

The USDA-FAS Soil Moisture System: Palmer Model, Data Assimilation, and Satellite Observations

a. Theoretical Basis

The two-layer Palmer Model used by the USDA-FAS is a bookkeeping water balance model that accounts for the water gained by precipitation and lost by evapotranspiration [10]. The top layer is assumed to have 2.54 cm available water-holding capacity at saturation, while the holding capacity of the lower layer varies depending on the depth of the bedrock. The model is driven by daily precipitation data and daily minimum and maximum temperature observations provided by the United States (U.S.) Air Force 557th Weather Wing (formerly known as the U.S. Air Force Weather Agency, AFWA). The AFWA dataset is derived using multiple sources, inducting remotely sensed observations and gauge data acquired from the World Meteorological Organization (WMO). The model is enhanced by adding a data assimilation unit, which allows the routine integration of satellite-based observations into the model using a one-dimensional (1D) ensemble Kalman filter approach (EnKF) [11,12]. The purpose of this modification is to improve the PM RZSM information by integrating the added values of the surface soil moisture retrievals to the model and examining their potential to correct for meteorological forcing uncertainty. A detailed description of Bayesian theory-based filtering, including the EnKF is beyond the scope of this paper; however, the methods are well established and documented [25–29].

The Kalman filter is a sequential Monte Carlo assimilation technique, where the model forecasts are optimally updated in response to the satellite observations via the Kalman gain (K). The operational implementation of EnKF requires some knowledge of the model uncertainty (Q) and the error of the satellite observations (R). Here, both of these parameters have been parameterized using some a priori knowledge. Given the above-discussed and well-established dependence of the model accuracy on the uncertainty of the rainfall data and the fact that the AFWA rainfall dataset is rain-gauge corrected, R has been modeled as a function of proximity to the WMO gauge station. Q, on the other hand, has been parameterized as a land cover type using published accuracy assessment analysis [30–36]. We discuss the implementation of the satellite-enhanced PM using remotely sensed observations acquired from two L-band missions, SMOS and SMAP. Full technical descriptions of these missions can be found in [13], respectively.

The corresponding SMOS and SMAP soil moisture estimates assimilated into the PM are derived using slightly different retrieval approaches; however, both systems and soil moisture products show similar performance and overall accuracy [37,38]. The global SMOS soil moisture data are operationally acquired from the National Oceanic and Atmospheric Administration (NOAA) Soil Moisture Operational Products System (SMOPS), which are distributed at 0.25° grid spacing (https://data.nodc.noaa.gov/cgi-bin/iso?id=gov.noaa.ncdc:C00994; last accessed on May 2018). SMAP offers a variety of soil moisture products (https://smap.jpl.nasa.gov/data/; last accessed on May 2018). This study applies the L3 passive-only SMAP soil moisture product. The data are routinely downloaded from the National Snow and Ice Data Center (https://nsidc.org/data/SPL3SMP/versions/4; last accessed on May 2018). SMAP is distributed in Equal-Area Scalable Earth (EASE)-grid 2 projection at 36-km grid spacing; therefore, the data have been pre-processed to match the model grid of 0.25°.

b. Operational Implementation

The satellite enhanced Palmer Model is set to run operationally on NASA's Global Inventory Modeling and Mapping Studies (GIMMS) Global Agricultural Monitoring (GLAM) system [22]. The model covers the Land Information System (LIS) domain (180°W–180°E, 90°N–60°S) at 0.25° [39]. The system generates various soil moisture products such as surface and root zone soil moisture [mm], profile soil moisture [%], and surface and root zone soil moisture anomalies [–]. The latter represents standardized anomalies, which are calculated using the following equation:

$$SMA = \frac{X_{SM-}\mu_{SM}}{\sigma_{SM}}$$

where, X_{SM} is the SMOS/SMAP soil moisture, μ_{SM} is the mean value, and σ_{SM} is the standard deviation of the SMOS/SMAP soil moisture. Each value shows the deviation of the current conditions relative to a long-term average standardized by the climatological standard deviation, where the climatology values are estimated based on the full data record of the satellite observation period over a 31-day moving window (e.g., climatology of a day of interest is calculated using the 15 days prior and 15 days after that day of year for the entire historical record). Negative anomaly values indicate that the current conditions are below average, while positive values indicate a surplus of water.

The system is executed daily as new AFWA, and satellite observations become available. However, SMOS and SMAP provide complete global coverage every 3 days; therefore, the output generated from the satellite-enhanced PM is binned to 3-day composites. Once a new 3-day composite product is produced, the data are operationally pushed to the USDA-FAS, and the data are automatically displayed at the agency's Crop Explorer web site. It should be noted that the SMOS- and SMAP-based systems are currently run independently and are expected to have slightly different climatologies given that each covers a different time period (SMOS: January 2010–present; SMAP: April 2015–present).

2.2. Ancillary Datasets

Several additional datasets have been used in this study to explore the relationship between RZSM anomalies and meteorological drought indices as a function of land cover variability. The Climate Hazards Group Infrared Precipitation with Station (CHIRPS) dataset, developed by the United States Geological Survey in collaboration with the Earth Resource Observation and Science center, is used to explore the spatial and temporal variability of precipitation with different land cover types. CHIRPS is generated by integrating satellite imageries and in situ gauge-collected observations. The daily rainfall data are distributed at 0.5° spatial resolution [40]. Vegetation-type information was obtained from the ESA's global land cover data developed by utilizing observations from the Medium Resolution Imaging Spectrometer collected by the Environmental Satellite [41]. The land cover map includes 22 land cover classes as defined by the Food and Agriculture Organization of the United Nations Land Cover Classification System.

The Standardized Precipitation Index (SPI) is a meteorological drought index used to assess different drought characteristics. It represents the standardized deviation of the observed cumulative precipitation relative to the long-term precipitation average. In this study, SPI at 3-, 6-, and 9-month scales was obtained from the International Research Institute for Climate and Society at Columbia University [21]. The SPI dataset was derived from the Climate Prediction Center's Outgoing Longwave Radiation blended gauge-based global daily precipitation data. The SPI was calculated by fitting a probability distribution to the long-term series of precipitation accumulation over the period of interest, where the resulting cumulative probability function is consequently transferred to a normal distribution. The monthly SPI data offers global coverage at spatial resolution of 1°.

The Normalized Difference Vegetation Index (NDVI) data were obtained from GIMMS system. This dataset is derived using the Moderate Resolution Imaging Spectroradiometer (MODIS) Terra surface reflectance products, which are provided by NASA Goddard Space Flight Center MODIS Adaptive Processing System [42]. We processed and ingested SPI and NDVI datasets as private assets in the GEE data catalog as those are not available in the GEE public data catalog.

All data products used in this study have been averaged to monthly composites and then resampled to 1° grid spacing to ensure comparable temporal and spatial resolutions among the different datasets.

3. Google Earth Engine Tools

We have developed several GEE tools, which enable easy processing, analysis, and visualization of the SMOS- and SMAP-based soil moisture data in the GEE platform. These tools can be arranged in three groups according to their functionality: (i) tools to process and ingest soil moisture data in the GEE data catalog, (ii) tools to explore spatial and temporal variation of soil moisture and precipitation as a function of land cover, and (iii) tools to estimate drought characteristics such as duration and intensity using soil moisture anomalies and intercomparing the latter against alternative drought indices. Detailed descriptions of the individual tools are given below.

3.1. Data Uploading Routine

The data upload routine has been designed to process, upload, and manage the SMOS- and SMAP-based soil moisture products in the GEE platform. This routine first converts the original soil moisture data stored in binary format into the georeferenced tagged image file format (GeoTIFF) as required by the GEE. Then, it creates a metadata file of the resulting imagery. The metadata is needed by the analysis routines, which are used to filter the data based on user-specified spatial and temporal information. Next, the GEE batch asset manager (https://github.com/tracek/gee_asset_manager) tool is used to upload a bulk amount of data automatically in the GEE (Figure 2). An alternative uploading option is to use the asset manager option in the GEE. However, the latter is time inefficient for large datasets as it allows the user to upload a single imagery at a time.

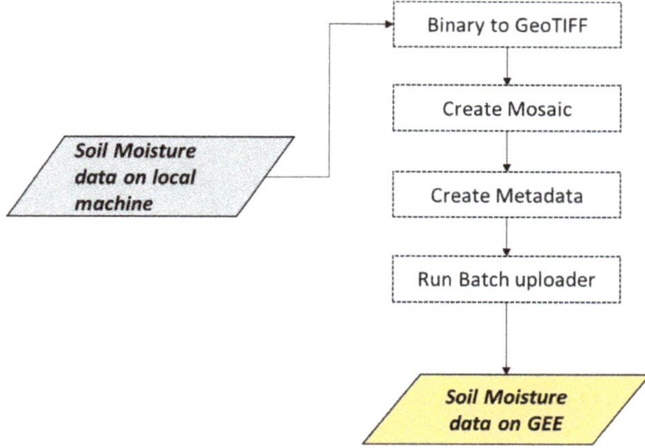

Figure 2. Ingestion of soil moisture datasets to the Google Earth Engine (GEE). The gray and gold boxes represent inputs and outputs, respectively. GeoTIFF, georeferenced tagged image file format.

3.2. Soil Moisture Exploration Routine

The soil moisture exploration routine has been specifically designed to assess the spatial and temporal variability of soil moisture from local to regional and global scales. This function first filters the soil moisture data based on user-specified temporal and spatial criteria using the GEE '*filter date*' and '*filter bounds*' functions. Next, the subset data are grouped by month using the *filter calendar range*' function and are aggregated from the original 3-day composites into monthly composites (Figure 3). Then, interactive monthly soil moisture plots are generated using the chart function by using the '*chart image series*' function in the GEE, which can be viewed and exported in multiple formats (i.e., comma-separated values, portable network graphics, etc.). The multi-annual image collection can be further reduced to a long-term average image representing mean soil moisture for the region of interest, which can be visualized through GEE Google Maps. This routine also enables an assessment of the variability of soil moisture data as a function of land cover type. The ESA land cover data is clipped based on the user-defined region of interest, and a histogram is plotted to estimate the major land cover types of the study region. Then, the monthly soil moisture values are filtered based on land cover class, and interactive plots are generated for additional analysis and visualization.

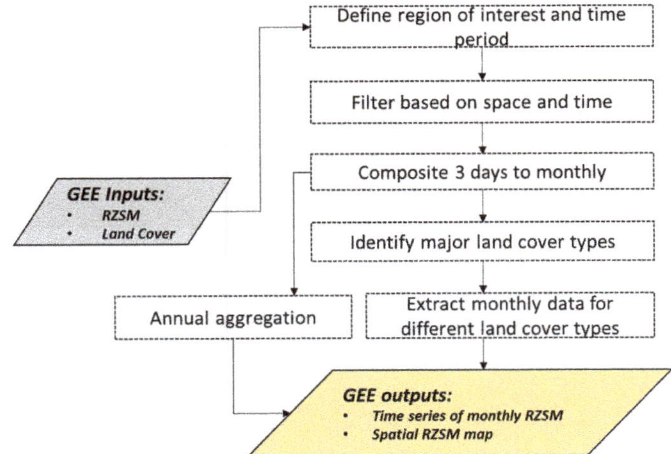

Figure 3. Data processing steps in the soil moisture exploration routine. The gray and gold boxes represent GEE inputs and outputs, respectively. The boxes identified by dotted lines represent the processes that run in the GEE server.

3.3. Drought Assessment Routine

The drought assessment routine has been developed using the GEE functionalities to compare various drought indicators based on specific drought characteristics, such as percentage of the month with drought conditions, maximum drought duration, drought severity, and intensity. In addition, any drought indicators have been developed to monitor, predict, and assess the severity of different drought types, which can be classified into two major categories of drought: meteorological and agricultural. Meteorological drought indicators are derived using precipitation data and have multi-scale features that identify different types of drought condition. As root zone soil moisture affects plant growth and productivity, RZSM anomalies are often used for quantifying and monitoring agricultural drought and capturing its impact on crop heath [4,43–45]. In this study, we used four drought indicators—SPI3, SPI6, SPI9 (meteorological drought indicators), and SMOS RZSM anomalies (agricultural drought indicator)—to assess drought conditions. Here, we focused on SMOS RZSM anomalies, as it has longer observation period compared with SMAP and is hence well suited to estimate drought characteristics and Pearson's correlation coefficient. Positive values of RZSM anomalies, SPI3, SPI6, and SPI9 are masked out to identify only months with drought conditions, the corresponding number is divided by the total number of months to calculate the percentage of months with drought conditions. Each product has been examined in terms of the following drought characteristics: drought duration (defined as the period during which the drought indices are continuously negative); drought severity (computed as the absolute value of the sum of all drought indices during a drought event); and drought intensity (calculated by dividing the severity by the drought duration) [46,47]. This routine also computes cross-correlations between agricultural- and meteorological-based drought indices and allows us to estimate the Pearson, Spearman, and lag correlation coefficients using the GEE *'reducerpearsonscorrelation'* function between the paired monthly time series of soil moisture anomalies and SPI, as well soil moisture and NDVI (Figure 4).

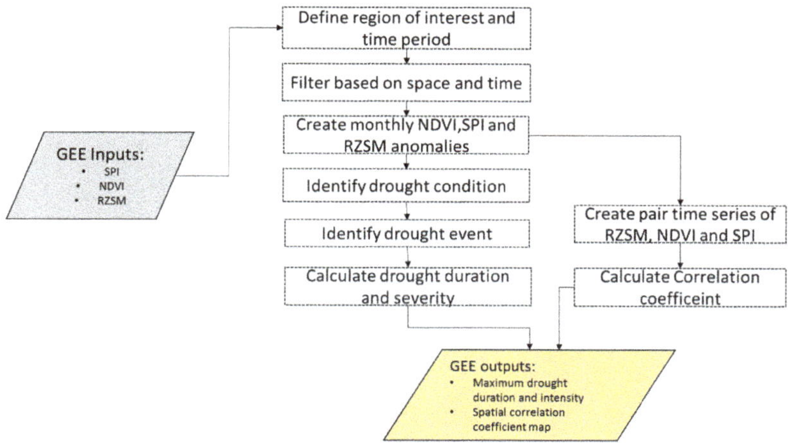

Figure 4. Data processing steps in the drought assessment routine using the GEE. The gray and gold boxes represent GEE inputs and outputs, respectively. The boxes identified by the dotted lines represent the process that run in the GEE server.

Developed tools are accessible through the links provided in the supplementary materials, though potential users are required to register to access (https://code.earthengine.google.com/). Once the link is clicked, the user is presented with the Google Earth Engine code editor, which is a web-based integrated development environment for the Earth Engine JavaScript Application Programming Interface (Figure 5). Then, the user can execute the program by clicking the 'run' button located above the JavaScript code editor panel, if it does not start automatically. Once this has been done, time series plots and a spatial map are displayed in the console tab and Google Maps respectively. The GEE outputs results can be exported by clicking on the run button in the tasks tab located in the right panel next to the code editor.

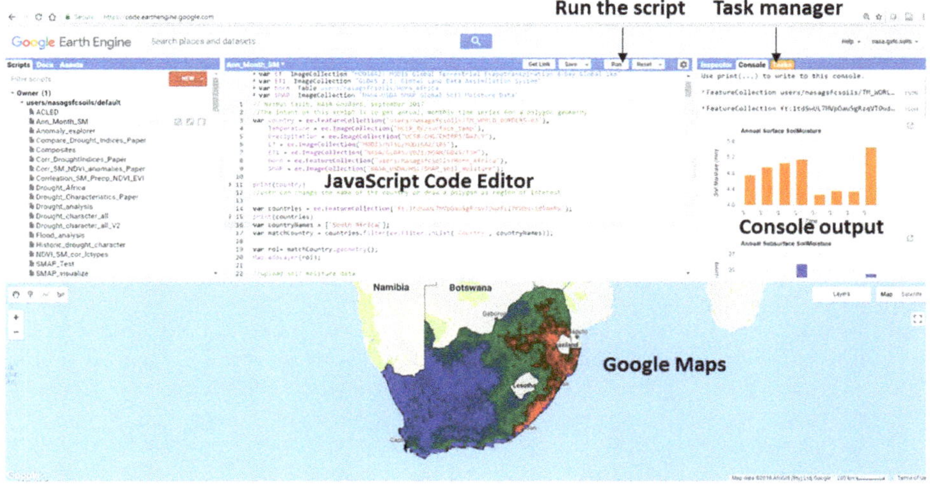

Figure 5. Components of Google Earth Engine code editor.

4. Example Applications

The GEE tools described in the previous section have been implemented to evaluate the spatial and temporal dynamics of soil moisture and precipitation and assess the ability of the drought indices described in the previous sections to capture the severity, duration, and intensity of drought events over South Africa and Ethiopia during 2010–2017.

Drought is common in South Africa and Ethiopia and occurs in all climate areas with varying degrees of intensity, spatial extent, and duration [48]. In recent years, the spatial extent and frequency of drought have increased in this area, causing significant water shortage, economic losses, and adverse social consequences [49]. Therefore, a better understanding of the climatology and drought characteristics over these areas is important in order to improve decision-making and aid activities aimed to mitigate the impact of drought. Our analysis is focused on the 2010–2017-time period, which was determined by the availability of the SMOS datasets. For this analysis, the soil moisture explorer routine is executed in the GEE code editor by clicking on the link provided in the supporting materials to generate spatial map and time series plots of the precipitation and soil moisture over South Africa. Then, we run the drought assessment routine to estimate the drought characteristics and the correlation among the different drought indices over South Africa. Next, we re-run both the soil moisture explorer and the drought assessment routine for Ethiopia by changing the country name inside the script. The output results of the GEE are imported into ArcGIS (Esri, Redlands, CA, USA) [50] to add a legend, scale, and proper color scheme and R [51] to generate boxplots of drought characteristics.

4.1. Spatial and Temporal Variability of Precipitation and Soil Moisture

We first examined the long-term spatial distribution of the precipitation and RZSM and then analyzed the variability of those variables with different land cover types. Spatial variability of rainfall and RZSM over South Africa and Ethiopia are shown in Figure 6. Both variables exhibit high regional variability. In South Africa, generally the mean annual precipitation increases from west to east with the maximum rainfall (680 mm) occurring over the Mpumalanga and KwaZulu-Natal provinces, while minimum rainfall (172 mm) falls over the western part of the country. The spatial variability captured by the RZSM reflects the precipitation variability showing wetter SM conditions in the east and drier conditions in the west (Figure 6, top row). The topographical variability significantly influences the spatial distribution of the precipitation and the soil moisture in Ethiopia. For example, the rainfall and soil moisture values are higher over the highland areas located in the central and northwestern portion of the country, while the lowland areas located in the eastern part of the country are associated with lower rainfall amounts (Figure 6, bottom row).

The monthly precipitation over Ethiopia and South Africa is driven mainly by the position of the intertropical convergence zone (ITCZ), which changes over the course of year [52,53]. A majority of the rainfall in Ethiopia falls during the summer seasons when the ITCZ is at its most northern position; however, the amount of rainfall also varies as a function of land cover (Figure 7). For example, forest, cropland, grassland, and shrub land show identical rainfall patterns with one main wet season (June–September) and a secondary wet season (February–May), where the highest rainfall occurs during the month of September (Figure 8). Over sparse vegetation, the major rainfall falls during the summer and winter season, as most of this land cover is located in the southern part of the country, where the rainfall timing is associated with ITCZ, which passes through the southern position of the equator at that time. The monthly RZSM follows the rainfall distribution reaching the wettest soil moisture conditions during the month of September. The position of ITCZ also results in two distinct seasons in South Africa—a wet and dry season roughly from November–April and May–October, respectively. The monthly soil moisture time series captures this seasonality, as seen in Figure 2. The monthly rainfall and soil moisture time series across South Africa vary with land cover, where regions covered by the grassland and sparse vegetation receive the highest and lowest amount of precipitation and soil moisture, respectively (Figure 8).

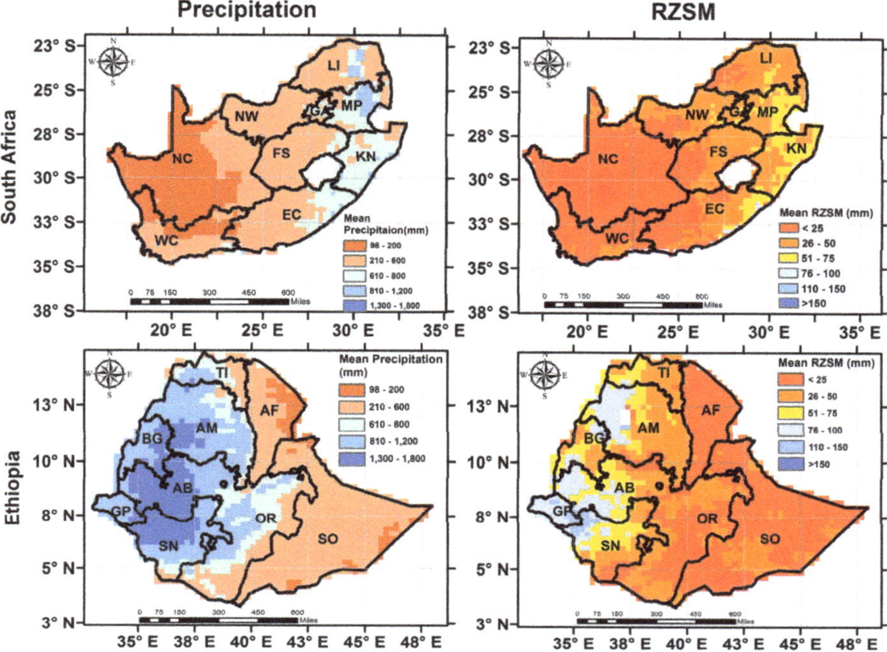

Figure 6. Spatial variability of precipitation (derived from CHIRPS precipitation data) and RZSM (derived from United States Department of Agriculture-Foreign Agricultural Services (USDA-FAS) soil moisture data) over South Africa (**top row**) and Ethiopia (**bottom row**) for the period of 2010–2017. The locations of the each province of South Africa (NC: Northern Cape, WC: Western Cape, EC: Eastern Cape, NW: North West, FS: Free State, KN: KwaZulu-Natal, MP: Mpumalanga, GA: Gauteng, and LI: Limpopo) and Ethiopia (SO: Somali, OR: Oromia, SN: Southern Nations, GP: Gambela Peoples, BG: Benshangul-Gumaz, AM: Amhara, TI: Tigray, AF: Afar, and AB: Addis Ababa) are also indicated.

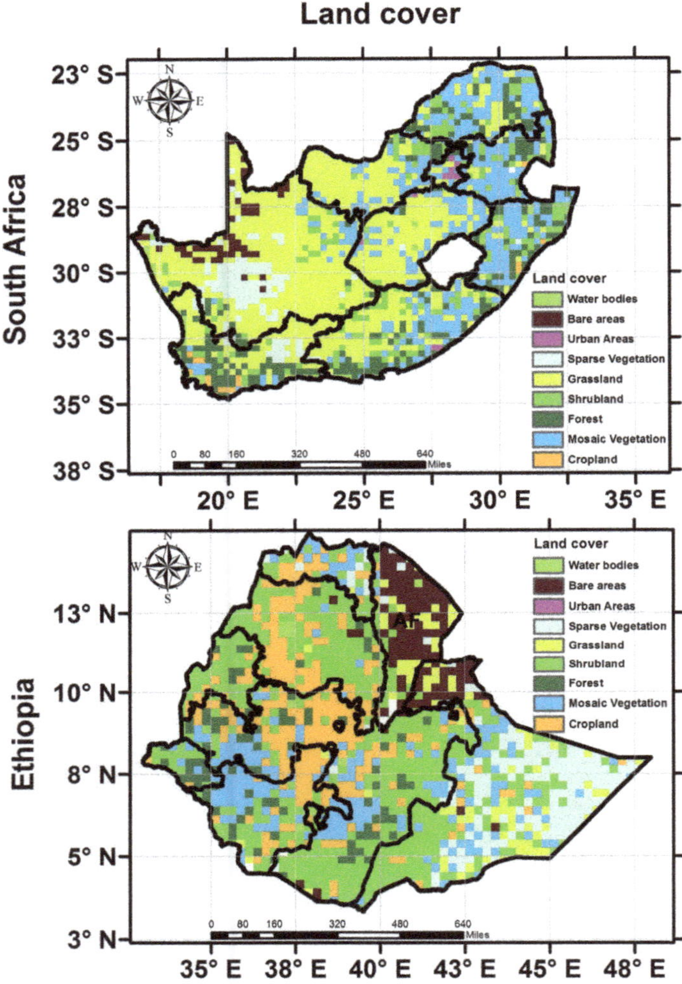

Figure 7. Land cover of South Africa (**top**) and Ethiopia (**bottom**).

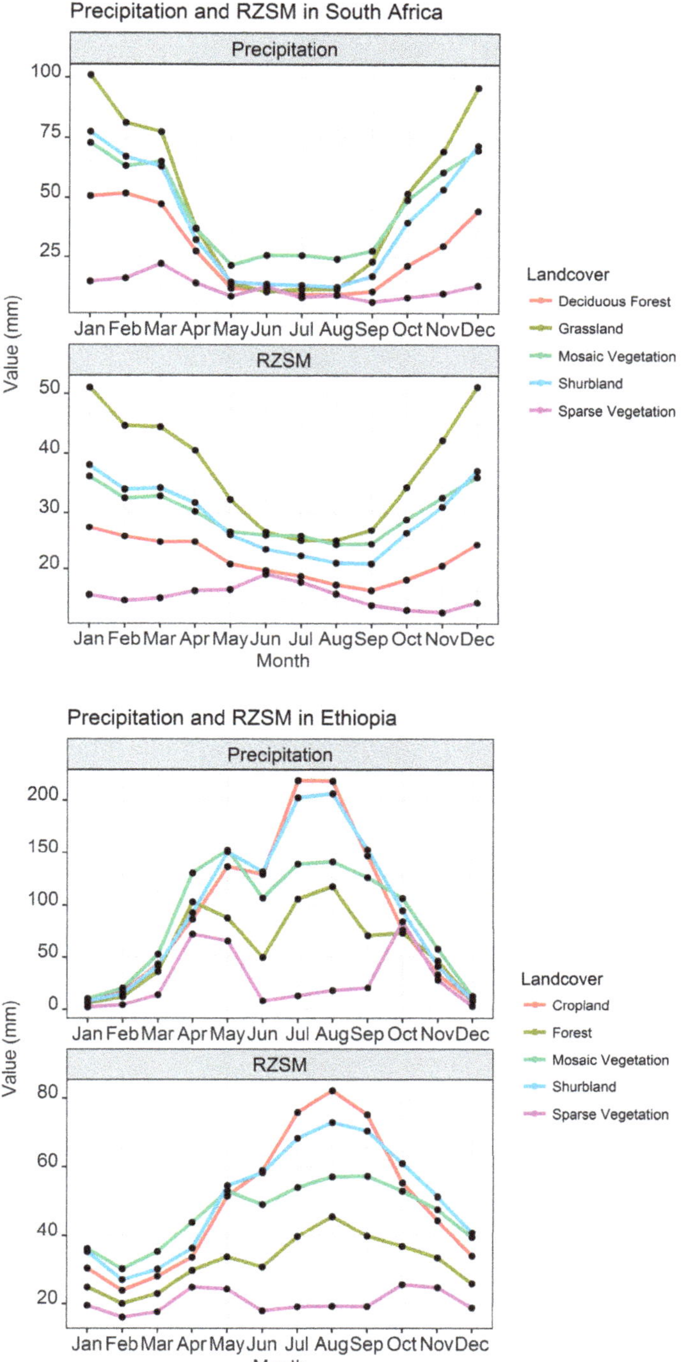

Figure 8. Monthly variation of soil moisture and rainfall for different land cover types over South Africa (**top**) and Ethiopia (**bottom**).

4.2. Comparison of Drought Characteristics

The RZSM anomalies indicated a higher percentage of months with drought conditions compared with SPI (SPI3, SPI6, and SPI9) over both study regions (Figure 9). Over South Africa, the average percentage of drought events identified in the RZSM anomaly data was 27%, which is 6% higher than the drought events captured by the SPI3. Additionally, among the rainfall-based drought indices, the SPI9 had the lowest percentage of months with drought events compared with the SPI3 and SPI6. This is in line with other studies [46], where the author found that agricultural-based drought indicators depict relatively larger values of drought months compared with the meteorological drought indices. The maximum drought duration varied among the different drought indicators. Based on our analysis, the maximum drought duration appeared to be higher in the meteorological-based indices than the agricultural-based indices. This is primarily because the meteorological-based drought indices integrate the drought condition over a longer period of time than do the agricultural-based drought indices [46]. The drought intensity was found to be higher in the agricultural-based drought variables and lower in the metrological-based drought indices. This example demonstrates the capability of the drought assessment tools, which can help to better assess the drought conditions.

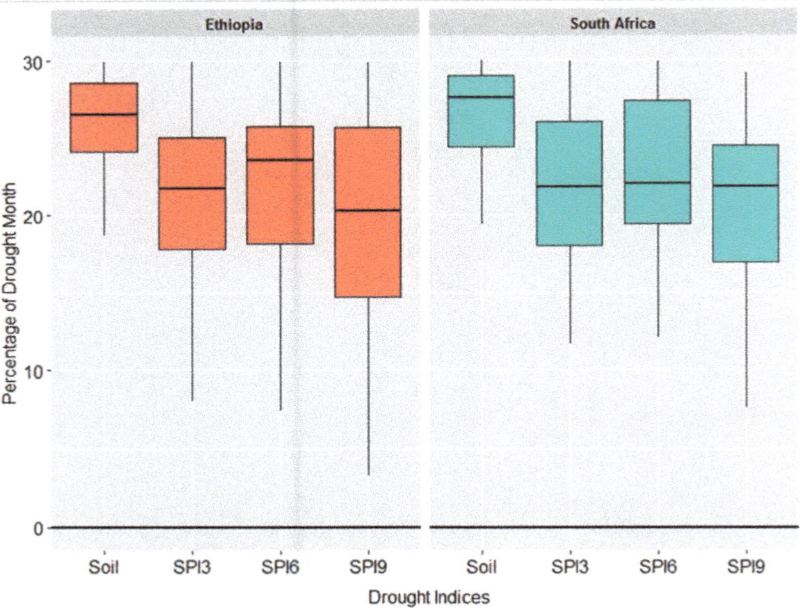

Figure 9. *Cont.*

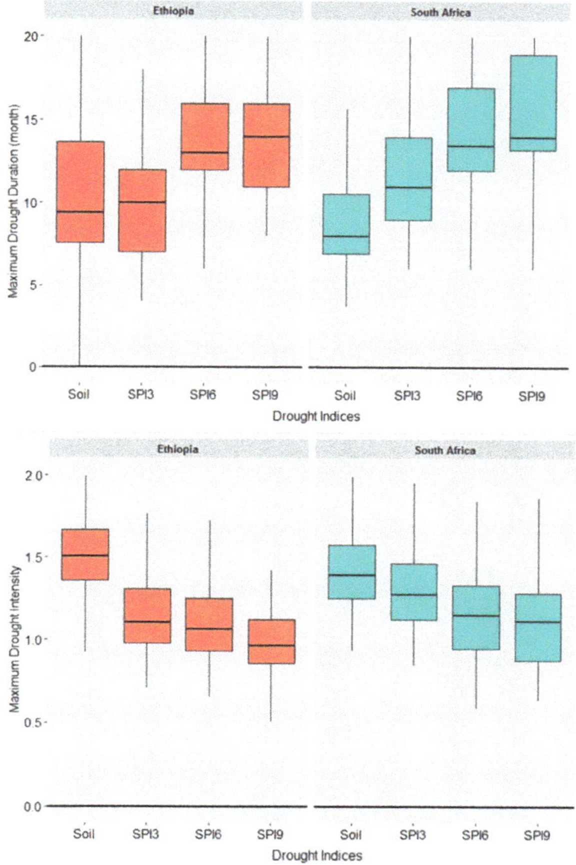

Figure 9. Comparison of percentage of months with drought conditions, maximum drought duration, and drought intensity over Ethiopia and South Africa for multiple drought indices. The center line of each boxplot depicts the median value (50th percentile) and the box encompasses the 25th and 75th percentiles of the sample data. The whiskers extend from q1 − 1.5 × (q3 − q1) to q3 + 1.5 × (q3 − q1), where q1 and q3 are the 25th and 75th percentiles of the sample data, respectively.

4.3. Correlation between Soil Moisture and Normalized Difference Vegetation Index Anomalies

Variations in RZSM substantially influence the vegetation dynamics (i.e., NDVI), which is a widely used vegetation index. Therefore, correlation analysis between RZSM and NDVI anomalies is important to understand the impact of changes in soil moisture on vegetation growth, which can be effectively utilized for early warning of time and areas of increased food insecurity [40,54]. The correlation of the RZSM and NDVI anomalies varied with the geographic location and the degree of lag time. The highest positive correlation coefficients and confidence level (i.e., p-value < 0.1) are observed when soil moisture change is concurrent or precedes the change in NDVI by one month. In most of the locations, NDVI and RZSM anomalies have positive correlations; however, some regions indicate negative correlation at higher lags due to coincidence of negative NDVI anomalies with positive soil moisture anomalies [55]. In South Africa, the semi-arid Western Cape and Eastern Cape show higher coefficients compared with other parts of the country as the vegetation growth in those regions has high reliance on root zone soil moisture [56,57]. No spatial variability in the lag correlation values was observed over Ethiopia (Figure 10). We further investigated the variation of the soil moisture and NDVI relationship as a function of major land cover types. The highest agreement was found over areas covered by grassland for both study areas, while the lowest agreement was achieved over the shrub-covered areas in Ethiopia and South Africa (Table 2). This is partly due to the fact that grassland roots are located on the shallow depths and are more sensitive to changes in soil moisture than deep rooted plants such as shrubs.

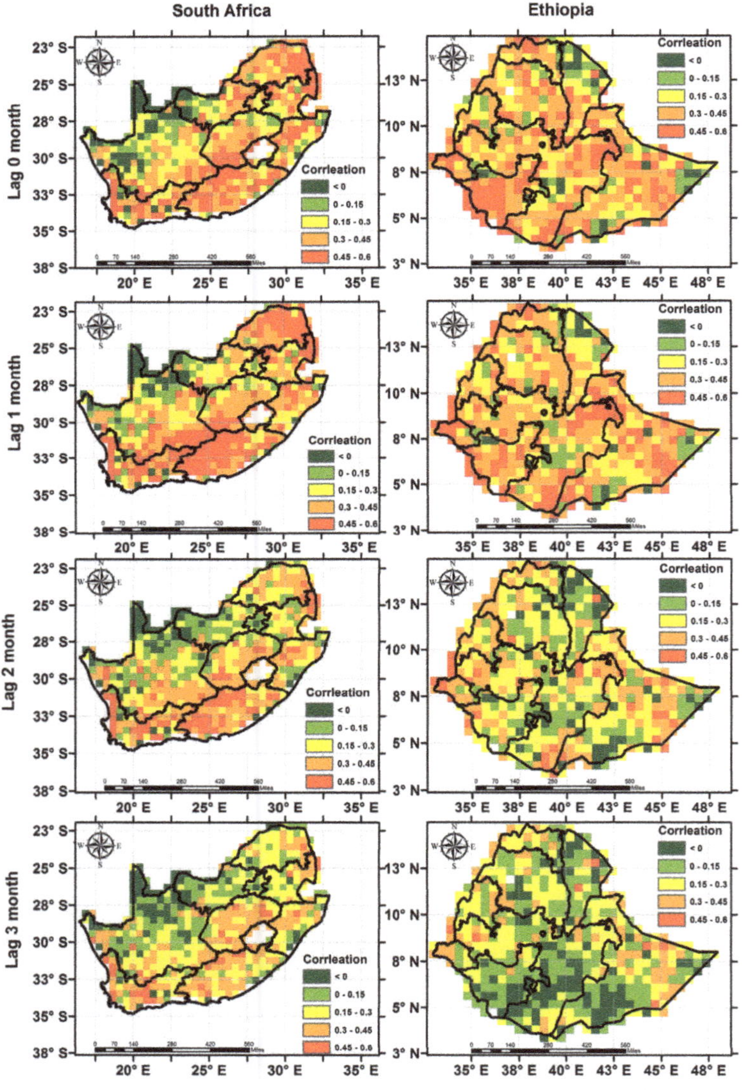

Figure 10. *Cont.*

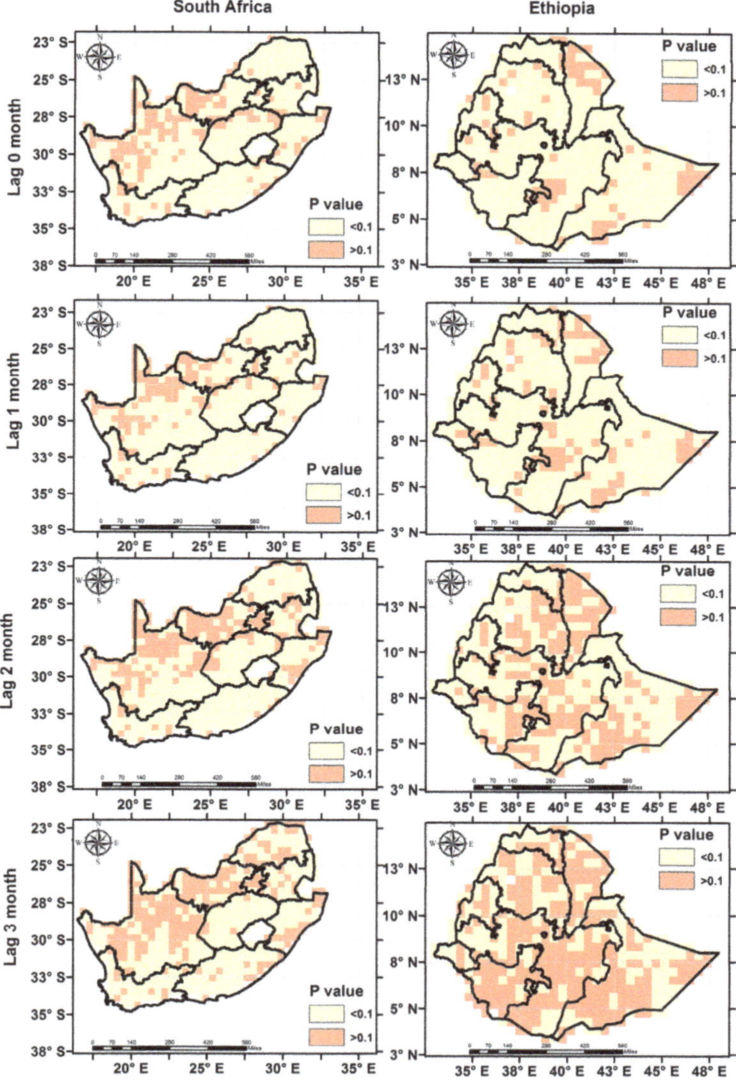

Figure 10. Correlation coefficient (**top**) and *p*-value (**bottom**) between RZSM and NDVI anomalies for different lag times.

Table 2. The Pearson's correlation coefficient computed between agricultural-based drought indices and meteorological-based drought indices for different land cover types.

	Ethiopia				South Africa			
Land Cover	Lag0	Lag1	Lag2	Lag3	Lag0	Lag 1	Lag2	Lag 3
Cropland	0.32	0.31	0.20	0.13	0.31	0.32	0.25	0.25
Grassland	0.34	0.32	0.21	0.11	0.37	0.38	0.28	0.23
Shrubland	0.32	0.30	0.20	0.11	0.24	0.29	0.23	0.18

4.4. Correlation among Soil Moisture Anomalies and Meteorological Based Indices

A correlation analysis was carried out between RZSM anomalies and meteorological-based drought indicators to evaluate how well the meteorological-based drought indicators represent agricultural-based drought. Such information could be used to help indicate times and areas that are likely to experience agricultural stress. It is envisaged that such approaches will improve drought monitoring and early warning systems that rely mostly on meteorological indicators [58]. The GEE-based inter-comparative analysis between soil moisture anomalies against SPI showed high agreement and alludes to the value of combining such datasets to compliment a regional drought assessment that incorporates both meteorological and agricultural drought. Over both study regions, SPI3 had higher correlation values compared with SPI6 and SPI9 (Figure 11), which indicates that SPI3 captures more of the agricultural drought. The performance of the meteorological drought indices varied spatially. In the case of South Africa, the correlation values were relatively higher and statistically significant (p-value < 0.1) in the Western Cape and Eastern Cape compared with the Northern Cape of the region. The spatial distribution of the correlation for all meteorological droughts in Ethiopia have similar pattern, where higher and lower correlation values are generally distributed over the northwest and northeast side of the country, respectively. The highest correlation between the soil moisture anomalies and meteorological drought indicators are associated with the cropland (Table 3), which is consistent with [59],who showed that a 3-month SPI has the highest correlation with vegetation growth on croplands of the midlatitude U.S. Great Plains.

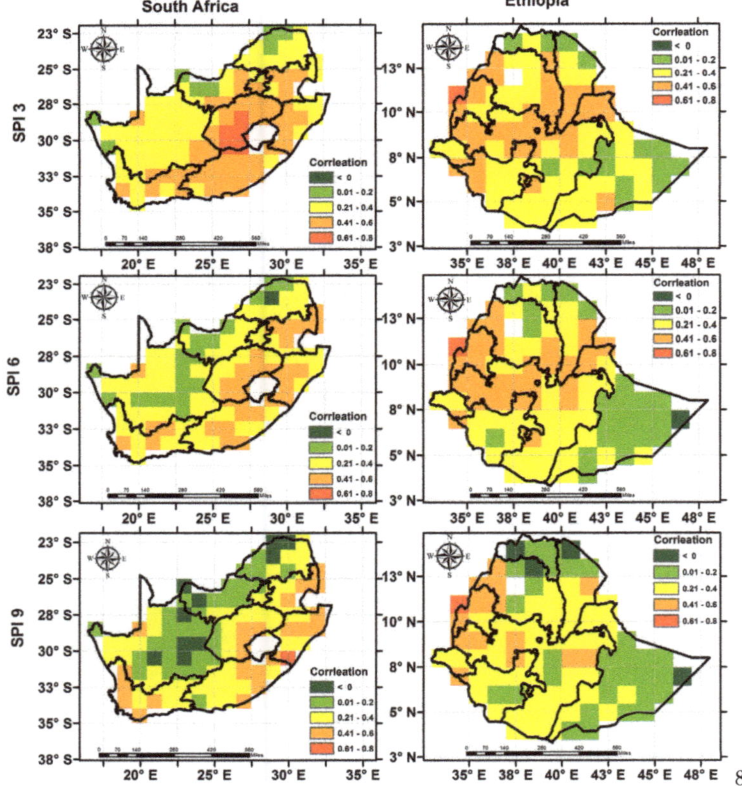

Figure 11. *Cont.*

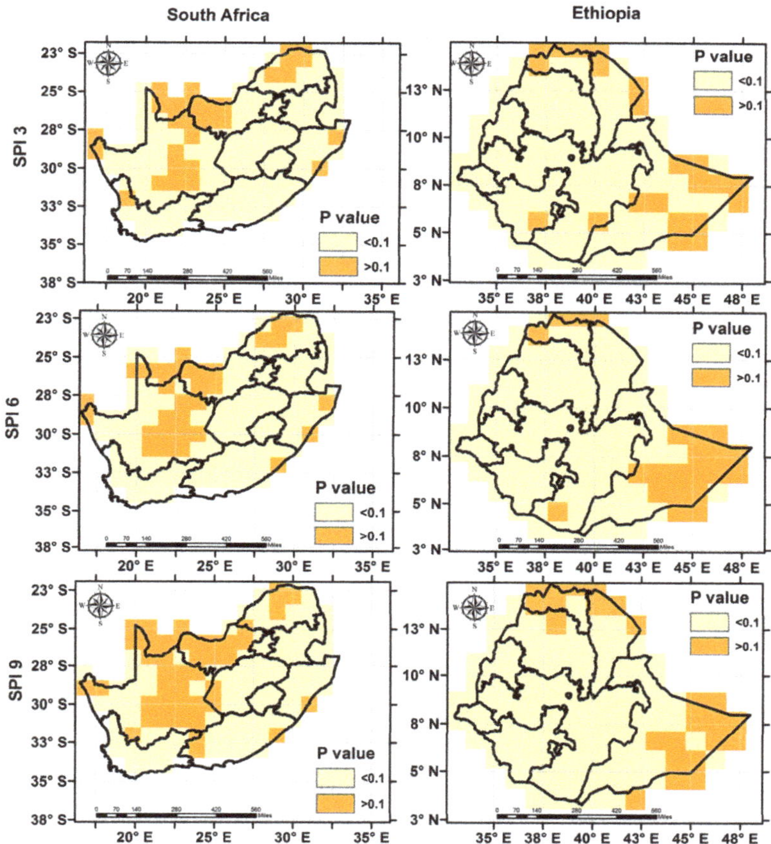

Figure 11. Spatial variation of Pearson correlation coefficients (**top**) and *p*-value (**bottom**) of RZSM anomalies with meteorological-based drought indices.

Table 3. The Pearson's correlation coefficient computed between agricultural-based drought indices and meteorological-based drought indices for different land cover types.

	Ethiopia			South Africa		
Land Cover	**SPI3**	**SPI6**	**SPI9**	**SPI3**	**SPI6**	**SPI9**
Cropland	0.41	0.42	0.32	0.43	0.39	0.34
Grassland	0.37	0.36	0.29	0.37	0.30	0.23
Shrubland	0.25	0.21	0.16	0.37	0.28	0.20

5. Discussion

Highest correlation of SPI3 with the RZSM anomalies indicated that short-time meteorological drought represents the agricultural drought better compare with long-term meteorological-based indicators such as SPI6 and SPI9. The impact of meteorological drought on vegetation is cumulative, meaning that vegetation does not respond instantaneously to the precipitation changes. The three-month SPI, which captures the precipitation pattern not only for the specific month of the interest, but also the previous two months, results in a highest correlation between SPI and soil moisture anomalies. On the contrary, the 12- and 6-month SPI values reflect precipitation patterns for annual and the entire growing season, respectively, and tend to diminish the variance in the precipitation data and smooth the SPI values results in lower correlation values [59]. The relationship between soil moisture and rainfall anomalies were also explored by Sims et al. [60] in North Carolina, where the authors suggest that SPI on a scale of 2–3 months yielded the highest correlation with soil moisture anomalies.

Our results indicate lower and higher correlation between RZSM anomalies and SPI-based indicators in the dry and wet regions, respectively, which could be related to the rainfall amount and soil types. The dominant soil types in the wet regions are clay and clay loam, which have higher water-holding capacities, which could result in slower response to the rainfall; therefore, soil moisture on a specific month would be more dependent on the previous month's rainfall. On the other hand, the arid region shows quicker response to the rainfall anomalies due to dry soil conditions and the limited water-holding capacity of the sandy soils that cover that region. Therefore, soil moisture in a specific month has a smaller dependence on previous months compared with the wet region [61].

In general, the land cover type has a significant impact on the relationship between RZSM anomalies and other drought indicators. For example, shrubland exhibits lower correlation values compared to the cropland, which could be due to the fact that cropland roots are located in the shallow depths and are more sensitive to changes in soil moisture than deep-rooted plants such as shrubs. Similar observations were made by Camberlin et al. [62] and Huber et al. [63] for Africa, by Li et al. [59] for China, and by Wang et al. [60] for the US central Great Plains. We also noticed a delayed response of NDVI to RZSM anomalies for the shrubland over South Africa, which might be related to soil texture and soil moisture amounts, as most of the shrubland is located in the wet region of the country characterized by more clay soils leading to slower response [64]. This is consistent with the findings of Wang et al. [65], who showed that the NDVI at humid sites takes a longer time to respond compared with the arid sites.

6. Conclusions

Soil moisture data are recognized as a fundamental physical variable that can be used to address science and resource-management questions requiring near real-time monitoring of the land–atmosphere boundary, including flood and drought monitoring and regional crop yield assessment. This study introduced new sets of near real-time global soil moisture data and demonstrates the potential of the GEE web-based tools and soil moisture data to assess regional drought conditions. In general, the meteorological drought indicator, SPI3, gives higher correlation values compared with SPI6 and SPI 9 and compared with the RZSM anomalies. When comparing the drought characteristics, RZSM anomalies exhibit relatively larger drought duration but smaller drought intensity compared with the meteorological-based drought indicators. The NDVI-RZSM anomalies are influenced by the vegetation cover, specifically shallow-rooted plants that are more sensitive to soil moisture changes compared with deep-rooted plants. The methods demonstrated here can be applied to other areas requiring early warning of food shortage or improved agricultural monitoring to help provide greater economic security within the agriculture sector.

Incorporating the global soil moisture data into the GEE data catalog enables users to efficiently and quickly acquire and process large amount of data. The available tools allow easy analysis, visualization, and interpretation of the data. To this end, these the GEE-based tools could enable scientists, policy makers, and the general public to explore spatial and temporal variation of soil moisture information and drought conditions for any location in the world with minimal data processing or data management. In addition, all the tools are easily transferable and can be used to explore spatial and temporal dynamics of other climate variables such as temperature and evapotranspiration. The GEE does not require any additional software installation, which helps to overcome compatibility limitations and allows user to access the available codes and data from any computer connected to the internet. This significantly increases the data and tools usability and applicability. The GEE tools and the soil moisture data are open source and are freely available, which can enable users to use, modify, and suggest future improvements for both the tools and the data.

Although the GEE offers many benefits, it has limitations as well. First, it requires basic knowledge of Python and JavaScript, and users with limited programming knowledge might have a steep learning curve. Second, users are sometimes required to export the analyzed results to perform additional analyses due to limited functionalities and plotting options in the GEE. Finally, debugging the code is challenging as the user-created algorithms run in the Google cloud distributed over many computers. Despite these limitations, the data distribution and processing approach offered by the GEE platform can be very beneficial, specifically for developing countries that are typically data-poor areas and lack high-performance data processing platforms for drought monitoring or crop forecasting.

Supplementary Materials: The SMOS and SMAP soil moisture data sets are available at https://explorer.earthengine.google.com/#detail/NASA_USDA%2FHSL%2Fsoil_moisture and https://explorer.earthengine.google.com/#detail/NASA_USDA%2FHSL%2FSMAP_soil_moisture respectively. The soil moisture exploration routine and drought assessment routine are available at https://code.earthengine.google.com/737906c2e5f814170e802859dbe94692, https://code.earthengine.google.com/1da21ee96e5f9ce92a076fc9784485bc and https://code.earthengine.google.com/074dad2e2ddcf24036b2dc0504363281.

Author Contributions: N.S., I.M., and J.B. designed the work. N.S. and I.M. undertook the data analysis. All the authors contributed equally to the final version of the paper.

Funding: This work is supported by the NASA Applied Sciences Program.

Acknowledgments: We would like to thank Simon Ilyushchenko for his support in adding our soil moisture data sets into the Google Earth Engine Public data catalog.

Conflicts of Interest: The authors declare no conflict of interest.

References

1. Tsang, L.; Jackson, T. Satellite Remote Sensing Missions for Monitoring Water, Carbon, and Global Climate Change [Scanning the Issue]. *Proc. IEEE* **2010**, *98*, 645–648. [CrossRef]
2. Mladenova, I.E.; Jackson, T.J.; Njoku, E.; Bindlish, R.; Chan, S.; Cosh, M.H.; Holmes, T.R.H.; de Jeu, R.A.M.; Jones, L.; Kimball, J.; et al. Remote monitoring of soil moisture using passive microwave-based techniques—Theoretical basis and overview of selected algorithms for AMSR-E. *Remote Sens. Environ.* **2014**, *144*, 197–213. [CrossRef]
3. McCabe, M.F.; Rodell, M.; Alsdorf, D.E.; Miralles, D.G.; Uijlenhoet, R.; Wagner, W.; Lucieer, A.; Houborg, R.; Verhoest, N.E.C.; Franz, T.E.; et al. The future of Earth observation in hydrology. *Hydrol. Earth Syst. Sci.* **2017**, *21*, 3879–3914. [CrossRef]
4. Eswar, R.; Das, N.N.; Poulsen, C.; Behrangi, A.; Swigart, J.; Svoboda, M.; Entekhabi, D.; Yueh, S.; Doorn, B.; Entin, J. SMAP Soil Moisture Change as an Indicator of Drought Conditions. *Remote Sens.* **2018**, *10*, 788.
5. Blyth, E.M.; Daamen, C.C. The accuracy of simple soil water models in climate forecasting. *Hydrol. Earth Syst. Sci.* **1997**, *1*, 241–248. [CrossRef]
6. Chakraborty, R.; Rahmoune, R.; Ferrazzoli, P. Use of passive microwave signatures to detect and monitor flooding events in Sundarban Delta. In Proceedings of the 2011 IEEE International Geoscience and Remote Sensing Symposium (IGARSS), Vancouver, BC, Canada, 24–29 July 2011; pp. 3066–3069.
7. Bourgeau-Chavez, L.L.; Kasischke, E.S.; Rutherford, M.D. Evaluation of ERS SAR data for prediction of fire danger in a Boreal region. *Int. J. Wildland Fire* **1999**, *9*, 183–194. [CrossRef]
8. Reichle, R.; Koster, R.D. Global assimilation of satellite surface soil moisture retrievals into the NASA catchment land surface model. *Geophys. Res. Lett.* **2005**, *32*, 2353–2364. [CrossRef]
9. Crow, W.T.; Zhan, X. Continental-Scale Evaluation of Remotely Sensed Soil Moisture Products. *IEEE Geosci. Remote Sens. Lett.* **2007**, *4*, 451–455. [CrossRef]
10. Palmer, W.C. *Meteorological Drought*; U.S. Weather Bureau Research Paper 45; Office of Climatology, US Department of Commerce: Washington, DC, USA, 1965.
11. Bolten, J.D.; Crow, W.T. Improved prediction of quasi-global vegetation conditions using remotely-sensed surface soil moisture. *Geophys. Res. Lett.* **2012**, *39*, L19406. [CrossRef]

12. Bolten, J.D.; Crow, W.T.; Zhan, X.; Jackson, T.J.; Reynolds, C.A. Evaluating the Utility of Remotely Sensed Soil Moisture Retrievals for Operational Agricultural Drought Monitoring. *IEEE J. Sel. Top. Appl. Earth Obs. Remote Sens.* **2010**, *3*, 57–66. [CrossRef]
13. Kerr, Y.H. *Soil Moisture and Ocean Salinity SMOS*; The European Space Agency: Paris, France, 1998.
14. Entekhabi, D.; Njoku, E.G.; O'Neill, P.E.; Kellogg, K.H.; Crow, W.T.; Edelstein, W.N.; Entin, J.K.; Goodman, S.D.; Jackson, T.J.; Johnson, J. The Soil Moisture Active Passive (SMAP) Mission. *Proc. IEEE* **2010**, *98*, 704–716. [CrossRef]
15. Gorelick, N.; Hancher, M.; Dixon, M.; Ilyushchenko, S.; Thau, D.; Moore, R. Google Earth Engine: Planetary-scale geospatial analysis for everyone. *Remote Sens. Environ.* **2017**, *202*, 18–27. [CrossRef]
16. Dong, J.; Xiao, X.; Menarguez, M.A.; Zhang, G.; Qin, Y.; Thau, D.; Biradar, C.; Moore, B. Mapping paddy rice planting area in northeastern Asia with Landsat 8 images, phenology-based algorithm and Google Earth Engine. *Remote Sens. Environ.* **2016**, *185*, 142–154. [CrossRef] [PubMed]
17. Boken, V.K.; Cracknell, A.P.; Heathcote, R.L. *Monitoring and Predicting Agricultural Drought: A Global Study*; Oxford University Press: Oxford, UK, 2005; p. 495.
18. Lobell, D.B.; Thau, D.; Seifert, C.; Engle, E.; Little, B. A scalable satellite-based crop yield mapper. *Remote Sens. Environ.* **2015**, *164*, 324–333. [CrossRef]
19. NASA-USDA Global Soil Moisture Data. Available online: https://explorer.earthengine.google.com/#detail/NASA_USDA%2FHSL%2Fsoil_moisture (accessed on 10 May 2018).
20. NASA-USDA SMAP Global Soil Moisture Data. Available online: https://explorer.earthengine.google.com/#detail/NASA_USDA%2FHSL%2FSMAP_soil_moisture (accessed on 10 May 2018).
21. International Research Institute for Climate and Society (IRI). Global Drought Analysis Tool. Available online: https://iridl.ldeo.columbia.edu/maproom/Global/Drought/Global/CPC_GOB/Analysis.html (accessed on 10 May 2018).
22. Global Agticultural Monitoring System. Available online: https://glam1.gsfc.nasa.gov/ (accessed on 10 May 2018).
23. CHIRPS Pentad: Climate Hazards Group InfraRed Precipitation with Station Data (Version 2.0 Final). Available online: https://explorer.earthengine.google.com/#detail/UCSB-CHG%2FCHIRPS%2FPENTAD. (accessed on 5 May 2018).
24. GlobCover: Global Land Cover Map. Available online: https://explorer.earthengine.google.com/#detail/ESA%2FGLOBCOVER_L4_200901_200912_V2_3 (accessed on 5 May 2018).
25. Evensen, G. The Ensemble Kalman Filter: Theoretical formulation and practical implementation. *Ocean Dyn.* **2003**, *53*, 343–367. [CrossRef]
26. De Wit, A.J.W.; van Diepen, C.A. Crop model data assimilation with the Ensemble Kalman filter for improving regional crop yield forecasts. *Agric. For. Meteorol.* **2007**, *146*, 38–56. [CrossRef]
27. Crow, W.T.; Kustas, W.P.; Prueger, J.H. Monitoring root-zone soil moisture through the assimilation of a thermal remote sensing-based soil moisture proxy into a water balance model. *Remote Sens. Environ.* **2008**, *112*, 1268–1281. [CrossRef]
28. Reichle, R.H. Data assimilation methods in the Earth sciences. *Adv. Water Resour.* **2008**, *31*, 1411–1418. [CrossRef]
29. Crow, W.; Ryu, D. A new data assimilation approach for improving hydrologic prediction using remotely-sensed soil moisture retrievals. *Hydrol. Earth Syst. Sci.* **2009**, *13*, 1–16. [CrossRef]
30. Panciera, R.; Walker, J.P.; Kalma, J.D.; Kim, E.J.; Saleh, K.; Wigneron, J.-P. Evaluation of the SMOS L-MEB passive microwave soil moisture retrieval algorithm. *Remote Sens. Environ.* **2009**, *113*, 435–444. [CrossRef]
31. Bitar, A.A.; Leroux, D.; Kerr, Y.H.; Merlin, O.; Richaume, P.; Sahoo, A.; Wood, E.F. Evaluation of SMOS Soil Moisture Products Over Continental U.S. Using the SCAN/SNO^{TEL} Network. *IEEE Trans. Geosci. Remote Sens.* **2012**, *50*, 1572–1586. [CrossRef]
32. Jackson, T.J.; Bindlish, R.; Cosh, M.H.; Zhao, T.; Starks, P.J.; Bosch, D.D.; Seyfried, M.; Moran, M.S.; Goodrich, D.C.; Kerr, Y.H.; et al. Validation of Soil Moisture and Ocean Salinity (SMOS) Soil Moisture Over Watershed Networks in the U.S. *IEEE Trans. Geosci. Remote Sens.* **2012**, *50*, 1530–1543. [CrossRef]
33. Van der Schalie, R.; Kerr, Y.H.; Wigneron, J.P.; Rodríguez-Fernández, N.J.; Al-Yaari, A.; de Jeu, R.A. Global SMOS Soil Moisture Retrievals from The Land Parameter Retrieval Model. *Int. J. Appl. Earth Obs. Geoinf.* **2016**, *45*, 125–134. [CrossRef]

34. Chan, S.K.; Bindlish, R.; O'Neill, P.E.; Njoku, E.; Jackson, T.; Colliander, A.; Chen, F.; Burgin, M.; Dunbar, S.; Piepmeier, J.; et al. Assessment of the SMAP Passive Soil Moisture Product. *IEEE Trans. Geosci. Remote Sens.* **2016**, *54*, 4994–5007. [CrossRef]
35. O'Neill, P.E.; Chan, S.; Njoku, E.G.; Jackson, T.J.; Bindlish, R. *SMAP Enhanced L2 Radiometer Half-Orbit 9 km EASE-Grid Soil Moisture, Version 1*; NASA National Snow and Ice Data Center Distributed Active Archive Center: Boulder, CO, USA, 2016.
36. Colliander, A.; Jackson, T.J.; Bindlish, R.; Chan, S.; Das, N.; Kim, S.B.; Cosh, M.H.; Dunbar, R.S.; Dang, L.; Pashaian, L.; et al. Validation of SMAP surface soil moisture products with core validation sites. *Remote Sens. Environ.* **2017**, *191*, 215–231. [CrossRef]
37. Burgin, M.S.; Colliander, A.; Njoku, E.G.; Chan, S.K.; Cabot, F.; Kerr, Y.H.; Bindlish, R.; Jackson, T.J.; Entekhabi, D.; Yueh, S.H. A Comparative Study of the SMAP Passive Soil Moisture Product With Existing Satellite-Based Soil Moisture Products. *IEEE Trans. Geosci. Remote Sens.* **2017**, *55*, 2959–2971. [CrossRef]
38. Al-Yaari, A.; Wigneron, J.P.; Kerr, Y.; Rodriguez-Fernandez, N.; O'Neill, P.E.; Jackson, T.J.; De Lannoy, G.J.M.; Al Bitar, A.; Mialon, A.; Richaume, P. Evaluating soil moisture retrievals from ESA's SMOS and NASA's SMAP brightness temperature datasets. *Remote Sens. Environ.* **2017**, *193*, 257–273. [CrossRef] [PubMed]
39. Kumar, S.V.; Peters-Lidard, C.D.; Tian, Y.; Houser, P.R.; Geiger, J.; Olden, S.; Lighty, L.; Eastman, J.L.; Doty, B.; Dirmeyer, P.; et al. Land information system: An interoperable framework for high resolution land surface modeling. *Environ. Model. Softw.* **2006**, *21*, 1402–1415. [CrossRef]
40. Funk, C.; Peterson, P.; Landsfeld, M.; Pedreros, D.; Verdin, J.; Shukla, S.; Husak, G.; Rowland, J.; Harrison, L.; Hoell, A.; et al. The climate hazards infrared precipitation with stations—A new environmental record for monitoring extremes. *Sci. Data* **2015**, *2*, 150066. [CrossRef] [PubMed]
41. Arino, O.; Ramos Perez, J.J.; Kalogirou, V.; Bontemps, S.; Defourny, P.; Van Bogaert, E. *Global Land Cover Map for 2009 (GlobCover 2009)*; European Space Agency (ESA): Paris, France; Université Catholique de Louvain (UCL): Louvain-la-Neuve, Belgium, 2012.
42. Tucker, C.J.; Pinzon, J.E.; Brown, M.E.; Slayback, D.A.; Pak, E.W.; Mahoney, R.; Vermote, E.F.; El Saleous, N. An extended AVHRR 8-km NDVI dataset compatible with MODIS and SPOT vegetation NDVI data. *Int. J. Remote Sens.* **2005**, *26*, 4485–4498. [CrossRef]
43. Mishra, A.; Vu, T.; Veettil, A.V.; Entekhabi, D. Drought monitoring with soil moisture active passive (SMAP) measurements. *J. Hydrol.* **2017**, *552*, 620–632. [CrossRef]
44. Martínez-Fernández, J.; González-Zamora, A.; Sánchez, N.; Gumuzzio, A. A soil water based index as a suitable agricultural drought indicator. *J. Hydrol.* **2015**, *522*, 265–273. [CrossRef]
45. Hunt Eric, D.; Hubbard Kenneth, G.; Wilhite Donald, A.; Arkebauer Timothy, J.; Dutcher Allen, L. The development and evaluation of a soil moisture index. *Int. J. Clim.* **2008**, *29*, 747–759. [CrossRef]
46. Bayissa, Y.; Maskey, S.; Tadesse, T.; van Andel, J.S.; Moges, S.; van Griensven, A.; Solomatine, D. Comparison of the Performance of Six Drought Indices in Characterizing Historical Drought for the Upper Blue Nile Basin, Ethiopia. *Geosciences* **2018**, *8*, 81. [CrossRef]
47. Tan, C.; Yang, J.; Li, M. Temporal-Spatial Variation of Drought Indicated by SPI and SPEI in Ningxia Hui Autonomous Region, China. *Atmosphere* **2015**, *6*, 1399–1421. [CrossRef]
48. Rouault, M.; Richard, Y. Intensity and spatial extent of droughts in southern Africa. *Geophys. Res. Lett.* **2005**, *32*, 297–318. [CrossRef]
49. Gebrehiwot, T.; van der Veen, A.; Maathuis, B. Spatial and temporal assessment of drought in the Northern highlands of Ethiopia. *Int. J. Appl. Earth Obs. Geoinf.* **2011**, *13*, 309–321. [CrossRef]
50. ESRI 2011. *ArcGIS Desktop: Release 10*; Environmental Systems Research Institute: Redlands, CA, USA, 2011.
51. R Core Team. *R: A Language and Environment for Statistical Computing*; R Foundation for Statistical Computing: Vienna, Austria, 2013.
52. Jury, M.R. Economic Impacts of Climate Variability in South Africa and Development of Resource Prediction Models. *J. Appl. Meteorol.* **2002**, *41*, 46–55. [CrossRef]
53. Nicholson, S.E. Climate and climatic variability of rainfall over eastern Africa. *Rev. Geophys.* **2017**, *55*, 590–635. [CrossRef]
54. Cash, B.A.; Rodó, X.; Ballester, J.; Bouma, M.J.; Baeza, A.; Dhiman, R.; Pascual, M. Malaria epidemics and the influence of the tropical South Atlantic on the Indian monsoon. *Nat. Clim. Chang.* **2013**, *3*, 502. [CrossRef]
55. Udelhoven, T.; Stellmes, M.; del Barrio, G.; Hill, J. Assessment of rainfall and NDVI anomalies in Spain (1989–1999) using distributed lag models. *Int. J. Remote Sens.* **2009**, *30*, 1961–1976. [CrossRef]

56. Herrmann, S.M.; Anyamba, A.; Tucker, C.J. Recent trends in vegetation dynamics in the African Sahel and their relationship to climate. *Glob. Environ. Chang.* **2005**, *15*, 394–404. [CrossRef]
57. Asoka, A.; Mishra, V. Prediction of vegetation anomalies to improve food security and water management in India. *Geophys. Res. Lett.* **2015**, *42*, 5290–5298. [CrossRef]
58. Bachmair, S.; Tanguy, M.; Hannaford, J.; Stahl, K. How well do meteorological indicators represent agricultural and forest drought across Europe? *Environ. Res. Lett.* **2018**, *13*, 034042. [CrossRef]
59. Ji, L.; Peters, A.J. A spatial regression procedure for evaluating the relationship between AVHRR-NDVI and climate in the northern Great Plains. *Int. J. Remote Sens.* **2004**, *25*, 297–311. [CrossRef]
60. Sims, A.P.; Niyogi, D.D.; Raman, S. Adopting drought indices for estimating soil moisture: A North Carolina case study. *Geophys. Res. Lett.* **2002**, *29*, 24-1. [CrossRef]
61. Richards, J.F. *Drought Assessment Tools for Agricultural Water Management in Jamaica*; McGill University Library: Montreal, QC, Canada, 2010.
62. Camberlin, P.; Martiny, N.; Philippon, N.; Richard, Y. Determinants of the interannual relationships between remote sensed photosynthetic activity and rainfall in tropical Africa. *Remote Sens. Environ.* **2007**, *106*, 199–216. [CrossRef]
63. Huber, S.; Fensholt, R.; Rasmussen, K. Water availability as the driver of vegetation dynamics in the African Sahel from 1982 to 2007. *Glob. Planet. Chang.* **2011**, *76*, 186–195. [CrossRef]
64. Jamali, S.; Seaquist, J.; Ardö, J.; Eklundh, L. Investigating temporal relationships between rainfall, soil moisture and MODIS-derived NDVI and EVI for six sites in Africa. In Proceedings of the 34th International Symposium on Remote Sensing of Environment, Sydney, Australia, 21 April 2011.
65. Wang, J.; Rich, P.M.; Price, K.P. Temporal responses of NDVI to precipitation and temperature in the central Great Plains, USA. *Int. J. Remote Sens.* **2003**, *24*, 2345–2364. [CrossRef]

Article

Multitemporal Cloud Masking in the Google Earth Engine

Gonzalo Mateo-García *, Luis Gómez-Chova, Julia Amorós-López, Jordi Muñoz-Marí and Gustau Camps-Valls

Image Processing Laboratory, University of Valencia, 46980 Paterna, Spain; luis.gomez-chova@uv.es (L.G.-C.); julia.amoros@uv.es (J.A.-L.); jordi.munoz@uv.es (J.M.-M.); gustau.camps@uv.es (G.C.-V.)

* Correspondence: gonzalo.mateo-garcia@uv.es

Received: 28 May 2018; Accepted: 5 July 2018; Published: 6 July 2018

Abstract: The exploitation of Earth observation satellite images acquired by optical instruments requires an automatic and accurate cloud detection. Multitemporal approaches to cloud detection are usually more powerful than their single scene counterparts since the presence of clouds varies greatly from one acquisition to another whereas surface can be assumed stationary in a broad sense. However, two practical limitations usually hamper their operational use: the access to the complete satellite image archive and the required computational power. This work presents a cloud detection and removal methodology implemented in the Google Earth Engine (GEE) cloud computing platform in order to meet these requirements. The proposed methodology is tested for the Landsat-8 mission over a large collection of manually labeled cloud masks from the Biome dataset. The quantitative results show state-of-the-art performance compared with mono-temporal standard approaches, such as FMask and ACCA algorithms, yielding improvements between 4–5% in classification accuracy and 3–10% in commission errors. The algorithm implementation within the Google Earth Engine and the generated cloud masks for all test images are released for interested readers.

Keywords: image time series; multitemporal analysis; change detection; cloud masking; Landsat-8; Google Earth Engine (GEE)

---

1. Introduction

Reliable and accurate cloud detection is a mandatory first step towards developing remote sensing products based on optical satellite images. Undetected clouds in the acquired satellite images hampers their operational exploitation at a global scale since cloud contamination affects most Earth observation applications [1]. Cloud masking of time series is thus a priority to obtain a better monitoring of the land cover dynamics and to generate more elaborated products [2].

Cloud detection approaches are generally based on the assumption that clouds present some useful features for their identification and discrimination from the underlying surface. On the one hand, a simple approach to cloud detection consists then in applying thresholds over a set of selected features, such as reflectance or temperature of the processed image, based on the physical properties of the clouds [3–6]. Apart from its simplicity, such approaches produce accurate results for satellite instruments that acquire enough spectral information, but it is challenging to adjust a set of thresholds that work at a global level. On the other hand, there is empirical evidence that supervised machine learning approaches outperform threshold-based ones in single scene cloud detection [1,7–9]. For instance, Ref. [1,7,9] show that neural networks are good candidates for cloud detection. However, they present practical limitations since they need a statistically significant, large collection of labeled images to learn from. This is because, in order to design algorithms capable of working globally over different types of surfaces and over different seasons, a huge number of image pixels labeled as cloudy or cloud free must be available to train the models. This labelling process usually requires a large amount of tedious manual work, which is also not exempt from errors. Furthermore, additional independent data has to be gathered to validate the performance of the algorithms, which increases the data requirements and dedication. In any case, both threshold and machine learning based cloud detection algorithms relying only on the information of the analyzed image are still far from being perfect and produce systematic errors specially over high reflectance surfaces such as urban areas, bright sand over coastlines, snow and ice covers [10].

In this complex scenario, including temporal information helps to distinguish clouds from the surface, since the latter usually remains stable over time. Cloud detection methodologies can thus be divided into monotemporal single scene and multitemporal approaches. Single scene approaches only use the information from a given image to build the cloud mask, while multitemporal approaches also exploit the information of previously acquired images, collocated over the same area, to improve the cloud detection accuracy. Multitemporal cloud detection is therefore an intrinsically easier problem because location and features of clouds vary greatly between acquisitions, whereas the surface is to a certain extent stable. However, multitemporal methods are computationally demanding, and the lack of accessibility to previous data usually hampers their operational application to most satellite missions. Therefore, in order to exploit the wealth of the temporal information, long-term missions with a granted access to the satellite images archive, and suitable computing platforms, are required. A clear example fulfilling these requirements is the Landsat mission from NASA [11], which provides global image data over land since 1972. For this reason we will focus here on Landsat images, although the methodology and the subsequent discussion can also be applied to other similar satellites [12].

There exists a wide variety of multitemporal approaches for cloud detection that have been applied to Landsat imagery [13–19]. In the Multitemporal Cloud Detection (MTCD) algorithm [14], the authors use a composite cloud-free image as reference, then they detect clouds by setting a threshold on the difference between the target and the reference in the blue band. In order to reduce false positives, they use an extra correlation criteria with at least 10 previous images. In Ref. [15], a previous spatially collocated cloud-free image from the same region is manually selected as the reference image. Then a set of thresholds over the reflectance in some Landsat bands (B1, B4 and B6) and over the difference in reflectance between the target and the reference image are set. The Temporal mask (TMask) algorithm [16] builds a pixel-wise time series regression to model the cloud-free reflectance of each pixel. It uses the FMask algorithm [5] to decide which pixels to include in such a regression model. Then, it applies a set of thresholds over the difference in reflectance between the estimated and the target image in Landsat bands B2, B4 and B5. The work presented in Ref. [17] is also based on FMask. In this case, they first remove one of the FMask tests to reduce over-detection, and compute the FMask cloud probability for each image in the time series. Afterwards, they compute the pixel-wise median FMask cloud probability and the standard deviation over the time series. Then, analyzed pixels are masked as cloudy if (a) the modified FMask says it is a cloud; or (b) if the cloud probability exceeds 3.5 standard deviations the median value. Recently, Ref. [19] proposed to also use a composite reference image and a set of thresholds over the difference in reflectance between the target and the reference in Landsat bands B3, B4 and B5. Thus the method is similar to the one presented in Ref. [14] but without the correlation criteria over the time series. Finally, in Ref. [18], we modeled the background surface from the three previous collocated cloud-free images using a non-linear kernel ridge regression that minimizes both prediction and estimation errors simultaneously. Then, the difference image between this background surface reference and the target is clustered and a threshold over the mean difference reflectance is applied to each cluster to decide if it belongs to a cloudy or cloud-free area. In summary, one can see how most of the multitemporal cloud detection schemes proposed in the literature cast the cloud detection problem as a change detection problem [20]: a reference image is built using cloud-free pixels and clouds are detected as particular changes over this reference. To decide whether the change is relevant enough, several thresholds are usually proposed based on heuristics.

Three main issues not properly addressed can be identified in all multitemporal approaches proposed so far:

- **Data access and image retrieval**. Most of the proposed methods assume that a sufficiently long time series of collocated images is available. It is worth pointing out that retrieving the images to build the time series in an easy and operational manner is technically difficult. We need access to the full catalog and powerful enough GIS software to select and co-register the images. We overcome this limitation using the Google Earth Engine (GEE) platform.
- **Computational cost**. Most of the proposed methods require a sufficiently long time series of images to operate: at least 15 in the case of TMask [16] and at least 10 in the case of MTCD [14]. This is a critical problem if the algorithm cannot be implemented using parallel computing techniques. We again solve this issue ensuring that our algorithm can be implemented on the GEE cloud computing platform.
- **Validation of results**. A consistent drawback in most cloud detection studies is the lack of quantitative validation over a large collection of independent images. On the one hand, as we have mentioned in the two previous points, if the multitemporal algorithm is computationally demanding and required images are hard to retrieve, it will be difficult to test the method over a large dataset. On the other hand, simultaneous collocated observations informing about the presence of clouds or independent datasets of manually annotated clouds are often not available. Therefore, without a comprehensive *ground truth*, validation of cloud detection results is usually limited to a visual inspection of the generated cloud masks. In this work we take advantage of the recently released Landsat-8 Biome Cloud Masks dataset [10], which contains manually generated cloud masks from 96 images from different Biomes around the world.

Therefore, we propose a multitemporal cloud detection algorithm that is also based on the hypothesis that surface reflectance smoothly varies over time, whereas abrupt changes are caused by the presence of clouds. Our proposed methodology extends the work we presented in Ref. [18]. In particular, the proposed methodology presented in this paper consists of four main steps. First, the surface background is estimated using few previous cloud-free images that are automatically retrieved from the Landsat archive stored in the GEE catalog. Then, the difference between the analyzed cloudy image (target) and the cloud-free estimated background (reference) is computed in order to enhance the changes due to the presence of clouds. This difference image is then processed to find homogeneous clusters corresponding to clouds and surface. Finally, the obtained clusters are labelled as cloudy or cloud-free areas by applying a set of thresholds on the difference intensity and on the reflectance of the representative clusters.

In addition, the surface background estimated from the previous cloud-free images can be also used to perform a *cloud removal* (or *cloud filling*) in the analyzed cloudy image [21,22]. Pixels masked as clouds can be replaced by the estimated surface background at these locations obtaining a completely cloud-free image [23,24]. The improved frequency of the satellite images time series can then be used to better monitor land cover dynamics and to generate more elaborated products.

The proposed algorithm is fully implemented in the GEE platform, which grants access to the complete Landsat-8 catalog, reducing the technical complexity of the multitemporal cloud detection and transferring the computational load to the GEE parallel computing infrastructure. The potential of the proposed approach is tested over 2661 500×500 patches extracted from the Biome dataset [10], and the obtained results are available online for the interested readers (http://isp.uv.es/projects/cdc/viewer_l8_GEE.html).

The rest of the paper is organized as follows. In Section 2 the Landsat-8 data, the GEE platform, and the Biome dataset are presented. In Section 3, we explain the proposed methodology for cloud detection and removal. Section 4 presents the evaluation of the proposed methodology. It shows the predictive power of the proposed variables over the dataset, the accuracy, commission and omission errors, some illustrative scenes with the proposed cloud mask, and the cloud removal errors. The algorithm implementation in the Google Earth Engine is briefly described in Section 5. Finally, Section 6 discusses the results and summarizes the conclusions.

2. Satellite Data and Ground Truth

2.1. Landsat-8 Data

The Landsat Program [11] consists of a series of Earth observation satellite missions jointly managed by NASA and the United States Geological Survey (USGS). Landsat is a unique resource with the world's longest continuously acquired image collection of the Earth's land areas at moderate to high resolution to support resource assessment, land-cover mapping, and to track inter-annual changes. It started with the first Landsat satellite launched in 1972, and is continued with both Landsat 7 and 8, which are still operational. Landsat 9 is expected to be launched in late 2020 ensuring the Landsat data continuity.

The Landsat-8 payload consists of two science instruments: the Operational Land Imager (OLI) and the Thermal InfraRed Sensor (TIRS), acquiring multispectral images with 11 spectral bands that cover from deep blue to the thermal infrared: B1—*Coastal and Aerosol* (0.433–0.453 μm), B2—*Blue* (0.450–0.515 μm), B3—*Green* (0.525–0.600 μm), B4—*Red* (0.630–0.680 μm), B5—*Near Infrared* or *NIR* (0.845–0.885 μm), B6—*Short Wavelength Infrared* or *SWIR* (1.560–1.660 μm), B7—*SWIR* (2.100–2.300 μm), B8—*Panchromatic* (0.500–0.680 μm), B9—*Cirrus* (1.360–1.390 μm), B10—*Thermal Infrared* or TIR (10.30–11.30 μm) and B11—*TIR* (11.50–12.50 μm). Note that the visible channels are B1-B4 (and B8), which are useful to distinguish the white and bright clouds. Additionally, Landsat-8 presents a band (B9) specifically designed to detect cirrus and high clouds.

2.2. Google Earth Engine Platform

The Google Earth Engine platform [25] is a cloud computing platform for geographical data analysis. It gives access to a full complete catalog of remote sensing products together with the capability to process these products quickly online through massive parallelization. The GEE data catalog includes data from Landsat 4, 5, 7 and 8 processed by the United States Geological Survey (USGS), several MODIS products, including global composites, recently imagery from Sentinel 1, 2 and 3 satellites, and many more. All data are pre-processed and geo-referenced, facilitating its direct use. In addition, user data in raster or vector formats can be uploaded (*ingested* using GEE terminology) and processed in the GEE. We took advantage of this feature for uploading the manual cloud masks used as ground truth in our experiments.

In this work, all required Landsat images were retrieved from the `LANDSAT/LC8_L1T_TOA_FMASK` *Image Collection* available in the GEE. These images consist of top of atmosphere (TOA) reflectance (calibration coefficients are included in metadata [26]). These products also include two additional bands: the quality assessment band (BQA) and the FMask cloud mask [5]. We use the cloud flag included in the BQA nominal product [11] to assess if previous images over each test site location are cloud free or not, which allows us to easily and automatically retrieve cloud-free images from the entire archive. In addition to the Automated Cloud Cover Assessment (ACCA) cloud masking algorithm in the BQA band presented in Ref. [27], the FMask [5] is used to benchmark the proposed cloud detection algorithm. Both algorithms are single-image approaches mainly based on combination of rules and thresholds over a set of spectral indexes.

The GEE computation engine offers both JavaScript and Python application programming interfaces (API), which allow to easily develop algorithms that work in parallel on the Google data computer facilities. The programming model is object oriented and based on the MapReduce paradigm [28]. On the one hand, the GEE engine is accessible through a web-based integrated development environment (IDE) using the JavaScript API. The web-based IDE allows the user to visualize images, results, tables and charts that can be easily exported. On the other hand, the Python API offers the same set of methods, which allow to make requests to the Engine and access the catalog, but without the visualization capabilities of the web-based IDE. However, we chose the Python API to develop our cloud detection scheme because it is easier to integrate with long running tasks, which are essential to run the full validation study in an automatic manner.

2.3. Cloud Detection Ground Truth

Validation of cloud detection algorithms is an extremely difficult task due to the lack of accurate simultaneous collocated information per pixel about the presence of clouds. In this scenario one is forced to manually generate a labeled dataset of annotated clouds, which is time consuming and always includes some uncertainties. Recent validation studies carried out for single scene cloud detection, e.g. for Landsat-8 [10] and for Proba-V [9], are extremely important efforts for the development and validation of cloud screening algorithms. The public dissemination of this data gives the opportunity to fairly benchmark the results on independent datasets and allows to quantify and analyze the cloud screening quality. This is the case of the Landsat 7 Irish dataset [3,29], the Landsat-8 SPARCS dataset [7,30], the Landsat-8 Biome dataset [10,31] or the Sentinel 2 Hollstein dataset [8]. In this work, we take advantage of the Landsat-8 Biome dataset [31] created in Ref. [10]. The Biome dataset consists of 96 Landsat-8 acquisitions (∼7500 × 7500 pixels approximately) from eight different biomes around the world, in which all pixels have been manually labeled. Figure 1 shows the geographic location of the 96 images that form the dataset.

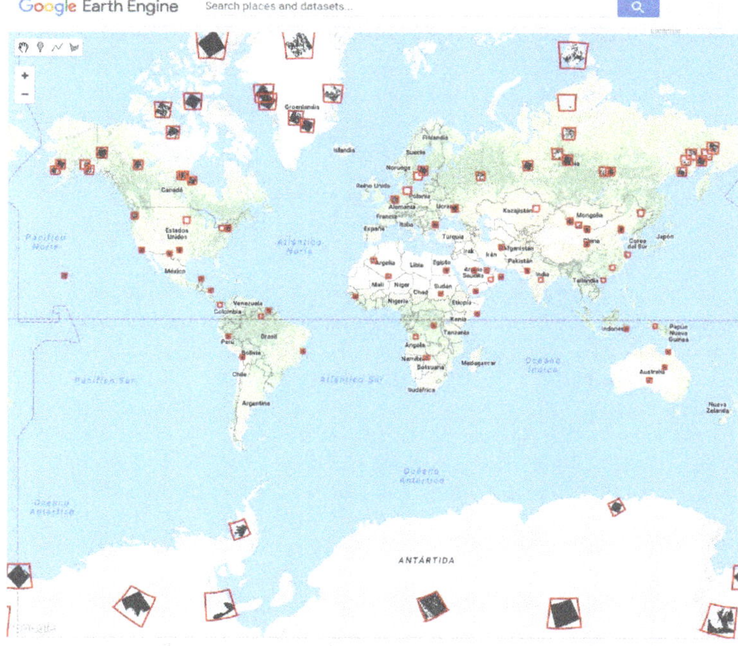

Figure 1. Geographic location of the 96 images from the Landsat-8 Biome dataset [31] ingested on the Google Earth Engine.

We add these cloud masks with the corresponding Landsat-8 products by ingesting this dataset in the GEE. Then, a few previous cloud-free images for each acquisition were automatically retrieved using the GEE API. From the original 96 Biome products, only 23 of them have enough (three) previous cloud-free images. This is mainly because unfortunately most of the labeled acquisitions selected for the Biome dataset are close to the launch of the Landsat-8 satellite. Therefore, we divided these 23 images in smaller patches of 500 × 500 pixels for our analysis, resulting in 2661 patches.

It is worth noting that validation studies of cloud detection algorithms over large datasets are scarce in the literature and, in the particular case of multitemporal cloud detection, the algorithms have been usually validated on a few images. The use of processing platforms such as the GEE make our study much more feasible.

3. Methodology

The proposed methodology for multitemporal cloud detection is based on our previous work [18]. It works under the assumption that surface reflectance is stable over time or at least follows smooth variations compared to the abrupt changes induced by the presence of clouds. Therefore, this work follows the widespread approach for cloud detection based on multitemporal background modeling with difference change detection extensively used in the remote sensing literature [13–19]. Figure 2 shows a diagram summarizing the proposed multitemporal cloud detection approach. The following sections describe the main methodological steps.

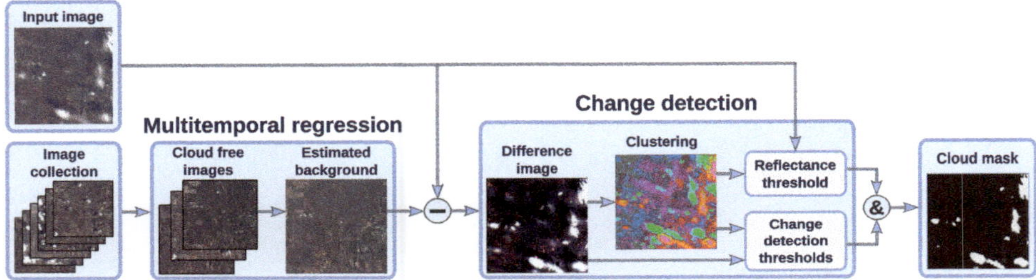

Figure 2. Multitemporal cloud detection scheme implemented on the Google Earth Engine platform.

3.1. Background Estimation

One of the main challenges of the background modeling step is to make it computationally scalable: previous attempts in Ref. [14,16] are computationally demanding, which make them difficult to apply in operational settings. In order to alleviate these problems, in this study we limit the proposed algorithm to work with only three previous collocated images for the surface background estimation. The key for this process to be fully operational is that the selection and retrieval of the three previous cloud-free collocated images has to be carried out automatically. We use the BQA band included in the Landsat products to discriminate if an image is cloud free; and, as we have mentioned, one of the main advantages of using the GEE Python API together with the Landsat image collection is that this step can be fully automated requiring no human intervention.

We call *pre-filtering* to the first image retrieval step, which consists of assessing if previous images are cloud free or not. Pre-filtering can be solved applying some rough cloud detection method, e.g., setting a threshold over the brightness or over the blue channel as proposed in Ref. [14], or taking advantage of automatic single scene cloud detection schemes if they exist for the given satellite. For this study we use the cloud flag from the Level 1 BQA band of Landsat-8 [27]. We consider an image cloud free if less than 10% of its pixels are masked as cloudy. This raises an important consideration on the design of the cloud detection scheme: it should be robust to errors on the pre-filtering method. An extremely inaccurate pre-filtering algorithm can undermine the performance of the method since cloudy pixels will be used to model the background surface. We will see that these methods are robust enough to work on situations where previous images have some clouds. It is worth pointing out that, since we limit the cloud cover to be less than 10% in each selected image and we assume that clouds are randomly located from one image to another, the probability that the same pixel is cloudy in all three images is expected to be really low.

The estimation of the background from the cloud-free image time series is one of the critical steps of the method. We compare four different background estimation methodologies presented in the literature, from simpler to more complex:

- **Nearest date**: It consists of taking the nearest cloud-free image in time as the background. This is the approach used in Ref. [15], however they rely on human intervention to assess that the image does not present any cloud.
- **Median filter**: It takes the pixel-wise median over time using the three previous cloud-free images. This is the approach suggested in TMask [16] for pixels where the time series is not long enough.
- **Linear regression**: It fits a standard regularized linear regression using the time series of the previous cloud-free images [18]. Similarly, TMask [16] used an iterative re-weighted least squares regression at pixel level, which mitigates the effect of eventual cloudy pixels in the time series.
- **Kernel regression**: The nonlinear version of the former method. It is based on a specific kernel ridge regression (KRR) formulation for change detection presented in Ref. [18].

3.2. Change Detection and Clouds Identification

Once the background is estimated, we use it as a cloud-free reference image to tackle the cloud detection as a change detection problem. Therefore, we compute the difference image between the cloudy target image and the estimated cloud-free reference, which is the base for most change detection methods [20].

However, we do not find changes by applying thresholds directly to the difference image, i.e., target minus estimated. Instead, we previously apply a k-means clustering algorithm over the difference image using all Landsat-8 bands. Afterwards, specific thresholds are applied at a cluster level, i.e., to some features computed over the pixels belonging to each cluster. In particular, we compute three different features for each cluster i: (a) the norm (intensity) of the difference reflectance image over the visible bands (B2, B3 and B4 for Landsat 8), we denote this quantity with α_i; (b) the mean of the difference reflectance image over visible bands, β_i; and (c) the norm of reflectance image over the visible bands, γ_i. A cluster is classified as cloudy if the three following tests over these features are satisfied: $\alpha_i \geq 0.04$, $\beta_i \geq 0$ and $\gamma_i \geq 0.175$.

The threshold 0.04 on the difference of reflectance image is ubiquitous in the existing literature. For instance, TMask [16] also suggested 0.04 for the B4 channel, MTCD [14] suggests 0.03 on the blue band weighted by the difference between the acquisition time of the image and the reference. The method proposed in Ref. [19] also used 0.04 in the B3 and B4 bands. In contrast, in our previous work [18] the threshold was higher (0.09) since we used the norm over all the reflectance bands. Here we select the norm as a more robust indicator but restricted to the visible bands (B2, B3 and B4). This threshold is intended to detect significant differences, i.e., with a sufficient intensity to be considered changes, while the other two conditions to be satisfied are specifically included to distinguish clouds from the rest of possible changes in the surface. On the one hand, clouds are usually brighter than the surface so clouds imply an increase in reflectance with respect to the reference background image. By imposing the temporal difference over the visible bands to be positive we exclude intense changes decreasing the reflectance, such as shadows, flooded areas, agricultural changes, etc. On the other hand, we also want to discard changes that increase the brightness but do not look like a cloud in the target image, e.g., agricultural crops. Therefore, we also impose that the norm of the top of atmosphere (TOA) reflectance over the visible bands is higher than 0.175 in order to consider that the cluster corresponds to a cloud. The norm of the visible reflectance bands is also used in Ref. [17] to distinguish potentially cloudy pixels, although in this work they set a lower threshold of 0.15 because they wanted to over-detect cloudy areas.

Modifying these thresholds will make the algorithm more or less cloud conservative. We believe that the subsequent user of the cloud mask should have some flexibility to choose to be more or less cloud conservative. For instance, applications like land use or land cover classification are less affected by the presence of semitransparent cirrus whereas for instance estimating the water content of canopy should be much more cloud conservative. Providing the receiver operator curve (ROC) [32] for the entire dataset allows the users to better select these thresholds in order to obtain a trade-off between commission (false positives) and omission (false negatives) errors for their particular application.

3.3. Remarks

One of the main differences of our proposal for cloud detection is the clustering step. We apply a k-means clustering over the difference image over all bands of the satellite. We fixed the number of clusters to 10; this number is related to the size of the image (500×500 pixels in the experiments) so if larger images are used this number should be increased. We tried however different numbers of clusters (5, 15 and 20) but we did not observe major differences in performance. The clustering step seeks to capture patterns over all the bands that cannot be captured with a single static threshold. For example, it is well known that the Thermal Infrared Bands (TIR, B10 and B11) have good predictive power for the cloud detection problem. However, setting a global threshold independently of location and season is very difficult since surface temperature greatly varies over places and surfaces. In addition, working with time series exacerbates this problem since the surface temperature might vary quite a lot with the date of the acquired image. Therefore, k-means clustering is intended to group similar patterns, e.g., in temperature, and pixels assigned to the same cluster will be classified afterwards to the same class (cloudy or clear). The clustering step simplifies the problem since instead of classifying pixels we have to classify clusters. However, it might introduce errors in mixed clusters where not all the pixels are purely from one of the two classes (cloudy or clear). In our case, if we classify each cluster according to its majority class using the ground truth, we obtain a classification error lower than 3% for all the proposed background estimation methods in the used dataset. This error can be considered a lower bound of the classification error for the presented results. Finally, it is worth mentioning that if we apply the thresholds directly over the difference image, i.e., without the clustering step, numerical accuracy is not significantly affected, but visual inspection showed less consistency on the masks and higher salt-and-pepper effects.

4. Experimental Results

This section contains the experimental results. First we describe an illustrative example where we show some intermediate results of the method, then the analysis over the full dataset is presented. For these results, we will first explore the parameters and the discriminative power of the multitemporal difference, then we will show the results over the complete dataset for cloud detection and cloud removal.

4.1. Cloud Detection Example

Figure 3 shows the cloud detection results for a cloudy image over Texas (USA). The top right corner shows the RGB composite of the acquired image included in the Biome dataset. We see that it contains several thin clouds scattered across the image. In the bottom left image, the manually labeled ground truth cloud mask (in yellow) from the Biome dataset overlay the RGB composite. We can see here that some very thin clouds are not included in the provided ground truth. The three top left images are the previous cloud-free images retrieved automatically with the GEE API from the GEE Landsat image collection. We see that the top left one is not completely free of clouds: this is because it has less than 10% of clouds according to the ACCA algorithm of the BQA band. The image of the bottom right corner corresponds to the cloud-free estimated background using the median method. We see that this estimation method is robust enough in this case since it has not been affected by the unscreened clouds present in the previous "cloud-free" images, and it correctly preserves other bright surfaces such as urban areas. Finally, we compare both the proposed and the Fmask cloud masks with the available Biome ground truth. The second image of the bottom starting from the left shows the differences between our proposed method and the ground truth. In white it shows the true positives (clouds) of the method with respect the ground truth, in orange the false negatives (omissions) and in blue the false positives (commissions). We see that the overall agreement is very high and most of the discrepancies are on the borders of the clouds. The image to the right corresponds to the differences between FMask and the ground truth, in this case we see that FMask missed some thin clouds in the bottom and in the left part of the image.

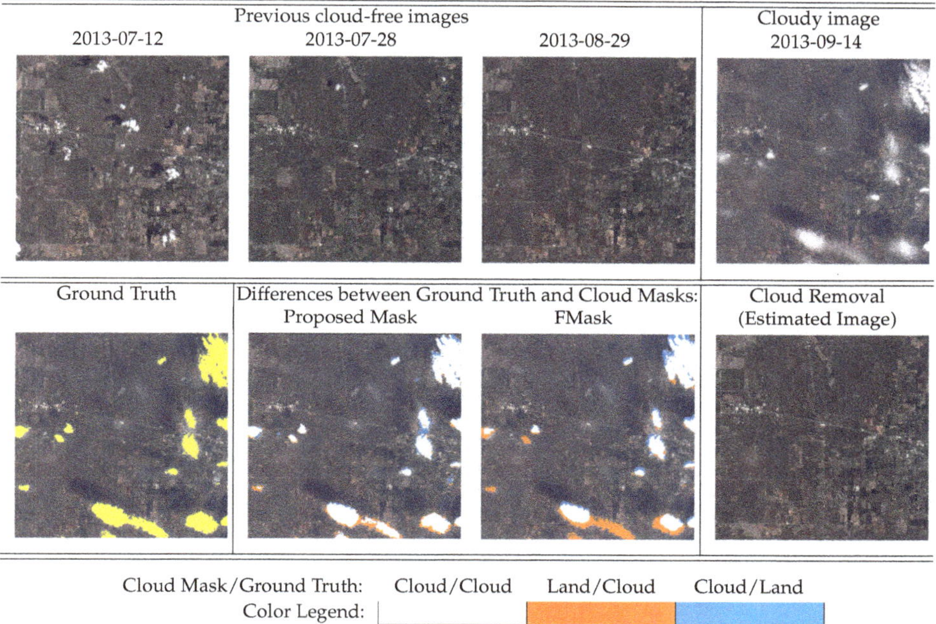

Figure 3. Illustration of the Cloud detection scheme. Comparison between the ground truth and the proposed cloud mask algorithm and the FMask. Discrepancies are shown in blue when the proposed method detects 'cloudy', and in orange when pixels are classified as 'cloud-free'.

4.2. Parameters and Errors Analysis

We evaluate the results in terms of commission errors, omission errors and overall accuracy. Table 1 contains the definition of these metrics. Generally, we can obtain a trade-off between commission and omission errors depending on the requirements. For instance, to reduce the omission error we can reduce the threshold over the reflectance which will make the algorithm more cloud conservative and, as a result, the commission error will increase. On the other hand, if we increase the threshold the commission error will decrease and the algorithm will be more clearly conservative and will probably raise the omission error.

First we want to demonstrate the discriminating capability of the norm of the difference image (α_i) for cloud detection. The receiver operator curve (ROC) shows the true positive rate vs. false positive rate trade-off as we vary the threshold over the predictive variable α_i. Figure 4 shows four curves corresponding to the four different background estimation methods (nearest date, median filter, Linear regression, and Kernel ridge regression). A cross is displayed for the case of the proposed threshold (0.04). We also show a cross indicating the TPR and FPR values for FMask [5] and for ACCA (BQA) [27] over this dataset. As we see, using the nearest date or the median filter to estimate the background and a single threshold over α we outperform FMask and ACCA (BQA) since those points are below the obtained ROC curves. This means that we can reduce commission error while having the same omission error as FMask, or reduce omission error while maintaining the same commission error as FMask.

Table 1. Validation metrics: True Positive Rate (TPR), False Positive Rate (FPR), Commission Error, Omission Error, Overall Accuracy.

True Positive Rate	TPR	$\frac{\text{cloudy pixels predicted as clouds}}{\text{cloudy pixels}}$
False Positive Rate (Commission Error)	FPR	$\frac{\text{clear pixels predicted as clouds}}{\text{clear pixels}}$
Omission Error	1- TPR	$\frac{\text{cloudy pixels predicted as clear}}{\text{cloudy pixels}}$
Overall Accuracy		$\frac{\text{cloudy pixels pred. as clouds+clear pixels pred. as clear}}{\text{total pixels}}$

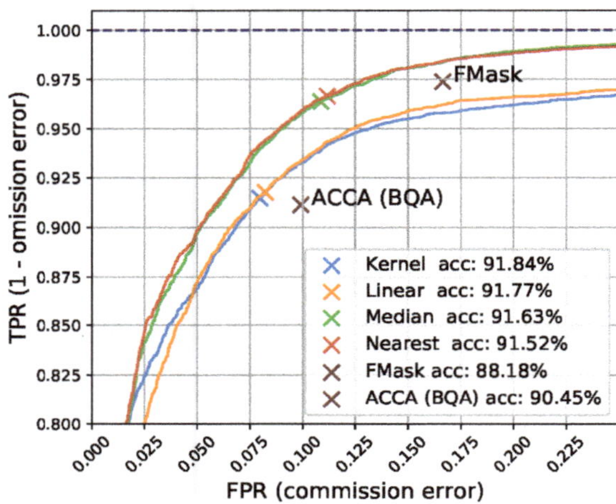

Figure 4. This figure shows ROC curves for the four proposed background estimation methods. Crosses show the TPR and FPR values for the proposed threshold (0.04) and for FMask [5] and ACCA (BQA) [27] on the same dataset.

It is worth mentioning that the simpler background estimation methods (Median and Nearest) have better performance in terms of cloud detection accuracy. In the case of the median, it is more robust to outliers (e.g., clouds contaminating the images used for the background estimation) than the linear or kernel regression approaches. For the nearest date, it might be because the closer in time the image is, the more similar it is to the target image in terms of surface changes. Nevertheless, we will see in the next sections that the kernel and linear regression methods obtain better results in terms of mean squared error in reflectance and, therefore, will be the more appropriate for the cloud removal task.

Figure 4 shows that using only a threshold over the norm difference has a very good performance on cloud detection. However, as we have mentioned in Section 3.2, by doing this we detect all high differences (changes) in reflectance. Whereas most of these differences are because of the presence of clouds, some of them are due to changes in the surface. Figure 5 shows an example of agricultural crops in Bulgaria. The image on the center shows that some of those fields are detected as clouds if we use only a threshold over the differences.

Figure 5. Landsat 8 image (LC81820302014180LGN00 006_013) acquired on 10 June 2013. Rural area in Bulgaria presenting crops misclassified as clouds when the threshold in reflectance is not applied.

In order to reduce these false positive cases we added an additional threshold applied directly over the reflectance of the cloudy image instead of over the difference image. In particular, we applied the threshold over the norm of the visible bands, γ. This is physically grounded since clouds have high reflectance on the visible spectral range. In addition, it has been exploited before in Ref. [17] as a measure of potentially cloudy pixels. Figure 6 confirms this approach. The left plot shows cluster centers colored in orange if the majority of their pixels are cloudy and in blue if most of them are clear. The X-axis shows the norm of the difference in reflectance, α, and the Y-axis shows the norm of the reflectance, γ. We can see that the threshold of 0.04 in α was correctly fixed and that 0.175 is a natural threshold in γ for this dataset. The right plot in Figure 6 shows the ROC curves with and without the extra threshold on reflectance γ. We show the ROC curves corresponding to the thresholds 0.15 and 0.175. We can see that the inclusion of this additional threshold (0.175) increases the overall accuracy from 91 to 94%. The threshold at 0.15 could be used instead of 0.175 for cloud conservative applications. Overall we see that by including this additional restriction (either in 0.175 or 0.15) the resulting algorithm is more accurate and less cloud conservative.

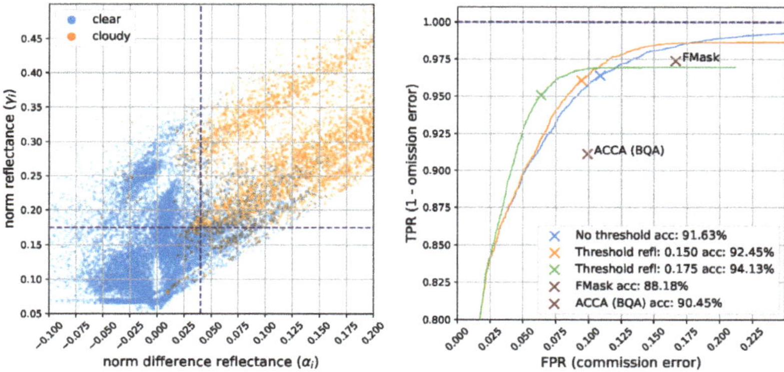

Figure 6. (**Left**) Scatter plot of the clusters. Norm of TOA reflectance of the visible bands on the Y-axis (γ) and norm of difference in reflectance on the X-axis (α). Each point corresponds to one of the 10 clusters from each of the 2661 image patches. Vertical and horizontal lines show the proposed thresholds. (**Right**) ROC curves with and without the extra threshold on reflectance γ. The median is used for background estimation in both cases.

4.3. Cloud Detection Results

Once the parameters and methodology have been fixed we analyze the cloud detection results over the whole dataset. Table 2 shows the cloud detection statistics using both the proposed thresholds for the four background estimation models and for the independent FMask and ACCA (BQA) cloud detection algorithms. We see that multitemporal methods yield higher overall accuracy than single scene methods. In addition, we see that the simpler background estimations, such as the median filter and the nearest date, yield a good trade-off in commission and omission errors. Figure 7 shows the mean accuracy and standard deviation over the patches for each of the 23 Landsat-8 acquisitions. We see here again that the multitemporal approach using the median as background estimator consistently outperforms FMask.

Table 2. Cloud detection statistics over all pixels of the used Landsat-8 Biome Dataset.

Method	Overall Accuracy	Kappa Statistic	Commission Error	Omission Error
FMask [5]	88.18%	0.7550	16.64%	2.62%
ACCA (BQA) [27]	90.45%	0.7933	9.90%	8.86%
Nearest date	94.18%	0.8733	6.31%	4.87%
Median filter	94.13%	0.8720	6.36%	4.94%
Linear regression	93.66%	0.8593	4.78%	9.32%
Kernel regression	93.56%	0.8572	4.82%	9.53%

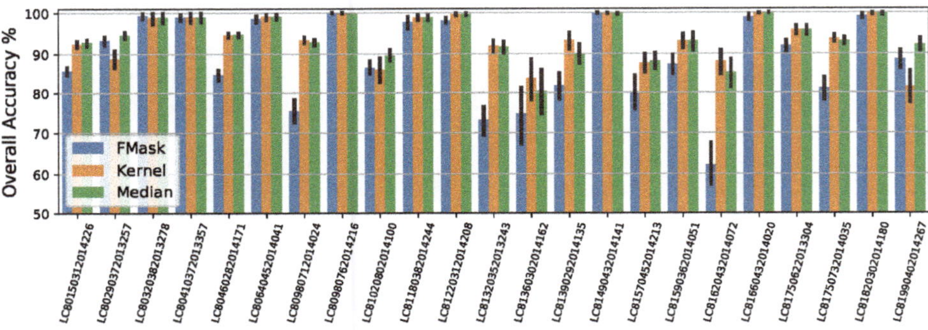

Figure 7. Average accuracy over the image patches for each of the 23 different Landsat-8 acquisitions selected from the Biome dataset.

Finally, Figure 8 shows some cherry-picked results of the proposed method using the median to estimate the background. Rows 1, 3 and 5 show some systematic errors of FMask over cities, coastal areas and riversides. Row 2 presents small errors in semitransparent clouds in the middle of the image that are not correctly labeled in the manual ground truth cloud mask but that our method correctly identifies. On the other hand, thin clouds on Rows 1 and 7 are misclassified by the proposed method whereas FMask identifies them correctly. Row 6 again shows errors in the ground truth labels. In this case, a path is falsely identified as a cloud. We found these errors specially over bright surfaces, which remind us that single scene cloud detection is challenging even for human experts. Actually, in some cases, we detected them only because we have previous images from the same location with which to compare. Interested readers can visually inspect cloud detection results and the comparison of both the proposed method and FMask [5] with the Biome ground truth, which are available online at http://isp.uv.es/projects/cdc/viewer_l8_GEE.html for all 2661 patches from the Biome dataset.

Figure 8. Patches of 500 × 500 pixels from Biome dataset. From left to right: RGB scene, RGB scene with ground truth cloud mask in yellow, differences between ground truth and the proposed cloud mask (using the median as background estimation), difference between ground truth and FMask, and estimated cloud-free image.

4.4. Cloud Removal

In addition to the cloud detection problem, we consider also the task of cloud removal (or cloud filling). Most land applications generally discard cloud contaminated pixels for the estimation of biophysical parameters. In cloudy areas, this causes a big amount of missing values in the processed time series that undermine the statistical significance of the subsequent analysis. In this subsection, we benchmark the different background estimation methodologies proposed in this paper and evaluate their suitability for cloud removal in a large dataset. In Ref. [18], we proposed to use previous images together with the current one to estimate the TOA reflectance of the cloud contaminated areas. This idea is also presented in Ref. [33] using more sophisticated methods, however, our proposal in Ref. [18] can be directly implemented in the GEE platform. We compare these linear and kernel based regression approaches with the two simplest baselines, i.e., using the latest available cloud-free pixel or the pixel-wise median filter. The performance of the cloud removal is quantified and evaluated in terms of the error between the estimated and actual background pixels in the cloud-free areas (since cloud contaminated pixels cannot be compared with the background). The accurate cloud mask and the posterior cloud removal provide cloud-free time series that allow a better monitoring of land cover dynamics and the generation of further remote sensing products. The last column in Figure 8 shows the estimated cloud-free image for some scenes. The plots show the estimated image where clouds have been removed. In fact, we show estimated values for the whole scene and not clouded areas only. We can see how the spatial and radiometric features are well preserved and no cloud residuals can be observed.

Quantitative results for the cloud removal are shown in Figure 9. It shows the distribution of the root mean square error on the 2661 patches separately for each spectral band. In the plots, the mid lines represent the mean RMSE for all the (cloud-free) image pixels, the boxes define the 25 and 75 percentiles of the RMSE distribution, and the vertical lines define the maximum and minimum RMSE values. We see here that the more sophisticated methods for background estimation (Linear and Kernel regression) perform better than the simpler ones (median and nearest). This confirms the results presented in Ref. [18] and shows that estimated reflectance is an accurate option to fill the gaps caused by cloud contamination.

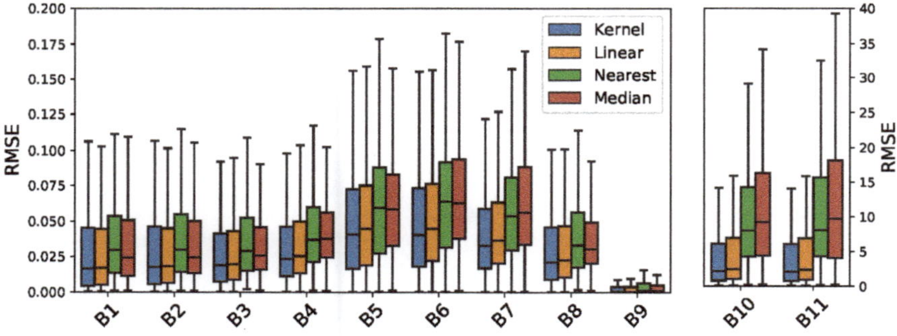

Figure 9. Root mean square error between the estimated and actual background pixels in the cloud-free areas. Distribution of the errors is shown separately for all Landsat bands for the four background estimation approaches.

5. Algorithm Implementation in the Google Earth Engine

The data in the GEE platform is organized in collections, usually composed of images or features. Images contain bands (spectra, masks, products, etc.), properties, and metadata. Features can contain any kind of information needed to process data, such as labels for supervised algorithms, polygons to define geographical areas, etc. Users can apply their own defined functions, or use the ones provided by the API, using an operation called *mapping*, which essentially applies a function over any given collection independently. This allows a straightforward processing of large amounts of images and data in parallel. Using this computational paradigm we implemented the full proposed cloud detection scheme using the Python API. In particular, given an input target image, we *map* and *filter* the Landsat-8 `LANDSAT/LC8_L1T_TOA_FMASK` *Image Collection*. Then the filtered collection is *reduced* to produce an *Image* which is the background estimation. With this reference image we compute the difference image and apply the k-means *clustering*. Finally, we apply the thresholds as defined in Section 3.

The manual cloud masks from the Biome dataset were ingested in the GEE. Therefore, the proposed methodology together with the comparison with the ground truth is implemented using only the Python API of the GEE platform. The developed code has been published in GitHub at https://github.com/IPL-UV/ee_ipl_uv. In that package we provide a function that computes the cloud mask following the proposed methodology for a given GEE image. In addition, some Python notebooks with examples that go step by step on the proposed methodology have been included in the software package.

Finally, as we have mentioned, in order to show the potential of the GEE platform the proposed algorithms have been tested over 2661 patches extracted from the Biome dataset. The obtained cloud masks can be inspected online for the whole dataset at http://isp.uv.es/projects/cdc/viewer_l8_GEE.html.

6. Discussion

In previous sections we presented a simple yet efficient multitemporal algorithm for cloud detection. The results show an overall increase in detection accuracy and commission error compared to state-of-the-art mono-temporal approaches such as FMask and ACCA. In addition, omission error could be reduced slightly more for the same commission error than FMask using a lower threshold in reflectance ($\gamma = 0.15$), as can be seen in Figure 6 (left). For cloud detection, it is normally taken for granted that commission errors are better than omissions, thus operational algorithms tend to overmask in order to avoid false negatives. However, we think that the proliferation of open access satellite image archives implies that in the future more advanced users will be interested in controlling by themselves the trade-off between commission and omission errors depending on their underlying application. To this end, we provide Table 3 as a guide to help in tuning the thresholds of the current algorithm, where we can see the selected combination providing the best trade-off highlighted in bold. From results shown in Tables 2 and 3, we can see that the proposed method presents improvements between 4–5% in classification accuracy and 3–10% in commission errors, compared with FMask and ACCA algorithms.

Table 3. Cloud detection statistics for different thresholds combinations over all selected pixels of the Landsat-8 Biome Dataset, using the *median* as background estimation method.

Thresholds		Overall Accuracy	Kappa Statistic	Commission Error	Omission Error
Difference (α)	Reflectance (γ)				
0.02	0.000	85.54%	0.7076	21.53%	0.95%
0.02	0.150	89.52%	0.7820	15.16%	1.54%
0.02	0.175	92.59%	0.8415	9.61%	3.21%
0.03	0.000	89.56%	0.7823	14.90%	1.93%
0.03	0.150	91.43%	0.8187	11.83%	2.35%
0.03	0.175	93.63%	0.8624	7.76%	3.74%
0.04	0.000	91.63%	0.8217	10.86%	3.63%
0.04	0.150	92.45%	0.8382	9.43%	3.95%
0.04	**0.175**	**94.13%**	**0.8720**	**6.36%**	**4.94%**
0.05	0.000	92.50%	0.8378	8.54%	5.53%
0.05	0.150	92.97%	0.8474	7.70%	5.76%
0.05	0.175	94.26%	0.8738	5.32%	6.54%
FMask		88.18%	0.7550	16.64%	2.62%
ACCA (BQA)		90.45%	0.7933	9.90%	8.86%

The proposed multitemporal methodology resembles popular multitemporal algorithms such as TMask [16] and MTCD [14] since all of them are based on background estimation and thresholds over the difference image. However, our methodology is simpler and requires less images in the time series to operate. For this reason we consider the current work as a baseline to evaluate trade-offs in processing performance for these more complex multitemporal schemes. It would be of great interest for the community to compare all these approaches in a common benchmark; unfortunately, to this end, we would need labeled images and common open-sourced versions of the algorithms to evaluate the models.

Obviously there are limitations to the proposed multitemporal methodology: for instance, it might fail in situations with sudden changes in the underlying surface, such as permanent snow in upper latitudes. The current dataset lacks these situations, hence we do not recommend its use in such cases.

In addition, another limitation of current and future works on cloud detection is the quality of the ground truth masks: for the Irish dataset [3], the work [34] estimated a mean overall disagreement of 7% over the manual cloud masks labelled by three different experts. The labelling procedure to create the Irish dataset [10] is similar to the Biome dataset that we use in the present work. Therefore, current overall errors are in line with the intrinsic error of human experts following the current labelling procedure. This indicates that in the future, in order to increase the performance, we should develop better labelling methods and provide results by cloud type and underlying surface.

7. Conclusions

In this work, we proposed a multitemporal cloud detection methodology that can be applied at a global scale using the GEE cloud computing platform. We applied the proposed approach to Landsat-8 imagery and we validated it using a large independent dataset of manually labelled images.

The approach is based on a simple multitemporal background modelling algorithm together with a set of tests applied over the segmented difference image, which has shown a high cloud detection power. Our principal findings and contributions can be summarized as follows. This approach outperforms single-scene threshold-based cloud detection approaches for Landsat such as FMask [5] and ACCA (BQA) [27]. We provided an evaluation of different background estimation methods and different variables and thresholds in terms of commission and omission errors. In particular, we showed that simple background detection models such as the median or the nearest cloud-free image are both accurate and robust for the cloud detection task. In addition, for the first time to the authors knowledge, a multitemporal cloud detection scheme is validated over a large collection of independent manually labelled images. The whole process has been implemented within the GEE cloud computing platform and a ready to use implementation has been provided. Compared to previous multitemporal open source implementations, our approach also includes the image retrieval and coregistration steps, which are essential for the operational use of the algorithm. The generated cloud masks can be inspected at http://isp.uv.es/projects/cdc/viewer_l8_GEE.html.

Future lines of research include the application to other optical multispectral satellites requiring accurate and automatic cloud detection. For example, the satellite constellations of Sentinel missions from the European Copernicus programme aim to optimize global coverage and data delivery. In particular, Sentinel-2 mission [12] acquires image time series with a high temporal frequency and unprecedented spatial resolution for satellite missions providing open access data at a global scale. Additionally, another line of research consists of using the multitemporal cloud masks as a proxy of a ground truth that can be used to train single scene supervised machine learning cloud detection algorithms. This approach has been recently successfully applied to image classification tasks [35] and would alleviate data requirements of machine learning methods.

Supplementary Materials: An interactive tool for the visualization of the validation results is available online at http://isp.uv.es/projects/cdc/viewer_l8_GEE.html. The code is published at https://github.com/IPL-UV/ee_ipl_uv.

Author Contributions: G.M.G. and L.G.C. conceived and designed the experiments; G.M.G. performed the experiments; G.M.G. and L.G.C. analyzed the data; J.M.M. contributed to develop the analysis tools; G.C.V. and J.A.L. contributed to write and review the manuscript.

Funding: This research has been funded by the Spanish Ministry of Economy and Competitiveness (MINECO-ERDF, TEC2016-77741-R and TIN2015-64210-R projects) and by Google Inc. under the Google Earth Engine (GEE) Award titled 'Cloud detection in the Cloud' granted to Luis Gómez-Chova. Camps-Valls was supported by the European Research Council (ERC) through the ERC Consolidator Grant SEDAL (project id 647423).

Conflicts of Interest: The authors declare no conflict of interest.

References

1. Gómez-Chova, L.; Camps-Valls, G.; Calpe, J.; Guanter, L.; Moreno, J. Cloud-Screening Algorithm for ENVISAT/MERIS Multispectral Images. *IEEE Trans. Geosci. Remote Sens.* **2007**, *45*, 4105–4118. [CrossRef]
2. Malenovsky, Z.; Rott, H.; Cihlar, J.; Schaepman, M.E.; García-Santos, G.; Fernandes, R.; Berger, M. Sentinels for science: Potential of Sentinel-1, -2, and -3 missions for scientific observations of ocean, cryosphere and land. *Remote Sens. Environ.* **2012**, *120*, 91–101. [CrossRef]
3. Irish, R.R.; Barker, J.L.; Goward, S.N.; Arvidson, T. Characterization of the Landsat-7 ETM+ Automated Cloud-Cover Assessment (ACCA) Algorithm. *Photogramm. Eng. Remote Sens.* **2006**, *72*, 1179–1188. [CrossRef]
4. Zhu, Z.; Woodcock, C.E. Object-based cloud and cloud shadow detection in Landsat imagery. *Remote Sens. Environ.* **2012**, *118*, 83–94. [CrossRef]
5. Zhu, Z.; Wang, S.; Woodcock, C.E. Improvement and expansion of the Fmask algorithm: Cloud, cloud shadow, and snow detection for Landsats 4–7, 8, and Sentinel 2 images. *Remote Sens. Environ.* **2015**, *159*, 269–277. [CrossRef]
6. Mei, L.; Vountas, M.; Gómez-Chova, L.; Rozanov, V.; Jäger, M.; Lotz, W.; Burrows, J.P.; Hollmann, R. A Cloud masking algorithm for the XBAER aerosol retrieval using MERIS data. *Remote Sens. Environ.* **2017**, *197*, 141–160. [CrossRef]
7. Hughes, M.J.; Hayes, D.J. Automated Detection of Cloud and Cloud Shadow in Single-Date Landsat Imagery Using Neural Networks and Spatial Post-Processing. *Remote Sens.* **2014**, *6*, 4907–4926. [CrossRef]
8. Hollstein, A.; Segl, K.; Guanter, L.; Brell, M.; Enesco, M. Ready-to-Use Methods for the Detection of Clouds, Cirrus, Snow, Shadow, Water and Clear Sky Pixels in Sentinel-2 MSI Images. *Remote Sens.* **2016**, *8*, 666. [CrossRef]
9. Iannone, R.Q.; Niro, F.; Goryl, P.; Dransfeld, S.; Hoersch, B.; Stelzer, K.; Kirches, G.; Paperin, M.; Brockmann, C.; Gómez-Chova, L.; et al. Proba-V cloud detection Round Robin: Validation results and recommendations. In Proceedings of the 2017 9th International Workshop on the Analysis of Multitemporal Remote Sensing Images (MultiTemp), Brugge, Belgium, 27–29 June 2017; pp. 1–8.
10. Foga, S.; Scaramuzza, P.L.; Guo, S.; Zhu, Z.; Dilley, R.D.; Beckmann, T.; Schmidt, G.L.; Dwyer, J.L.; Joseph Hughes, M.; Laue, B. Cloud detection algorithm comparison and validation for operational Landsat data products. *Remote Sens. Environ.* **2017**, *194*, 379–390. [CrossRef]
11. Loveland, T.R.; Dwyer, J.L. Landsat: Building a strong future. *Remote Sens. Environ.* **2012**, *122*, 22–29. [CrossRef]
12. Drusch, M.; Bello, U.D.; Carlier, S.; Colin, O.; Fernandez, V.; Gascon, F.; Hoersch, B.; Isola, C.; Laberinti, P.; Martimort, P.; et al. Sentinel-2: ESA's Optical High-Resolution Mission for GMES Operational Services. *Remote Sens. Environ.* **2012**, *120*, 25–36. [CrossRef]
13. Wang, B.; Ono, A.; Muramatsu, K.; Fujiwara, N. Automated Detection and Removal of Clouds and Their Shadows from Landsat TM Images. *IEICE Trans. Inf. Syst.* **1999**, *82*, 453–460.
14. Hagolle, O.; Huc, M.; Pascual, D.V.; Dedieu, G. A multi-temporal method for cloud detection, applied to FORMOSAT-2, VENμS, LANDSAT and SENTINEL-2 images. *Remote Sens. Environ.* **2010**, *114*, 1747–1755. [CrossRef]
15. Jin, S.; Homer, C.; Yang, L.; Xian, G.; Fry, J.; Danielson, P.; Townsend, P.A. Automated cloud and shadow detection and filling using two-date Landsat imagery in the USA. *Int. J. Remote Sens.* **2013**, *34*, 1540–1560. [CrossRef]
16. Zhu, Z.; Woodcock, C.E. Automated cloud, cloud shadow, and snow detection in multitemporal Landsat data: An algorithm designed specifically for monitoring land cover change. *Remote Sens. Environ.* **2014**, *152*, 217–234. [CrossRef]

17. Frantz, D.; Röder, A.; Udelhoven, T.; Schmidt, M. Enhancing the Detectability of Clouds and Their Shadows in Multitemporal Dryland Landsat Imagery: Extending Fmask. *IEEE Geosci. Remote Sens. Lett.* **2015**, *12*, 1242–1246. [CrossRef]
18. Gómez-Chova, L.; Amorós-López, J.; Mateo-García, G.; Muñoz-Marí, J.; Camps-Valls, G. Cloud masking and removal in remote sensing image time series. *J. Appl. Remote Sens.* **2017**, *11*, 015005. [CrossRef]
19. Candra, D.S.; Phinn, S.; Scarth, P. Cloud and cloud shadow removal of landsat 8 images using Multitemporal Cloud Removal method. In Proceedings of the 2017 6th International Conference on Agro-Geoinformatics, Fairfax, VA, USA, 7–10 August 2017; pp. 1–5.
20. Bovolo, F.; Bruzzone, L. The Time Variable in Data Fusion: A Change Detection Perspective. *IEEE Geosci. Remote Sens. Mag.* **2015**, *3*, 8–26. [CrossRef]
21. Melgani, F. Contextual reconstruction of cloud-contaminated multitemporal multispectral images. *IEEE Trans. Geosci. Remote Sens.* **2006**, *44*, 442–455. [CrossRef]
22. Lin, C.H.; Tsai, P.H.; Lai, K.H.; Chen, J.Y. Cloud Removal From Multitemporal Satellite Images Using Information Cloning. *IEEE Trans. Geosci. Remote Sens.* **2013**, *51*, 232–241. [CrossRef]
23. Hu, G.; Li, X.; Liang, D. Thin cloud removal from remote sensing images using multidirectional dual tree complex wavelet transform and transfer least square support vector regression. *J. Appl. Remote Sens.* **2015**, *9*, 095053. [CrossRef]
24. Chen, B.; Huang, B.; Chen, L.; Xu, B. Spatially and Temporally Weighted Regression: A Novel Method to Produce Continuous Cloud-Free Landsat Imagery. *IEEE Trans. Geosci. Remote Sens.* **2017**, *55*, 27–37. [CrossRef]
25. Gorelick, N.; Hancher, M.; Dixon, M.; Ilyushchenko, S.; Thau, D.; Moore, R. Google Earth Engine: Planetary-scale geospatial analysis for everyone. *Remote Sens. Environ.* **2017**, *202*, 18–27. [CrossRef]
26. Chander, G.; Markham, B.L.; Helder, D.L. Summary of current radiometric calibration coefficients for Landsat MSS, TM, ETM+, and EO-1 ALI sensors. *Remote Sens. Environ.* **2009**, *113*, 893–903. [CrossRef]
27. Scaramuzza, P.L.; Bouchard, M.A.; Dwyer, J.L. Development of the Landsat Data Continuity Mission Cloud-Cover Assessment Algorithms. *IEEE Trans. Geosci. Remote Sens.* **2012**, *50*, 1140–1154. [CrossRef]
28. Dean, J.; Ghemawat, S. MapReduce: Simplified Data Processing on Large Clusters. In Proceedings of the Sixth Symposium on Operating System Design and Implementation (OSDI'04), San Francisco, CA, USA, 6–8 December 2004; pp. 137–150.
29. U.S. Geological Survey. *L7 Irish Cloud Validation Masks*; Data Release; U.S. Geological Survey: Reston, VA, USA, 2016.
30. U.S. Geological Survey. *L8 SPARCS Cloud Validation Masks*; Data Release; U.S. Geological Survey: Reston, VA, USA, 2016.
31. U.S. Geological Survey. *L8 Biome Cloud Validation Masks*; Data Release; U.S. Geological Survey: Reston, VA, USA, 2016.
32. Fawcett, T. An introduction to ROC analysis. *Pattern Recognit. Lett.* **2006**, *27*, 861–874. [CrossRef]
33. Meng, F.; Yang, X.; Zhou, C.; Li, Z. A Sparse Dictionary Learning-Based Adaptive Patch Inpainting Method for Thick Clouds Removal from High-Spatial Resolution Remote Sensing Imagery. *Sensors* **2017**, *17*, 2130. [CrossRef] [PubMed]
34. Oreopoulos, L.; Wilson, M.J.; Várnai, T. Implementation on Landsat Data of a Simple Cloud-Mask Algorithm Developed for MODIS Land Bands. *IEEE Geosci. Remote Sens. Lett.* **2011**, *8*, 597–601. [CrossRef]
35. Mahajan, D.; Girshick, R.; Ramanathan, V.; He, K.; Paluri, M.; Li, Y.; Bharambe, A.; van der Maaten, L. Exploring the Limits of Weakly Supervised Pretraining. *arXiv* **2018**, arXiv:1805.00932.

 *remote sensing*

Article

Historical and Operational Monitoring of Surface Sediments in the Lower Mekong Basin Using Landsat and Google Earth Engine Cloud Computing

Kel N. Markert [1,2,*], Calla M. Schmidt [3], Robert E. Griffin [4], Africa I. Flores [1,2], Ate Poortinga [5,6], David S. Saah [5,6,7], Rebekke E. Muench [1,2], Nicholas E. Clinton [8], Farrukh Chishtie [6,9], Kritsana Kityuttachai [10], Paradis Someth [10], Eric R. Anderson [1,2], Aekkapol Aekakkararungroj [6,9] and David J. Ganz [11]

1. Earth System Science Center, The University of Alabama in Huntsville, 320 Sparkman Dr., Huntsville, AL 35805, USA; africa.flores@nsstc.uah.edu (A.I.F.); rem0016@uah.edu (R.E.M.); eric.anderson@nsstc.uah.edu (E.R.A.)
2. SERVIR Science Coordination Office, NASA Marshall Space Flight Center, 320 Sparkman Dr., Huntsville, AL 35805, USA
3. Environmental Science Department, University of San Francisco, 2130 Fulton St., San Francisco, CA 94117, USA; cischmidt@usfca.edu
4. Department of Atmospheric Science, The University of Alabama in Huntsville, 320 Sparkman Dr., Huntsville, AL 35805, USA; robert.griffin@nsstc.uah.edu
5. Spatial Informatics Group, LLC, 2529 Yolanda Ct., Pleasanton, CA 94566, USA; apoortinga@sig-gis.com (A.P.); dssaah@usfca.edu (D.S.S.)
6. SERVIR-Mekong, SM Tower, 24th Floor, 979/69 Paholyothin Road, Samsen Nai Phayathai, Bangkok 10400, Thailand; farrukh.chishtie@adpc.net (F.C.); aekkapol.a@adpc.net (A.A.)
7. Geospatial Analysis Lab, University of San Francisco, 2130 Fulton St., San Francisco, CA 94117, USA
8. Google, Inc., 1600 Amphitheatre Parkway, Mountain View, CA 94043, USA; nclinton@google.com
9. Asian Disaster Preparedness Center, SM Tower, 24th Floor, 979/69 Paholyothin Road, Samsen Nai Phayathai, Bangkok 10400, Thailand
10. Technical Support Division, Mekong River Commission Secretariat, P.O. Box 6101, 184 Fa Ngoum Road, Unit 18, Ban Sithane Neua, Sikhottabong District, Vientiane 01000, Lao PDR; kritsana@mrcmekong.org (K.K.); someth@mrcmekong.org (P.S.)
11. RECOFTC—The Center for People and Forests, P.O. Box 1111, Kasetsart Post Office Bangkok 10903, Thailand; david.ganz@recoftc.org

* Correspondence: km0033@uah.edu or kel.markert@nasa.gov; Tel.: +1-256-961-7484

Received: 15 April 2018; Accepted: 7 June 2018; Published: 8 June 2018

Abstract: Reservoir construction and land use change are altering sediment transport within river systems at a global scale. Changes in sediment transport can impact river morphology, aquatic ecosystems, and ultimately the growth and retreat of delta environments. The Lower Mekong Basin is crucial to five neighboring countries for transportation, energy production, sustainable water supply, and food production. In response, countries have coordinated to develop programs for regional scale water quality monitoring that including surface sediment concentrations (SSSC); however, these programs are based on a limited number of point measurements and due to resource limitations, cannot provide comprehensive insights into sediment transport across all strategic locations within the Lower Mekong Basin. To augment in situ SSSC data from the current monitoring program, we developed an empirical model to estimate SSSC across the Lower Mekong Basin from Landsat observations. Model validation revealed that remotely sensed SSSC estimates captured the spatial and temporal dynamics in a range of aquatic environments (main stem of Mekong river, tributary systems, Mekong Floodplain, and reservoirs) while, on average, slightly underestimating SSSC by about 2 mg$\cdot$L^{-1} across all settings. The operational SSSC model was developed and implemented using Google Earth Engine and Google App Engine was used to host an online application that allows users, without any knowledge of remote sensing, to access SSSC data across the region. Expanded access to

SSSC data should be particularly helpful for resource managers and other stakeholders seeking to understand the dynamics between surface sediment concentrations and land use conversions, water policy, and energy production in a globally strategic region.

Keywords: lower mekong basin; landsat collection; suspended sediment concentration; online application; google earth engine

---

1. Introduction

Human activities such as agriculture, forestry, and urbanization are increasing sediment transport in rivers globally, while reservoir construction is simultaneously decreasing the total sediment flux to coastal environments [1]. Because sediment transport by rivers impacts channel morphology, aquatic ecosystems, reservoir storage capacity, and ultimately the growth or retreat of delta environments [2]; monitoring changes in sediment concentration and transport is critical to effective basin management. Timely information on total suspended matter is critical for land managers to assess the effects of a wide range of issues caused by poor water quality. Unfortunately, collection of reliable suspended sediment concentration data at the spatial and temporal resolution necessary for effective basin management and planning can often be prohibitively time consuming and expensive in large rivers.

The Mekong River is the largest trans-boundary river basin in Asia that covers an area of 795,000 km^2, and has an annual discharge of 475 km^3 [3]. Each year the Mekong delivers approximately 160 million tons of sediment to the South China Sea [4]. The Mekong River hydrology is dominated by the seasonality of snowmelt runoff into the northern headwaters on the Tibetan Plateau, and the seasonal monsoon in the lower basin. The lowest flows are between February and April, with peak discharge between August and September. The Mekong River Basin is rapidly developing and due to increasing demands for hydropower and freshwater, reservoir construction has accelerated in recent decades [5] with potential to alter sediment transport in the region [6,7]. Changes in the Mekong river sediment supply, most likely caused by dam retention of sediment and channel-bed sand extraction in the Mekong delta, are suspected to be the cause of erosion patterns observed in the Mekong delta [2]. However, while changes in sediment transport have been observed on other major Asian rivers such as the Indus, Yellow, and Yangtze Rivers, limited field data has made it difficult to detect changes in sediment discharge on the Lower Mekong River, despite the fact that the Lower Mekong Basin is experiencing similar pressures of population growth, land use change, infrastructure development, and reservoir construction [8].

The most comprehensive field dataset for suspended sediments in the Lower Mekong has been collected by The Mekong River Commission (MRC). Suspended sediment is a water quality constituent of particular concern to MRC because suspended sediments increase turbidity and can influence the transport of particle bound contaminants such as nutrients, organic compounds, pesticides, and trace metals. In addition, suspended sediments are critical to the accumulation of wetland soils in the Mekong Delta [9]. The MRC has an extensive water quality monitoring network, with 132 stations across five countries in the Lower Mekong Basin (Thailand, Vietnam, Laos PDR, Cambodia, and Myanmar) (Figure 1). At a number of stations samples have been collected monthly since 1985, however there are significant gaps in the records at numerous stations. While the MRC dataset on suspended matter provides invaluable information on water quality in the Lower Mekong Basin, it consists of single point measurements taken from the surface of the river at a single point in the river cross section, and therefore the dataset provides limited understanding of the spatial distribution of suspended sediments along the river.

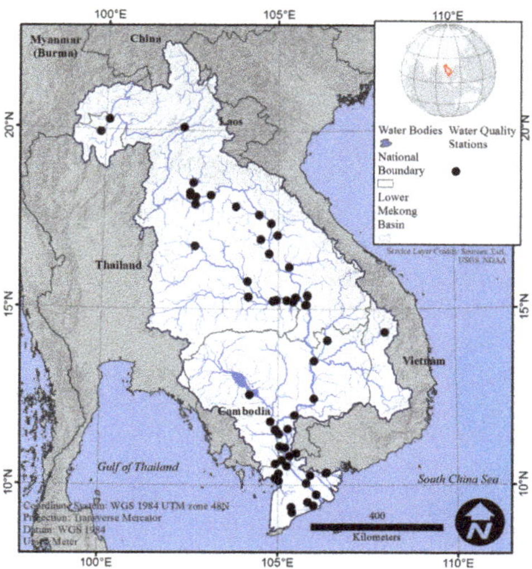

Figure 1. Study area map of the Lower Mekong Basin highlighting geopolitical boundaries, the river systems in the region, and the water quality monitoring stations included in the MRC water quality database.

Remote sensing tools can provide spatial and temporal resolution for surface suspended sediment concentration (SSSC) in large rivers that are not available from traditional in situ measurements [10–13]. The retrieval of SSSC from remote sensing systems relies on the optical properties (transmittance, absorption and scattering) of water and the dissolved and suspended constituents in the water. Suspended solids are responsible for most of the scattering in an aquatic system, whereas chlorophyll-a (chl-a) and colored dissolved matter are mainly responsible for absorption [14]. A variety of techniques have been used to estimate SSSC from different remote sensing systems [15–17]. There is a wide body of research on assessing water quality analytical optical modeling using in situ inherent optical properties [18–20]. Unfortunately, these approaches are often complex, iterative and location specific, leading researchers to explore the extent to which empirical models can provide robust estimates of water quality parameters. Methods used to relate in situ data to the satellite observations through statistical relationships include simple linear regression, non-linear regressions, principal component analysis, and neural networks. Previous studies have shown that SSSC is well correlated with the first four bands of the Landsat sensors [21–23] and the use of a single band from the sensor series, provided the band is chosen appropriately, can provide a robust estimation of SSSC [24–29]. Moreover, other studies have also illustrated the utility of band ratios from Landsat sensors to estimate SSSC [11,30,31]. Sensors from the Landsat satellite series (TM, ETM+, OLI) are the most commonly used remote sensing platforms for estimating SSSC [32].

Because the relationship between SSSC and surface reflectance is a function of sediment mineralogy, color, and grain size distribution [33], empirical models perform best when calibrated with local in situ observations. This is particularly important in riverine SSSC studies because there is often considerable spatial and temporal variability in these parameters. Previous efforts to estimate SSSC and TSS from remote sensing data sources in the Lower Mekong Basin have focused on the main stem of the Mekong River or on the Mekong Delta, but not both. For example, Suif et al. [34] developed an empirical model using the near-infrared band on Landsat TM as well as the blue, green, and red bands in a multiple linear regression along the Mekong River. In another recent study, Duc et al. [35] found a strong correlation between the 1st principle component of Landsat TM and ETM+ imagery and SSSC in the Mekong Delta. While these studies provide models to estimate SSSC accurately from remotely sensed data, the models were not validated at regional scale and therefore, need careful consideration before broad use. Other studies, such as Bui et al. [36] have used soil erosion modeling to estimate sediment transport dynamics within watersheds in Southeast Asia. Although, modeling approaches produce suspended sediment load (SSL) information irrespective of remote sensing inputs, the models require detailed parameterization and quality in situ data to accurately estimate SSL, which can limit application in data sparse regions.

Cloud based remote sensing platforms such as Google Earth Engine (GEE) offer exciting new opportunities to provide policy makers with high resolution near real-time SSSC data through a simple web interface without the need for expensive software, technical expertise or other resource demanding solutions. The goals of this study were to (1) calibrate a regional remote sensing model of SSSC in the Lower Mekong which augments the spatial and temporal resolution of existing field records and (2) develop an online application that allows users, without any expert knowledge of remote sensing, to monitor and analyze trends in SSSC for decision making within the region. We accomplished these goals by utilizing the extensive MRC SSSC dataset across the entire Lower Mekong Basin to develop an empirical model for estimating SSSC from satellite data. The model includes 8 main stem gauging stations on the Mekong River, and 36 additional gauging stations from tributary systems, the Mekong Floodplain, and reservoirs. The model is accessible through an operational online application that allows users to monitor and analyze trends in SSSC from continuously updated remote sensing datasets in a timely manner. This novel web application can provide actionable SSSC data to decision makers throughout the region.

2. Data and Methods

In this study we correlated in situ SSSC measurements with coinciding Landsat observations to create an empirical model to estimate SSSC in the Lower Mekong Basin. Due to large data volume and processing needs, GEE [37] was used to facilitate data processing. GEE allows users to run algorithms on georeferenced imagery, vector data, and other precomputed value-added products stored on Google's cloud-based infrastructure in an easy-to-use manner. GEE was used in this study to (1) query all Landsat observations over the Mekong Basin that coincide with in situ measurements; (2) extract the spectra to develop an empirical SSSC model and (3) provide back-end processing for an online suspended sediment application. Figure 2 displays the overarching workflow used in this study.

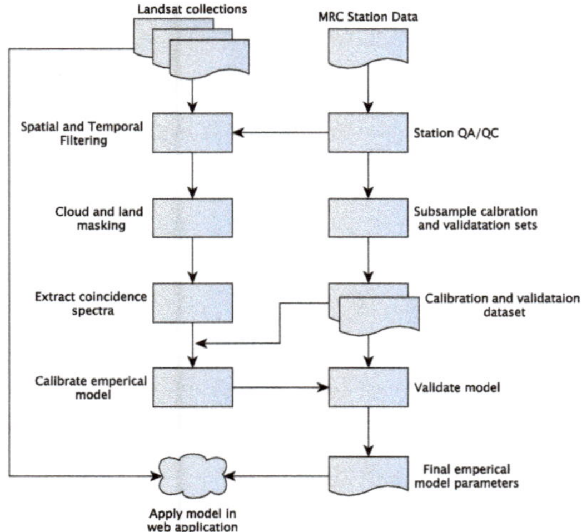

Figure 2. General workflow schematic for building SSSC empirical model from satellite imagery.

2.1. Field Measurement of Suspended Sediment Concentration

The reference dataset used to create and validate the SSSC model was acquired from the MRC data portal (http://portal.mrcmekong.org/). The MRC established a water quality monitoring network in 1985 to measure discharge and collect monthly water quality samples for a wide suite of constituents including cations, anions, nutrients and SSSC [38]. SSSC samples are collected near the surface (0.3–0.5 m) using a bottle, and therefore likely underestimate SSSC given that suspended sediment concentration typically increases with depth. Nevertheless, these surface samples provide an excellent comparison with remotely sensed optical properties which are most representative of shallow depths. Although this is the best available water quality dataset for the Lower Mekong Basin, it does include uncertainties in sampling technique consistency and quality from earlier collection years. At many stations, SSSC records are not complete, with data gaps ranging from months to years. For a thorough discussion of SSSC data reliability in the Lower Mekong Basin see Walling [8].

2.2. Landsat Collection Data

Data from the Landsat TM (from both Landsat 4 and 5 satellites), ETM+ (Landsat 7) and OLI (Landsat 8) sensors were used to estimate SSSC for water bodies in the Lower Mekong Basin. The Landsat 4, 5, 7, and 8 satellites are each in a sun-synchronous orbit each with a 16-day revisit time. Landsat 4 had an operating lifetime from 1982-1994, Landsat 5's lifetime was from 1984–2011, and Landsat 7 has been active since its launch in 1999. Landsat 8, the most recent satellite in the series, was launched in 2013. While any two Landsat satellites were in operation, there was an 8-day offset of data acquisition between the two satellites increasing the temporal resolution. The TM, ETM+, and OLI sensors collect spectral channel data in the visible, near-infrared (NIR), and short-wave infrared (SWIR) portions of the electromagnetic spectrum at 30 $\times$ 30 m resolution. Landsat data collected from 1985 to 2011 from Landsat 4, 5, and 7 were used in this study to maximize the number of Landsat observations that coincide with the in situ data from MRC. All Landsat sensors are used within the online applications.

2.3. Data Preprocessing

The in situ data were provided as a table with geographic latitude and longitude along with suspended sediment concentration for each collection time. To reduce the influence of the channel bottom or upwelling that occurs next to the bank and in shallow waters, station locations were filtered to ensure that only stations at least 60 m (two Landsat pixels) from the bank of the waterbody were included in further analysis. To account for changes in river morphology during the time from 1985–2011, the shoreline was dynamically calculated from the European Commission's Joint Research Center (JRC) Monthly Water History v1.0 image collection (JRC/GSW1_0/MonthlyHistory) [39] for each in situ collection date. Landsat or in situ data that was collected within 60m of the dynamic shoreline was not used in the analysis.

Next, the Landsat collections within GEE were queried to identify scene IDs that overlap MRC stations within one day of a in situ collection date. The precomputed surface reflectance (SR) Landsat collections were used in this analysis (LANDSAT/LT04/C01/T1_SR; LANDSAT/LT05/C01/T1_SR; LANDSAT/LE07/C01/T1_SR); these Landsat data collections have been converted from raw digital numbers to Top of Atmosphere (TOA) reflectance using the methods and band specific irradiance values from Chander et al. [40].

Atmospheric correction is an important process for the remote sensing of water quality as water-leaving radiance constitutes a small fraction of the total energy measured by the sensor, with the main contribution coming from the atmosphere [41]. Studies have found that image-based [42–44], site-specific [41,45,46], and radiative transfer model [47–49] atmospheric correction methods can provide adequate retrievals of surface reflectance for water quality mapping. The Landsat data used in this study have been atmospherically corrected using the Earth Resources Observation and Science (EROS) Center Science Processing Architecture (ESPA) surface reflectance processing system. The ESPA surface reflectance processing system uses the Landsat ecosystem disturbance adaptive processing system (LEDAPS) algorithm [50] to atmospherically correct data from Landsat 4–7. The LEDAPS processing algorithm is built upon the 6S atmospheric correction model [51,52]. The 6S atmospheric correction model is a single layer radiative transfer model that enables accurate simulations of satellite observations that account for the elevation of targets, includes the modeling of a realistic molecular/aerosol/mixed atmosphere, allows for the retrieval values from Lambertian or anisotropic ground surfaces, and includes the calculation of gaseous absorption [53]. Furthermore, the 6S model is a widely used and heavily documented radiative transfer code that has been rigorously validated [54,55] and applied for remote sensing of water quality [56–58] making it a suitable atmospheric correction procedure.

Included in the ESPA surface reflectance processing is the C Function of the Mask (CFMask) algorithm [59] used to map cloud, cloud confidence, cloud shadow, and snow/ice pixels in Landsat scenes. The CFMask is an implementation of the FMask algorithm [60] written in C programming language. The CFMask code is a multi-pass algorithm that first labels pixels based on a decision trees classifier; it then uses scene-wide statistics to validate or discard the initial pixel labels.The cloud shadow mask is created by iteratively projecting clouds to the ground with multiple cloud heights. Pixels flagged as cloud or cloud shadow were masked in the Landsat data collection using the CFMask pixel QA band. To ensure that the extracted image spectra were water, the JRC water mask [39] was used to extract water only pixels for analysis. The JRC Monthly Water History v1.0 image collection (JRC/GSW1_0/MonthlyHistory) was temporally filtered for the month coinciding with each individual Landsat scene and used to mask land pixels. The JRC data is only available from 16 March 1984 to 18 October 2015 which affects data processing outside of the JRC date range. Thus, the CFMask QA band was used to extract pixels flagged as water as a secondary/backup algorithm for Landsat scenes that fall out of the JRC data availability.

After the MRC station-satellite acquisition coincidence check and data masking, the image spectra were extracted from the coinciding and masked Landsat scenes. After the preprocessing and spectra extraction, a natural logarithmic transform was applied to both the atmospherically corrected image spectra and in situ SSSC measurements to reduce skewness and make the distributions more Gaussian for linear statistical modeling. The log transformed image spectra and SSSC measurements were then used to create an empirical model to estimate SSSC from Landsat imagery.

2.4. Statistical Methods

The total sample of image spectra and SSSC measurements was sub-setted into calibration and validation samples; 70% of the data were used for calibration and 30% were used for validation. The sub-setting was completed using a Monte Carlo approach where the validation data was randomly selected. The index of the samples selected for validation were stored for 10,000 iterations. The final validation sample was selected by finding 30% of the data indices that were selected the most from all Monte Carlo simulations, data not selected for the validation sample were used within the calibration sample.

To determine the optimal band or combination of bands for the model [32], the spectral data from visible and NIR bands and all possible visible and NIR band ratios were correlated to the calibration sample. Only the bands and band ratios that had an absolute linear correlation greater than 0.50 were used to test for the best covariate for estimating SSSC. The selected bands were then used to derive an empirical model of SSSC from the calibration dataset. Five statistical models were optimized between the image spectra and SSSC measurements using the Scientific Python (SciPy) module [61,62] in a local Python environment; these models include: (1) linear; (2) exponential; (3) 2nd order polynomial; (4) 3rd order polynomial and (5) 4th order polynomial functions. Each model was tested on the selected bands using the following objective functions: coefficient of determination (R^2), sum of square error (SSE), and significance (p). The resulting model fit statistics were used to rank the best performing model based on the calibration dataset. The best performing model was then applied on the validation dataset and statistically analyzed to understand the accuracy and errors associated with estimating SSSC using the specified approach.

2.5. Online Application

The preprocessing methodology and selected empirical model of SSSC was implemented in an online application. Google's App Engine was used to host the application that relies on the GEE backend to process the imagery. The framework for requesting data, performing spatial calculations, and serving the information in a browser is provided in Figure 3. The web interface relies on the Google App Engine technology, using elements of HTML, CSS and JavaScript (or Clojure). Requests from the front-end are made by a call from JavaScript to the Python script using Asynchronous JavaScript And XML (AJAX). The GEE Python library handles requests to GEE and receives the result. The information returned to the JavaScript is displayed in the browser. Spatial information is displayed with the Google Maps Application Programming Interface (API) and graphical data is displayed with the Google Visualization API.

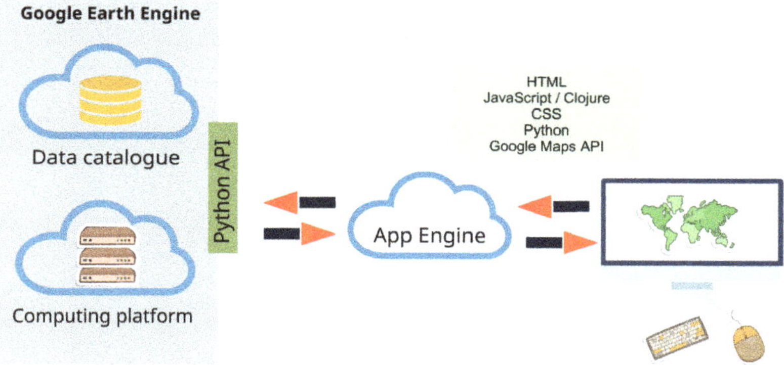

Figure 3. The infrastructure for spatial application development provided by Google. GEE consists of a cloud-based data catalogue and computing platform. The App Engine framework was used to serve the data to a web browser and communicate with the GEE using the Python API. Figure reused with permission from Poortiga et al. [63].

3. Results

3.1. Model Calibration and Validation

From the original 132 MRC water quality stations, a total of 44 stations met our quality control criteria for having collections >60 m from the bank. (Figure 4). For the 44 selected MRC stations there were relatively few cloud-free, coinciding Landsat observations within one day of in situ sampling. Out of a total 24,749 in situ samples from the MRC dataset, a total of 118 Landsat observations met the criteria for inclusion in the empirical model across the 44 stations (see Appendix A for detailed information on the stations used). Figure 4 displays the spatial, temporal, and SSSC concentration distributions from the acquired coincidence samples. Overall, in situ samples and coincident Landsat observations were recovered from a variety of locations throughout the basin, but coincident observations were most common along the Mekong mainstem and larger tributary systems (Figure 4a). In total, the calibration dataset includes 8 stations located in the the Mekong River main stem, 11 stations located in tributary systems, 22 stations from the Mekong Floodplain (which constitutes all stations downstream from Kratie [35]), and 3 stations located within reservoirs

(Table A1). The calibration and validation dataset includes 118 Landsat observations coincident with in situ SSSC measurements, 66 of these instances come from the dry season (December to May) , while 52 occurred during the wet season (June to November), providing a good representation of both the wet and dry seasons for statistical analysis (Figure 4b). The range of in situ SSSC values in the calibration dataset is 1.0–1155.0 mg$\cdot$L^{-1} for the dry season and 8.0–655.0 mg$\cdot$L^{-1} for the wet season (Figure 4c).

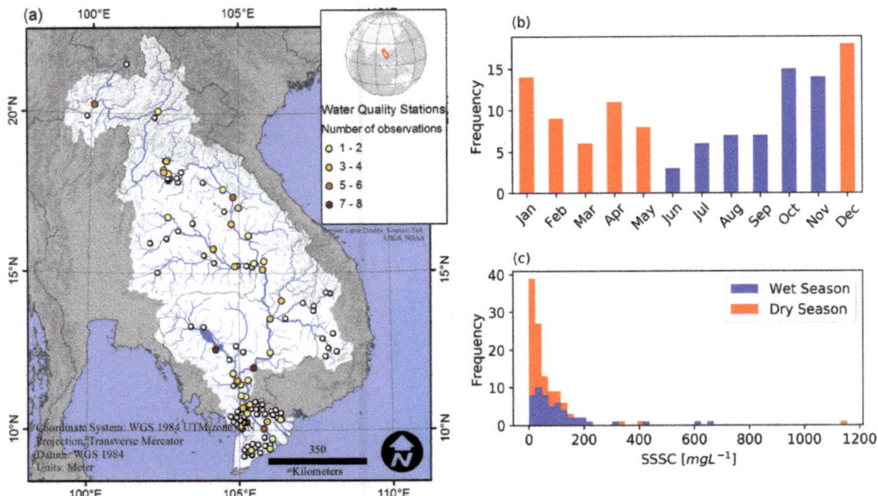

Figure 4. (**a**) Map highlighting the locations of water quality stations included in the calibration and validation dataset. Stations are coded by number of coincident Landsat 4, 5 and 7 observations from 1985 to 2011 for each station. Unused stations left out due to bank proximity or lack of coincidence Landsat data are shown in white; (**b**) Temporal distribution of Landsat observations coincident with in situ SSSC measurements in the calibration dataset (n = 189); (**c**) Distribution of in situ SSSC values in the calibration dataset.

The statistical correlation analysis between SSSC and the Landsat spectra yielded five candidate bands for the empirical model creation and further analysis based on our criteria (Table A2). The bands and band ratios with a correlation coefficient over 0.5 with the SSSC calibration dataset included: $\rho_{Green}, \rho_{Red}, \rho_{NIR}, \frac{\rho_{Blue}}{\rho_{Red}}, \frac{\rho_{Blue}}{\rho_{NIR}}$ and $\frac{\rho_{Green}}{\rho_{Red}}$. The $\frac{\rho_{Green}}{\rho_{Red}}$ band ratio yielded the highest absolute correlation of 0.73. Empirical model fitting found that a 4th order polynomial fit with the $\frac{\rho_{Red}}{\rho_{Green}}$ band ratio yielded the best fit ; however, to avoid overfitting we used an exponential model with similar fit statistics to estimate SSSC (Table A3). The fitting procedure yielded the following equation to derive SSSC from Landsat $\frac{\rho_{Red}}{\rho_{Green}}$ band ratio:

$$y = 1.904 \times e^{1.448 \cdot x + 0.630} \quad (1)$$

where y is the estimated ln(SSSC) and x is $\ln(\frac{\rho_{Red}}{\rho_{Green}})$. Equation (1) was applied to the validation sample subset and to determine the accuracy of the model (Figure 5). Model error statistics were calculated to quantify the performance of the model (Figure 5, Table 1) by converting model output from log space to actual SSSC values. The validation dataset was split into wet and dry season observations in order to assess model performance under high flow and low flow conditions (Figure 5b,c). Model error statistics such as correlation coefficient (R) and Nash-Sutcliffe model efficiency coefficient (NSE) are well within acceptable model range for modeling of water quality parameters according to Moriasi et al. [64]. However, the relative error (RE) for the entire validation dataset of the model is approximately 43%, with a RE of 44% and 41% for dry season and wet season samples respectively. While this RE is above the desired value RE of 35% set by NASA's Ocean Biology and Biogeochemistry Program [65], we are confident in our model because this relative error corresponds to a root-mean-square error (RMSE) between observed and modeled SSSC values of 17.50 mg·L^{-1} for the dry season validation dataset which ranges from 4–133 mg·L^{-1} and a RMSE of 22.50 mg·L^{-1} for the wet season validation dataset which ranges from 6.0–255.5 mg·L^{-1}. The bias of the model is −1.63 mg·L^{-1} for the entire validation dataset, indicating an underestimation of SSSC.

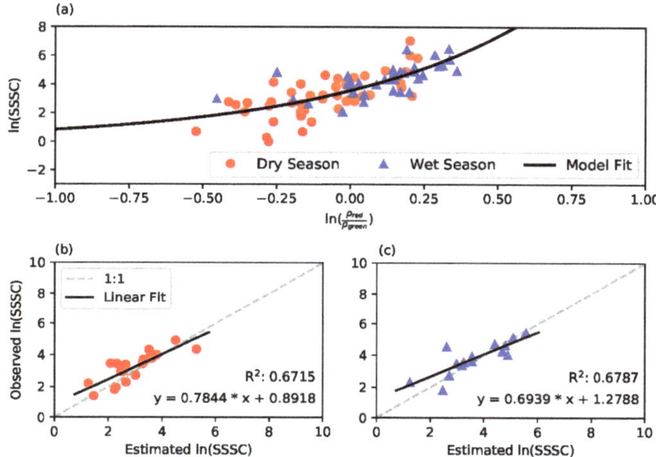

Figure 5. Model calibration using the $\frac{\rho_{Red}}{\rho_{Green}}$ band ratio with a exponential fit from Equation (1) (**a**) and validation (**b**,**c**) plots of statistical modeling for the dry (**b**) and wet (**c**) seasons.

Table 1. Error statistics for validation of empirical model compared to the validation dataset.

Season	R [-]	Bias [mg·L^{-1}]	RMSE [mg·L^{-1}]	RE [%]	NSE [-]	Validation Data Range [mg·L^{-1}]
Dry	0.82	−3.73	17.50	43.95	0.54	4.0–133.0
Wet	0.82	1.17	22.50	41.65	0.52	6.0–225.5
Total	0.84	−1.63	19.64	42.96	0.58	4.0–225.5

Previous studies of suspended sediment in the Lower Mekong Basin (using both in situ and remote sensing approaches) have focused largely on the mainstem water quality stations [8,27,34]. In this study we used all available data (mainstem and tributaries) to develop a empirical model of SSSC that can be applied throughout the entire basin. To assess model performance across a range of settings in the Lower Mekong Basin, we compared monthly median SSSC from the entire in situ dataset and all remotely sensed estimates (independent of coincidence) at the 44 stations that met our quality control criteria. To make the comparison, stations were classified into four groups that represent different environments: (a) the mainstem of the Mekong river; (b) tributary river systems; (c) Mekong Floodplain region and (d) reservoirs (Table A1). For all location groups the model accurately captures temporal dynamics of SSSC (Figure 6). However, in general, median monthly SSSC values measured in situ are higher than modeled SSSC monthly medians, particularly in August and September in the main stem and Mekong Floodplain locations. Interestingly, the model does an excellent job estimating monthly median SSSC in reservoirs throughout the year, while the largest difference in monthly median SSSC is observed in the Mekong Floodplain stations. This could indicate that optical properties of shallow water are influenced by additional factors (e.g., channel bottom, aquatic plants, chlorophyll concentration) not included in our model.

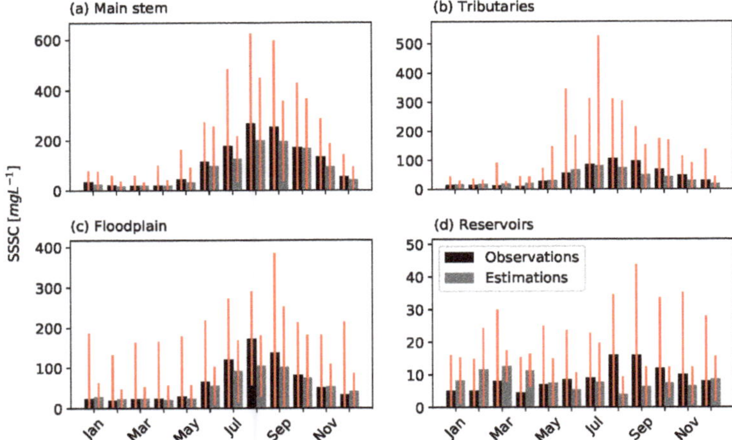

Figure 6. Monthly median estimates of SSSC from 1985–2011 against observed monthly median SCC for the 44 water quality stations used in this study. The stations were classified into four groups: (**a**) mainstem of Mekong river; (**b**) tributary river systems; (**c**) the Mekong floodplain region and (**d**) reservoirs. Red bars indicate the standard deviation of each month from all stations in the group.

3.2. Web Application

Landsat derived SSSC values are available through a simplified web-interface (Figure 7, https://mekong-ssc.appspot.com) that utilizes GEE and AppSpot technologies. Users define a time period of interest, season to process (wet season, dry season, or all months), and a region of interest. Then the application calculates the mean SSSC value for each Landsat pixel in the region of interest from all of the cloud free Landsat observations of that pixel within the selected time period. Users also have the option to generate time series charts through the web application. Calculated SSSC time series values are displayed in the web application and the calculated values are available for download. Users can download the chart as an image or the raw time series data as a CSV file. The time series option displays the mean SSSC concentration calculated for all pixels in the defined region for each time step available.

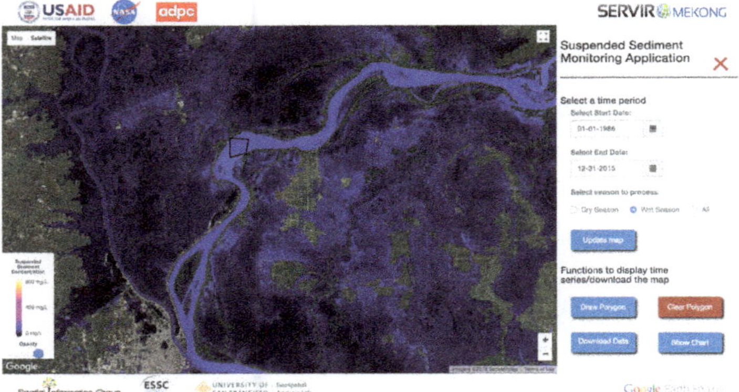

Figure 7. A screenshot of the Mekong Suspended Sediment Monitoring web application (https://mekong-ssc.appspot.com). To download a dataset, users set their desired parameters in the right panel. The region of interest can be a polygon drawn directly on the map. A mean composite for the time period is processed on the fly and is available for users to download directly.

4. Discussion

4.1. Improved Spatio-Temporal Resolution and Coverage

Our results are generally consistent with previous work which finds that exponential models using band ratios (as opposed to individual bands) are the most successful at estimating SSSC values over a broad range in concentration [16]. The 19 mg·L^{-1} RMSE of our model represents an improvement from the previously published models for the Mekong. For example, Suif et al. [34] report an RMSE ranging from 50.2 to 109.7 mg·L^{-1} for the Mekong main stem, and Wackerman et al. [17] report an RMSE of 34 mg·L^{-1} for an empirical model of SSCC in the Mekong Delta. Given that these studies were calibrated using fewer in situ observations from a smaller region within the Mekong Basin, our results illustrate that a regionally applicable model can be developed with sufficient in situ observations (n = 118 samples in this study).

The operational model developed in this study considerably increases the availability of SSSC data in the Lower Mekong Basin, making it possible to estimate SSSC for large stretches of the Mekong River and its tributaries. One impact of improved spatial coverage is that cities not located near MRC stations can now monitor SSSC through the web application (Figure 8). For example, MRC currently has water quality stations near the cities of Vientiane, Nong Khai, and Nakhon Phanom along the border of Thailand and Laos, but the city of Bueng Kan, which is located 185 km downstream of the nearest MRC station does not have a source of local water quality data. Through the online tool monthly estimates of SSSC can easily be downloaded for this region. The time series function in the online tool additionally allows users to explore how SSSC may be changing in their region.

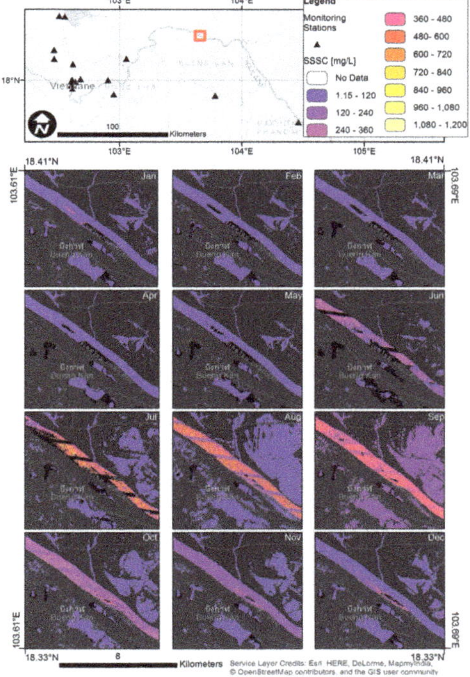

Figure 8. Estimated monthly average SSSC from 1985–2011 near Bueng Kan city located 185 km downstream of the nearest MRC station. Top map displays northeast Thailand bordering Laos with surrounding MRC water quality monitoring stations and an indication of the monthly SSSC maps. The SSSC scale in the legend refers to the monthly maps below.

In addition to expanding spatial coverage, the model also extends temporal coverage for many parts of the Mekong Basin. For example, SSSC data was collected at the Ban Chai Buri station located at the Songkhram River in Thailand (17.6422 °N, 104.4615 °E) from 2004–2011, but now with remotely sensed SSSC estimates this record is extended to the Landsat series availability (1985-present). Previous efforts to detect changes in sediment concentrations in the Mekong Basin have been difficult in part due to the limited time series of SSSC records at many stations [8,66]. The larger SSSC dataset available through this GEE web application should facilitate future investigations of SSSC trends, allowing users to compare local SSSC estimates for time periods before and after dam construction, major land use changes, or other issues of regional interest.

4.2. Model Limitations

While our model produces a spatially and temporally extensive SSSC dataset, there are limitations to remote sensing models of SSSC that must be carefully considered when analyzing the model results. First, temporal resolution is limited both by satellite repeat cycle, and by cloud cover. Cloud cover during the wet season is particularly problematic because this is when discharge and SSSC is the highest in the Lower Mekong Basin. In our study the calibration dataset includes 52 Landsat overpasses coincident with in situ observation in the wet season, but only 9 of these samples have concentrations over 200 $mg \cdot L^{-1}$ (Figure 4). As a result, the range of estimated SSSC values produced by the model is smaller than the range of SSSC values in the entire MRC database, and in particular we are missing the highest values during July and August, when in situ measurements of >200 $mg \cdot L^{-1}$ SSSC are not uncommon. Second, remote sensing techniques measure the optical properties of the top 1–2 meters of the water column, and therefore in shallow water reflectance from the bottom will significantly modify results [67]. We suspect that this is one of the reasons we see the worst match between the MRC dataset and our remotely sensed SSSC dataset for stations downstream of Kratie in the Mekong Floodplain. In this study we attempted to limit interference from the channel bottom by filtering scenes to insure they were at least 60 m from the channel edge. Development of additional filtering algorithms may be required to improve model performance in the region.

Remote Sensing of SSSC concentration can aid regional efforts to assess of water clarity, habitat conditions, nutrient transport and channel morphodynamics, but SSSC estimates cannot be used directly for sediment transport modeling because surface water contains only a fraction of the suspended sediment present over the entire water column. Moreover, the discrepancy between SSSC and depth integrated sediment concentrations will be more significant during the wet season. Particularly at high discharge, bedload transport and course material in the lower water column may constitute a significant fraction of the total sediment load. An additional constraint is that from remote sensing of SSSC alone, it is not possible to differentiate if an increase in SSSC results from mixing of sediment from the lower water column into the upper water column or an increase of suspended sediment in the whole water column. In recognition that depth integrated samples are required for robust sediment transport analysis, the MRC created the Discharge and Sediment Monitoring Program (DSMP) which collected depth integrated samples at 15 locations between 2009 and 2013 through a combination of isokinetic samplers and Acoustic Doppler Current Profilers [68]. Unfortunately, the DSMP dataset is not large enough to calibrate a remote sensing model, but future work could focus on developing rating curves to integrate this information into the model.

4.3. Implications and Future Work

The presented web-application and delivery of SSSC data derived from satellite imagery to users on an operational basis is the first of its kind. With the advent of cloud and web-based applications, such as GEE and Google AppSpot, customizable geospatial data products produced operationally are more widely available. These publicly available applications ultimately allow users greater flexibility to provide input data, filter spatially and temporally, and modify algorithm parameters to provide relevant information for analysis. The web application presented here, along with other such applications (e.g., Robinson et al. [69]), are examples of the changing paradigm from serving static geospatial products to dynamic data products.

There is great potential for this model to be adapted to estimate concentrations of additional water quality constituents, particularly those that are associated with suspended sediments. Future work on this system will focus on utilizing additional sensors, such as Sentinel-2, to improve temporal resolution of satellite observations. Sentinel-2 has been shown to provide high quality estimates of suspended sediments and other water quality parameters [70] and data from Sentinel-2′ has already been successfully integrated with Landsat for monitoring water quality [46]. Furthermore, although remote sensing relies on optically active water constituents, recent studies have explored the utility of remote sensing technologies for optically inactive water quality parameters. For example, Wu et al. [71] used Landsat TM data to empirically estimate total phosphorus concentrations in a riverine environment. The fidelity of such methodologies for estimating other optically inactive water quality parameters, such as nitrogen or pH, has not yet been explored in the Lower Mekong Basin. Expansion of the model and web application to include both optically active and inactive water quality parameters will ultimately provide a more holistic view of water quality and aquatic ecosystem health within the Mekong Basin.

5. Conclusions

The objective of this study was to increase spatial-temporal density of SSSC data in the Lower Mekong Basin by integrating remote sensing observations with in situ measurements. The model developed in this study allows for a consistent and reliable indicator of surface sediment concentrations in an operational near real-time environment. Expanded access to sediment concentration data should be particularly helpful to resource managers interested in the dynamics between sediment concentrations and land use conversions, water policy, and energy production in a globally strategic region.

Supplementary Materials: The online suspended sediment concentration application is available at: http://mekong-ssc.appspot.com/; All source code for the website and processing is available at: https://github.com/KMarkert/mekong-ssc-gae/.

Author Contributions: K.N.M., R.E.G., A.I.F., C.M.S. and D.S.S. designed the methodology; K.K. and P.S. (MRC) provided in situ water quality data and analysis, K.N.M., C.M.S., R.E.M. performed the data processing; K.N.M., A.P., N.E.C. and C.M.S. designed and developed the web application, K.N.M., C.M.S., D.S.S., and A.P. wrote the manuscript; all authors discussed and reviewed the manuscript.

Acknowledgments: The authors would like to thank the MRC for supplying the field data used in this study. Also, the authors would like to thank the Google Earth Engine team for their support and allowing access and use of the Earth Engine platform. Thanks goes to the three anonymous reviewers for their comments that improved the quality of the manuscript. Support for this work was provided through the joint US Agency for International Development (USAID) and National Aeronautics and Space Administration (NASA) initiative SERVIR, particularly through the NASA Applied Sciences Capacity Building Program, NASA Cooperative Agreement NNM11AA01A.

Conflicts of Interest: The authors declare no conflict of interest.

Abbreviations

The following abbreviations are used in this manuscript:

AJAX	Asynchronous JavaScript And XML
API	Application Programming Interface
CFMask	C Function of Mask
chl-a	Chlorophyll-a
CSS	Cascading Style Sheets
EROS	Earth Resources Observation and Science
ESPA	EROS Center Science Processing Architecture

ETM+	Enhanced Thematic Mapper Plus
GEE	Google Earth Engine
HTML	Hypertext Markup Language
JRC	Joint Research Center
LEDAPS	Landsat Ecosystem Disturbance Adaptive Processing System
MRC	Mekong River Commission
NSE	Nash-Sutcliffe model efficiency coefficient
OLI	Operational Land Imager
R	Correlation Coefficient
R^2	Coefficient of Determination
RE	Relative Error
RMSE	Root Mean Square Error
SciPy	Scientific Python
SR	Surface Reflectance
SSSC	Surface Suspended Sediment Concentration
SSE	Sum of Square Error
SSL	Suspended Sediment Load
TM	Thematic Mapper
TOA	Top of Atmosphere
TSS	Total Suspended Solids
QA	Quality Assurance

Appendix A. MRC Station Information

Detailed information describing the stations used in this study is presented in Table A1.

Table A1. Table describing the 44 stations used in the study.

Station ID	Name	Lat.	Lon.	Waterbody	Type	Observations
H010501	Chiang Sean	20.2755	100.090986	Mekong River	Main stem	5
H013101	Nakhon Phanom	17.399339	104.800227	Mekong River	Main stem	6
H013401	Savannakhet	16.559887	104.743416	Mekong River	Main stem	3
H013801	Khong Chiam	15.32027	105.5	Mekong River	Main stem	2
H013900	Pakse	15.120617	105.78276	Mekong River	Main stem	3
H014501	Stung Treng	13.547263	106.015905	Mekong River	Main stem	4
H014901	Kratie	12.47827	106.015	Mekong River	Main stem	2
H019801	Chroy Chang Var	11.58483	104.9425	Mekong River	Main stem	4
H019802	Kampong Cham	12.001609	105.46783	Mekong River	Floodplain	8
H019804	My Thuan	10.277645	105.906339	Mekong River	Floodplain	4
H019805	My Tho	10.3444	106.35056	Mekong River	Floodplain	2
H019806	Neak Luong	11.59511	105.28694	Mekong River	Floodplain	1
H019807	Krom Samnor	11.069384	105.208978	Mekong River	Floodplain	1
H020101	Phnom Penh Port	11.57316	104.93167	Tonle Sap River	Floodplain	3
H020102	Prek Kdam	11.81319	104.8	Tonle Sap River	Floodplain	1
H020106	Kampong Luong	12.579697	104.213025	Tonle Sap Lake	Floodplain	7
H029812	Dai Ngai	9.733668	106.075538	Bassac River	Floodplain	2
H033401	Takhmao	11.564623	104.934383	Bassac River	Floodplain	3
H033402	Koh Khel	11.456477	105.039326	Bassac River	Floodplain	2
H033403	Khos Thom	11.105372	105.061034	Mekong River	Floodplain	3
H039801	Chau Doc	10.710065	105.124479	Bassac River	Floodplain	5
H039803	Can Tho	10.053218	105.800404	Bassac River	Floodplain	5
H039805	My Tho	10.35145	106.368236	Mekong River	Floodplain	1
H100101	Ban Hat Kham	20.084313	102.258378	Nam Ou River	Tributary	1
H230102	Tha Ngon	18.133752	102.621123	Nam Ngum River	Tributary	1
H230199	Nam Ngum at Damsite	18.53229	102.55333	Nam Ngum Reservoir	Reservoir	4
H231801	Nam Souang	18.25215	102.55333	Souang River	Tributary	2
H231901	Nam Houm	18.178159	102.55333	Nam Houm Reservoir	Reservoir	4

Table A1. *Cont.*

Station ID	Name	Lat.	Lon.	Waterbody	Type	Observations
H290103	Ban Chai Buri	17.641781	104.461561	Nam Songkhram	Tributary	1
H320101	Se Bang Fai	17.076623	104.983587	Se Bang Fai	Tributary	2
H350101	Ban Keng Done	16.18774	105.316627	Se Bang Hieng	Tributary	2
H370104	Yasothon	15.783679	104.138624	Nam Chi	Tributary	4
H370299	Nam Pong Dam	16.77213	102.618581	Nam Pong	Reservoir	2
H380103	Ubon	15.223357	104.861663	Nam Mun	Tributary	4
H380128	Mun (Khong Chiam)	15.32194	105.51	Mekong River	Main stem	1
H390104	Souvanna Khili	15.385382	105.823818	Se Done	Tributary	1
H430102	Siempang	14.12097	106.3933	Se Kong	Tributary	3
H450101	Lumphat	13.552984	106.528211	Sre Pok	Tributary	1
H988102	Tan Thanh	10.81751	105.59028	Hong Ngu Canal	Floodplain	2
H988114	Tu Thuong	10.825895	105.339373	Tu Thuong Canal	Floodplain	2
H988202	My Thanh	9.429292	105.998322	My Thanh Canal	Floodplain	2
H988214	Phuoc Sinh	9.38372	105.38333	Quan Lo-Phung Hiep	Floodplain	2
H988302	Ba The	10.54331	105.25694	Kinh Ba The Canal	Floodplain	2
H988314	Soc Xoai	10.13242	105.02889	Rach Gia-Ha Tien	Floodplain	2

Appendix B. Statistical Exploration Results

Resulting statistical data from correlation analysis between remotely sensed wave length bands (Table A2) and the comparison between statistical models for estimating SSC (Table A3).

Table A2. Correlation matrix for the log spectral values of each band and ln(SSSC).

	ρ_{Blue}	ρ_{Green}	ρ_{Red}	ρ_{NIR}	$\frac{\rho_{Blue}}{\rho_{Green}}$	$\frac{\rho_{Blue}}{\rho_{Red}}$	$\frac{\rho_{Blue}}{\rho_{NIR}}$	$\frac{\rho_{Green}}{\rho_{Red}}$	$\frac{\rho_{Green}}{\rho_{NIR}}$	$\frac{\rho_{Red}}{\rho_{NIR}}$	SSSC [mg·L^{-1}]
ρ_{Blue}	1										
ρ_{Green}	0.911	1									
ρ_{Red}	0.826	0.961	1								
ρ_{NIR}	0.750	0.75	0.809	1							
ρ_{Blue}/ρ_{Green}	0.023	0.432	0.522	0.240	1						
ρ_{Blue}/ρ_{Red}	0.367	0.677	0.827	0.588	0.840	1					
ρ_{Blue}/ρ_{NIR}	0.230	0.345	0.468	0.816	0.334	0.544	1				
ρ_{Green}/ρ_{Red}	0.537	0.722	0.886	0.720	0.578	0.928	0.590	1			
ρ_{Green}/ρ_{NIR}	0.234	0.183	0.275	0.762	0.069	0.221	0.917	0.380	1		
ρ_{Red}/ρ_{NIR}	0.101	0.274	0.283	0.335	0.445	0.367	0.581	0.246	0.803	1	
SSSC [mg L^{-1}]	0.396	0.557	0.666	0.654	0.486	0.705	0.616	0.727	0.447	0.001	1

Table A3. Statistical modeling results for each combination of bands/ratios selected and fitting functions.

Band	Model	R^2	SSE	p
ρ_{Green}	Linear	0.305	98.84	<0.01
	Exponential	0.324	96.16	<0.01
	2nd order Polynomial	0.332	94.91	<0.01
	3rd order Polynomial	0.364	90.45	<0.01
	4th order Polynomial	0.372	89.37	<0.01
ρ_{Red}	Linear	0.412	83.71	<0.01
	Exponential	0.453	77.93	<0.01
	2nd order Polynomial	0.468	75.67	<0.01
	3rd order Polynomial	0.472	75.08	<0.01
	4th order Polynomial	0.476	74.50	<0.01

Table A3. *Cont.*

Band	Model	R^2	SSE	p
ρ_{NIR}	Linear	0.384	87.52	<0.01
	Exponential	0.370	89.74	<0.01
	2nd order Polynomial	0.385	87.52	<0.01
	3rd order Polynomial	0.386	87.35	<0.01
	4th order Polynomial	0.388	87.00	<0.01
ρ_{Red}/ρ_{Blue}	Linear	0.430	81.15	<0.01
	Exponential	0.430	81.15	<0.01
	2nd order Polynomial	0.437	80.08	<0.01
	3rd order Polynomial	0.474	74.79	<0.01
	4th order Polynomial	0.476	74.55	<0.01
ρ_{Blue}/ρ_{NIR}	Linear	0.306	98.51	<0.01
	Exponential	0.289	101.14	<0.01
	2nd order Polynomial	0.315	97.51	<0.01
	3rd order Polynomial	0.322	96.47	<0.01
	4th order Polynomial	0.323	96.29	<0.01
ρ_{Red}/ρ_{Green}	Linear	0.463	76.45	<0.01
	Exponential	0.494	71.94	<0.01
	2nd order Polynomial	0.502	70.90	<0.01
	3rd order Polynomial	0.504	70.57	<0.01
	4th order Polynomial	0.505	70.46	<0.01

References

1. Syvitski, J.P.M.; Voosmarty, C.J.; Kettner, A.J.; Green, P. Impact of Humans on the Flux of Terrestrial Sediment to the Global Coastal Ocean. *Science* **2005**, *308*, 376–380, doi:10.1126/science.1109454. [CrossRef] [PubMed]
2. Anthony, E.J.; Brunier, G.; Besset, M.; Goichot, M.; Dussouillez, P.; Nguyen, V.L. Linking rapid erosion of the Mekong River delta to human activities. *Sci. Rep.* **2015**, *5*, 14745. [CrossRef] [PubMed]
3. Mekong River Commission. *State of the Basin Report: 2005*; Mekong River Commission: Phnom Penh, Cambodia, 2005.
4. Milliman, J.; Syvitski, J.P.M. Geomorphic/Tectonic Control of Sediment Discharge to the Ocean: The Importance of Small Mountainous Rivers. *J. Geol.* **1992**, *100*, 525–544, doi:10.1086/629606. [CrossRef]
5. Kondolf, G.M.; Rubin, Z.K.; Minear, J.T. Dams on the Mekong: Cumulative sediment starvation. *Water Resour. Res.* **2014**, *50*, doi:10.1002/ 2013WR014651. [CrossRef]
6. Kummu, M.; Varis, O. Sediment-related impacts due to upstream reservoir trapping, the Lower Mekong River. *Geomorphology* **2007**, *85*, 275–293. [CrossRef]
7. Kummu, M.; Lu, X.; Wang, J.J.; Varis, O. Basin-wide sediment trapping efficiency of emerging reservoirs along the Mekong. *Geomorphology* **2010**, *119*, 181–197, doi:10.1016/j.geomorph.2010.03.018. [CrossRef]
8. Walling, D.E. The Changing Sediment Load of the Mekong River. *Ambio* **2008**, *37*, 150–157. [CrossRef]
9. Mekong River Commission. *State of the Basin Report: 2010*; Mekong River Commission: Phnom Penh, Cambodia, 2010.
10. Ritchie, J.C.; Zimba, P.V.; Everitt, J.H. Remote Sensing Techniques to Assess Water Quality. *Photogramm. Eng. Remote Sens.* **2003**, *69*, 695–704. [CrossRef]
11. Wang, F.; Han, L.; Kung, H.T.; van Arsdale, R. Applications of Landsat-5 TM imagery in assessing and mapping water quality in Reelfoot Lake, Tennessee. *Int. J. Remote Sens.* **2006**, *27*, 5269–5283. [CrossRef]
12. Park, E.; Latrubesse, E.M. Modeling suspended sediment distribution patterns of the Amazon River using MODIS data. *Remote Sens. Environ.* **2014**, *147*, 232–242. [CrossRef]
13. Umar, M.; Rhoads, B.L.; Greenberg, J.A. Use of multispectral satellite remote sensing to assess mixing of suspended sediment downstream of large river confluences. *J. Hydrol.* **2018**, *556*, 325–338. [CrossRef]
14. Myint, S.W.; Walker, N.D. Quantification of surface suspended sediments along a river dominated coast with NOAA AVHRR and Sea WiFS measurements: Louisiana, USA. *Int. J. Remote Sens.* **2002**, *23*, 3229–3249, doi:10.1080/01431160110104700. [CrossRef]

15. Matthews, M.W. A current review of empirical procedures of remote sensing in Inland and near-coastal transitional waters. *Int. J. Remote Sens.* **2011**, *32*, 6855–6899, doi:10.1080/01431161.2010.512947. [CrossRef]
16. Long, C.M.; Pavelsky, T.M. Remote sensing of suspended sediment concentration and hydrologic connectivity in a complex wetland environment. *Remote Sens. Environ.* **2013**, *129*, 197–209. [CrossRef]
17. Wackerman, C.; Hayden, A.; Jonik, J. Deriving spatial and temporal context for point measurements of suspended-sediment concentration using remote-sensing imagery in the Mekong Delta. *Cont. Shelf Res.* **2017**, *147*, 231–245, doi:10.1016/j.csr.2017.08.007. [CrossRef]
18. Dekker, A.; Vos, R.; Peters, S. Comparison of remote sensing data, model results and in situ data for total suspended matter (TSM) in the southern Frisian lakes. *Sci. Total Environ.* **2001**, *268*, 197–214. [CrossRef]
19. Laanen, M.L. *Yellow Matters: Improving the Remote Sensing of Coloured Dissolved Organic Matter in Inland Freshwaters*; Water Insight B.V.: Wageningen, The Netherlands, 2007.
20. Tilstone, G.H.; Peters, S.W.; van der Woerd, H.J.; Eleveld, M.A.; Ruddick, K.; Schönfeld, W.; Krasemann, H.; Martinez-Vicente, V.; Blondeau-Patissier, D.; Röttgers, R.; et al. Variability in specific-absorption properties and their use in a semi-analytical ocean colour algorithm for MERIS in North Sea and Western English Channel Coastal Waters. *Remote Sens. Environ.* **2012**, *118*, 320–338. [CrossRef]
21. Cox, R.M.; Forsythe, R.D.; Vaughan, G.E.; Olmsted, L. Assessing water quality in Catawba River reservoirs using Landsat Thematic Mapper satellite data. *Lake Reserv. Manag.* **1998**, *14*, 405–416. [CrossRef]
22. Dekker, A.G.; Vos, R.; Peters, S. Analytical algorithms for lake water TSM estimation for retrospective analyses of TM and SPOT sensor data. *Int. J. Remote Sens.* **2002**, *23*, 15–35. [CrossRef]
23. Brezonik, P.; Menken, K.D.; Bauer, M. Landsat-based remote sensing of lake water quality characteristics, including chlorophyll and colored dissolved organic matter (CDOM). *Lake Reserv. Manag.* **2005**, *21*, 373–382. [CrossRef]
24. Curran, P.; Hansom, J.; Plummer, S.; Pedley, M. Multispectral remote sensing of nearshore suspended sediments: A pilot study. *Int. J. Remote Sens.* **1987**, *8*, 103–112. [CrossRef]
25. Novo, E.; Hansom, J.; Curran, P. The effect of viewing geometry and wavelength on the relationship between reflectance and suspended sediment concentration. *Int. J. Remote Sens.* **1989**, *10*, 1357–1372. [CrossRef]
26. Hellweger, F.; Schlosser, P.; Lall, U.; Weissel, J. Use of satellite imagery for water quality studies in New York Harbor. *Estuar. Coast. Shelf Sci.* **2004**, *61*, 437–448. [CrossRef]
27. Fleifle, A. Suspended Sediment Load Monitoring Along the Mekong River from Satellite Images. *J. Earth Sci. Clim. Chang.* **2013**, *4*, doi:10.4172/2157-7617.1000160. [CrossRef]
28. Papoutsa, C.; Retalis, A.; Toulios, L.; Hadjimitsis, D. Defining the Landsat TM/ETM+ and chris/proba spectral regions in which turbidity can be retrieved in inland water bodies using eld spectroscopy. *Int. J. Remote Sens.* **2014**, *35*, 1674–1692. [CrossRef]
29. Overeem, I.; Hudson, B.D.; Syvitski, J.P.M.; Mikkelsen, A.B.; Hasholt, B.; van den Broeke, M.R.; Noël, B.P.Y.; Morlighem, M. Substantial export of suspended sediment to the global oceans from glacial erosion in Greenland. *Nat. Geosci.* **2017**, *10*, 859, doi:/10.0.4.14/ngeo3046. [CrossRef]
30. Nechad, B.; Ruddick, K.; Park, Y. Calibration and validation of a generic multisensor algorithm for mapping of total suspended matter in turbid waters. *Remote Sens. Environ.* **2010**, *110*, 854–866. [CrossRef]
31. Feng, L.; Hu, C.; Chen, X.; Song, Q. Influence of the Three Gorges Dam on total suspended matters in the Yangtze Estuary and its adjacent coastal waters: Observations from MODIS. *Remote Sens. Environ.* **2014**, *140*, 779–788. [CrossRef]
32. Gholizadeh, M.H.; Melesse, A.M.; Reddi, L. A Comprehensive Review on Water Quality Parameters Estimation Using Remote Sensing Techniques. *Sensors* **2016**, *16*, 1298, doi:10.3390/s16081298. [CrossRef] [PubMed]
33. Bowers, D.; Binding, C. The optical properties of mineral suspended particles: A review and synthesis. *Estuar. Coast. Shelf Sci.* **2006**, *67*, 219–230, doi:10.1016/j.ecss.2005.11.010. [CrossRef]
34. Suif, Z.; Fleifle, A.; Yoshimura, C.; Saavedra, O. Spatio-temporal patterns of soil erosion and suspended sediment dynamics in the Mekong River Basin. *Sci. Total Environ.* **2016**, *568*, 933–945, doi:10.1016/j.scitotenv.2015.12.134. [CrossRef] [PubMed]
35. Dang, T.D.; Cochrane, T.A.; Arias, M.E. Quantifying sediment dynamics in mega deltas using remote sensing data: A case study of the Mekong floodplains. *Int. J. Appl. Earth Obs.* **2018**, *68*, 105–115. [CrossRef]

36. Bui, Y.T.; Orange, D.; Visser, S.M.; Hoanh, C.T.; Laissus, M.; Poortinga, A.; Tran, D.T.; Stroosnijder, L. Lumped surface and sub-surface runoff for erosion modeling within a small hilly watershed in northern Vietnam. *Hydrol. Process.* **2014**, *28*, 2961–2974, doi:10.1002/hyp.9860. [CrossRef]
37. Gorelick, N.; Hancher, M.; Dixon, M.; Ilyushchenko, S.; Thau, D.; Moore, R. Google Earth Engine: Planetary-scale geospatial analysis for everyone. *Remote Sens. Environ.* **2017**, *202*, 18–27. [CrossRef]
38. Mekong River Commission. *Hydrological/Water Quality Database*; Mekong River Commission: Phnom Penh, Cambodia, 2011.
39. Pekel, J.F.; Cottem, A.; Gorelick, N.; Belward, A.S. High-resolution mapping of global surface water and its long-term changes. *Nature* **2016**, *540*, 418–422, doi:10.1038/nature20584. [CrossRef] [PubMed]
40. Chander, G.; Markham, L.; Halder, D.L. Summary of current radiometric calibration coefficients for Landsat MSS, TM, ETM+, and EO-1 ALI sensors. *Remote Sens. Environ.* **2009**, *113*, 893–903. [CrossRef]
41. Dash, P.; Walker, N.; adn Eurico D'Sa, D.M.; Ladner, S. Atmospheric Correction and Vicarious Calibration of Oceansat-1 Ocean Color Monitor (OCM) Data in Coastal Case 2 Waters. *Remote Sens.* **2012**, *4*, 1716–1740, doi:10.3390/rs4061716. [CrossRef]
42. Tian, L.; Wai, O.W.H.; Chen, X.; Liu, Y.; Feng, L.; Li, J.; Huang, J. Assessment of Total Suspended Sediment Distribution under Varying Tidal Conditions in Deep Bay: Initial Results from HJ-1A/1B Satellite CCD Images. *Remote Sens.* **2014**, *6*, 9911–9929, doi:10.3390/rs6109911. [CrossRef]
43. Barrett, D.C.; Frazier, A.E. Automated Method for Monitoring Water Quality Using Landsat Imagery. *Water* **2016**, *8*, 257, doi:10.3390/w8060257. [CrossRef]
44. Liu, H.; Li, Q.; Shi, T.; Hu, S.; Wu, G.; Zhou, Q. Application of Sentinel 2 MSI Images to Retrieve Suspended Particulate Matter Concentrations in Poyang Lake. *Remote Sens.* **2017**, *9*, 761, doi:10.3390/rs9070761. [CrossRef]
45. Carswell, T.; Costa, M.; Young, E.; Komick, N.; Gower, J.; Sweeting, R. Evaluation of MODIS-Aqua Atmospheric Correction and Chlorophyll Products of Western North American Coastal Waters Based on 13 Years of Data. *Remote Sens.* **2017**, *9*, 1063, doi:10.3390/rs9101063. [CrossRef]
46. Page, B.P.; Kumar, A.; Mishra, D.R. A novel cross-satellite based assessment of the spatio-temporal development of a cyanobacterial harmful algal bloom. *Int. J. Appl. Earth Obs.* **2018**, *66*, 69–81. [CrossRef]
47. Watanabe, F.S.Y.; Alcântara, E.; Rodrigues, T.W.P.; Imai, N.N.; Barbosa, C.C.F.; da Silva Rotta, L.H. Estimation of Chlorophyll-a Concentration and the Trophic State of the Barra Bonita Hydroelectric Reservoir Using OLI/Landsat-8 Images. *Int. J. Environ. Res. Public Health* **2015**, *12*, 10391–10417. [CrossRef] [PubMed]
48. Rotta, L.H.; Alcântara, E.H.; Watanabe, F.S.; Rodrigues, T.W.; Imai, N.N. Atmospheric correction assessment of SPOT-6 image and its influence on models to estimate water column transparency in tropical reservoir. *Remote Sensing Applications: Society and Environment* **2016**, *4*, 158–166, doi:10.1016/j.rsase.2016.09.001. [CrossRef]
49. Pahlevan, N.; Schott, J.R.; Franz, B.A.; Zibordi, G.; Markham, B.; Bailey, S.; Schaaf, C.B.; Ondrusek, M.; Greb, S.; Strait, C.M. Landsat 8 remote sensing reflectance (Rrs) products: Evaluations, intercomparisons, and enhancements. *Remote Sens. of Environ.* **2017**, *190*, 289–301, doi:10.1016/j.rse.2016.12.030. [CrossRef]
50. Masek, J.G.; Vermote, E.F.; Saleous, N.; Wolfe, R.; Hall, F.G.; Huemmrich, F.; Gao, F.; Kutler, J.; Lim, T.K. A Landsat surface reflectance data set for North America, 1990-2000. *IEEE Geosci. Remote Sci.* **2006**, *3*, 68–72. [CrossRef]
51. Vermote, E.F.; Saleous, N.; Justice, C.O.; Kaufman, Y.J.; Privette, J.L.; Remer, L.; Roger, J.C.; Tanre, D. Atmospheric correction of visible to middle-infrared EOS-MODIS data over land surfaces: Background, operational algorithm, and validation. *J. Geophys. Res.* **1997**, *102*, 17131–17141. [CrossRef]
52. Vermote, E.F.; Saleous, N.; Justice, C.O. Atmospheric correction of MODIS data in the visible to middle infrared: First results. *Remote Sens. Environ.* **2002**, *83*, 97–111. [CrossRef]
53. Vermote, E.F.; Tanre, D.; Deuze, J.L.; Herman, M.; Morcrette, J.J. Second Simulation of the Satellite Signal in the Solar Spectrum, 6S: An Overview. *IEEE Trans. Geosci. Remote Sens.* **1997**, *35*, 675–686. [CrossRef]
54. Kotchenova, S.Y.; Vermote, E.F.; Mataresse, R.; Frank, J.K., Jr. Validation of a vector version of the 6S radiative transfer code for atmospheric correction of satellite data. Part I: Path Radiance. *Appl. Opt.* **2006**, *45*, 6726–6774. [CrossRef]
55. Kotchenova, S.Y.; Vermote, E.F. Validation of a vector version of the 6S radiative transfer code for atmospheric correction of satellite data. Part II: Homogeneous Lambertian and anisotropic surfaces. *Appl. Opt.* **2007**, *46*, 4455–4464. [CrossRef] [PubMed]

56. Potes, M.; Costa1, M.J.; Salgado, R. Satellite remote sensing of water turbidity in Alqueva reservoir and implications on lake modelling. *Hydrol. Earth Syst. Sci.* **2012**, *16*, 1623–1633, doi:10.5194/hess-16-1623-2012. [CrossRef]
57. Giardino, C.; Bresciani, M.; Cazzaniga, I.; Schenk, K.; Rieger, P.; Braga, F.; Matta, E.; Brando, V.E. Evaluation of Multi-Resolution Satellite Sensors for Assessing Water Quality and Bottom Depth of Lake Garda. *Sensors* **2014**, *14*, 24116–24131, doi:10.3390/s141224116. [CrossRef] [PubMed]
58. Shang, P.; Shen, F. Atmospheric Correction of Satellite GF-1/WFV Imagery and Quantitative Estimation of Suspended Particulate Matter in the Yangtze Estuary. *Sensors* **2016**, *16*, 1997, doi:10.3390/s16121997. [CrossRef] [PubMed]
59. Foga, S.; Scaramuzza, P.L.; Guo, S.; Zhu, Z.; Dilley, R.D.; Beckmann, T.; Schmidt, G.L.; Dwyer, J.L.; Hughes, M.J.; Laue, B. Cloud detection algorithm comparison and validation for operational Landsat data products. *Remote Sens. Environ.* **2017**, *194*, 379–390. [CrossRef]
60. Zhu, Z.; Wang, S.; Woodcock, C.E. Improvement and expansion of the Fmask algorithm: cloud, cloud shadow, and snow detection for Landsats 4–7, 8, and Sentinel 2 images. *Remote Sens. Environ.* **2015**, *159*, 269–277, doi:10.1016/j.rse.2014.12.014. [CrossRef]
61. Oliphant, T.E. Python for Scientific Computing. *Comput. Sci. Eng.* **2007**, *9*, 10–20, doi:10.1109/MCSE.2007.58. [CrossRef]
62. Millman, K.J.; Aivazis, M. Python for Scientists and Engineers. *Comput. Sci. Eng.* **2011**, *13*, 9–12. [CrossRef]
63. Poortinga, A.; Clinton, N.; Saah, D.; Cutter, P.; Chishtie, F.; Markert, K.N.; Anderson, E.R.; Troy, A.; Fenn, M.; Tran, L.H.; Bean, B.; Nguyen, Q.; Bhandari, B.; Johnson, G.; Towashiraporn, P. An operational Before-After-Control-Impact (BACI) designed platform for vegetation monitoring at planetary scale. *Remote Sens.* **2018**, *10*, 760, doi:10.3390/rs10050760. [CrossRef]
64. Moriasi, D.N.; Arnold, J.G.; Van Liew, M.W.; Bingner, R.L.; Harmel, R.D.; Veith, T.L. Model evaluation guidelines for systematic quantification of accuracy in watershed simulations. *Trans. ASABE* **2007**, *50*, 885–900. [CrossRef]
65. McCain, C.; Hooker, S.; Feldman, G.; Bontempi, P. Satellite data for ocean biology, biogeochemistry, and climate research. *Eos Trans. Am. Geophys. Union* **2006**, *87*, 337–343. [CrossRef]
66. Wang, J.; Lu, X.X.; Kummu, M. Sediment load estimates and variations in the Lower Mekong River. *River Res. Appl.* **2011**, *27*, 33–46, doi:10.1002/rra.1337. [CrossRef]
67. Volpe, V.; Silvestri, S.; Marani, M. Remote sensing retrieval of suspended sediment concentration in shallow waters. *Remote Sens. Environ.* **2011**, *115*, 44–54, doi:10.1016/j.rse.2010.07.013. [CrossRef]
68. Koehnken, L. *Discharge Sediment Monitoring Project (DSMP) 2009–2013 Summary and Analysis of Results: Final Report*; Mekong River Commission: Vientiane, Lao PDR, 2014.
69. Robinson, N.P.; Allred, B.W.; Jones, M.W.; Moreno, A.; Kimball, J.S.; Naugle, D.E.; Erikson, T.A.; Richardson, A.D. A Dynamic Landsat Derived Normalized Difference Vegetation Index (NDVI) Product for the Conterminous United States. *Remote Sens.* **2017**, *9*, 863, doi:10.3390/rs9080863. [CrossRef]
70. Toming, K.; Kutser, T.; Laas, A.; Sepp, M.; Paavel, B.; Noges, T. First Experiences in Mapping Lake Water Quality Parameters with Sentinel-2 MSI Imagery. *Remote Sens.* **2016**, *8*, doi:10.3390/rs8080640. [CrossRef]
71. Wu, C.; Wu, J.; Qi, J.; Zhang, L.; Huang, H.; Lou, L.; Chen, Y. Empirical estimation of total phosphorus concentration in the mainstream of the Qiantang River in China using Landsat TM data. *Remote Sens. Environ.* **2010**, *31*, 2309–2324. [CrossRef]

Letter

Mapping Mining Areas in the Brazilian Amazon Using MSI/Sentinel-2 Imagery (2017)

Felipe de Lucia Lobo [1,2,*], Pedro Walfir M. Souza-Filho [1,3], Evlyn Márcia Leão de Moraes Novo [2], Felipe Menino Carlos [4] and Claudio Clemente Faria Barbosa [4]

1. Instituto Tecnológico Vale, Rua Boaventura da Silva, 955, Belém, PA 66055-090, Brazil; pedro.martins.souza@itv.org
2. Remote Sensing Division, National Institute for Space Research (INPE), Av. dos Astronautas 1758, São José dos Campos, SP 12227-010, Brazil; evlyn.novo@inpe.br
3. Geosciences Institute, Universidade Federal do Pará, Rua Augusto Correa 1, Belém, PA 66075-110, Brazil
4. Image Processing Division, National Institute for Space Research (INPE), Av. dos Astronautas 1758, São José dos Campos, SP 12227-010, Brazil; felipe.carlos@fatec.sp.gov.br (F.M.C.); claudio.barbosa@inpe.br (C.C.F.B.)

* Correspondence: felipe.lobo@inpe.br; Tel.: +55-12-3208-6810

Received: 27 June 2018; Accepted: 20 July 2018; Published: 25 July 2018

Abstract: Although mining plays an important role for the economy of the Amazon, little is known about its attributes such as area, type, scale, and current status as well as socio/environmental impacts. Therefore, we first propose a low time-consuming and high detection accuracy method for mapping the current mining areas within 13 regions of the Brazilian Amazon using Sentinel-2 images. Then, integrating the maps in a GIS (Geography Information System) environment, mining attributes for each region were further assessed with the aid of the DNPM (National Department for Mineral Production) database. Detection of the mining area was conducted in five main steps. (a) MSI (MultiSpectral Instrument)/Sentinel-2A (S2A) image selection; (b) definition of land-use classes and training samples; (c) supervised classification; (d) vector editing for quality control; and (e) validation with high-resolution RapidEye images (Kappa = 0.70). Mining areas derived from validated S2A classification totals 1084.7 km^2 in the regions analyzed. Small-scale mining comprises up to 64% of total mining area detected comprises mostly gold (617.8 km^2), followed by tin mining (73.0 km^2). The remaining 36% is comprised by industrial mining such as iron (47.8), copper (55.5) and manganese (8.9 km^2) in Carajás, bauxite in Trombetas (78.4) and Rio Capim (48.5 km^2). Given recent events of mining impacts, the large extension of mining areas detected raises a concern regarding its socio-environmental impacts for the Amazonian ecosystems and for local communities.

Keywords: small-scale mining; industrial mining; google engine; image classification; land-use cover change

---

1. Introduction

Brazil is one of the leading countries in mineral production, 40% of which comes from the Amazonian states [1] generating large financial compensation for local municipalities and states. At the same time, these activities have profound impacts on the Amazon regional economy, and cause intense social conflicts [2] and environmental impacts, such as water contamination [3]. For example, a recent leak of toxic mining debris discharged by a bauxite mining company has contaminated several communities in Barcarena (Pará State) with high levels of lead, aluminum, sodium, and other toxins detected in drinking water up to 2 km downstream [3]. Another example regards gold exploitation in the Tapajós River watershed, where water has been contaminated with mercury and impacted by the siltation process due to discharges of artisanal gold-mining tailings since the 1950s [4]. Moreover, recent publications have demonstrated the influence of mining project/activities on the Land Use Cover Change (LUCC) caused by other land-use covers, such as pasture and agribusiness, indicating its importance to territory expansion as well [5,6].

Although mining has an important role for the economy of the Amazon, little is known about its attributes such as area, type, scale, status and socio/environmental impacts. One of the reasons for this absence of information is due the lack of accurate mapping of the mining area [7,8]. This lack is explained by the nature of the mining activity characterized by a diversity of techniques and scale of exploitation. Mining in the Amazon ranges from small-scale with rudimentary techniques (water jets and rafts), more mechanized exploitation with pit loaders and cyanide tanks (in the case of gold extraction) where miners organized themselves into cooperatives, to large-scale mining characterized by high mechanization at an industrial scale (ports, pipelines, roads, etc.) [9,10]. This diversity of techniques and exploitation scale causes different types of land cover, which includes barren soil, land pits, water bodies, degraded and recovering areas [7]. Moreover, a significant amount of small-scale mining occurs in small areas (<10,000 m^2) within forest land which are only detectable by medium- to high-resolution images ($\leq$20 m). Therefore, both high spectral variability and high frequency of small-scale mining areas often causes misclassification by image classifiers and interpreters because the visible features of the mines are similar to many other land cover and land-use changes such as clearance for agriculture or cattle farms [8]. Another challenge to map mining areas is that the stream network in which mining takes place can comprise of clear water, but also of highly turbid water and riverbanks, with spectral signatures similar to those of bare soil [8]. One more aspect regarding the criteria for including a given land use to the mining class area is that different land uses such as ports, airstrips, access roads, can be considered to be direct LUCC by mining activities, and therefore added to the exploitation area [7].

Considering all these difficulties and challenges, maps of mining areas are sporadic. To overcome that, research efforts have been made to map mining areas around the world using different remote sensing data and classification methods [11,12]. A recent review article indicates that classification of mining area is improved when using high-resolution imagery and applies machine learning, such as Support Vector Machine (SVM), object-oriented or decision tree to classify mining areas. To detect small-scale mining in Malaysia, for example, some researchers have applied high-resolution imagery [8] and provided high accuracy (89%) maps using object-oriented/SVM classification to detect several mining-related land-cover uses. The high accuracy, however, relied on time-consuming image processing methods. Also, the swath widths of high-resolution images are usually narrow, preventing their application to large-scale areas. For the Brazilian Amazon, the only mining area estimation publicly available is based on visual interpretation of Landsat imagery (30 m) and vector editing (TerraClass Project [13], TC2014). This product, despite the satisfactory location of mining occurrence, lacks details and is subject to misinterpretation. Moreover, due to the use of a time-demanding method, it cannot be updated frequently [12].

Given the economic and environmental importance of mining exploitation around the world, and particularly in the Amazon, the development of rapid and accurate methods for classifying mining areas is key to understanding the influence of mining activities on the regional environment [7]. In this context, this research aims to map current active mining areas (open pits) at a large scale (Brazilian Amazon) and integrate attributes regarding the mining scale and ore exploited. The first objective of this paper is to map current operational mining areas within the main mining regions of the Brazilian Amazon using Sentinel-2A images (S2A) applying a low time-consumption mapping method based on GEE (Google Earth Engine). In the face of a lack of information about the type of mineral exploited and the scale (industrial or small-scale), the second objective is the integration of data provided by Brazilian National Department for Mineral Production—DNPM (license status, mineral type among other information) to the mining map. This integration will be carried out in a GIS (Geographic Information

System) database and will allow computation of the area occupied by each mining category, providing key information for the environmental management of mining activities.

2. Materials and Methods

2.1. Mining Sites in the Brazilian Amazon

Small-scale and large-scale mining are present within numerous regions in the Brazilian Amazon, exploiting several ores such as iron, manganese, bauxite, tin, gold, nickel, and copper. Besides the different ore types, the techniques applied in the exploitation also vary considerably. Industrial iron mining in Carajás, for example, applies high-end technology throughout the whole process of surveying, exploiting, and recovering the area, resulting in high yields with minimum environmental impacts. On the other hand, small-scale gold mining in the Tapajós River, for example, uses low-end technology in ore processing (water jets and dredges) with low yields and high impact levels on land degradation and water contamination (sediment and mercury) [14–17].

To cover the variety of mining activities in the Amazon, 13 main regions for ore exploitation were considered for analysis, including both small-scale and large-scale that present specific characteristics in terms of socio-economic and territorial aspects (Figure 1). The areas were selected based on literature review [10] and other databases [1,13]. Although other areas in the Brazilian Amazon present relevant mining activities, such as the state of Roraima and north of the Amazon State (diamond and gold mining), they were not included either due to the lack of cloud-free images or to the mining process, mainly in rivers in rafts and dredges, which are hardly detectable by satellite imagery [2,18].

Figure 1 shows the study area location and the 38 Sentinel-2A tiles used for mining detection. All the images were selected with data range from July to September 2017, with exception of Taboca (AM) and Serra do Navio (AP), due to a lack of a more recent cloud-free images.

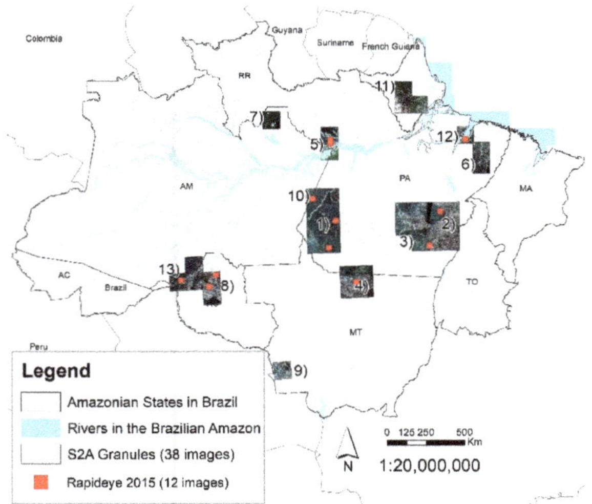

Figure 1. Thirteen mining regions in the Brazilian Amazon evaluated in this study: (1) Tapajós River; (2) Carajás; (3) Rio Xingu; (4) Peixoto de Azevedo; (5) Trombetas/Juruti; (6) Capim River; (7) Mina Taboca; (8) Ariquemes; (9) Pontes e Lacerda; (10) Amanã River; (11) Amapá; (12) Barcarena; (13) Madeira River. Indication of 38 Sentinel-2A images used for mapping mining areas and 12 RapidEye images used for validation purposes. Federal States within the Amazon region: Acre (AC), Amapá (AP), Amazonas (AM), Maranhão (MA), Mato Grosso (MT), Pará (PA), Rondônia (RO), Roraima (RR), and Tocantins (TO).

Based on a literature review [1,13,18], Table 1 provides more general information about each study area including type of ore exploited, Federal Units-State, and whether it is a large-scale (industrial) or small-scale mining (locally called "garimpos").

Table 1. Information of the mining sites regarding to location, Sentinel-2 tiles and acquisition date, ore exploited and mining scale. Mining scale varies from small-scale ("garimpos" with rudimentary techniques), medium (mines with improved techniques, pitloaders, cyanide tanks), industrial scale (large mining operations with high mechanization which can include ports, pipelines, ore refinery, etc.). Federal Units: AM—Amazonas, AP—Amapá, MT—Mato Grosso, PA—Pará, RO—Rondônia.

#	Mining Sites	F.U.	Sentinel-2 Granule	Date	RapidEye Tile_Date	Ore	Mining Scale
1	Tapajós region	PA	T21 MWQ, MVP, MWP, MVN, MWN, MVM, MWM	19 July 2017	2135416_2015-06-07 2136118_2015-07-23 2136712_2015-07-19	Gold	Small-Medium
2	Carajás region	PA	T22MEU, MFU, MET	13 and 20 July 2017	2235415_2014-07-30 2236318_2015-08-01	Iron, Copper, Manganese, Nickel	Industrial
3	Xingu/Rio Fresco	PA	T22MCU, MDU, MCT, MDT, MET, MDS, MES, MFS	13 and 20 July 2017		Gold/Tin/Nickel	Small/Industrial
4	Teles Pires/Peixoto de Az.	MT	T21LXK, LYK, LXJ, LYJ	6 July 2017	2134423_2015-07-28	Gold	Small-scale
5	Trombetas/Juruti	PA	T21MWU, MWT	1 December 2017 29 July 2017	2138317_2015-09-21 2138217_2015-09-21	Bauxite	Industrial
6	Rio Capim	PA	T22MHB, MHC	20 July 2017		Kaolinite	Industrial/Pipeline
7	Mina Taboca	AM	T20MRE	3 September 2017		Tin	Industrial
8	Ariquemes region	RO	T20LMQ, LMP	18 and 25 July 2017	2034714_2014-07-28 2034814_2015-07-02	Tin	Small/Industrial
9	Pontes e Lacerda	MT	T20LRJ	19 July 2017		Gold	Small/Industrial
10	Amanã	PA/AM	T21MVQ	19 July 2017		Gold	Small
11	Amapá/Serra do Navio	AP	T22NCF, NCG	18 November 2016		Manganese/Gold	Industrial
12	Barcarena	PA	T22MGD	20 July 2017	2238325_2015-06-28	Bauxite/Kaolinite	Industrial/Port
13	Rio Madeira	RO/AM	T20LLR, LLQ, LKQ	18 and 25 July 2017	2034705_2015-06-24	Gold	Small-scale

2.2. Classification and Validation of Mining Areas

To address the first objective, the mapping process was conducted in five main steps, three of which were carried out in the GEE platform and two in ArcGIS (ESRI, Redlands, CA, USA). In GEE platform:

(a) The first step was to select current (2017) cloud-free images from the Sentinel-2 database within the study areas (Figure 2a). A selection script was applied based on criteria of date and cloud percentage (<20%) to identify the 38 Sentinel-2 granules used (see GEE script at https://goo.gl/2S8zSp). The S2A image database available on GEE and used in this study is in digital number (DN) and not submitted to atmospheric correction.

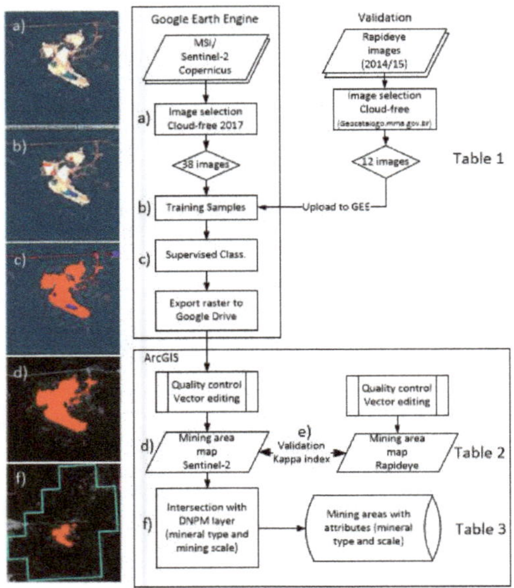

Figure 2. Flowchart with examples of steps taken to address objective 1 (**a**–**e**) and objective 2 (**f**). In GEE (Google Earth Engine): (**a**) Images selection; (**b**) training samples; (**c**) supervised classification followed by exportation. In a Geographical Information System—GIS software: (**d**) vector editing for quality control, (**e**) validation of mining areas with RapidEye classification. (**f**) Identification of mineral type as well as scale by intersecting validated mining map with mining licenses (National Department for Mineral Production—DNPM) layer.

(b) The second step was to define the land-cover classes and select training samples [17]. Initially, four land-use classes were used as reference for the classification process: dense forest, clear-cut deforestation (usually pastures with regular geometry), water bodies and barren soil (open-pit mining areas) (Figure 3a). For each study area, at least 10 small-sized polygons (~5 × 5 pixel) were used as training samples for each class. Although mining areas can include several land-use cover types, the criteria used in this study was to select only currently active areas, i.e., open-pit mining areas that present high albedo levels in comparison to other classes (Figure 3b). The forest class includes dark dense vegetation areas only and for the water class, natural rivers, and lakes, preferentially of low turbidity, were sampled for training process. Clear-cut contains areas of pasture with regular shape with little or non-vegetation (Figure 3a). These classes show spectral difference between each other (Figure 3b). For example, band 4 shows values up to 400 (DN) for water and forest and increased values for clear-cut (1500) and mining (2900), whereas band 6 shows low values for water (200), intermediate level for vegetation (1800) and clear-cut (2000) and high values for mining (3100). The spectral difference among the classes observed allowed proceeding with the classification process.

(c) The third step was to apply a supervised classification available on GEE to the selected images (Figure 3c). The classification method used for mining detection was the Classification and Regression Trees (CART), a non-parametric classifier that does not require any a priori statistical assumptions regarding the distribution of data. Recent research has compared ten machine-learning image classification methods implemented on GEE, such as CART and Random Forest, to map wetlands in Indonesia [19], and indicated that CART presented the mapping accuracy (96%) demonstrating that CART provides reliable outputs for mapping land-use cover. CART is a pixel-based classifier that uses the DN levels from the training samples (polygons) to create a decision tree that classifies each pixel of the image (20 m). Based on Figure 3c, the bands 3 (550 nm), 4 (665 nm), 6 (740 nm), 8 (1600 nm) and

11 (2200 nm) were used by CART to classify S2A imagery into mining, water, forest, and clear-cut areas. To avoid extensive misclassification with inactive mining areas that includes some vegetation mostly where small-scale mining occurs, the mining areas considered for this research included only active mining areas, i.e., barren soil (active exploitation areas) and associated water bodies. Therefore, for the following steps, only the mining and water classes were taken into consideration by removing forest and clear-cut classes from the raster classification (Figure 3c). The resulting layer was exported to Tiff file type into Google Drive. Unfortunately, the export option only provides image exportation up to 1 million pixels, hindering the export of large areas (such as the whole granule extension). This issue is even more relevant in the case of MSI/S2A imagery (10 m). To overcome this issue, the exportation step was conducted with a spatial resolution of 20 m and using subsets to areas of mining occurrence (Region of Interest, ROI).

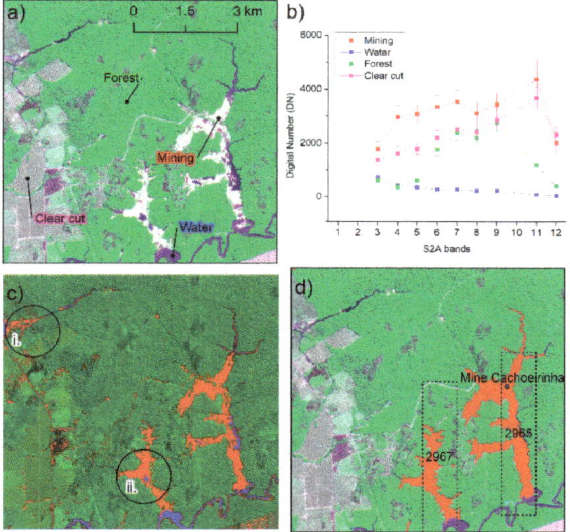

Figure 3. Illustration of the classification process for an industrial mining in Itapuã do Oeste/RO: (**a**) Sample selection for four classes used (Mining in red, water in blue, forest in green and clear-cut areas in magenta). (**b**) Average and standard deviation of Sentinel-2A spectra for 25 pixels of each class, bands 1 (443 nm), 2 (490 nm) and 10-Cirrus (1375 nm) were omitted due to high atmospheric effects. (**c**) Classification result from GEE based on CART (Classification and Regression Trees), only mining areas in red and water areas in blue are shown: (i) indicates commission areas within mining class; and (ii) indicates water bodies derived from mining activities and, therefore, integrated into mining class. (**d**) Final mining area map after vector edition with indication of Mining License Areas (dashed rectangles) used for type of ore identification (see Table 2 for details on Mining License).

Table 2. Example of information related to the Exploitation License given by DNPM (National Department for Mineral Production). Process numbers refer to the areas indicated in Figure 3d. In this case, two industrial companied have the license to exploit tin in Itapuã do Oeste (RO) with 485 ha each.

Process N.	Year	Area (ha)	Status	Last Update	Name	Ore	Usage	Federal Unit	Municipality
2965	1965	485	Active	6 February 2018	Indústria e Comércio S A	Tin (Sn)	Not informed	RO	Itapuã do Oeste
2967	1965	485	Active	7 February 2018	Ltd.a	Tin (Sn)	Not informed	RO	Itapuã do Oeste

In GIS environment: for data quality control (d), the exported Tiff files, from both S2A and RapidEye classifications, were then collated in a GIS software and submitted to vector editing at 1:25,000 scale to eliminate major miss-classification of mining areas. For example, in Figure 3(ci), areas that were committed into mining class were manually removed from vector layer, whereas water bodies that are directly derived from mining operation were re-classified into mining class (Figure 3(cii)). This procedure is dependent on the user's interpretation to distinguish mining areas from those of other uses, which can lead to omission and commission errors (in particular small-scale mining). To support the user's visual interpretation, several ancillary data were used to minimize these errors such as such as roads, water bodies, limits of gold-mining districts, mining sites and protected areas [7]. This information helped to identify whether barren soil is an open-pit (mining area) or a clear-cut for other land use, for example. Moreover, the capacity of identifying mining areas is more precise using high spatial resolution imagery. Thus, for validation purposes, 12 RapidEye images (5 m) available at geocatalogo.mma.gov.br [20] acquired in 2014/2015 were uploaded to GEE and submitted to the same classification procedure developed for MSI/Sentinel-2 data (Table 1). New independent training samples were selected for each class using RapidEye bands 2 (560 nm), 3665 nm), 4 (710 nm) and 5 (805 nm) for CART classification. Although the time gap between S2A 2017) and RapidEye images may introduce commission errors as mining tends to expand from 2015 and 2017, RapidEye is the only high-resolution imagery freely available which can be used for validation. Other high-resolution images, such as Ikonos and WorldView, are available but under purchase only, which would make the mapping very expensive due to the wide extent of the study area.

Therefore, as part of the validation step (e), the results of Sentinel-2 classification (20 m) were compared to the classification derived from RapidEye imagery (5 m), the latter being the reference map (Figure 2e). Digital classification of RapidEye images were resampled to 20 m and used as ground truth for the S2A classification via confusion matrix (Kappa Index). To minimize spatial errors due to georeferencing, all the maps used for validation were projected to UTM (Universal Transverse Mercator) SAD-1969 Brazil coordinate system.

2.3. Identification of Mineral Type and Mining Scale

To address the second objective, validated mining maps derived from S2A images were subject to identification of mineral exploited, mining scale (whether is an industrial or small-scale mining) followed by tabulation of mining area in km^2. To do so, the validated Mining areas map (2017) was intersected with the territorial mining license layer controlled by Brazilian National Department for Mineral Production (DNPM) [21], which contains information about the mineral, license status and the owner (Figure 2f and Table 2). This information along with ancillary data allowed identification of mineral exploited and scale of mining activities.

3. Results

3.1. Validation of Mining Areas in 2017

In terms of total mining area, even with the time gap of approximately 2 years, both sensors provided quite similar results when examining individual tiles (Figure 4). In terms of map accuracy, Figure 4 shows a better agreement in areas of large-scale mining (>0.70) than those of small-scale mining (<0.70). For example, for Carajás (Figure 4a), characterized by industrial mining activities, Kappa index was 0.86. However, for Tapajós (Figure 4b), containing extensive small-scale mining, the index drops to 0.62. It is interesting to note that for industrial regions, mining concentrated in large areas (presented by few polygons) and in regions dominated by small-scale, a more diffuse land occupation, usually along the river network, was observed. The overall accuracy is above 98%, taking into consideration the 12 RapidEye images (see Table 1) with Kappa index of 0.70 (Table 3).

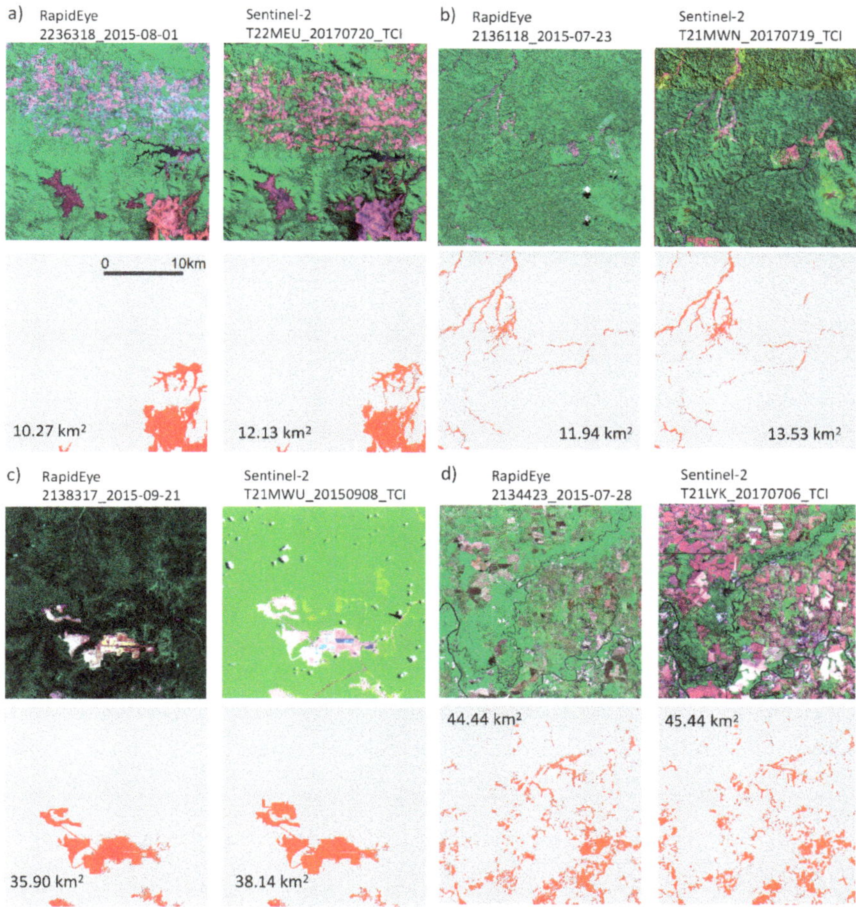

Figure 4. Evaluation of mining classification using S2A images having RapidEye classification maps (25 × 25 km) as ground truth. (**a**) Carajás region (Industrial), Kappa = 0.86; (**b**) Tapajós region (Small-scale), Kappa = 0.62; (**c**) Trombetas region (Industrial), Kappa = 0.87; and (**d**) Peixoto de Azevedo/Telez Pires region (Small-scale), Kappa = 0.59. RapidEye RGB (3, 5, 2) composition. S2A RGB (4, 8, 3) composition.

Table 3. Confusion matrix of 12 RapidEye tiles and their Sentinel (S2A) equivalent area. Only two classes were considered, mining and not mining areas. Overall accuracy is above 98% and overall Kappa Index is 0.70.

		RapidEye—2014/2015		
		Not	**Mining**	**Total**
S2A-2017	**Not**	7138.63	42.46	6637.09
	Mining	73.17	165.75	238.91
	Total	7218.30	208.20	7426.54

Commission error was larger (73.17 km^2) than omission error (42.46 km^2) and the total area detected by both imagery data was 165.75 km^2. The results show that the methodology proposed for S2A data provided satisfactory results, giving support to the analysis of total mining distribution and type of mineral exploited.

3.2. Total Mining Area, Type of Mineral and Mining Scale in 2017

As a result of mining areas derived from S2A images previously validated with RapidEye imagery, a total of 1084.7 km^2 were mapped considering the 13 regions analyzed. Table 4 shows the total mining area per study area including type of mineral exploited.

Table 4. Mining area in km^2 mapped with S2A images acquired in 2017 distributed over the study areas and specified by mineral type. For gold and tin, percentage of small-scale mining is shown between brackets. Federal Unit (F.U.).

N	Study Area	F.U.	Gold (% SS)	Tin (% SS)	Baux.	Iron	Cop.	Kaol.	Mang.	Nickel	Sand	Clay	TOTAL (SS%)
1	Rio Tapajós	PA	345.8 (100)										345.8 (100)
2	Carajás	PA	24.2 (100)			47.8	55.1		8.9	4.2	0.7		140.9 (17)
3	Rio Xingu	PA	92.1 (100)	37.8 (100)			2.6			1.9			134.4 (97)
4	Peixoto de A.	MT	116.7 (100)										116.7 (100)
5	Tromb./Juruti	PA			78.4								78.4
6	Rio Capim	PA			48.5			12.2					60.6
7	Mina Taboca	AM		48.8 (0)									48.8 (0)
8	Ariquemes	RO		45.8 (50)									45.8 (50)
9	Pontes e L.	MT	27.7 (50)										27.7 (50)
10	Amanã	AM	27.5 (100)										27.5 (100)
11	Amapá	AP	1.7 (0)			22.1							23.8 (0)
12	Barcarena	PA			12.0			2.3			4.5	0.6	19.4
13	Rio Madeira	RO	1.3 (100)	13.5 (100)									14.8 (100)
	TOTAL (km^2)		637.0 (97)	146.0 (50)	138.9	69.8	57.7	14.5	8.9	6.1	5.2	0.6	1084.7 (64)

The total area exploited by gold mining totals 637.0 km^2, which is equivalent to 58.7% of the mining area in the Amazon Region. Industrial gold mining takes place in Pontes e Lacerda (13.2 km^2) and Amapá (1.7 km^2). However, small-scale gold mining is responsible for 97% of the total area distributed over Tapajós (345.8), Peixoto de A. (116.7), Xingu (92.1), Amanã (27.5), Pontes e Laderda (13.5), and Madeira (1.3 km^2). Small-scale also comprises 50% of total tin exploitation, mostly in Rio Xingu (37.8), Ariquemes (22.9), and Rio Madeira (13.5 km^2). The remaining 50% of tin mining areas are related to industrial activities in Taboca (48.8) and Ariquemes (22.8 km^2). Overall, small-scale mining, including both gold and tin, comprises 64% of total mining area detected.

For the other minerals in Table 4, only industrial mining exploits them. Industrial bauxite mining, for example, takes place in Trombetas (78.4 km^2), Rio Capim (48.5 km^2) and Barcarena (12.0 km^2), totaling 138.9 km^2. Iron mining is present in Carajás (47.8 km^2) and Amapá (22.1 km^2), at a total of 69.8 km^2. Copper is exploited in Carajás (55.1 km^2) and Xingu (2.6 km^2). Kaolin is exploited at Rio Capim (12.2 km^2) and sent by pipeline to Barcarena (2.3 km^2 of infra-structure). Manganese mining occurs in Carajás only (8.9 km^2). For nickel, mining is found in Carajás (4.2 km^2) and Rio Xingu (1.9 km^2). Sand exploitation is found in Barcarena (4.5 km^2) and Carajás (0.7 km^2), whereas clay exploitation only occurs in Barcarena (0.6 km^2). Figure 5a shows the exploited area per different mining activities.

According to Figure 5, the majority (74%) of the mining areas identified in this study, mostly small-scale gold mining, is within the State of Pará boundaries. In this state, bauxite is the second most exploited mineral, followed by copper, iron, tin and others. For Mato Grosso state, only gold mining was identified, which has 9% out of 144.5 km^2 as industrial mining. The remaining area is composed of small-scale gold mining. In Rondônia and Amazonas states, tin mining is predominant. However, in Rondônia, tin is exploited equally by small-scale (50%) and industrial mining (50%), whereas in Amazonas only one industrial mining site (Mina Taboca) was identified. In Amapá, a large-scale exploitation of iron comprises most of the mining areas (93%).

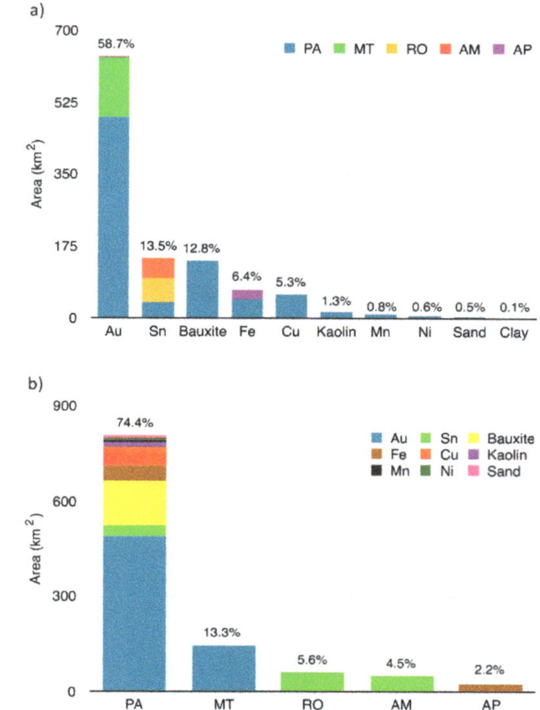

Figure 5. Distribution of mining area (km^2) based on mineral types (**a**): Au—Gold; Fe—Iron; Mn—Manganese; Sn—Tin; Cu—Copper; Ni—Nickel; and Federal Units (**b**): AM—Amazonas, AP—Amapá, MT—Mato Grosso, PA—Pará, RO—Rondônia.

4. Discussion

4.1. Method for Mapping Mining Areas and Constrains

Considering that the only mining area estimation for the Brazilian Amazon currently available is based on visual interpretation (TC2014), the first contribution of this research was to develop a cost-effective method, in terms of time and resources, to map mining areas in the Brazilian Amazon with satisfactory performance. Created in GEE environment, the mapping process allows quick access and selection of S2A images, which usually would take hours for selecting and downloading S2A images from the usgs.gov database for example, can be now done in minutes [22]. In addition, the classification step is much faster than on stand-alone image processing. For example, in the cloud (GEE) a supervised classification takes a few seconds, whereas in GIS software it usually takes several minutes or hours, depending on the computer configuration. Finally, employing the GEE for the classification suffers from the limitation of restricting the size of exports and thus, preventing the exportation of large areas in 10 m resolution.

On the other hand, the use of CART to classify mining areas presented some issues. Because CART is a pixel-based algorithm, some mining areas where not fully classified, leaving isolated pixels within a mining area, for example. One way to minimize it was to apply a low-pass filter in GIS environment. Another issue regards commission errors. Even though a careful selection of training samples were conducted for four main classes, i.e., mining (open-pits), water (natural water bodies), forest (dense dark vegetation) and clear-cut (pasture with regular shape), commission errors were often identified, mostly commissions of barren soil of other land-cover uses, such as pasture (Figure 3c). Given the high spectral variability of mining areas (barren soil, water, recovery areas, etc.) [7], a refined evaluation of what areas/polygons were or were not mining areas was necessary based on ancillary information to improve the map precision. The use of ancillary data such as mining sites, mining license, information about other land-cover uses (such as aquaculture, cattle regions, agriculture) as well as literature review was fundamental to identify the main mining areas, and is encouraged to be used on research for mining area detection. Therefore, a fully automatic method that detects mining areas with high precision is yet to be developed. Overall, the method presented here shows a progress toward a quick and accurate method for mining detection in large scale, such as the Brazilian Amazon.

4.2. Validation of Mining Areas with RapidEye Imagery

Besides being a quick method for mapping mining areas, the accuracy between mining maps derived from S2A in comparison to those from RapidEye confirms that the proposed method provides satisfactory results. As shown in Figure 4, the accuracy is higher among large-scale areas than for small-scale mining. The difference in Kappa index occurs because of two reasons: (1) the more concentrated mining areas in industrial mining allows easier detection in comparison to small-scale where a diffuse distribution of small polygons is observed; (2) Also, on small-scale, the LUCC is more dynamic, expanding or recovering quicker than industrial activities. Therefore, in the case of a 2-year gap, the commission error is more likely to occur.

The mining detection improvement using the proposed methods is more evident when we compared the currently available mining areas (TC2014 database) to RapidEye classification (2014/2015) used as ground truth (Figure 6).

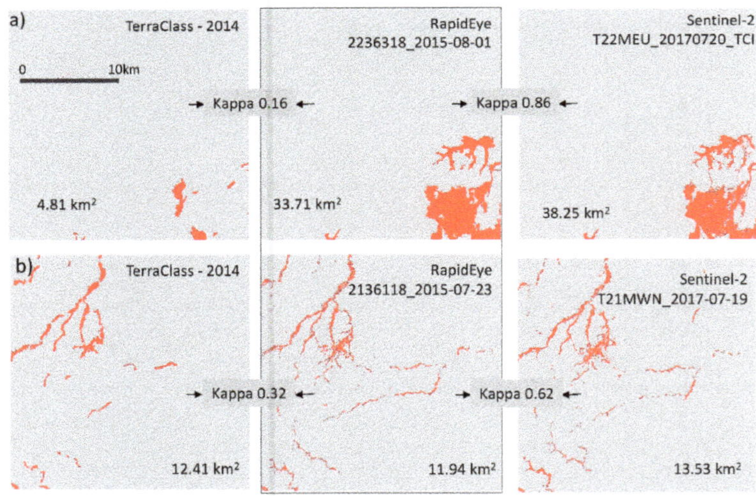

Figure 6. Comparison of mining areas between current available data (TerraClass 2014), map from RapidEye (taken as ground truth and limited to validation tiles, 25 × 25 km) and Sentinel-2A (mining maps for all study areas). In Carajás region (**a**), accuracy measured by Kappa index shows an improvement from 0.16 (TC2014) to 0.86 (S2A), due to a large omission error in TC2014. For small-scale mining in Tapajós (**b**), Kappa increased from 0.32 to 0.62. S2A classification were able to detect small polygons, whereas TC2014, due to a coarser classification, lacks detail.

For industrial mining in Carajás, for example, TC2014 detected only 4.81 km^2, resulting in a weak agreement with RapidEye classification (Kappa = 0.16), whereas S2A classification showed a Kappa index of 0.86. In the case of small-scale in Tapajós, the total area among all classifications were similar but the TC22014 classification lacks in detecting very small polygons (Figure 6). The lack of details in TC2014 maps can be attributed to (1) misclassification during visual interpretation and (2) to a lower spatial resolution (30 m) of Landsat images used for TC2014 maps, as compared to RapidEye (5 m) and Sentinel-2 (20 m) used in this study.

4.3. Total Area, Type of Mineral and Scale

The second main contribution of this research is the identification of scale and mineral type for each mining area mapped. According to Table 4, some minerals such as iron, manganese, copper, and nickel are exploited by industrial mining only, even though it comprises approximately a third of the total mining area mapped in this research. Industrial mining in the Brazilian Amazon is responsible for approximately 14% of total national annual exportation [1]. For some minerals such as bauxite, copper, kaolin, manganese and tin, Amazonian production totals 90% of national production. To illustrate its economic importance, the Financial Compensation for Mining Exploitation [1] indicates that incomes from Amazonian mining exportation yielded, approximately, US$ 531 million in 2016, of which 96% are related to the state of Pará, mostly from iron, bauxite and copper exploitation, which corresponds to 90% of total exportation in Pará.

While industrial mining has a significant financial importance and usually follows legal procedures for mining exploitation and the commitment of recovering degraded areas, several socio-environment impacts have recently been reported. For example, water contamination by bauxite exploitation in Barcarena [3], and by nickel exploitation in the Cateté river nearby Xikrin Indigenous Land [4].

Table 4 also shows that mining areas in the Brazilian Amazon is mostly occupied by small-scale gold mining, in particular Tapajós, Xingu/Fresco and Peixoto de Azevedo (MT). Including tin mining, small-scale comprises up to 64% of total mining area detected in this study. According to Enriquez [18], approximately 150,000 people are directly involved in small-scale mining in the Brazilian Amazon. This activity creates a large amount of income (~US$ 10 million per month for the regional economy [2]) which is rarely subject to taxes or control because most of the area is informally or illegally installed.

The large extension of small-scale mining raises a concern regarding its socio-environment impacts for the Amazonian ecosystems and for local people, since it usually does not follow environmental protocols to recover degraded areas. In fact, recent studies have demonstrated the impact of small-scale gold mining in the Tapajós River and its tributaries (Crepori, Novo and Tocantinzinho rivers) where intense water siltation occurs [12,16,17]. The impacts also include geomorphological changes in the riverbed and landscapes, causing severe impacts on benthic [23] and fish communities [24–27], not to mention the mercury contamination used in the gold amalgamation process [24,28–30] which is usually illegally imported [31].

5. Conclusions

This research presents a quick and efficient classification method developed in GEE to map mining areas in the Brazilian Amazon. The methods rely on quick access to S2A database, supervised classification of mining and other land-cover uses, exportation to GIS software, followed by vector editing to assure the product's quality control. Besides reducing image processing time, the methods provided accurate maps when compared to classifications derived from high-resolution RapidEye imagery (overall Kappa equals 0.70).

The main novelty of this research is the identification of scale and mineral type for each mining area mapped. Based on this identification, we concluded that small-scale mining (gold and tin) comprises up to 64% of total mining area detected in this study, which raises a concern regarding its socio-environment impacts for the Amazonian ecosystems and for local people, since it usually does not follow environmental protocols to recover degraded areas. The remaining 36% of mining areas in the Brazilian Amazon is comprised of industrial mining, which is responsible for most iron, manganese, copper, and nickel production. While industrial mining has a significant financial importance and usually follows legal procedures for mining exploitation and the commitment of recovering degraded areas, several socio-environment impacts have recently been reported and are raising a concern regarding its socio-environment impacts.

Given the satisfactory results, the methods presented here will be applied to map deforestation far beyond operational mining boundaries. Furthermore, we will carry out a historical mapping from satellite images, such as Landsat, to support further investigation of water quality and on LUCC in areas under the influence of both industrial and small-scale mining in the Brazilian Amazon.

Author Contributions: F.d.L.L. participated in every step of this article from planning, data processing, analysis, and writing the manuscript as the first author. P.W.M.S.-F. and E.M.L.d.M.N. were the co-supervisors of this research and were fundamental to define the research approach and data analysis. F.M.C. developed the GEE script and C.C.F.B., head of LabISA (The Instrumentation Laboratory for Aquatic Systems) where this research was conducted, collaborated with the methodological approach.

Funding: This research was funded by Coordenação de Aperfeiçoamento de Pessoal de Nível Superior (CAPES, Brazil) and Instituto Tecnológico da Vale (ITV, Brazil), Project 'Uso Sustentável de Recursos Naturais em Regiões Tropicais' (15024016001P1), as a Post-PhD Fellowship for F.d.L.L. Trainee Scholarship for F.M.C. funded by CNPq (Brazilian National Council for Scientific and Technological Development, Process: 135114/2017-9).

Acknowledgments: Research developed in a partnership between the ITV (Instituto Tecnológico da Vale) and INPE (Instituto Nacional de Pesquisas Espaciais) in Brazil. We would like to thank both reviewers for their insightful comments, which enabled us to improve the paper. Thanks to Lauren Pansegrouw for English revision.

Conflicts of Interest: The authors declare no conflict of interest.

References

1. Pinheiro, W.F.; Ferreira Filho, O.B.; Neves, C.A.R. *Anuário Mineral Brasileiro: Principais Substâncias Metálicas*; DNPM: Brasília, Brazil, 2016; p. 31.
2. Fernandes, F.R.C.; Alamino, R.d.C.J.; Araújo, E.R. *Recursos Minerais e Comunidade: Impactos Humanos, Socioambientais e Econômicos*; CETEM/MCTI: Rio de Janeiro, Brazil, 2014; Volume 1.
3. Nathanson, M. Norsk Hydro Accused of Amazon Toxic Spill, Admits 'Clandestine Pipeline'. *Mongabay*, 27 February 2018. Available online: https://news.mongabay.com/2018/02/norsk-hydro-accused-of-amazon-toxic-spill-admits-clandestine-pipeline/ (accessed on 5 June 2018).
4. Hofmeister, N.; Silva, J.C.D. The River is Dead: Is a Mine Polluting the Water of Brazil´s Xikrin Tribe? *Gaurdian*. 2018. Available online: https://www.theguardian.com/global-development/2018/may/15/brazil-xikrin-catete-river-amazon (accessed on 18 May 2018).
5. Sonter, L.J.; Herrera, D.; Barrett, D.J.; Galford, G.L.; Moran, C.J.; Soares-Filho, B.S. Mining drives extensive deforestation in the Brazilian Amazon. *Nat. Commun.* **2017**, *8*, 1013. [CrossRef] [PubMed]
6. Sonter, L.J.; Moran, C.J.; Barrett, D.J.; Soares-Filho, B.S. Processes of land use change in mining regions. *J. Clean. Product.* **2014**, *84*, 494–501. [CrossRef]
7. Chen, W.; Li, X.; He, H.; Wang, L. A review of fine-scale land use and land cover classification in open-pit mining areas by remote sensing techniques. *Remote Sens.* **2017**, *10*, 15. [CrossRef]
8. Isidro, C.; McIntyre, N.; Lechner, A.; Callow, I. Applicability of earth observation for identifying small-scale mining footprints in a wet tropical region. *Remote Sens.* **2017**, *9*, 945. [CrossRef]
9. Veiga, M.M.; Hinton, J.J. Abandoned artisanal gold mines In the Brazilian Amazon: A legacy of mercury pollution. *Nat. Res. Forum* **2002**, *26*, 15–26. [CrossRef]
10. Cuchierato, G. 200 maiores minas brasileiras. *Minérios & Minerales*, November/December 2016; p. 45.
11. Garai, D.; Narayana, A.C. Land use/land cover changes in the mining area of Godavari coal fields of Southern India. *Egypt. J. Remote. Sens. Space Sci.* **2018**. [CrossRef]
12. Lobo, F.; Costa, M.; Novo, E.; Telmer, K. Distribution of artisanal and small-scale gold mining in The Tapajós River Basin (Brazilian Amazon) over the past 40 years and relationship with water siltation. *Remote Sens.* **2016**, *8*, 579. [CrossRef]
13. Almeida, C.A.D.; Coutinho, A.C.; Esquerdo, J.C.D.M.; Adami, M.; Venturieri, A.; Diniz, C.G.; Dessay, N.; Durieux, L.; Gomes, A.R. High spatial resolution land use and land cover mapping of the Brazilian Legal Amazon in 2008 using landsat-5/tm and modis data. *Acta Amazon.* **2016**, *46*, 291–302. [CrossRef]

14. Coelho, M.C.N.; Wanderley, L.J.; Costa, R.C. Small scale gold mining in the xxi century. Examples in the south-west brazilian amazon. *Anuário do Instituto de Geociências-UFRJ.* **2016**, *39*, 5–14. [CrossRef]
15. Monteiro, M.D.A.; Coelho, M.C.N.; Cota, R.G.; Barbosa, E.J.D.S. Ouro, empresas e garimpeiros na amazônia: O caso emblemático de Serra Pelada. *Revista Pós Ciências Sociais-UFMA (Universidade Federal do Maranhão)* **2010**, *7*, 28.
16. Lobo, F.; Costa, M.; Novo, E.; Telmer, K. Effects of small-scale gold mining tailings on the underwater light field in the Tapajós River Basin, Brazilian Amazon. *Remote Sens.* **2017**, *9*, 861. [CrossRef]
17. Lobo, F.L.; Costa, M.P.F.; Novo, E.M. Time-series analysis of Landsat-MSS/TM/OLI images over Amazonian waters impacted by gold mining activities. *Remote Sens. Environ.* **2015**, *157*, 170–184. [CrossRef]
18. Enríquez, M.A. Mineração na Amazônia. *Parcerias Estratégicas.* **2014**, *19*, 155–198.
19. Farda, N.M. Multi-temporal land use mapping of coastal wetlands area using machine learning in Google Earth Engine. *IOP Conf. Ser. Earth Environ. Sci.* **2017**, *98*, 012042. [CrossRef]
20. MMA. *Geocatalogo Mma-Rapideye Imagery*; MMA: Brasília, Brazil, 2018. Available online: http://geocatalogo.mma.gov.br/index.jsp (accessed on 14 February 2018).
21. Departamento Nacional de Produção Mineral (DNPM). *Sistema De Informações Geográficas Da Mineração-Sigmine*; DNPM: Brasília, Brazil, 2018; Volume 1. Available online: http://sigmine.dnpm.gov.br/webmap/ (accessed on 20 February 2018).
22. Parente, L.; Ferreira, L. Assessing the spatial and occupation dynamics of the Brazilian pasturelands based on the automated classification of MODIS images from 2000 to 2016. *Remote Sens.* **2018**, *10*, 606. [CrossRef]
23. Couceiro, S.R.M.; Hamada, N.; Forsberg, B.R.; Padovesi-Fonseca, C. Trophic structure of macroinvertebrates in Amazonian streams impacted by anthropogenic siltation. *Austral Ecol.* **2011**, *36*, 628–637. [CrossRef]
24. Nevado, J.J.B.; Martin-Doimeadios, R.C.R.; Bernardo, F.J.G.; Moreno, M.J.; Herculano, A.M.; do Nascimento, J.L.M.; Crespo-Lopez, M.E. Mercury in the Tapajos River Basin, Brazilian Amazon: A review. *Environ. Int.* **2010**, *36*, 593–608. [CrossRef] [PubMed]
25. Sampaio da Silva, D.; Lucotte, M.; Paquet, S.; Davidson, R. Influence of ecological factors and of land use on mercury levels in fish in the Tapajós River Basin, Amazon. *Environ. Res.* **2009**, *109*, 432–446. [CrossRef] [PubMed]
26. Dorea, J.G.; Barbosa, A.C. Anthropogenic impact of mercury accumulation in fish from the Rio Madeira and Rio Negro rivers (amazonia). *Biol. Trace Element Res.* **2007**, *115*, 243–254. [CrossRef] [PubMed]
27. Boudou, A.; Maury-Brachet, R.; Coquery, M.; Durrieu, G.; Cossa, D. Synergic effect of gold mining and damming on mercury contamination in fish. *Environ. Sci. Technol.* **2005**, *39*, 2448–2454. [CrossRef] [PubMed]
28. Rabitto, I.d.S.; Bastos, W.R.; Almeida, R.; Anjos, A.; de Holanda, Í.B.B.; Galvão, R.C.F.; Neto, F.F.; de Menezes, M.L.; dos Santos, C.A.M.; de Oliveira Ribeiro, C.A. Mercury and ddt exposure risk to fish-eating human populations in Amazon. *Environ. Int.* **2011**, *37*, 56–65. [CrossRef] [PubMed]
29. Coelho-Souza, S.A.; Guimarães, J.R.D.; Miranda, M.R.; Poirier, H.; Mauro, J.B.N.; Lucotte, M.; Mergler, D. Mercury and flooding cycles in the Tapajós River Basin, Brazilian Amazon: The role of periphyton of a floating macrophyte (paspalum repens). *Sci. Total Environ.* **2011**, *409*, 2746–2753. [CrossRef] [PubMed]
30. Telmer, K.; Veiga, M. Chapter 6: World Emissions of Mercury from Artisanal and Small Scale Gold Mining. In *Mercury Fate and Transport in the Global Atmosphere*; Pirrone, N., Mason, R., Eds.; Springer Science, Switzerland AG: Basel, Switzerland, 2009; pp. 131–172.
31. NSCTV. Operacao faz apreensao de Carga de Mercurio no Porto de Itajai. 2018. Available online: https://g1.globo.com/sc/santa-catarina/noticia/operacao-faz-apreensao-de-carga-de-mercurio-no-porto-de-itajai.ghtml (accessed on 12 June 2018).

Technical Note

Estimating Satellite-Derived Bathymetry (SDB) with the Google Earth Engine and Sentinel-2

Dimosthenis Traganos [1], Dimitris Poursanidis [2,*], Bharat Aggarwal [1], Nektarios Chrysoulakis [2] and Peter Reinartz [3]

1 German Aerospace Center (DLR), Remote Sensing Technology Institute, Rutherfordstraße 2, 12489 Berlin, Germany; Dimosthenis.Traganos@dlr.de (D.T.); 851.bharat@gmail.com (B.A.)
2 Foundation for Research and Technology—Hellas (FORTH), Institute of Applied and Computational Mathematics, N. Plastira 100, Vassilika Vouton, 70013 Heraklion, Greece; zedd2@iacm.forth.gr
3 German Aerospace Center (DLR), Earth Observation Center (EOC), 82234 Weßling, Germany; peter.reinartz@dlr.de
* Correspondence: dpoursanidis@iacm.forth.gr, Tel.: +30-2810-391774

Received: 7 May 2018; Accepted: 30 May 2018; Published: 1 June 2018

Abstract: Bathymetry mapping forms the basis of understanding physical, economic, and ecological processes in the vastly biodiverse coastal fringes of our planet which are subjected to constant anthropogenic pressure. Here, we pair recent advances in cloud computing using the geospatial platform of the Google Earth Engine (GEE) with optical remote sensing technology using the open Sentinel-2 archive, obtaining low-cost in situ collected data to develop an empirical preprocessing workflow for estimating satellite-derived bathymetry (SDB). The workflow implements widely used and well-established algorithms, including cloud, atmospheric, and sun glint corrections, image composition and radiometric normalisation to address intra- and inter-image interferences before training, and validation of four SDB algorithms in three sites of the Aegean Sea in the Eastern Mediterranean. Best accuracy values for training and validation were $R^2 = 0.79$, RMSE = 1.39 m, and $R^2 = 0.9$, RMSE = 1.67 m, respectively. The increased accuracy highlights the importance of the radiometric normalisation given spatially independent calibration and validation datasets. Spatial error maps reveal over-prediction over low-reflectance and very shallow seabeds, and under-prediction over high-reflectance (<6 m) and optically deep bottoms (>17 m). We provide access to the developed code, allowing users to map bathymetry by customising the time range based on the field data acquisition dates and the optical conditions of their study area.

Keywords: satellite-derived bathymetry; image composition; pseudo-invariant features; sun glint correction; empirical; spatial error; Google Earth Engine; low cost in situ; Sentinel-2; Mediterranean

1. Introduction

Bathymetry is important for understanding how global Earth processes interact as they influence the flow of the sea water carrying heat, salt, nutrients, and pollutants. Bathymetry also aids in understanding the propagation of energy from undersea seismic events that impact navigation and commerce, and shape habitats for marine life, especially in coastal areas [1,2]. Coastal areas are under constant pressure due to intense anthropogenic activities such as urbanisation, exploitation of natural resources, and climate change-induced natural hazards (e.g., coastal erosion due to sea level changes) [1]. The littoral zone of this interface is spatially complex and determines biodiversity-related processes as increasing bathymetric values decrease light penetration and cause changes in habitat compositions and the depth zonation of biota [2]. Studies of the coastal zone—including the modelling of tsunami expansion and wave height estimations, sea-level change scenarios, risk assessment, and coastal habitat mapping—require the availability or the creation of updated bathymetric data of high (10 m) to very high resolution (2 m) [3–6]. Up to today, numerous researchers have mapped coastal bathymetry with a range of different tools and methods of variable spatial and temporal resolution, but with increased cost. Globally, bathymetry has been extracted by the inversion of the spaceborne geoid data from Geosat and the European Remote Sensing satellite ERS-1 [7] at 12-km spatial resolution as well as from a combination of sound navigation and ranging (SONAR) multibeam sounding data from marine agencies and an improved gravity model from CryoSat-2 and Jason-1 at a 500-m resolution [8]. With respect to currently open access data, Landsat missions, especially Landsat 8 Operational Land Imager (OLI) due to the availability of its coastal/aerosol band 1 centred at a 443-nm wavelength featuring high water penetration) open a new window in coastal satellite-derived bathymetry (SDB) due to their spatial (30-m) and temporal (16-day) resolution. With these characteristics, the Landsat program allows for the selection of proper images (cloud-free or atmospheric images, surface and water column conditions) and the testing of hypertemporal approaches for monitoring changes in seabed morphology [9–11]. The more recently launched Copernicus Sentinel-2 twin-satellite mission has created a new era in terrestrial and marine monitoring due to its high spatial resolution of 10 m, availability of a coastal/aerosol band at 443 nm (60-m spatial resolution), quick revisit time of 5 days, and more significantly, its open and free data access policy. Focusing on SDB, studies have employed Sentinel-2 data in shallow inland waters [12] and semi-closed bays [13,14] with promising results, but not at large scales, with notable differences in the water column and seabed composition. The aforementioned optical sensor technology, machine learning algorithms, and cloud computing system infrastructures provide an unprecedented environment for high spatiotemporal large-scale analysis of natural and anthropogenic ecosystems and associated biophysical variables. Among the available cloud systems, Google Earth Engine (GEE) [15] has attracted the attention of environmental scientists due to its unique components. GEE is a cloud-based geospatial computing platform which offers a petabyte-scale archive of freely available optical satellite imagery. Among characteristics, it features the whole archive of Landsat, the first three Sentinel missions, and full Moderate Resolution Imaging Spectroradiometer (MODIS) data and products. Researchers have utilised it for country-scale vegetation metrics [16], continental-scale mapping of croplands [17], and global land surface temperature [18] and albedo [19] estimation.

Here, we have developed an empirical preprocessing workflow within GEE to estimate four satellite-derived bathymetry algorithms using Sentinel-2 images at three different locations in the Eastern Mediterranean. The preprocessing chain of Sentinel-2 data combines a plethora of simple, widely used, and well-established cloud, atmospheric, and sun glint corrections with a seasonal 10-m image composition approach (median composites). The latter approach opts to reduce the effects of variable water surface conditions, water quality, and related optical properties such as sun glint and sky glint, whitecaps, turbidity, and sedimentation etc. Furthermore, we implement in situ data acquired by a low-cost methodology in the training and validation of the SDB which shows, in turn, promising results for a time- and cost-efficient epoch for wide-scale SDB. We perform the training and validation of the four empirical models in different optically shallow environments to reduce the statistical bias according to the first law of geography by Tobler [20]—spatially neighbouring observations tend to be more similar than the ones further apart. Moreover, we employ a pseudo-invariant-feature (PIF) approach to normalise the differences in the reflectance ranges between the pre-processed composites used in the training and validation steps [21]. Finally, we map the model residuals of the SDB maps to unveil possible over- or under-prediction patterns. Our proposed method can be widely adapted in a cloud computing environment for estimating large-scale coastal bathymetry given, naturally, the availability of relevant open access in situ data.

2. Materials and Methods

2.1. Study Sites and In Situ Data

Figure 1 shows our three study sites; the first study site is Samaria National Park in the southern part of West Crete, Greece, in the Eastern Mediterranean. The selected area covers the east part of the National Park. Mixed habitats cover the seabed here, while depths vary from smooth areas to steep ridges and provide a challenging seascape for space-borne coastal bathymetry. Seagrass meadows reach depths of 40 meters [22]. Bathymetric data was collected during the period 2012–2015 by means of a 5-m inflatable boat using a Lowrance High Definition System (HDS)-5 single-beam echosounder with the HST-WSU 83/200 kHz Skimmer Transducer. The Global Positioning System (GPS) was positioned on top of the roll bar, directly above the transducer to record the exact position. Data was recorded with a frequency of 1 Hz. Preprocessing of Lowrance files was performed using DrDepth software. Data was imported in ArcMap 10.5 in comma separated value (CSV) format and converted into a three-dimensional shapefile for further use. The second study site is Apokoronas area (Obros Gyalos) in West Crete, close to Georgioupolis Bay. It is a remote area with a mixed bottom—shallow rocky reefs followed by bright white sand substrate, ending with seagrass meadows which are followed by a mixed pebble/sand substrate. In this location, the first diving park in Crete is under development, acting also as a protected marine area. Bathymetric data was collected following the aforementioned methodology. The third study site is at Nea Moudania (Thermaikos) in the SouthEast part of the greater Thermaikos Gulf, NorthWest Aegean Sea, Greece. Several types of human activity, including agriculture, aquaculture, industry, tourism, fishing, and trade directly affect the coastal system of Thermaikos. The seascape is mainly made by soft substrates followed by dense continuous seagrass meadows down to approximately 16.5 m. All the aforementioned habitats have been verified by snorkelling and diving. Bathymetric data was collected between 10 and 13 July 2016, utilising the Garmin Fishfinder 160C with an 80/200 kHz (dual beam) sonar. The different transducers in use are due to the availability of the systems at the employed boats. Data was corrected for position, adding the distance from the transducer to the GPS in postprocessing. We calculate the mean depth values that fall within each Sentinel-2 10-m pixel prior to the analysis. Both bathymetry and Sentinel-2 are referenced with respect to World Geodetic System (WGS) 84 (G1762). Additionally, as the whole Aegean basin is a principally tideless environment (mean tide amplitude of a few cm), we assume an identical vertical datum for all the images within the implemented median composite (as seen in Section 2.2; 1 August–31 December 2016) and the singlebeam echosounder data.

2.2. Sentinel-2 Data Preprocessing

A plethora of factors affect the state of the atmosphere (e.g., haze, aerosols, and clouds), sea surface (e.g., sun glint, sky glint, and white caps) and water column (e.g., sedimentation, turbidity and variable optical properties) hindering the remote sensing of the optically shallow extent (where part of the water surface remote signal contains a bottom signal). A suitable preprocessing approach which precedes the estimation of bathymetry from remotely sensed images should address and correct most of these impeding factors. Google Earth Engine's open petabyte-scale satellite image archive allows unprecedented, on the fly, multi-image processing in the cloud. This offers the opportunity to address the hindrances which are inherent in the nature of optically shallow water remote sensing. As such, we implement a preprocessing workflow (Figure 2) that features a combination of metadata information, widely used algorithms in the field of coastal aquatic remote sensing, image composition (median composite), image normalisation, and smoothing. We applied the preprocessing workflow on Sentinel-2 (S2) Level-1C (L1C) top of atmosphere (TOA) data which is the standard S2 archive in GEE (ImageCollection ID: COPERNICUS/S2). Seven S2 bands (coastal aerosol: b1, blue: b2, green: b3, red: b4, near infrared (NIR): b8, shortwave infrared (SWIR) 1: b11, quality assurance (QA) 60 band) form this data input which spans the period between 1 August and 31 December 2016, which is close chronologically to the acquisition of all in situ bathymetry data and is when the water column is better stratified in the study areas [23]. Our preprocessing workflow consists of seven steps:

1. We employ the QA60 bitmask band which contains cloud information to mask out opaque and cirrus clouds and scale S2 L1C TOA data by 10,000 (S2 quantification value).
2. We use a classification and regression tree (CART) classifier [24] on a b3-b8-b11 composite to mask the terrestrial environment. It is noteworthy that although the classifier is trained with selected aquatic and terrestrial points (35 and 32, respectively) around Crete Island only (Figure 1d), it is utilised in all three sites.
3. To atmospherically correct the masked for clouds and land images, we implement a modified dark pixel subtraction (DPS) method after [25] which subtracts the mean radiance of optically deep-water pixels (>40 m) to address path radiance and two standard deviations to address sensor-related noise in all bands.
4. We employ the so-called image composition where a new pseudo-image is created using—in our case—the median values of the already pre-processed images [16]. This approach aims to further reduce image artefacts which have not been corrected by the previous preprocessing steps. In fact, 79 tiles (34 SGE, 34 SGD, 35 SKV and 35 SKU; also called granules—100 × 100 km^2 ortho images projected in Universal Transverse Mercator UTM/WGS84 [26]) form the Samaria National Park and Apokoronas pseudo-image, while 18 tiles (34 TFK) form the Thermaikos pseudo-image.
5. We apply the sun glint correction algorithm from [27] to the median composite. Following a user-defined set of pixels that represents sun glint of variable intensity (two polygons in the south of tSamaria National Park site in South Crete), the algorithm equals the sun glint-corrected/sun glint-polluted median composite reduced by the product of the slope of the regression of NIR b8 against b1–b4 and the difference between b8 and its minimum value. We should state here that in [28], the preprocessing step included the mean and not the minimum NIR signal over optically deep water. The two last steps of our preprocessing chain are performed in a GIS environment (ArcMap 10.5) as we export the sun glint-corrected median composites from the previous step for all three areas of interest.
6. We implement the pseudo-invariant features (PIF) [21] to radiometrically normalise b2 and b3 bands which are used in the validation of the SDB models (Apokoronas and Thermaikos composites; Figure 1a,b) to the b2 and b3 of the National Samaria Park composite (Figure 1c). This technique was developed to quantitatively transform a subject multispectral image to the reference multispectral image as if they were sensed under the same atmospheric conditions, and in the case of a coastal study, the same water surface and column conditions. The PIF-based composite normalisation is employed here to decrease the spectral differences which caused high RMSEs in our first SDB validation experiments (results are not shown here). Figure 3 displays the location of the selected PIF features (44 in total for each site), with shallow sands as the bright features and *Posidonia oceanica* seagrasses (National Samaria Park and Thermaikos) and brown algae (Apokoronas; mainly *Cystoseira* sp.) as the dark features. Figure S1 shows the linear equations which are utilised in the radiometric normalisations. We select these specific features because they occupy the extremes (high, low) of the observed reflectance range in all three sites following the recommendations that can be found in [29]; the underlying assumption here is that the PIF changes little between the same tiles (=same site) and different composites (=different site).
7. We apply a 3 × 3 low pass filter to the normalised S2 b2 and b3 bands to further reduce noise before the training and validation of the empirical SDB models.

2.3. Empirical Satellite-Derived Bathymetries (SDB)

Generally, empirical SDB methods require certain bands in the visible wavelength, with blue and green being the most widely used (as independent variables), and a set of known in situ depths (the dependent variable) as the only inputs in simple or multiple linear regressions which lead to bathymetry estimations in a given area. Here, we implement and compare four different empirical approaches to derive bathymetry from the pre-processed GEE Sentinel-2 composites. To be consistent in our comparison with the other SDB models and the existing SDB literature, we exploit only the blue and green S2 wavelengths (b2: 496.6 and b3: 560 nm for S2A) in the estimations due to their reasonable water penetration. The training of all four models takes place in the National Samaria Park, while the validation was performed in two different areas: the Apokoronas area and the Thermaikos Gulf (Figure 1). Table 1 provides information on the size of the survey area, number of used in situ points, and their depth range. Given the different depth range of the in situ points in the two validation sites (0–25 m in Apokoronas and 0–12 m in Thermaikos), we trained two different sub-models for each one of the four SDB methods to increase the accuracy of the estimated SDB as it is trained and validated using the same depth range. The first and oldest approach is the one proposed by the author of [28] (hereafter Lyzenga85) which assumes a linear relationship between the log-transformed bands and known depth via (multiple) linear regression. The produced coefficients of the regression are then used to train the Lyzenga85 SDB model which forms as:

$$z = h_0 + h_i X_i + h_j X_j \tag{1}$$

where z is the satellite-derived bathymetry, h_0, h_i, and h_j are the coefficients (intercept and slopes), and X_i and X_j are the independent variables (the radiance in the blue and green bands, respectively). For the 0–25-m depth range, the Lyzenga85 model features an R^2 of 0.79 and RMSE of 2.46 m and takes the form:

$$z = -27.85 + 4.95\mathrm{b2} - 14.13\mathrm{b3} \tag{2}$$

while for the 0–12-m depth range, the model exhibits an R^2 of 0.69 and RMSE of 1.39 m:

$$z = -7.76 + 4.76\mathrm{b2} - 8.71\mathrm{b3} \tag{3}$$

The second and third selected approaches follow two modifications of the proposal by [30]—hereafter referred to as modified Stumpf03 and Traganos17 (after its first use in [31]), respectively—concerning the empirical relationship between the ratio of the log-transformed green band to the log-transformed blue band and water depth. The modified Stumpf03 is simply the multiplicative inverse of the original ratio; it employs the blue to green ratio instead of the original green to blue ratio:

$$z = m_1 \frac{\ln(n\mathrm{b2})}{\ln(n\mathrm{b3})} - m_0 \tag{4}$$

where m_1 and m_0 are the slope and y-intercept, set by the linear regression between the ratio and bathymetry, and n is a fixed constant (1000 in all experiments related to the approach in [31]) to assure the linear response of the logarithmic ratio with depth and that it will remain positive at all points. The m_1 and m_0 values are 44.39 and 33.17 for the 25-m training set (R^2 = 0.59, RMSE = 3.49 m) and 20.37 and 12.16 for the 12-m training set (R^2 = 0.5, RMSE = 1.78 m), respectively. The Traganos17 SDB algorithm is essentially the ratio of log-transformed blue to log-transformed green (x) without the n constant of Equation (4) and has shown more accurate results over low-reflectance bottoms (e.g., seagrasses and algae) than the original algorithm in [30], which was primarily tested in and tuned with high-reflectance bottoms (e.g., sand, coral reefs). The exponential equation:

$$z = 4416.3\mathrm{e}^{-6.12x} \tag{5}$$

is implemented to estimate bathymetry in the Apokoronas area (R^2 = 0.67, RMSE = 3.65 m) up to a 25-m depth, while the linear equation with m_1 and m_0 values of 20.37 and 12.16, respectively, derives bathymetry of up to a 12-m depth in the Thermaikos Gulf (R^2 = 0.56, RMSE = 1.68 m). The fourth and final SDB empirical approach is the one developed by the authors of [32] (hereafter Dierssen03), which takes the log-transformed ratio of the blue to green median composite (x) here (green to red band in the original paper) to map bathymetry. The exponential equation:

$$z = 9.46e^{1.52x} \tag{6}$$

is used to estimate the bathymetry in Apokoronas ($R^2 = 0.66$, RMSE = 3.59 m) in the 0–25-m depth range and the linear equation:

$$z = 7.05 \ln\left(\frac{b2}{b3}\right) + 8.31 \tag{7}$$

derives the bathymetry in Thermaikos in the 0–12 m depth range ($R^2 = 0.53$, RMSE = 1.72 m).

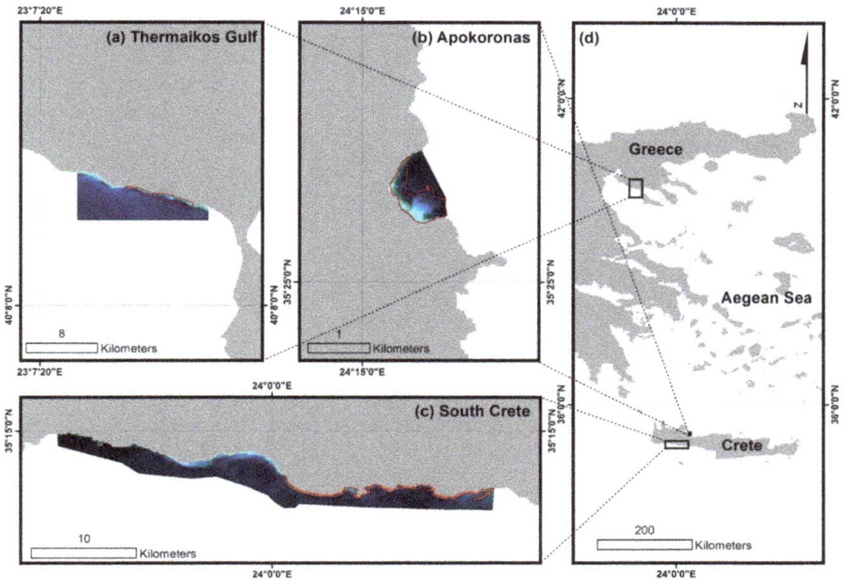

Figure 1. Locations of selected study sites for bathymetry model analysis: (**a**) Thermaikos Gulf; (**b**) Obros Gyalos in the Apokoronas area; (**c**) National Samaria Park, South Crete; and (**d**) the Aegean Sea, Greece. The red polygons indicate the extent of the area that is used for the actual analysis. All depicted images are median Sentinel-2 composites using the preprocessing workflow from Figure 2.

2.4. Spatial Distribution of Model Residuals

As our studied seabed varies from bright shallow sand to dark seagrass meadows and algae-covered bottoms, we expect related errors in the resulting SDB maps. To further distil the spatial information from the SDB maps, we calculate the distribution of model residuals. We estimate the latter as the remainder of the subtraction of the in situ bathymetry data from the most accurate trained and validated SDB maps (Lyzenga85 in all three areas). Following the suggestions by the authors of [33], spatial error maps are derived by Kriging interpolation (ArcMap Spatial Analyst tool) of the remaining points to the whole extent of the training and validation sites to unveil over- and under-prediction patterns of the SDB models. The Kriging interpolation algorithm utilises a spherical semivariogram model and a variable search radius of 12 points to "capture both the deterministic and autocorrelated variation in the residual surface" [33].

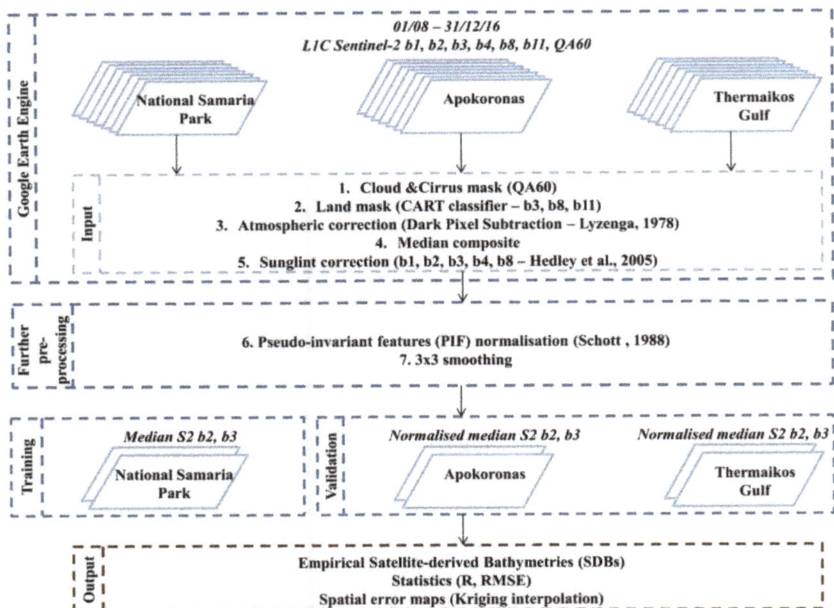

Figure 2. The methodological workflow of the present study. We employ seagrass and algae beds as dark features and shallow sand as bright features; We perform the training of the satellite-derived models in the National Samaria Park sites using two different depth ranges to capture the representative depth range in the validation sites: 0–25 m in Apokoronas and 0–12 m in Thermaikos.

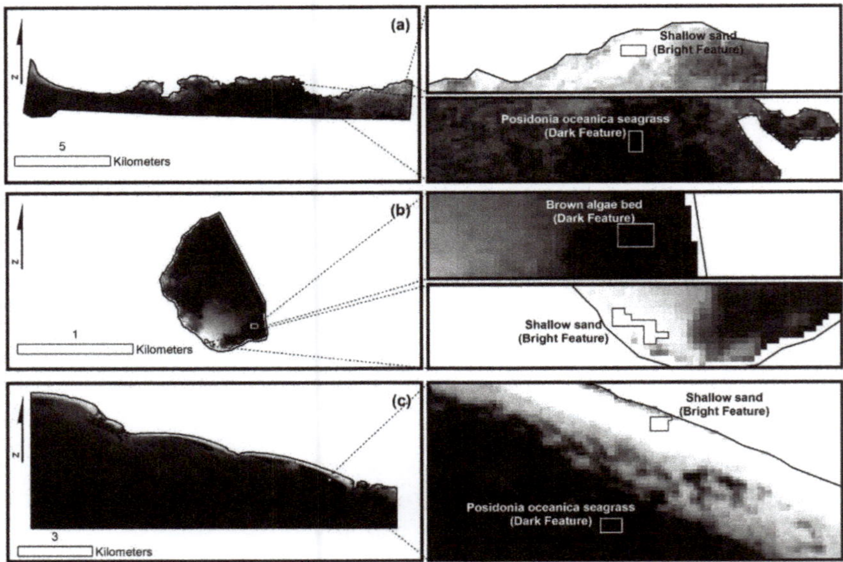

Figure 3. Location of used pseudo-invariant features (zoomed right panels) in the normalisation of the Apokoronas and Thermaikos composites (**b**,**c**) to the National Samaria Park composite (**a**).

Table 1. Survey area and number of related in situ points used for training and validation of the satellite-derived bathymetry models. The Thermaikos Gulf has a lower number of validation points as the initially available dataset is much smaller in comparison to that of Apokoronas.

Survey Site	National Samaria Park (Training)	Apokoronas (Validation)	Thermaikos Gulf (Validation)
Survey area (km^2)	46.3	1.3	20.6
Number of in situ points	4978 (25-m models) 3230 (12-m models)	1557	53
Depth range and intervals of in situ points (m)	0–25: 0–5 (675), 5–10 (1998), 10–15 (1247), 15–20 (659), 20–25 (399) 0–12: 10–12 (556)	0–25: 0–5 (57), 5–10 (301), 10–15 (428), 15–20 (517), 20–25 (254)	0–12 0–5 (25), 5–10 (26), 10–15 (2)

3. Results

3.1. SDB Estimations

Figure 4 shows the pre-processed median S2 composites and their respective satellite-derived bathymetry maps of the training (Figure 4a,b) and validation sites (Figure 4b,c,e,f). Maximum SDB depths are 27 m in National Samaria Park, 23 m in Apokoronas, and 16 m in the Thermaikos Gulf (minimum depth of Thermaikos is 1.7 m, 0 in the two other sites). All statistical results are collectively given in Tables S1 and S2. In terms of validation, the Lyzenga85 model achieves the best accuracies, explaining 90% of the variation inside the validation dataset at Apokoronas (Figure 5—1557 points; RMSE = 1.67 m) and 86% of the variation inside the validation dataset at Thermaikos (Figure 6—53 points; RMSE = 4.1 m). More specifically, with regard to the Apokoronas area, the Lyzenga85 approach features approximately twelve-fold higher R^2 and five-fold smaller RMSEs than the average R^2 and RMSE values of 0.073 and 8.84 m, respectively, for the modified Stumpf03, Traganos17, and Dierssen03 approaches. On the other hand, with regard to the Thermaikos area, the Lyzenga85 model explains 2.6 times better the variation within the validation dataset than the average 33% of the other three models, but exhibits a 1.49-m greater RMSE than their average value of 2.6 m. The utilisation of the Traganos17 SDB model yields the lowest RMSE of 2.49 m here (R^2 = 0.52).

3.2. Methodological Gains Using a PIF Normalisation and 3 × 3 Smoothing

Figure 5 demonstrates the effectiveness of including pseudo-invariant-feature-based normalisation and a 3 × 3 low pass filter over the remotely-sensed composite before the validation of the SDB models. All four panels in Figure 5 are the validation plots of the model with the highest accuracy (Lyzenga85) in the Apokoronas area which differ in terms of the incorporated preprocessing methods. The validation on a pre-processed composite utilising only the five first steps of Figure 2 (Figure 5a) results in an R^2 of 0.89 and a RMSE of 3.79 m, which is nearly halved to 1.92 m (same R^2) by normalising the pre-processed composite prior to the validation (Figure 5b). The application of 3 × 3 smoothing without the normalisation (Figure 5c) produces the highest RMSE of 4.79 m (nearly same R^2 of 0.88), whereas with normalisation (Figure 5d), we observe that R^2 rises marginally by 0.02, and the RMSE drops to 1.67 m in comparison to step 3.

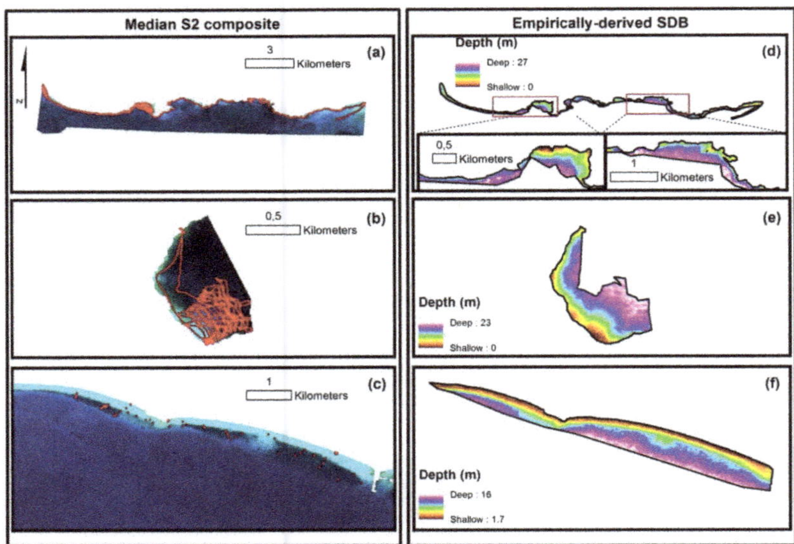

Figure 4. Pre-processed median Sentinel-2 composites using the workflow of Figure 2 on the left panels (**a–c**) and related best empirical satellite-derived bathymetry (SDB) estimates on the right (**d–f**): The National Samaria Park composite and related SDB estimates in panels (**a,d**); Apokoronas composite and related SDB estimates in panels (**b,e**); Thermaikos composite and related SDB estimates in panels (**c,f**). Red points depict the employed in situ points in the training (**a**) and validation (**b,c**) of the SDB models here. Best algorithm (highest R^2, lowest RMSE) in all three cases is Lyzenga85 model applied on 3 × 3 smoothed and normalised median Sentinel-2 composites using the shown pseudo-invariant (PIF) features in the zoomed maps of Figure 3b.

3.3. Spatial Distribution of Model Residuals

Figure 7 depicts the spatial variation of the model residuals of the SDB maps following training by the Lyzenga85 model. In conjunction with the validation plots in Figures 5 and 6, the three figures reveal potential geographical- (horizontal) and depth-related (vertical) patterns of over- or under-prediction (always as a reference to the used in situ data) and the relation of the latter to the presence of specific habitats. There are several distinctive over- and under-prediction SDB patterns. First, the National Samaria Park area (training site) features an under-prediction and over-prediction tendency in its westernmost and easternmost parts, respectively (Figure 7a). There are also model residuals related to under-prediction over shallow sand and rocks of less than ~6 m and more than ~21 m (Figures 3a, 4d and 7a). Second, the Apokoronas area (first validation site) shows variable prediction trends across variable depth zones—the Lyzenga85 model underpredicts bathymetry in approximately the first 6 m over the sand-covered seabed and in the algae-covered seabed deeper than approximately 17 m (Figures 3b, 4e and 7b). The latter is further justified by the validation plot (4) of Figure 5 which displays an under-estimation of depth towards the deeper seabed in Apokoronas. The remainder of the bottom of the same site—which is dominated by algae habitats—depicts an over-prediction of SDB (Figure 7b). Third, the residuals in the Thermaikos Gulf (second validation site) are all related to over-prediction, which ranges between 3 and 4.9 m. While the westernmost and easternmost Thermaikos' seabed, covered with dense *P. oceanica* seagrass and sparse *Cymodocea nodosa* seagrass, exhibit the greatest degree of over-prediction, the central Thermaikos, covered by a smaller *P. oceanica* meadow, lays out a smaller degree of over-prediction with a rising tendency from west to east (away from the centre of the composite) (Figure 7c). Last but not least, we note medium over-prediction values (off-white colour) over the sandy bottom in the northwestern and southeastern areas. This is greater than for the studied central Gulf.

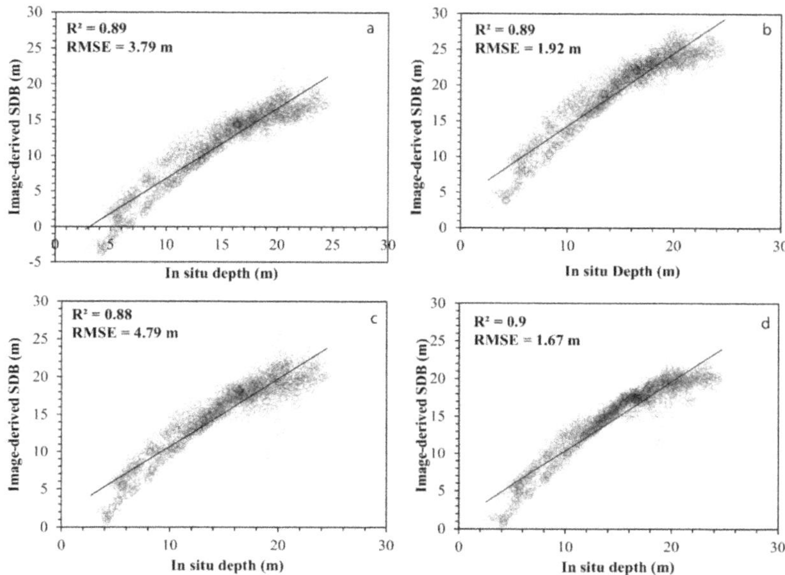

Figure 5. Validation plots of in situ depth points (x-axis) against image-derived bathymetries (y-axis) implementing the Lyzenga85 model which displays the best accuracy from the median Sentinel-2 composites of Apokoronas site for successive methodological steps including: (**a**) the initial median Sentinel-2 composite; (**b**) a normalised median Sentinel-2 composite using the shown pseudo-invariant (PIF) features in the zoomed maps of Figure 3b; (**c**) a 3 × 3 smoothed median Sentinel-2 composite; (**d**) a 3 × 3 smoothed and normalised median Sentinel-2 composite which shows the best accuracy among the four models of the present study in the Apokoronas site.

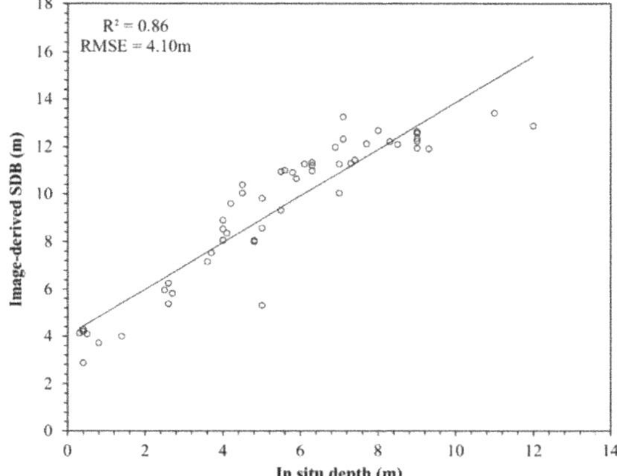

Figure 6. Validation plot of 53 in situ depth points (x-axis) against image-derived bathymetries (y-axis) employing the Lyzenga85 model on the 3 × 3 smoothed and normalised median Sentinel-2 composite of Thermaikos site.

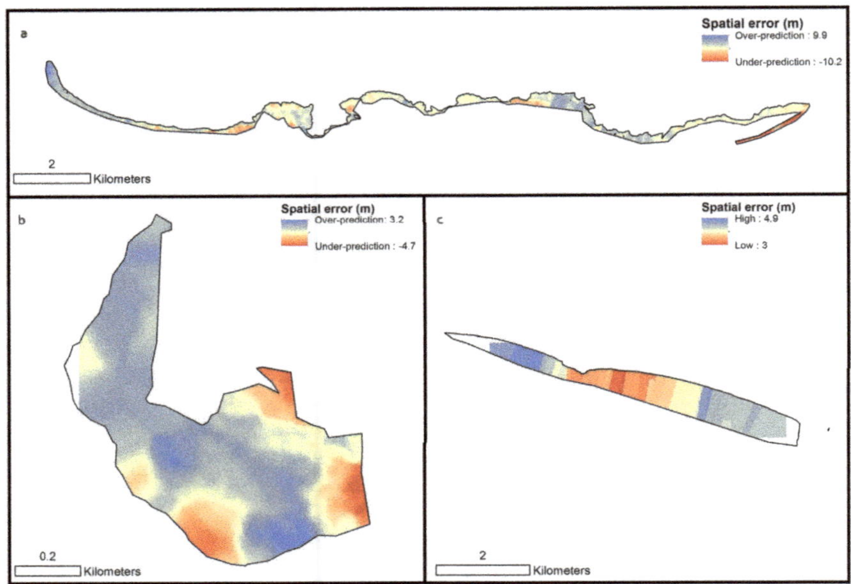

Figure 7. Spatial distribution of model residuals of the SDB maps with the highest accuracy. (**a**) the training site (National Samaria Park, 25-m maximum depth); (**b**,**c**) the two validation sites—Apokoronas in panel b (25-m maximum depth) and Thermaikos in panel c (12-m maximum depth). Note that the upper and lower left ramps indicate over- and under-prediction (as the difference between satellite-derived and in situ bathymetry), while the lower right ramp depicts high to low spatial error (both over-prediction).

3.4. SDB Sensitivity to Variation in Seabed Habitat and Reflectance

To further explore the performance of the SDB models with bottom habitat variability, we examine the two most widely implemented empirical algorithms, Lyzenga85 and Stumpf03, and the blue median reflectance composite (490 nm) along two different transects (0–20-m depth range) in the three study areas (Figure 8). All references to panels in this paragraph are as given for those in Figure 8. All transects run seaward and nearly perpendicularly to the coastline.

Generally, the better performance of Lyzenga85 in comparison to the modified Stumpf03 is exemplified by the smoother variation of the former model with changes in seabed habitats and thus more blue reflectance than in the latter model. Stumpf03 and blue reflectance appear to be more dependent across the majority of transects; at the same time, the Lyzenga85 model exhibits a smaller habitat interdependence (as manifested by variations of blue reflectance), especially over dark-reflectance bottom areas (seagrass and algae) as seen in the upper left (240–500 m), middle right, and lower left panels. Additionally, in comparison to Stumpf03, Lyzenga85 seems to underpredict bathymetry over sandy and rocky habitats, except for the deeper sand (b2 < ~0.015) (upper left and right, middle left, and lower left panels).

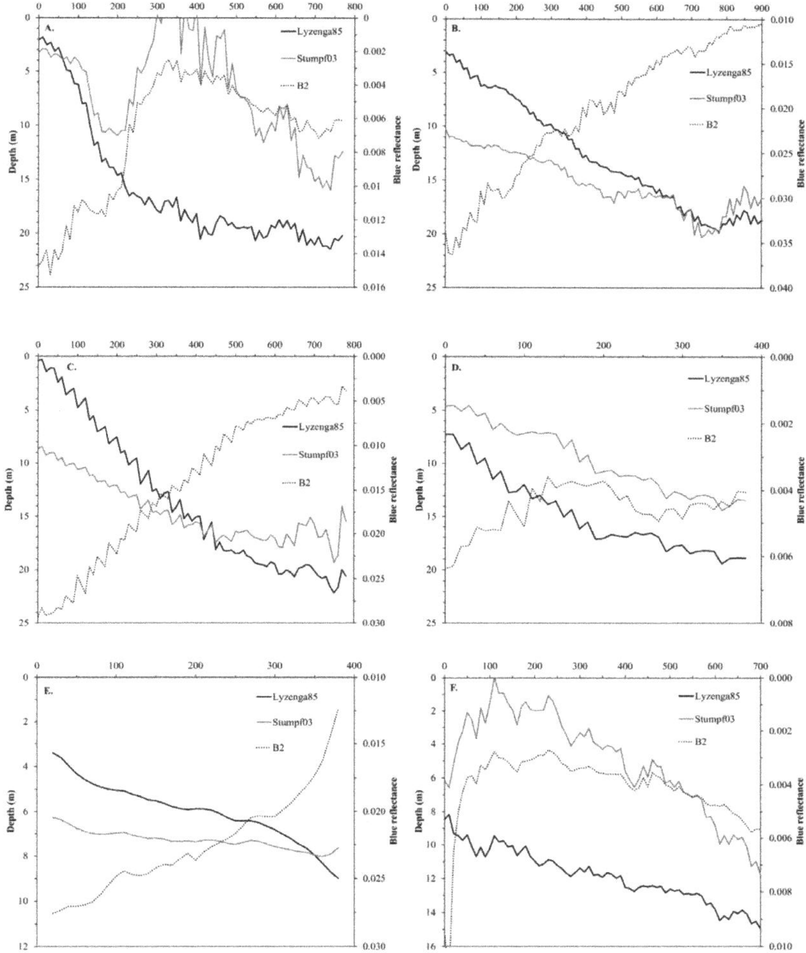

Figure 8. Satellite-derived bathymetry profiles using most accurate Lyzenga85 and modified Stumpf03 models, and a 3 × 3 smoothed and normalised median blue Sentinel-2 reflectance composite (490 nm). (**A**) Samaria National Park: profile over seabed with rocks, sand, and *Posidonia oceanica* seagrass, successively; (**B**) Samaria National Park: profile over sand; (**C**,**D**) the Apokoronas site: profiles over sand and algae-covered beds, respectively; (**E**,**F**) Thermaikos site: profiles over sand and *P. oceanica* seagrass, respectively.

4. Discussion

4.1. Cloud-Based SDB and Model Performance

Cloud-based geospatial analysis platforms, such as the herein utilised Google Earth Engine, provide an unprecedented opportunity for large-scale, on the fly preprocessing, processing, and analysis of vital to the coastal marine environment open data. Here, we developed a preprocessing workflow within GEE and a GIS environment which implements a plethora of widely used and well established algorithms in the remote sensing of optically shallow habitats (seagrasses, corals, algae etc.)

prior to the calculation of four empirical satellite-derived bathymetries [28,30–32] with Sentinel-2 along with their statistics and model residuals.

During the last three decades, a series of empirical, semi-analytical, and analytical methods have been developed for the estimation of optically shallow bathymetry from remote sensing images [34]. Here, we focus on the empirical methodologies which employ linear- or ratio-based statistical relationships between the log-transformed, water-penetrating bands (most frequently at blue and green wavelengths) and acoustic-derived in situ depth data to calculate SDB. Empirical algorithms assume a priori homogeneous water column and bottom composition; this is rarely the case, however, in a typical coastal Aegean benthic site with several seagrasses neighbouring sands and rocks with algal communities within the optical depth limit of the herein used Sentinel-2. Therefore, the non-unique nature of our selected study sites somehow breaches the assumptions of the empirical approaches. Nevertheless, we examine them here due to their historical significance, simple and widespread utilisation, and good accuracy. Among the investigated models in both training and validation in the two depth ranges (0–12 and 0–25 m), the Lyzenga85 model [28] exhibits the highest R^2 and smaller RMSE values on average (0.81 and 2.4 m respectively) followed by the Traganos17 model [31] with an R^2 of 0.45 and RMSE of 4.03 m. The empirical model of Stumpf03 features the poorest overall performance with an R^2 of 0.3 and RMSE of 4.41 m (Table S1 and S2). All models are tuned in the National Samaria Park over primarily sand- and rocky-covered seabed with sparse seagrass patches and validated over the mainly sandy bottom with a few algae in Apokoronas and the dense seagrass beds with few sands and algae in the Thermaikos Gulf (Figure 4). In comparison to unpublished SDB results in the South Cretan training site using the Dierssen03 [33] empirical band ratio of a super-resolved (60 to 10 m) Sentinel-2 coastal aerosol band 1 to green band 3 tuned with the same in situ depth data with the present study (depth range of 0–30 m), the herein Dierssen03 model produces a lower lower R^2 (reduced by 0.19), but also 0.72-m smaller RMSE.

On the other hand, previously published SDB results using Sentinel-2 in the validation site of the Thermaikos [31] with the original Traganos17 algorithm tuned in the same waters at depths between 0 and 20 m explained 1.76 times better the variation, with a 1.19-m smaller RMSE in comparison to the present Traganos17 results. In yet another SDB application in the Thermaikos site [34] with 5-m RapidEye imagery, the use of Dierssen03 empirical ratio exhibited a 0.49 lower R^2 with a decreased RMSE of 0.09 m in comparison to the herein validated Dierrsen03 model. In the previous approaches, training and validation data points originated from the same site and thus were characterised by high autocorrelation according to the first law of geography by Tobler which could have caused statistical bias; here, we attempt to lower this bias by employing two independent in situ data sets in two different sites to validate the calibrated empirical models.

4.2. On the Importance of Image Composition, PIF and the Spatial Distribution of Errors

The ability to create remotely sensed image composites by calculating the median of every pixel in the image across a given time range mitigates in an inter-image fashion intra-image issues related to water surface and water quality conditions, and optical properties; issues like waves, sun glint and sky glint, temporal sedimentation due to land-based rainfall runoffs, or resuspension of seabed sediments due to intensive wave activity for long periods tend to degrade the quality of single images and impede SDB applications. The selected time range affects the composite quality—a short time range could possibly cause data gaps while a long one could amplify the artificiality of the resulting pseudo-image. Here, we selected the time period between 1 August and 31 December 2016, which is chronologically closer to the acquisition of field bathymetry data within the Sentinel-2 lifespan and satisfies the criterion of a better-stratified water column for our study region (Aegean Sea) [23]. Future users of the herein proposed workflow are naturally expected to choose the time range based on the acquisition dates of their in situ data, and the optimum surface and water column conditions of their study area. It is worth mentioning that the temporal difference between data acquisition (2012–2015) and image composition (2016) in our study is not expected to obstruct the SDB estimations because all three sites are relatively protected from both weather phenomena and coastal processes, hence featuring a stable seabed. While the present work is within a coastal marine environment with almost no influence from tides, when the used methods and the proposed approaches are adopted at areas with tides that affect the results, a correction of the tide using measurements from tide gauges at fixed stations is a mandatory step in the satellite derived bathymetry process.

The radiometric normalization of the median image composites using the pseudo-invariant-feature approach [21] allows the successful development of our approach. The implemented image composites for validation here originate in sites which are approximately 26 (Apokoronas) and 562 km (Thermaikos) (Figure 1) away from the training site (National Samaria Park). This necessitates the utilisation of the PIF normalisation method to correct the two former composites to the reflectance range of the latter composite. The yielded better statistics following utilisation of PIF and 3 × 3 low pass in combination with the smaller spread of regressed values (Figure 5d) manifest the importance of these two preprocessing steps. The image composition using median values prior to the application of PIF increases the reliability of the subsequent SDB estimations due to the decrease of cross-image composite comparison issues [35]. To the best of our knowledge, this study is the first to employ PIF as a preprocessing step for SDB calculations. Previous uses of PIF normalisation within an optically coastal shallow setting have been confided to its preprocessing implementation for coral bleaching detection [29,36].

The depiction of the spatial distribution of model residuals (Figure 7) is integral here. There are two inherent assumptions in the empirical nature of SDB estimations; first, in situ observations employed in the training and validation of a given SDB model are independent in between them, and second, model residuals feature a normal distribution and random location. The presence of spatial dependence shown by the selection of observations from the same image (and thus in close proximity) adheres to the first law of geography by Tobler but violates the assumed freedom and location of in situ measurements, heightening the standard error and broadly lowering the statistical confidence of the SDB models. On one hand, the origination of both training and validations datasets from the same site is common practice in optically shallow bathymetry derivation studies due to practical reasons (usually the high cost of acquiring such data in the field). On the other hand, the existence of three in situ single beam-derived bathymetry datasets allows us to confront the aforementioned violation, subsequently restricting statistical bias in the results.

The spatial error maps unveil specific over- and under-prediction patterns. Demonstrated more pronouncedly in the National Samaria Park and Apokoronas sites (upper and lower left panel in Figure 7), we identify that mainly algae-covered beds relate to an over-prediction tendency because they comprise a low-reflectance habitat. In the very shallow pixels of Apokoronas, there might be an over-correction by the sun glint algorithm found in [26] due to the interaction of light photons with the seabed in the NIR wavelength which may have decreased reflectance composite values and added to the over-prediction trends. On the contrary, sandy and rocky bottoms reflect more photons, hence producing under-prediction patterns. Generally, the number of photons which reflect on the seabed decreases, and reaches zero past a certain depth limit as averaged by the herein image composition. At this point, the remote sensing signal would contain information arising only from the water column and not the water bottom, therefore being unable to estimate bathymetry. This could have caused the visible under-prediction patterns towards deeper areas independent of the underlying habitat as manifested by the fourth panel in Figure 5 (beyond ~22 m), and the upper and lower left panel in Figure 7 (red colours).

We also show the importance of using the same depth ranges of in situ data for both training and validation of our SDB models, which increase R^2 and decrease RMSE. Implementation of the 0–25 m training in National Samaria Park, validated in the Apokoronas site model in the Thermaikos site and spanning depths of up to 12 m, produces a 0.06 smaller R^2 value and a nearly two-fold smaller RMSE than the final trained Lyzenga85 model within 0–12 m (Figure 6). This has been also mentioned in [34]: "*The optimal performance of bathymetric estimates is like to be achieved when points covering a fully-representative range of depths are present in both datasets*".

4.3. In Situ Data Collection, Crowdsourced Information and an Outlook for SDB

A vital feature of the present study is the use of low-cost tools for the collection of in situ bathymetric data. Usually, such data are collected by expensive state-of-the-art equipment such as airborne lidar data [37] or multibeam echosounder systems [38]. While these tools derive very high-resolution bathymetries, both their spatial coverage and updating of the derived bathymetry are limited and particularly costly at the shallow coastal zone which is practically inaccessible for large boats carrying the acoustic equipment and where the multibeam swath is narrow. Additionally, airborne campaigns require special licenses that are difficult to obtain. Here, we employ commercial off-the-shelf (COTS) solutions with a total cost of less than 1500 euros, which can be utilised at any region of interest.

Another method of in situ data collection is the involvement of crowdsourcing/citizen science data collection methodology [39]. Since 1963, the Cooperative Charting Program between NOAA's Office of Coast Survey and the United States Power Squadron (USPS) has triggered the submission of bathymetric point data to cartographers via the postal service for chart application. Up to today, several initiatives based on the voluntary participation of boat owners with installed hydrographic equipment have been released [35,40], resulting in the provision of bathymetric maps and information on bottom type and vegetation cover in some cases. Most recently, the Nippon Foundation General Bathymetric Chart of the Oceans (GEBCO) Seabed 2030 project was launched at the United Nations Oceans Conference in 2017 with the goal to map the entirety of the world's ocean seabed by 2030. To achieve this unquestionably ambitious goal, the project aims to create a new fleet of research vessels by employing millions of fishing boats, thousands of cargo, cruise, and passenger ships, and private yachts to acquire crowdsourcing multibeam echosounder data [41]. Nonetheless, the extraction of raw data (XYZ) from the existing databases is not yet permitted for the public, but companies and research projects—such as the H2020 BASE-platform (https://base-platform.com/)—have reached an agreement and the first SDB results utilising crowdsourcing data are now available. The accessibility of such raw data is crucial towards a new era for global coastal bathymetry applications using the present proposed methodology. The empirical nature of our methodology, however, raises the computational demands within the GEE due to the estimation of the regressions between the image composition values and water depth—large scales in both space and time could cause the GEE to create a time-out (GEE error message).

In addition to the envisioned new era for large-scale bathymetry, we discuss five near-future endeavours which could directly or indirectly succeed and/or improve the present methodology and study:

1. The implementation of the estimated spatial error parameter as a postprocessing step to increase the statistical accuracy of the empirically derived SDB models following the work in [34].
2. The utilisation of best-available-pixel (BAP) composition, which employs a series of pixel-based scores related to distance to clouds and shadow masks, atmospheric opacity, day of the year etc., instead of image composition across a relevant time series to the user's study region [42].
3. The use of the proposed preprocessing workflow to conduct sea- to basin-wide habitat mapping and monitoring.
4. The incorporation of radiative transfer-based optimisations following the semi-analytical inversion method of [43] or machine learning methods [10,44] for the derivation of bathymetry.
5. The fusion of the Copernicus Sentinel-1 (also available in GEE) and the Sentinel-2 open and free image archive for the development of 10-m topobathymetric digital elevation models (DEMs)—seamless merged elevation products of terrestrial and underwater topography [45] useful for numerous Earth science applications including mapping and modelling of inundation, sediment transport, sea-level rise, and storm surge [46].

5. Conclusions

The present study proposes a complete preprocessing chain in Google Earth Engine—including simple and well-established algorithms in the remote sensing of optically shallow habitats—which could be easily implemented through the online provided code to estimate satellite-derived bathymetry using Sentinel-2 data and low-cost field data. The user can select a suitable time range of available Sentinel-2 images according to the available in situ depth data and optimum surface and water column conditions of the area of interest. Here, we train four SDB models utilising a pre-processed median composite of 79 S2 tiles in SW Crete (Eastern Mediterranean) and validate them employing 79- and 18-tile composites in NW Crete and the NW Aegean Sea, 26 and 562 km away, respectively. Given the good accuracies of the calibrated model in the two validation sites (R^2 up to 0.9 and RMSE as low as 1.67 m) despite the large horizontal distance, there is emerging potential for upscaling the herein developed SDB model to the whole optically shallow extent of the Aegean and Ionian Seas, hence further exploiting the capability of GEE for big data analysis. To this end, one particular challenge would surely be the existence of relevant field bathymetric data to validate the accuracy of the upscaling effort. The innovation of this work lies mainly in the fact that it is the first that implements the inter-image approach of image composition within GEE to address single-image issues like atmospheric, surface, and water column interferences which obstruct space-borne approaches in the field of aquatic remote sensing. Moreover, in comparison to their exclusion, the inclusion of PIF-based normalisation to match the reflectance range of the training composite to the validation composites along with a 3×3 smoothing to filter remaining noise prior to the SDB calculation notably improve the methodology as manifested by the increased SDB accuracies. On the other hand, regressions which lie in the heart of the empirical models (e.g., sun glint correction, PIF normalisation, SDB estimations) decrease processing time in GEE and could possibly create related time-out errors. This led us to conduct the radiometric normalisation and the SDB calculations outside GEE in this study for the sake of efficiency. In the near future, we aim to integrate these two preprocessing and analysis steps within the GEE platform in addition to adapting the code to also use Landsat 8 images as input. All in all, due to intense anthropogenic activities, coastal ecosystems are on the verge of significant degradation of their ecosystem services; however, Google Earth Engine and Sentinel-2 have created the perfect storm in the last three years for an unprecedented, global-scale, high spatiotemporal mapping of bathymetry and, more broadly, of the immensely vital optically shallow benthos which can in turn empower physical understanding, management and conservation practices.

Supplementary Materials: The following are available online at http://www.mdpi.com/2072-4292/10/6/859/s1.

Author Contributions: D.T. and D.P. conceived the idea, collected the in-situ data, processed the satellite-derived bathymetries, and wrote the paper; B.A. developed the preprocessing code in Google Earth Engine; P.R. and N.C. supervised the development of the present project, from start to finish.

Acknowledgments: D.P. and N.C. are supported by the European H2020 Project 641762 ECOPOTENTIAL: *Improving future ecosystem benefits through Earth Observations*. D.T. is supported by a DLR-DAAD Research Fellowship (No. 57186656).

Conflicts of Interest: The authors declare no conflict of interest.

References

1. Paterson, D.M.; Hanley, N.D.; Black, K.; Defew, E.C.; Solan, M. (Eds.) Biodiversity, ecosystems and coastal zone management: Linking science and policy. Theme Section. *Mar. Ecol. Prog. Ser.* **2011**, *434*, 201–301. [CrossRef]
2. Robertson, E. Crowd-Sourced Bathymetry Data via Electronic Charting Systems. ESRI Ocean GIS Forum, 2016. Available online: http://proceedings.esri.com/library/userconf/oceans16/papers/oceans_12.pdf (accessed on 20 April 2018).
3. Li, R.; Liu, J.-K.; Felus, Y. Spatial Modeling and Analysis for Shoreline Change Detection and Coastal Erosion Monitoring. *Mar. Geod.* **2010**, *24*, 1–12. [CrossRef]
4. Omira, R.; Baptista, M.A.; Leone, F.; Matias, L.; Mellas, S.; Zourarah, B.; Miranda, J.M.; Carrilho, F.; Cherel, J.P. Performance of coastal sea-defense infrastructure at El Jadida (Morocco) against tsunami threat: Lessons learned from the Japanese 11 March 2011 tsunami. *Nat. Hazards Earth Syst. Sci.* **2013**, *13*, 1779–1794. [CrossRef]

5. Roelfsema, C.; Kovacs, E.; Ortiz, J.C.; Wolff, N.H.; Callaghan, D.; Wettle, M.; Ronan, M.; Hamylton, S.M.; Mumby, P.J.; Phinn, S. Coral reef habitat mapping: A combination of object-based image analysis and ecological modelling. *Remote Sens. Environ.* **2018**, *208*, 27–41. [CrossRef]
6. Wang, J.; Yi, S.; Li, M.; Wang, L.; Song, C. Effects of sea level rise, land subsidence, bathymetric change and typhoon tracks on storm flooding in the coastal areas of Shanghai. *Sci. Total Environ.* **2018**, *621*, 228–234. [CrossRef] [PubMed]
7. Sandwell, D.T.; Smith, W.H. Marine gravity anomaly from Geosat and ERS 1 satellite altimetry. *J. Geophys. Res. Solid Earth* **1997**, *102*, 10039–10054. [CrossRef]
8. Olson, C.J.; Becker, J.J.; Sandwell, D.T. A new global bathymetry map at 15 arcsecond resolution for resolving seafloor fabric: SRTM15_PLUS. In Proceedings of the AGU Fall Meeting Abstracts, San Francisco, CA, USA, 15–19 December 2014.
9. Pe'eri, S.; Madore, B.; Nyberg, J.; Snyder, L.; Parrish, C.; Smith, S. Identifying bathymetric differences over Alaska's North Slope using a satellite-derived bathymetry multi-temporal approach. *J. Coast. Res.* **2016**, *76*, 56–63. [CrossRef]
10. Misra, A.; Vojinovic, Z.; Ramakrishnan, B.; Luijendijk, A.; Ranasinghe, R. Shallow water bathymetry mapping using Support Vector Machine (SVM) technique and multispectral imagery. *Int. J. Remote Sens.* **2018**. [CrossRef]
11. Pacheco, A.; Horta, J.; Loureiro, C.; Ferreira, Ó. Retrieval of nearshore bathymetry from Landsat 8 images: A tool for coastal monitoring in shallow waters. *Remote Sens. Environ.* **2015**, *159*, 102–116. [CrossRef]
12. Dörnhöfer, K.; Göritz, A.; Gege, P.; Pflug, B.; Oppelt, N. Water Constituents and Water Depth Retrieval from Sentinel-2A—A First Evaluation in an Oligotrophic Lake. *Remote Sens.* **2016**, *8*, 941. [CrossRef]
13. Chybicki, A. Mapping South Baltic Near-Shore Bathymetry Using Sentinel-2 Observations. *Pol. Mar. Res.* **2017**, *24*, 15–25. [CrossRef]
14. Chybicki, A. Three-Dimensional Geographically Weighted Inverse Regression (3GWR) Model for Satellite Derived Bathymetry Using Sentinel-2 Observations. *Mar. Geod.* **2017**, *41*, 1–23. [CrossRef]
15. Gorelick, N.; Hancher, M.; Dixon, M.; Ilyushchenko, S.; Thau, D.; Moore, R. Google Earth Engine: Planetary-scale geospatial analysis for everyone. *Remote Sens. Environ.* **2017**, *202*, 18–27. [CrossRef]
16. Robinson, N.P.; Allred, B.W.; Jones, M.O.; Moreno, A.; Kimball, J.S.; Naugle, D.E.; Erickson, T.A.; Richardson, A.D. A Dynamic Landsat Derived Normalized Difference Vegetation Index (NDVI) Product for the Conterminous United States. *Remote Sens.* **2017**, *9*, 863. [CrossRef]
17. Xiong, J.; Thenkabail, P.S.; Tilton, J.C.; Gumma, M.K.; Teluguntla, P.; Oliphant, A.; Congalton, R.G.; Yadav, K.; Gorelick, N. Nominal 30-m Cropland Extent Map of Continental Africa by Integrating Pixel-Based and Object-Based Algorithms Using Sentinel-2 and Landsat-8 Data on Google Earth Engine. *Remote Sens.* **2017**, *9*, 1065. [CrossRef]
18. Parastatidis, D.; Mitraka, Z.; Chrysoulakis, N.; Abrams, M. Online Global Land Surface Temperature Estimation from Landsat. *Remote Sens.* **2017**, *9*, 1208. [CrossRef]
19. Chrysoulakis, N.; Mitraka, Z.; Gorelick, N. Exploiting satellite observations for global surface albedo trends monitoring. *Theor. Appl. Climatol.* **2018**. accepted.
20. Tobler, W.R. A computer movie simulating urban growth in the detroit region. *Econ. Geogr.* **1970**, *46*, 234–240. [CrossRef]
21. Schott, J.R.; Salvaggio, C.; Vochok, W.J. Radiometric scene normalization using pseudo-invariant features. *Remote Sens. Environ.* **1988**, *26*, 1–16. [CrossRef]
22. Poursanidis, D.; Topouzelis, K.; Chrysoulakis, N. Mapping coastal marine habitats and delineating the deep limits of the Neptune's seagrass meadows using Very High Resolution Earth Observation data. *Int. J. Remote Sens.* **2018**. accepted.
23. Tanhua, T.; Hainbucher, D.; Schroeder, K.; Cardin, V.; Álvarez, M.; Civitarese, G. The Mediterranean Sea system: A review and an introduction to the special issue. *Ocean Sci.* **2013**, *9*, 789–803. [CrossRef]
24. Breiman, L.; Friedman, J.H.; Olshen, R.A.; Stone, C.J. *Classification and Regression Trees;* Chapman & Hall/CRC: Boca Raton, FL, USA, 1984.
25. Armstrong, R.A. Remote sensing of submerged vegetation canopies for biomass estimation. *Int. J. Remote Sens.* **1993**, *14*, 621–627. [CrossRef]
26. European Space Agency (ESA). *SENTINEL-2 User Handbook;* ESA: Paris, France, 2015; p. 64.

27. Hedley, J.D.; Harborne, A.R.; Mumby, P.J. Technical note: Simple and robust removal of sun glint for mapping shallow-water benthos. *Int. J. Remote Sens.* **2005**, *26*, 2107–2112. [CrossRef]
28. Lyzenga, D.R. Shallow-water bathymetry using combined lidar and passive multispectral scanner data. *Int. J. Remote Sens.* **1985**, *6*, 115–125. [CrossRef]
29. Elvidge, C.D.; Dietz, J.B.; Berkelmans, R.; Andréfouët, S.; Skirving, W.; Strong, A.E.; Tuttle, B.T. Satellite observation of Keppel Islands (Great Barrier Reef) 2002 coral bleaching using IKONOS data. *Coral Reefs* **2004**, *23*, 461–462. [CrossRef]
30. Stumpf, R.P.; Holderied, K.; Sinclair, M. Determination of water depth with high-resolution satellite imagery over variable bottom types. *Limnol. Oceanogr.* **2003**, *48*, 547–556. [CrossRef]
31. Traganos, D.; Reinartz, P. Mapping Mediterranean seagrasses with Sentinel-2. *Mar. Pollut. Bull.* **2017**. [CrossRef] [PubMed]
32. Dierssen, H.M.; Zimmerman, R.C.; Leathers, R.A.; Downes, T.V.; Davis, C.O. Ocean color remote sensing of seagrass and bathymetry in the Bahamas Banks by high- resolution airborne imagery. *Limnol. Oceanogr.* **2003**, *48*, 444–455. [CrossRef]
33. Hamylton, S.M.; Hedley, J.D.; Beaman, R.J. Derivation of High-Resolution Bathymetry from Multispectral Satellite Imagery: A Comparison of Empirical and Optimisation Methods through Geographical Error Analysis. *Remote Sens.* **2015**, *7*, 16257–16273. [CrossRef]
34. Traganos, D.; Reinartz, P. Interannual Change Detection of Mediterranean Seagrasses Using RapidEye Image Time Series. *Front. Plant Sci.* **2018**, *9*. [CrossRef] [PubMed]
35. TeamSurv, 2018. Available online: https://www.teamsurv.com/ (accessed on 28 Match 2018).
36. Hedley, J.D.; Roelfsema, C.M.; Chollett, I.; Harborne, A.R.; Heron, S.F.; Weeks, S.; Skirving, W.J.; Strong, A.E.; Eakin, C.M.; Christensen, T.R.L.; et al. Remote Sensing of Coral Reefs for Monitoring and Management: A Review. *Remote Sens.* **2016**, *8*, 118. [CrossRef]
37. Saylam, K.; Hupp, J.R.; Averett, A.R.; Gutelius, W.F.; Gelhar, B.W. Airborne lidar bathymetry: Assessing quality assurance and quality control methods with Leica Chiroptera examples. *Int. J. Remote Sens.* **2018**, *39*, 2518–2542. [CrossRef]
38. Ierodiaconou, D.; Schimel, A.C.G.; Kennedy, D.; Rattray, A. Combining pixel and object based image analysis of ultra-high resolution multibeam bathymetry and backscatter for habitat mapping in shallow marine waters. *Mar. Geophys. Res.* **2018**, *39*, 271. [CrossRef]
39. International Hydrographic Organization. *Guidance on Crowdsourced Bathymetry*; IHO, Monaco Cedex, 2018; p. 55. Available online: https://www.iho.int/iho_pubs/draft_pubs/CSB-Guidance_Document-Ed1.0.0.pdf (accessed on 20 April 2018).
40. BioBase, 2018. Available online: https://www.cibiobase.com/ (accessed on 28 Match 2018).
41. Nippon Foundation-GEBCO, 2018. Available online: https://seabed2030.gebco.net/ (accessed on 2 May 2018).
42. White, J.C.; Wulder, M.A.; Hobart, G.W.; Luther, J.E.; Hermosilla, T.; Griffiths, P.; Coops, N.C.; Hall, R.J.; Hostert, P.; Dyk, A.; et al. Pixel-based image compositing for large-area dense time series applications and science. *Can. J. Remote Sens.* **2014**, *40*, 192–212. [CrossRef]
43. Lee, Z.P.; Carder, K.L.; Mobley, C.D.; Steward, R.G.; Patch, J.S. Hyperspectral remote sensing for shallow waters: 2. Deriving bottom depths and water properties by optimization. *Appl. Opt.* **1999**, *38*, 3831–3843. [CrossRef] [PubMed]
44. Danilo, C.; Melgani, F. Wave period and coastal bathymetry using wave propagation on optical images. *IEEE Trans. Geosci. Remote Sens.* **2016**, *54*. [CrossRef]
45. Collin, A.; Hench, J.L.; Pastol, Y.; Planes, S.; Thiault, L.; Schmitt, R.J.; Holbrook, S.J.; Davies, N.; Troyer, M. High resolution topobathymetry using a Pleiades-1 triplet: Moorea Island in 3D. *Remote Sens. Environ.* **2018**, *208*, 109–119. [CrossRef]
46. Poursanidis, D.; Chrysoulakis, N. Remote Sensing, natural hazards and the contribution of ESA Sentinels missions. *Remote Sens. Appl. Soc. Environ.* **2017**, *6*, 25–38. [CrossRef]

Technical Note

Mean Composite Fire Severity Metrics Computed with Google Earth Engine Offer Improved Accuracy and Expanded Mapping Potential

Sean A. Parks [1,*], Lisa M. Holsinger [1], Morgan A. Voss [2], Rachel A. Loehman [3] and Nathaniel P. Robinson [4]

1 Aldo Leopold Wilderness Research Institute, Rocky Mountain Research Station, US Forest Service, 790 E. Beckwith Ave., Missoula, MT 59801, USA; lisamholsinger@fs.fed.us
2 Department of Geography, University of Montana, Missoula, MT 59812, USA; morgan.voss@umontana.edu
3 Alaska Science Center, US Geological Survey, 4210 University Drive, Anchorage, AK 99508, USA; rloehman@usgs.gov
4 W.A. Franke College of Forestry and Conservation & Numerical Terradynamic Simulation Group, University of Montana, Missoula, MT 59812, USA; Nathaniel.Robinson@umontana.edu
* Correspondence: sean_parks@fs.fed.us; Tel.: +1-406-542-4182

Received: 4 May 2018; Accepted: 4 June 2018; Published: 5 June 2018

Abstract: Landsat-based fire severity datasets are an invaluable resource for monitoring and research purposes. These gridded fire severity datasets are generally produced with pre- and post-fire imagery to estimate the degree of fire-induced ecological change. Here, we introduce methods to produce three Landsat-based fire severity metrics using the Google Earth Engine (GEE) platform: The delta normalized burn ratio (dNBR), the relativized delta normalized burn ratio (RdNBR), and the relativized burn ratio (RBR). Our methods do not rely on time-consuming a priori scene selection but instead use a mean compositing approach in which all valid pixels (e.g., cloud-free) over a pre-specified date range (pre- and post-fire) are stacked and the mean value for each pixel over each stack is used to produce the resulting fire severity datasets. This approach demonstrates that fire severity datasets can be produced with relative ease and speed compared to the standard approach in which one pre-fire and one post-fire scene are judiciously identified and used to produce fire severity datasets. We also validate the GEE-derived fire severity metrics using field-based fire severity plots for 18 fires in the western United States. These validations are compared to Landsat-based fire severity datasets produced using only one pre- and post-fire scene, which has been the standard approach in producing such datasets since their inception. Results indicate that the GEE-derived fire severity datasets generally show improved validation statistics compared to parallel versions in which only one pre-fire and one post-fire scene are used, though some of the improvements in some validations are more or less negligible. We provide code and a sample geospatial fire history layer to produce dNBR, RdNBR, and RBR for the 18 fires we evaluated. Although our approach requires that a geospatial fire history layer (i.e., fire perimeters) be produced independently and prior to applying our methods, we suggest that our GEE methodology can reasonably be implemented on hundreds to thousands of fires, thereby increasing opportunities for fire severity monitoring and research across the globe.

Keywords: burn severity; change detection; Landsat; dNBR; RdNBR; RBR; composite burn index (CBI); MTBS

---

1. Introduction

The degree of fire-induced ecological change, or fire severity, has been the focus of countless studies across the globe [1–5]. These studies often rely on gridded metrics that use pre- and post-fire imagery to estimate the amount of fire-induced change; the most common metrics are the delta normalized burn ratio (dNBR) [6], the relativized delta normalized burn ratio (RdNBR) [7], and the relativized burn ratio (RBR) [8]. These metrics generally have a high correspondence ($r^2 \geq 0.65$) to field-based measures of fire severity [9–12], making them an attractive alternative to expensive and time-consuming collection of post-fire field data. These satellite-inferred fire severity metrics are often produced using Landsat Thematic Mapper (TM), Enhanced Thematic mapper Plus (ETM+), and Operational Land Imager (OLI) imagery due to their combined temporal depth (1984-present) and global coverage, although they can be produced from other sensors such as the Moderate Resolution Imaging Spectroradiometer (MODIS) [13] and Sentinal2A [14].

However, producing satellite-inferred fire severity datasets can be challenging, particularly if severity data are needed for a large number of fires (>~20) or over broad spatial extents. For example, expertise in remote sensing technologies and software is necessary, indicating the need for a remote-sensing specialist or a substantial investment of time to learn such technologies and software. Furthermore, fire severity datasets have traditionally been produced using one pre-fire and one post-fire Landsat image [15,16], which requires careful attention to scene selection. Image selection can be time consuming in terms of identifying scenes with no clouds covering the fire of interest and avoiding scenes affected by a low sun angle and those with mismatched phenology between pre- and post-fire conditions [6,17]. Even when careful attention to image selection has been achieved, some images (those from Landsat ETM+ acquired after 2003) and the resulting gridded severity datasets will have missing data due to the failure of the Scan Line Corrector [18].

Challenges in producing satellite-inferred severity datasets have likely hampered development of regional to national fire severity products in many countries. The exception is in the United States (US), where Landsat-derived severity metrics have been produced for all 'large' fires (those ≥400 ha in the western US and ≥250 ha in the eastern US) that have occurred since 1984 [19]. This effort, undertaken by the US government, is called the Monitoring Trends in Burn Severity (MTBS) program and has mapped the perimeter and severity of over 20,000 fires. The MTBS program has provided data for numerous scientific studies ranging from those involving <10 fires [20–22] to those involving >1000 fires [2,23,24] and for topics such as fuel treatment effectiveness, climate change impacts, and time series analyses [25–28]. The fire severity datasets produced by the MTBS program have clearly advanced wildland fire research in the US. Although some studies involving the trends, drivers, and distribution of satellite-inferred fire severity are evident outside of the US [4,5,15,29,30], the number and breadth of such studies are relatively scarce and restricted compared to those conducted in the US. We suggest that, if spatially and temporally comprehensive satellite-inferred severity metrics were more widely available in other countries or regions, opportunities for fire severity monitoring and research would increase substantially.

In this paper, we present methods to quickly and easily produce Landsat-derived fire severity metrics (dNBR, RdNBR, and RBR). These methods are implemented within the Google Earth Engine (GEE) platform. As opposed to the standard approach in which one pre-fire and one post-fire Landsat scene are identified and used to produce these fire severity datasets, we use a mean compositing approach in which all valid pixels (e.g., cloud-free) over a pre-specified date range are stacked and the mean value for each pixel over each stack is calculated. Consequently, there is no need for a priori scene selection, which substantially speeds up the time necessary to produce fire severity datasets. The main caveat, however, is that a fire history GIS dataset (i.e., polygons of fire perimeters) must be available and produced independent of this process. Where fire history datasets are currently available or can easily be generated, our methods provide a means to produce satellite-inferred fire severity products similar to those distributed by the MTBS program. We also validate the severity metrics produced with our GEE methodology by evaluating the correspondence of dNBR, RdNBR, and RBR to a field-based measure of severity and measure the classification accuracy when categorized as low, moderate, and high severity. These validations were conducted on 18 fires in the western US [8] and were compared to parallel validations of fire severity datasets using one pre-fire and post-fire scene.

Code and a sample fire history GIS dataset are provided to aid users in replicating and implementing our methods.

2. Materials and Methods

2.1. Processing in Google Earth Engine

We produced the following Landsat-based fire severity metrics for each of the 18 fires that are described in Section 2.2; the perimeter of each fire was obtained from the MTBS program [19]. All fire severity metrics are based on the normalized burn ratio (NBR; Equation (1)) and include the: (i) Delta normalized burn ratio (dNBR; Equation (2)) [6]; (ii) relativized delta normalized burn ratio (RdNBR; Equation (3)) [7]; and (iii) relativized burn ratio (RBR; Equation (4)) [8]. These are produced using Landsat TM, ETM+, and OLI imagery.

$$NBR = \left(\frac{NIR - SWIR}{NIR + SWIR}\right) \quad (1)$$

$$dNBR = \left(NBR_{prefire} - NBR_{postfire}\right) \times 1000 \quad (2)$$

$$RdNBR = \begin{cases} \frac{dNBR}{\left|NBR_{prefire}\right|^{0.5}}, & \left|NBR_{prefire}\right| \geq 0.001 \\ \frac{dNBR}{|0.001|^{0.5}}, & \left|NBR_{prefire}\right| < 0.001 \end{cases} \quad (3)$$

$$RBR = \frac{dNBR}{NBR_{prefire} + 1.001} \quad (4)$$

where *NIR* (Equation (1)) is the near infrared band and *SWIR* (Equation (1)) is the shortwave infrared band. The $NBR_{prefire}$ qualifier in *RdNBR* (Equation (3)) is necessary because the equation fails when $NBR_{prefire}$ equals zero and produces very large values when it approaches zero.

Within GEE, mean pre- and post-fire NBR values (Equation (1)) across a pre-specified date range (termed a 'mean composite') were calculated per pixel across the stack of valid pixels (e.g., cloud- and snow-free pixels). For fires that occurred in Arizona, New Mexico, and Utah, the date range is April through June; for all other fires, the date range is June through September (Figure 1). These date ranges are based on various factors including the fire season, expected snow cover, expected cloud cover and latitude. We used the Landsat Surface Reflectance Tier 1 datasets, which among the bands, includes a quality assessment mask to identify those pixels with clouds, shadow, water, and snow. This mask is produced by implementing a multi-pass algorithm (called 'CFMask') based on decision trees and is described in detail by Foga et al. [31]. As such, pixels identified as cloud, shadow, water, and snow were excluded when producing the mean composite pre- and post-fire NBR. The resulting pre- and post-fire NBR mean composite images are then used to calculate dNBR, RdNBR, and RBR (Equations (2)–(4)). Our mean compositing approach renders the need for a priori scene selection unnecessary.

We also produced alternative versions of each severity metric in which we account for potential phenological differences between pre- and post-fire imagery, also known as the '$dNBR_{offset}$' [6]. The $dNBR_{offset}$ is the average dNBR of pixels outside the burn perimeter (i.e., unburned) and is intended to account for differences between pre- and post-fire imagery that arise due to varying conditions in phenology or precipitation between respective time periods. Incorporating the $dNBR_{offset}$ is advisable when making comparisons among fires [7,8]. For each fire, we determined the $dNBR_{offset}$ by calculating the mean dNBR value across all pixels located 180 m outside of the fire perimeter; informal testing indicated that a 180 m distance threshold adequately quantifies dNBR differences among unburned pixels. A simple subtraction of the fire-specific $dNBR_{offset}$ from each dNBR raster incorporates the $dNBR_{offset}$ [17]. The dNBR (with the offset) is then used to produce RdNBR and RBR (Equations (3) and (4)).

2.2. Validation

We aimed to determine whether our GEE methodology (specifically the mean compositing method) produced Landsat-based fire severity datasets with equivalent or higher validation statistics than severity datasets produced using one pre-fire and one post-fire scene (i.e., the standard approach since these metrics were introduced). This validation has three components (described below), all of which rely on 1681 field-based severity plots covering 18 fires in the western US that burned between 2001 and 2011; these are the same plots and fires that were originally evaluated by Parks et al. [8] (Figure 1) (Table 1). The field data represent the composite burn index (CBI) [6], which rates factors such as surface fuel consumption, soil char, vegetation mortality, and scorching of trees. CBI is rated on a continuous scale from zero to three, with CBI = 0 reflecting no change due to fire and CBI = 3 reflecting the highest degree of fire-induced ecological change. The fires selected by Parks et al. [8] and used in this study (Table 1) met the following criteria: (i) They had at least 40 field-based CBI plots; and (ii) at least 15% of the plots fell into each class representing low, moderate, and high severity. Of the 1681 field-based CBI plots, 30% are considered low severity (CBI < 1.25), 41% are moderate severity (CBI $\geq$ 1.25 and < 2.25), and 29% are high severity (CBI $\geq$ 2.25).

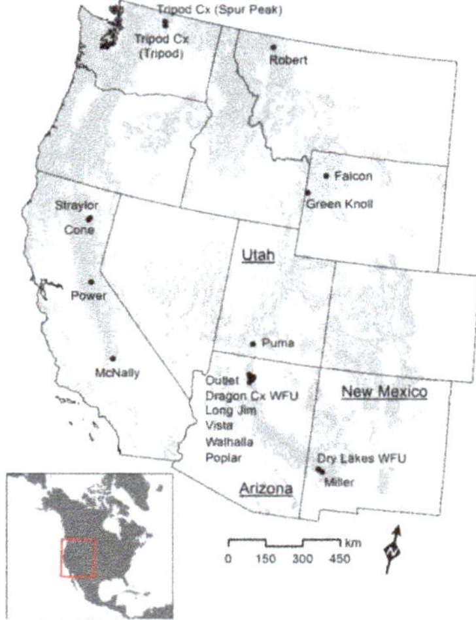

Figure 1. Location and names of the 18 fires included in the validation of the delta normalized burn ratio (dNBR), relativized delta normalized burn ratio (RdNBR), and relativized burn ratio (RBR). Forested areas in the western United States (US) are shown in gray shading. Inset shows the study area in relation to North America.

The first validation evaluates the correspondence of each severity metric to the CBI data for each fire. Exactly following Parks et al. [8], we extracted GEE-derived dNBR, RdNBR, and RBR values using bilinear interpolation and then used nonlinear regression in the R statistical environment [32] to evaluate the performance of each severity metric. Specifically, we quantified the correspondence of each severity metric (the dependent variable) to CBI (the independent variable) as the coefficient of determination, which is the R^2 of a linear regression between predicted and observed severity values. We conducted this analysis for each fire and reported the mean R^2 across the 18 fires. We then conducted a parallel analysis but used MTBS-derived severity datasets. This parallel analysis allows for a robust comparison of severity datasets produced using one pre-fire and one post-fire image (e.g., MTBS-derived metrics) with the mean compositing approach as achieved with GEE. This validation was conducted on the severity metrics without and with the $dNBR_{offset}$.

Table 1. Summary of fires analyzed in this study; this table is from originally Parks et al. 2014 [8].

				Historical Fire Regime [33]		
Fire Name	Year	Number of plots	Overstory species (in order of prevalence)	Surface	Mixed	Replace
Tripod Cx (Spur Peak) [1]	2006	328	Douglas-fir, ponderosa pine, subalpine fir, Engelmann spruce	80–90%	<5%	5–10%
Tripod Cx (Tripod) [1]	2006	160	Douglas-fir, ponderosa pine, subalpine fir, Engelmann spruce	>90%	<5%	<5%
Robert [2]	2003	92	Subalpine fir, Engelmann spruce, lodgepole pine, Douglas-fir, grand fir, western red cedar, western larch	5–10%	30–40%	40–50%
Falcon [3]	2001	42	Subalpine fir, Engelmann spruce, lodgepole pine, whitebark pine	0%	30–40%	60–70%
Green Knoll [3]	2001	54	Subalpine fir, Engelmann spruce, lodgepole pine, Douglas-fir, aspen	0%	20–30%	70–80%
Puma [4]	2008	45	Douglas-fir, white fir, ponderosa pine	20–30%	70–80%	0%
Dry Lakes Cx [3]	2003	49	Ponderosa pine, Arizona pine, Emory oak, alligator juniper	>90%	0%	0%
Miller [5]	2011	94	Ponderosa pine, Arizona pine, Emory oak, alligator juniper	80–90%	5–10%	0%
Outlet [6]	2000	54	Subalpine fir, Engelmann spruce, lodgepole pine, ponderosa pine, Douglas-fir, white fir	30–40%	5–10%	50–60%
Dragon Cx WFU [6]	2005	51	Ponderosa pine, Douglas-fir, white fir, aspen, subalpine fir, lodgepole pine	60–70%	20–30%	5–10%
Long Jim [6]	2004	49	Ponderosa pine, Gambel oak	>90%	0%	0%
Vista [6]	2001	46	Douglas-fir, white fir, ponderosa pine, aspen, subalpine fir	20–30%	70–80%	0%
Walhalla [6]	2004	47	Douglas-fir, white fir, ponderosa pine, aspen, subalpine fir, lodgepole pine	60–70%	20–30%	<5%
Poplar [6]	2003	108	Douglas-fir, white fir, ponderosa pine, aspen, subalpine fir, lodgepole pine	20–30%	20–30%	40–50%
Power [7]	2004	88	Ponderosa/Jeffrey pine, white fir, mixed conifers, black oak	>90%	0%	0%
Cone [7]	2002	59	Ponderosa/Jeffrey pine, mixed conifers	80–90%	<5%	<5%
Straylor [7]	2004	75	Ponderosa/Jeffrey pine, western juniper	>90%	0%	<5%
McNally [7]	2002	240	Ponderosa/Jeffrey pine, mixed conifers, interior live oak, scrub oak, black oak	70–80%	10–20%	0%

Composite burn index (CBI) data sources: [1] Susan Prichard, US Forest Service, Pacific Northwest Research Station; [2] Mike McClellan, Glacier National Park; [3] Zack Holden, US Forest Service, Northern Region; [4] Joel Silverman, Bryce Canyon National Park; [5] Sean Parks, US Forest Service, Rocky Mountain Research Station, Aldo Leopold Wilderness Research Institute; [6] Eric Gdula, Grand Canyon National Park; [7] Jay Miller, US Forest Service, Pacific Southwest Region.

Our second validation is nearly identical to that described in the previous paragraph but plot data from all 18 fires was combined (*n* = 1681). That is, instead of evaluating on a per-fire basis, we evaluated the plot data from all fires simultaneously. Following Parks et al. [8], this evaluation used a five-fold cross-validation. That is, five evaluations were conducted with 80% of the plot data used to train each nonlinear model and the remaining 20% used to test each model. The resulting coefficients of determination (R^2) and standard errors for the five testing datasets were averaged.

The third validation evaluates the classification accuracy when categorizing the satellite- and field-derived severity datasets into three discrete classes representing low, moderate, and high severity. To do so, we grouped the CBI plot data into severity classes using well-recognized CBI thresholds: Low severity corresponds to CBI values ranging from 0–1.24, moderate severity from 1.25–2.24, and high severity from 2.25–3.0 [7]. We then identified thresholds specific to each metric (with and without incorporating the $dNBR_{offset}$) corresponding to the low, moderate, and high CBI thresholds using nonlinear regression models as previously described. However, the nonlinear models used to produce low, moderate, and high severity thresholds for this evaluation used all 1681 plots combined and did not use the cross-validated versions. We measured the classification accuracy (i.e., the percent correctly classified) with 95% confidence intervals using the 'caret' package [34] in the R statistical environment [32]. We also produced confusion matrices for each severity metric and report the user's and producer's accuracy for each severity class (low, moderate, and high).

Finally, it is worth noting that we did not directly use the fire severity datasets distributed by the MTBS program. Our reasoning is that the MTBS program does not distribute the RBR. Furthermore, the MTBS program incorporates the $dNBR_{offset}$ into the RdNBR product but does not distribute RdNBR without the $dNBR_{offset}$. The MTBS program does, however, distribute the imagery used to produce each fire severity metric. In order to make valid comparisons to the GEE-derived datasets, we opted to use the pre- and post-fire imagery distributed by the MTBS program to produce dNBR, RdNBR, and RBR, with and without the $dNBR_{offset}$, for each of the 18 fires. All processing of MTBS-derived fires was accomplished with the 'raster' package [35] in the R statistical environment [32].

2.3. Google Earth Engine Implementation and Code

We provide a sample code and a geospatial fire history layer to produce a total of six raster datasets (dNBR, RdNBR, and RBR; with and without the $dNBR_{offset}$) for each of the 18 previously described fires. This code produces severity datasets that are clipped to a bounding box representing the outer extent of each fire. We designed the code to use imagery from one year before and one year after each fire occurs and to use a pre-specified date range for image selection for each fire, as previously described. These parameters can easily be modified to suit the needs of different users, ecosystems, and fire regimes.

3. Results

Using GEE, we were able to quickly produce dNBR, RdNBR, and RBR (with and without the $dNBR_{offset}$) for the 18 fires analyzed. The entire process was completed in approximately 1 h; fires averaged about 15,000 hectares in size and ranged from 723–60,000 hectares. This timeframe included a few minutes of active, hands-on time and about 60 min of GEE computational processing. This timeframe should be considered a very rough estimate, however, because GEE processing time varies widely among fires (larger fire sizes require more computational processing) and because production time depends on available resources shared among users within GEE's cloud-based computing platform [36]; nonetheless, processing time is very fast with fairly low investment in terms of human labor.

The mean compositing approach, in conjunction with the exclusion pixels classified as cloud, shadow, snow, and water, resulted in a variable number of valid Landsat scenes used in producing each pre- and post-fire NBR image. The average number of stacked pixels used to produce pre- and post-fire NBR was about 11. This varied by fire and ranged from 2–20 for pre-fire NBR and from 6–20 for post-fire NBR.

Our first validation, in which correspondence between CBI and each severity metric was computed independently for each fire, shows that there is not a substantial improvement between the MTBS- and GEE-derived dNBR and RBR (Table 2). For RdNBR, however, the GEE-derived severity metrics show a sizeable improvement, on average, over the MTBS-derived metrics.

Table 2. Mean R^2 of the correspondence between CBI and each MTBS- and GEE-derived fire severity metric across the 18 fires. MTBS: Monitoring Trends in Burn Severity; GEE: Google Earth Engine. The correspondence between CBI and the severity metrics were computed for each of the 18 fires and the mean R^2 is reported here.

	Mean R^2 without $dNBR_{offset}$		**Mean R^2 with $dNBR_{offset}$**	
	MTBS-Derived	**GEE-Derived**	**MTBS-Derived**	**GEE-Derived**
dNBR	0.761	0.768	0.761	0.768
RdNBR	0.736	0.782	0.751	0.782
RBR	0.784	0.791	0.784	0.790

When the correspondence between CBI and each severity metric for 1681 plots covering 18 fires was evaluated simultaneously using a five-fold cross-validation (our second evaluation), the R^2 was consistently higher for the GEE-derived fire severity datasets as compared to the MTBS-derived datasets (Table 3; Figure 2). Furthermore, the inclusion of the $dNBR_{offset}$ increased the correspondence to CBI for all fire severity metrics (Table 3). All terms in the nonlinear regressions for all severity metrics (those with and without the $dNBR_{offset}$) were statistically significant ($p < 0.05$) in all five folds of the cross-validation.

Table 3. R^2 of the five-fold cross-validation of the correspondence between CBI and each MTBS- and GEE-derived fire severity metric for 1681 plots across 18 fires; standard error shown in parentheses. The values characterize the average of five folds and represent the severity metrics excluding and including the $dNBR_{offset}$.

	R^2 without $dNBR_{offset}$ (Standard Error)		**R^2 with $dNBR_{offset}$ (Standard Error)**	
	MTBS-Derived	**GEE-Derived**	**MTBS-Derived**	**GEE-Derived**
dNBR	0.630 (0.026)	0.660 (0.025)	0.655 (0.026)	0.682 (0.025)
RdNBR	0.616 (0.026)	0.723 (0.024)	0.661 (0.027)	0.732 (0.024)
RBR	0.683 (0.025)	0.722 (0.024)	0.714 (0.025)	0.739 (0.024)

The GEE-derived fire severity datasets also provided a consistent improvement over the comparable MTBS-derived datasets in terms of overall classification accuracy (Table 4); this is evident regardless of whether or not the $dNBR_{offset}$ was incorporated. Inclusion of the $dNBR_{offset}$ provided additional improvement for the most part (Table 4). The only exception is for the GEE-derived RdNBR, in which the classification accuracy was identical with and without the $dNBR_{offset}$ (Table 4). The confusion matrices for each fire severity metric (with and without the $dNBR_{offset}$) indicate that the user's and producer's accuracies are usually higher with the GEE-derived metrics compared to the MTBS-derived metrics (Tables 5 and 6). The thresholds we used to classify plots as low, moderate, or high severity are shown in Table 7; these may be useful for others who implement our GEE methodology and want to classify the resulting datasets.

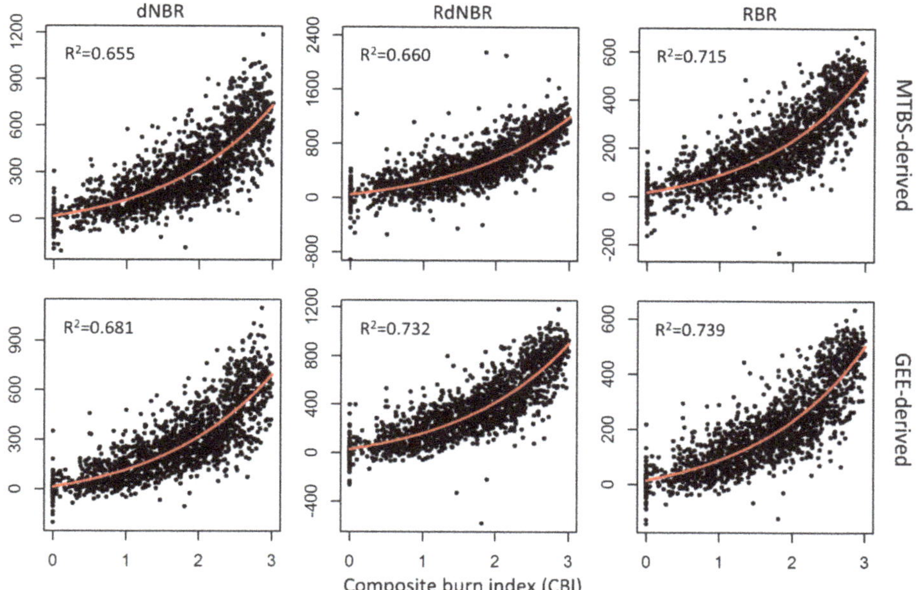

Figure 2. Plots show each MTBS- (top row) and GEE-derived (bottom row) severity metric and the corresponding field-based CBI. All severity metrics include the $dNBR_{offset}$. Red lines show the modeled fit of the nonlinear regressions for all 1681 plots. The model fits and the resulting R^2 shown here were not produced using cross-validation and therefore may differ slightly from the results shown in Table 3.

Table 4. Classification accuracy (percent correctly classified) and 95% confidence intervals (CI) for the three fire severity metrics (with and without the $dNBR_{offset}$). Each fire severity metric is classified into categories representing low, moderate, and high severity based on index-specific thresholds (see Table 7) and compared to the same classes based on composite burn index thresholds.

		Without $dNBR_{offset}$		With $dNBR_{offset}$	
		Accuracy (%)	95% CI	Accuracy (%)	95% CI
dNBR	MTBS-derived	69.6	67.3–71.8	70.2	68.0–72.4
	GEE-derived	71.3	69.0–73.4	71.7	69.5–73.9
RdNBR	MTBS-derived	71.4	69.2–73.5	73.6	71.4–75.6
	GEE-derived	73.9	71.8–76.0	73.9	71.8–76.0
RBR	MTBS-derived	72.4	71.1–74.5	73.5	71.4–75.6
	GEE-derived	73.5	71.4–75.6	74.1	72.0–76.2

Table 5. Confusion matrices for classifying as low, moderate, and high severity using the severity metrics computed without the $dNBR_{offset}$. Confusion matrices for MTBS-derived metrics are on the left and confusion matrices for GEE-derived metrics are on the right. UA: user's accuracy; PA: producer's accuracy.

		Reference CBI Class						**Reference CBI Class**			
		Low	Mod.	High	UA			Low	Mod.	High	UA
Classified using MTBS-derived dNBR	Low	401	159	18	69.4	Classified using GEE-derived dNBR	Low	407	139	13	72.8
	Mod.	91	412	114	66.8		Mod.	87	438	123	67.6
	High	5	124	357	73.5		High	3	118	353	74.5
	PA	80.7	59.3	73.0			PA	81.9	63.0	72.2	
		Reference CBI class						**Reference CBI class**			
		Low	Mod.	High	UA			Low	Mod.	High	UA
Classified using MTBS-derived RdNBR	Low	366	142	7	71.1	Classified using GEE-derived RdNBR	Low	396	130	7	74.3
	Mod.	119	451	99	67.4		Mod.	97	470	105	69.9
	High	12	102	383	77.1		High	4	95	377	79.2
	PA	73.6	64.9	78.3			PA	79.7	67.6	77.1	
		Reference CBI class						**Reference CBI class**			
		Low	Mod.	High	UA			Low	Mod.	High	UA
Classified using MTBS-derived RBR	Low	380	127	12	73.2	Classified using GEE-derived RBR	Low	403	130	9	74.4
	Mod.	113	462	102	68.2		Mod.	90	464	111	69.8
	High	4	106	375	77.3		High	4	101	369	77.8
	PA	76.5	66.5	76.7			PA	81.1	66.8	75.5	

Table 6. Confusion matrices for classifying as low, moderate, and high severity using the severity metrics computed with the $dNBR_{offset}$. Confusion matrices for MTBS-derived metrics are on the left and confusion matrices for GEE-derived metrics are on the right. UA: user's accuracy; PA: producer's accuracy.

		Reference CBI Class						**Reference CBI Class**			
		Low	Mod.	High	UA			Low	Mod.	High	UA
Classified using MTBS-derived dNBR	Low	397	156	13	70.1	Classified using GEE-derived dNBR	Low	402	141	10	72.7
	Mod.	98	425	118	66.3		Mod.	92	451	126	67.4
	High	2	114	358	75.5		High	3	103	353	76.9
	PA	79.9	61.2	73.2			PA	80.9	64.9	72.2	
		Reference CBI class						**Reference CBI class**			
		Low	Mod.	High	UA			Low	Mod.	High	UA
Classified using MTBS-derived RdNBR	Low	378	133	5	73.3	Classified using GEE-derived RdNBR	Low	386	122	7	75.0
	Mod.	112	467	92	69.6		Mod.	108	478	103	69.4
	High	7	95	392	79.4		High	3	95	379	79.5
	PA	76.1	67.2	80.2			PA	77.7	68.8	77.5	
		Reference CBI class						**Reference CBI class**			
		Low	Mod.	High	UA			Low	Mod.	High	UA
Classified using MTBS-derived RBR	Low	390	135	6	73.4	Classified using GEE-derived RBR	Low	386	123	7	74.8
	Mod.	105	460	97	69.5		Mod.	107	481	103	69.6
	High	2	100	386	79.1		High	4	91	379	80.0
	PA	78.5	66.2	78.9			PA	77.7	69.2	77.5	

Table 7. Threshold values for each fire severity metric corresponding to low (CBI = 0–1.24), moderate (CBI = 1.25–2.25), and high severity (CBI = 2.25–3).

		MTBS-Derived			GEE-Derived		
		Low	Moderate	High	Low	Moderate	High
Excludes $dNBR_{offset}$	dNBR	≤186	187–429	≥430	≤185	186–417	≥418
	RdNBR	≤337	338–721	≥722	≤248	249–544	≥545
	RBR	≤134	135–303	≥304	≤135	136–300	≥301
Includes $dNBR_{offset}$	dNBR	≤165	166–440	≥411	≤159	160–392	≥393
	RdNBR	≤294	295–690	≥691	≤212	213–511	≥512
	RBR	≤118	119–289	≥289	≤115	116–282	≥283

4. Discussion

The Google Earth Engine (GEE) methodology we developed to produce Landsat-based measures of fire severity is an important contribution to wildland fire research and monitoring. For example, our methodology will allow those who are not remote sensing experts, but have some familiarity with GEE, to quickly produce fire severity datasets (Figure 3). This benefit is due to the efficiency and speed of the cloud-based GEE platform [37,38] and because no a priori scene selection is necessary. Furthermore, compared to the standard approach in which only one pre- and post-fire scene are used, the GEE mean composite fire severity datasets exhibit higher validation statistics in terms of the correspondence (R^2) to CBI and higher classification accuracies for most severity classes. This suggests that mean composite severity metrics more accurately represent fire-induced ecological change, likely because the compositing method is less biased by pre- and post-fire scene mismatch and image characteristics inherent in standard processing. The computation and incorporation of the $dNBR_{offset}$ within GEE further improves, for the most part, the validation statistics of all metrics.

The improvements in the validation statistics of the GEE-derived severity metrics over the MTBS-derived severity metrics, when evaluated on a per-fire basis, are more or less negligible for dNBR and RBR (see Table 2). This suggests that if practitioners and researchers are interested in only one fire [20,39], it does not matter if dNBR or RBR are produced using the mean compositing approach or using one pre-fire and one post-fire image (e.g., MTBS). However, if RdNBR is the preferred severity metric, our results show that the mean compositing approach substantially outperforms (on average) RdNBR when produced using one pre-fire and one post-fire scene. It is also worth noting that the improvements in the validation statistics of the GEE-derived severity metrics over the MTBS-derived severity metrics, when all plots are evaluated simultaneously, are not statistically significant in most cases. That is, the overall classification accuracy of the GEE-derived metrics overlap the 95% confidence intervals of the MTBS-derived metrics in all comparisons except that of RdNBR without the $dNBR_{offset}$ (Table 4). Although the user's and producer's accuracy is oftentimes higher for the GEE-derived severity metrics (Tables 5 and 6), this is not always the case for all severity classes. In particular, the producer's accuracy (but not the user's accuracy) is generally higher for the MTBS-derived metrics when evaluating the high severity class. Nevertheless, the modest improvement in most validation statistics of the GEE-derived metrics, together with the framework and code we distribute in this study, will likely provide the necessary rationale and tools for producing fire severity datasets in counties that do not have national programs tasked with producing such datasets (e.g., MTBS in the United States).

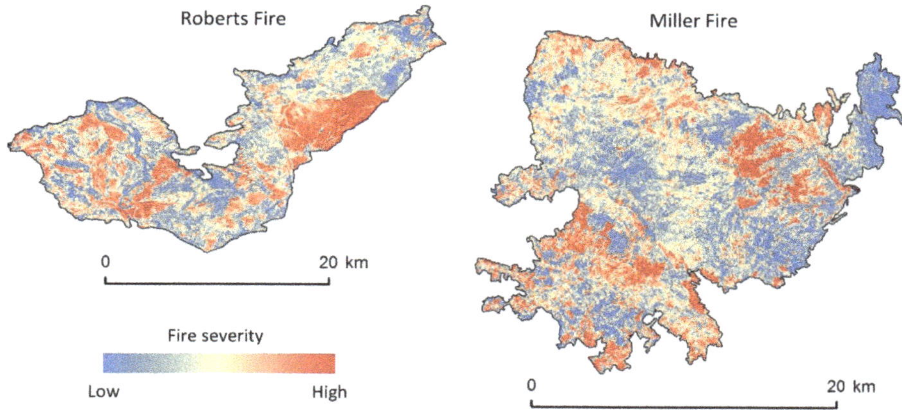

Figure 3. Example shows the RBR (includes the $dNBR_{offset}$) for two of the fires (Roberts and Miller) we evaluated. See Figure 1 to reference the locations of these fires.

The Monitoring Trends in Burn Severity (MTBS) program in the US, which produces and distributes Landsat-based fire severity datasets [19], has enabled scientists to conduct research involving hundreds to thousands of fires [2,24,40,41]. Outside of the US, where programs similar to MTBS do not exist, most fire severity research is limited to only a handful of fires, the exceptions being Fang et al. [15] (n = 72 fires in China) and Whitman et al. [42] (n = 56 fires in Canada). We suggest that the GEE methodology we developed will allow users in regions outside of the US to efficiently produce fire severity datasets for hundreds to thousands of fires in their geographic areas of interest, thereby providing enhanced opportunities for fire severity monitoring and research. Although fire history datasets (i.e., georeferenced fire perimeters) are a prerequisite for implementing our GEE methodology, such datasets have already been produced and used in scientific studies in Portugal [43], Spain [44], Canada [45], portions of Australia [46], southern France [47], the Sky Island Mountains of Mexico [48], and likely elsewhere. Therefore, the GEE methods developed here provide a common platform for assessing fire-induced ecological change and can provide more opportunities for fire severity monitoring and research across the globe.

The fires we analyzed primarily burned in conifer forests and were embedded within landscapes comprised of similar vegetation. As such, our approach to incorporating the $dNBR_{offset}$ that used pixels in a 180 m 'ring' around the fire perimeter may not be appropriate everywhere and we urge caution in landscapes in which fires burn vegetation that is not similar to that of the surrounding lands. For example, our methods for calculating and implementing the $dNBR_{offset}$ would not be appropriate if a fire burned a forested patch that was surrounded by completely different vegetation such as shrubland or agriculture. In such cases, we recommend that fire severity datasets exclude the $dNBR_{offset}$ as it may not improve burn assessments. Similarly, the low, moderate, and high severity thresholds identified in this study (Table 7) are likely only applicable to forested landscapes in the western US, and other thresholds may be more suitable to other regions of the globe and in different vegetation types. Finally, our choice of developing post-fire imagery from the period one-year after the fire may not be appropriate for all ecosystems. Arctic tundra ecosystems, for example, might be better represented by imagery derived immediately after the fire or after snowmelt but prior to green-up the year following the fire [49]. The GEE approach can be easily modified to select dates that best suit each ecosystem.

5. Conclusions

In this paper, we present practical and efficient methodology for producing three Landsat-based fire severity metrics: dNBR, RdNBR, and RBR. These methods rely on Google Earth Engine and provide expanded potential in terms of fire severity monitoring and research in regions outside of the US that do not have a dedicated program for mapping fire severity. In validating the fire severity metrics, our goal was not to compare and contrast individual metrics (e.g., dNBR vs. RBR) [11,12] nor to critique products produced by the MTBS program. Instead, we aimed to evaluate differences between the GEE-based mean compositing approach to the standard approach in which one pre-fire and post-fire Landsat scene are used to produce severity datasets. The GEE-based severity datasets generally achieved higher validation statistics in terms of correspondence to field data and overall classification accuracy. The inclusion of the $dNBR_{offset}$ provided additional improvements in these validation statistics for most fire severity metrics regardless of whether they were MTBS- or GEE-derived. This provides further evidence that inclusion of the $dNBR_{offset}$ should be considered when multiple fires are of interest [8,17]. Our evaluation included fires over a large spatial extent (the western US) and with varied fire regime attributes, ranging from those that are predominantly surface fire regimes to those that are stand-replacing regimes. Consequently, the higher validation statistics reported here for the GEE-derived composite-based fire severity datasets should provide researchers and practitioners with increased confidence in these products.

Author Contributions: S.A.P. conceived of the study, conducted the statistical validations, and wrote the paper. L.M.H. aided in designing the study, developed GEE code, and contributed to manuscript writing. R.A.L. aided in designing the study and contributed to manuscript writing. M.A.V. and N.P.R. developed GEE code and contributed to manuscript writing.

Funding: This research was partially funded by an agreement between the US Geological Survey and US Forest Service.

Acknowledgments: Any use of trade names is for descriptive purposes only and does not imply endorsement by the US Government.

Conflicts of Interest: The authors declare no conflict of interest.

Code Availability: The code to implement our methods is available here: https://code.earthengine.google.com/c76157be827be2f24570df50cca427e9. The code is set up to run on the 18 fires highlighted in this paper (Figure 1) and will produce dNBR, RdNBR, and RBR with and without the $dNBR_{offset}$.

References

1. Parks, S.A.; Parisien, M.A.; Miller, C.; Dobrowski, S.Z. Fire activity and severity in the western US vary along proxy gradients representing fuel amount and fuel moisture. *PLoS ONE* **2014**, *9*, e99699. [CrossRef] [PubMed]
2. Dillon, G.K.; Holden, Z.A.; Morgan, P.; Crimmins, M.A.; Heyerdahl, E.K.; Luce, C.H. Both topography and climate affected forest and woodland burn severity in two regions of the western US, 1984 to 2006. *Ecosphere* **2011**, *2*, 130. [CrossRef]
3. Veraverbeke, S.; Lhermitte, S.; Verstraeten, W.W.; Goossens, R. The temporal dimension of differenced Normalized Burn Ratio (dNBR) fire/burn severity studies: The case of the large 2007 Peloponnese wildfires in Greece. *Remote Sens. Environ.* **2010**, *114*, 2548–2563. [CrossRef]
4. Fernández-Garcia, V.; Santamarta, M.; Fernández-Manso, A.; Quintano, C.; Marcos, E.; Calvo, L. Burn severity metrics in fire-prone pine ecosystems along a climatic gradient using Landsat imagery. *Remote Sens. Environ.* **2018**, *206*, 205–217. [CrossRef]
5. Fang, L.; Yang, J.; White, M.; Liu, Z. Predicting Potential Fire Severity Using Vegetation, Topography and Surface Moisture Availability in a Eurasian Boreal Forest Landscape. *Forests* **2018**, *9*, 130. [CrossRef]
6. Key, C.H.; Benson, N.C. Landscape assessment (LA). In *FIREMON: Fire Effects Monitoring and Inventory System*; General Technical Report RMRS-GTR-164-CD; U.S. Department of Agriculture, Forest Service, Rocky Mountain Research Station: Fort Collins, CO, USA, 2006.
7. Miller, J.D.; Thode, A.E. Quantifying burn severity in a heterogeneous landscape with a relative version of the delta Normalized Burn Ratio (dNBR). *Remote Sens. Environ.* **2007**, *109*, 66–80. [CrossRef]

8. Parks, S.A.; Dillon, G.K.; Miller, C. A new metric for quantifying burn severity: The relativized burn ratio. *Remote Sens.* **2014**, *6*, 1827–1844. [CrossRef]
9. Holden, Z.A.; Morgan, P.; Evans, J.S. A predictive model of burn severity based on 20-year satellite-inferred burn severity data in a large southwestern US wilderness area. *For. Ecol. Manag.* **2009**, *258*, 2399–2406. [CrossRef]
10. Wimberly, M.C.; Reilly, M.J. Assessment of fire severity and species diversity in the southern Appalachians using Landsat TM and ETM+ imagery. *Remote Sens. Environ.* **2007**, *108*, 189–197. [CrossRef]
11. Veraverbeke, S.; Lhermitte, S.; Verstraeten, W.W.; Goossens, R. Evaluation of pre/post-fire differenced spectral indices for assessing burn severity in a Mediterranean environment with Landsat Thematic Mapper. *Int. J. Remote Sens.* **2011**, *32*, 3521–3537. [CrossRef]
12. Soverel, N.O.; Perrakis, D.D.B.; Coops, N.C. Estimating burn severity from Landsat dNBR and RdNBR indices across western Canada. *Remote Sens. Environ.* **2010**, *114*, 1896–1909. [CrossRef]
13. Beck, P.S.A.; Goetz, S.J.; Mack, M.C.; Alexander, H.D.; Jin, Y.; Randerson, J.T.; Loranty, M.M. The impacts and implications of an intensifying fire regime on Alaskan boreal forest composition and albedo. *Glob. Chang. Biol.* **2011**, *17*, 2853–2866. [CrossRef]
14. Mallinis, G.; Mitsopoulos, I.; Chrysafi, I. Evaluating and comparing Sentinel 2A and Landsat-8 Operational Land Imager (OLI) spectral indices for estimating fire severity in a Mediterranean pine ecosystem of Greece. *GISci. Remote Sens.* **2018**, *55*, 1–18. [CrossRef]
15. Fang, L.; Yang, J.; Zu, J.; Li, G.; Zhang, J. Quantifying influences and relative importance of fire weather, topography, and vegetation on fire size and fire severity in a Chinese boreal forest landscape. *For. Ecol. Manag.* **2015**, *356*, 2–12. [CrossRef]
16. Cansler, C.A.; McKenzie, D. How robust are burn severity indices when applied in a new region? Evaluation of alternate field-based and remote-sensing methods. *Remote Sens. Mol.* **2012**, *4*, 456–483. [CrossRef]
17. Key, C.H. Ecological and sampling constraints on defining landscape fire severity. *Fire Ecol.* **2006**, *2*, 34–59. [CrossRef]
18. Picotte, J.J.; Peterson, B.; Meier, G.; Howard, S.M. 1984–2010 trends in fire burn severity and area for the conterminous US. *Int. J. Wildl. Fire* **2016**, *25*, 413–420. [CrossRef]
19. Eidenshink, J.C.; Schwind, B.; Brewer, K.; Zhu, Z.-L.; Quayle, B.; Howard, S.M. A project for monitoring trends in burn severity. *Fire Ecol.* **2007**, *3*, 3–21. [CrossRef]
20. Kane, V.R.; Cansler, C.A.; Povak, N.A.; Kane, J.T.; McGaughey, R.J.; Lutz, J.A.; Churchill, D.J.; North, M.P. Mixed severity fire effects within the Rim fire: Relative importance of local climate, fire weather, topography, and forest structure. *For. Ecol. Manag.* **2015**, *358*, 62–79. [CrossRef]
21. Stevens-Rumann, C.; Prichard, S.; Strand, E.; Morgan, P. Prior wildfires influence burn severity of subsequent large fires. *Can. J. For. Res.* **2016**, *46*, 1375–1385. [CrossRef]
22. Prichard, S.J.; Kennedy, M.C. Fuel treatments and landform modify landscape patterns of burn severity in an extreme fire event. *Ecol. Appl.* **2014**, *24*, 571–590. [CrossRef] [PubMed]
23. Parks, S.A.; Holsinger, L.M.; Panunto, M.H.; Jolly, W.M.; Dobrowski, S.Z.; Dillon, G.K. High-severity fire: Evaluating its key drivers and mapping its probability across western US forests. *Environ. Res. Lett.* **2018**, *13*, 044037. [CrossRef]
24. Keyser, A.; Westerling, A. Climate drives inter-annual variability in probability of high severity fire occurrence in the western United States. *Environ. Res. Lett.* **2017**, *12*, 065003. [CrossRef]
25. Arkle, R.S.; Pilliod, D.S.; Welty, J.L. Pattern and process of prescribed fires influence effectiveness at reducing wildfire severity in dry coniferous forests. *For. Ecol. Manag.* **2012**, *276*, 174–184. [CrossRef]
26. Wimberly, M.C.; Cochrane, M.A.; Baer, A.D.; Pabst, K. Assessing fuel treatment effectiveness using satellite imagery and spatial statistics. *Ecol. Appl.* **2009**, *19*, 1377–1384. [CrossRef] [PubMed]
27. Parks, S.A.; Miller, C.; Abatzoglou, J.T.; Holsinger, L.M.; Parisien, M.-A.; Dobrowski, S.Z. How will climate change affect wildland fire severity in the western US? *Environ. Res. Lett.* **2016**, *11*, 035002. [CrossRef]
28. Miller, J.D.; Safford, H.D.; Crimmins, M.; Thode, A.E. Quantitative evidence for increasing forest fire severity in the Sierra Nevada and southern Cascade Mountains, California and Nevada, USA. *Ecosystems* **2009**, *12*, 16–32. [CrossRef]
29. Whitman, E.; Parisien, M.-A.; Thompson, D.K.; Hall, R.J.; Skakun, R.S.; Flannigan, M.D. Variability and drivers of burn severity in the northwestern Canadian boreal forest. *Ecosphere* **2018**, *9*. [CrossRef]

30. Ireland, G.; Petropoulos, G.P. Exploring the relationships between post-fire vegetation regeneration dynamics, topography and burn severity: A case study from the Montane Cordillera Ecozones of Western Canada. *Appl. Geogr.* **2015**, *56*, 232–248. [CrossRef]
31. Foga, S.; Scaramuzza, P.L.; Guo, S.; Zhu, Z.; Dilley, R.D.; Beckmann, T.; Schmidt, G.L.; Dwyer, J.L.; Hughes, M.J.; Laue, B. Cloud detection algorithm comparison and validation for operational Landsat data products. *Remote Sens. Environ.* **2017**, *194*, 379–390. [CrossRef]
32. R Core Team. *R: A Language and Environment for Statistical Computing*; R Foundation for Statistical Computing: Vienna, Austria, 2016; Available online: https://www.r-project.org/ (accessed on 1 July 2017).
33. Rollins, M.G. LANDFIRE: A nationally consistent vegetation, wildland fire, and fuel assessment. *Int. J. Wildl. Fire* **2009**, *18*, 235–249. [CrossRef]
34. Kuhn, M. Caret package. *J. Stat. Softw.* **2008**, *28*, 1–26.
35. Hijmans, R.J.; van Etten, J.; Cheng, J.; Mattiuzzi, M.; Sumner, M.; Greenberg, J.A.; Lamigueiro, O.P.; Bevan, A.; Racine, E.B.; Shortridge, A.; et al. *Package 'Raster'*; R. Package, 2015.
36. Gorelick, N.; Hancher, M.; Dixon, M.; Ilyushchenko, S.; Thau, D.; Moore, R. Google Earth Engine: Planetary-scale geospatial analysis for everyone. *Remote Sens. Environ.* **2017**, *202*, 18–27. [CrossRef]
37. Kennedy, R.E.; Yang, Z.; Gorelick, N.; Braaten, J.; Cavalcante, L.; Cohen, W.B.; Healey, S. Implementation of the LandTrendr Algorithm on Google Earth Engine. *Remote Sens.* **2018**, *10*, 691. [CrossRef]
38. Robinson, N.P.; Allred, B.W.; Jones, M.O.; Moreno, A.; Kimball, J.S.; Naugle, D.E.; Erickson, T.A.; Richardson, A.D. A Dynamic Landsat Derived Normalized Difference Vegetation Index (NDVI) Product for the Conterminous United States. *Remote Sens.* **2017**, *9*, 863. [CrossRef]
39. Lydersen, J.M.; Collins, B.M.; Brooks, M.L.; Matchett, J.R.; Shive, K.L.; Povak, N.A.; Kane, V.R.; Smith, D.F. Evidence of fuels management and fire weather influencing fire severity in an extreme fire event. *Ecol. Appl.* **2017**, *27*, 2013–2030. [CrossRef] [PubMed]
40. Reilly, M.J.; Dunn, C.J.; Meigs, G.W.; Spies, T.A.; Kennedy, R.E.; Bailey, J.D.; Briggs, K. Contemporary patterns of fire extent and severity in forests of the Pacific Northwest, USA (1985–2010). *Ecosphere* **2017**, *8*, e01695. [CrossRef]
41. Stevens, J.T.; Collins, B.M.; Miller, J.D.; North, M.P.; Stephens, S.L. Changing spatial patterns of stand-replacing fire in California conifer forests. *For. Ecol. Manag.* **2017**, *406*, 28–36. [CrossRef]
42. Whitman, E.; Batllori, E.; Parisien, M.-A.; Miller, C.; Coop, J.D.; Krawchuk, M.A.; Chong, G.W.; Haire, S.L. The climate space of fire regimes in north-western North America. *J. Biogeogr.* **2015**, *42*, 1736–1749. [CrossRef]
43. Fernandes, P.M.; Loureiro, C.; Magalhães, M.; Ferreira, P.; Fernandes, M. Fuel age, weather and burn probability in Portugal. *Int. J. Wildl. Fire* **2012**, *21*, 380–384. [CrossRef]
44. Trigo, R.M.; Sousa, P.M.; Pereira, M.G.; Rasilla, D.; Gouveia, C.M. Modelling wildfire activity in Iberia with different atmospheric circulation weather types. *Int. J. Climatol.* **2016**, *36*, 2761–2778. [CrossRef]
45. Parisien, M.-A.; Miller, C.; Parks, S.A.; Delancey, E.R.; Robinne, F.-N.; Flannigan, M.D. The spatially varying influence of humans on fire probability in North America. *Environ. Res. Lett.* **2016**, *11*, 075005. [CrossRef]
46. Price, O.F.; Penman, T.D.; Bradstock, R.A.; Boer, M.M.; Clarke, H. Biogeographical variation in the potential effectiveness of prescribed fire in south-eastern Australia. *J. Biogeogr.* **2015**, *42*, 2234–2245. [CrossRef]
47. Fox, D.M.; Carrega, P.; Ren, Y.; Caillouet, P.; Bouillon, C.; Robert, S. How wildfire risk is related to urban planning and Fire Weather Index in SE France (1990–2013). *Sci. Total Environ.* **2018**, *621*, 120–129. [CrossRef] [PubMed]
48. Villarreal, M.L.; Haire, S.L.; Iniguez, J.M.; Montaño, C.C.; Poitras, T.B. Distant Neighbors: Recent wildfire patterns of the Madrean Sky Islands of Southwestern United States and Northwestern México. *Fire Ecol.* **2018**, in press.
49. Kolden, C.A.; Rogan, J. Mapping wildfire burn severity in the Arctic tundra from downsampled MODIS data. *Arct. Antarct. Alp. Res.* **2013**, *45*, 64–76. [CrossRef]

MDPI
St. Alban-Anlage 66
4052 Basel
Switzerland
Tel. +41 61 683 77 34
Fax +41 61 302 89 18
www.mdpi.com

Remote Sensing Editorial Office
E-mail: remotesensing@mdpi.com
www.mdpi.com/journal/remotesensing

www.ingramcontent.com/pod-product-compliance
Lightning Source LLC
Chambersburg PA
CBHW051706210326
41597CB00032B/5385

* 9 7 8 3 0 3 8 9 7 8 8 4 8 *